Developments in Geotechnical Engineering

Series editors

Braja M. Das, Henderson, USA
Nagaratnam Sivakugan, Townsville, Australia

More information about this series at http://www.springer.com/series/13410

V.V.S. Rao · G.L. Sivakumar Babu
Editors

Forensic Geotechnical Engineering

Springer

Editors
V.V.S. Rao
Nagadi Consultants Pvt Limited
Chennai, Tamil Nadu
India

G.L. Sivakumar Babu
Department of Civil Engineering
Indian Institute of Science
Bangalore, Karnataka
India

ISSN 2364-5156 ISSN 2364-5164 (electronic)
Developments in Geotechnical Engineering
ISBN 978-81-322-2376-4 ISBN 978-81-322-2377-1 (eBook)
DOI 10.1007/978-81-322-2377-1

Library of Congress Control Number: 2015938088

Springer New Delhi Heidelberg New York Dordrecht London

Printed on acid-free paper

Springer (India) Pvt. Ltd. is part of Springer Science+Business Media (www.springer.com)

Foreword

In my programme of candidature for ISSMGE President, in 2005, I have emphasized "that it is my profound desire to bring a better balance and harmonisation of geotechnical activities for the different regions and more visibility for ISSMGE.

The diversity of all 81 ISSMGE Member Societies, is our great richness and a source of inspiration. It is our great challenge, but also a unique opportunity, to rethink ISSMGE, due to the changes in basic societal needs, to reach a new model. We need to recognize the importance of dialogue to work together and pursue perfection to reach optimum solutions for the problems facing us.

It is important to explore a new format of the Technical Committees (TCs) in order to turn them more attractive not only for academicians, but also for practitioners. The universality of the knowledge and the need for a recurring demand on the TCs to explore the best ways for research, design, construction and assessment of safe geotechnical structures, in order to reach the expectations and targets of our Society during the third millennium".

Within this framework I have Dr. V.V.S. Rao's proposal to create TC 40 "Forensic Geotechnical Engineering".

During our discussions to prepare the Terms of Reference I have emphasized that TCs are the spine of ISSMGE, a forum of discussion contributing towards the advancement of knowledge in geotechnical engineering and an opportunity to interact with the Society. It was important to communicate, to share experiences, to compare methodologies and to monitor results. The benefits of an open dialogue between academicians, researchers, practitioners, contractors and owners are huge.

The TCs should disseminate their work, through ISSMGE conferences, or workshops and satellite conferences.

At the end of the tenure of 4 years, each TC should produce an ISSMGE bulletin or book.

Engineering is an ambiguous word, in some languages an engineer is a person mastering engines, in Latin, an engineer is a person with genius, something like art, that is, skill plus intelligence. If the Earth is viewed as a big engine, a set of parts linked into intricate mechanisms, the first acceptance is right, but a bit of genius is required to master its complexity.

Bacon (1620) quoted that nature to be mastered, must be obeyed.

A good geotechnical engineer is one who knows the limits of his experience on problems and soil conditions comparable with his current assignment and makes appropriate extrapolations. He knows what he knows and uses it confidently. More importantly, he knows what he does not know, seeks available knowledge, and then proceeds, fully acknowledging his limitations and uncertainties.

Forensic geotechnical engineering is the investigation of geotechnical materials, products or components that fail (ultimate limit state) or do not operate or function as intended (serviceability limit state), causing personal injury or property loss. The consequences of failure are dealt with by the law of structure reliability. Generally, the purpose of a forensic engineering investigation is to identify the cause or causes of failure with a view to improve performance or life of a structure, or to assist a court of law in determining the facts of an incident or accident.

This book integrates 35 contributions from experts of all round the world, namely from the 6 Regions that compose ISSMGE: Africa, Asia, Australasia, Europe, North America and South America.

In a geotechnical world that is transforming and advancing everyday, we face new challenges. This book explores new ways, discusses our boundaries of responsibility, our ethical and legal environment, enhances the appeal of the ISSMGE to the practitioner engineers, promotes our Society to other professionals, to decision-makers and will be a continuous source of inspiration for today and tomorrow.

The lessons learned from the case studies addressed in this book remind us that the arrogance of our convictions should be subordinated to the study of alternative solutions, to the effort to render compatible divergent interests and opinions and the need to not forget past lessons. Forensic Geotechnical Engineering is simultaneously complex and a great challenge, but we need to humbly recognize the existence of uncertainties, the importance of dialogue and pursue the perfection to reach the optimum solution.

The purpose of this book to promote a cross-fertilization between academicians and practitioners was totally achieved. It is my hope that the spirit of cooperation fostered by this book will encourage additional projects and will contribute to the advancement of the knowledge in forensic geotechnical engineering, quoting Goldsmith's memorable lines:

"Hope, like the glimmering taper's light
Adorns and cheers the way
And still, as darker grows the night
Emits a brighter ray".

I would like to express on behalf of the ISSMGE, our deep gratitude to the Indian Geotechnical Society for the support and cooperation to host this Technical Committee, since 2005.

Last but not least, I would like to address to the Chairman of TC 302, Dr. V.V.S. Rao and to all members of this TC a word of praise and gratitude for their devotion and skill to overcome the difficulties and to organise very successfully all the activities, and also a message of hope that TC 302 will allow us to develop a feeling of universal responsibility and with the joint cooperation of decision-makers, general public, researchers, professors, designers and contractors to create the ambition to serve better our Society, quoting Julia Carney's memorable lines:

"Little drops of water, little grains of
Sand,
Make the mighty ocean and the pleasant
Land.
So the little minutes, humble though
They be,
Make the mighty ages of eternity".

This journey was an unforgettable experience for us. We learned that we have an enormous amount of choice to create the beautiful lives of our dreams.

We are very grateful. You help us in so many ways that words cannot express.

Prof. Pedro Seco e Pinto
President, ISSMGE (2009–2013)

Preface

Forensic geotechnical engineering is about analysing the causes of failure, understanding what went wrong, why and when from the geotechnical point of view. The role of a forensic geotechnical engineer is often complex and needs to be comprehensive to satisfy technical and legal perspectives. Often, the geotechnical engineer is required either as a technical expert to understand the failure or as expert witness in the Court of Law. To develop guidelines in Forensic Geotechnical Engineering, learning from failures is an essential step and is the hallmark of good engineering practice. In this edited volume on advances in Forensic Engineering, a number of technical contributions from experts and professionals in this area are included. The work is the outcome of deliberations at various conferences in the area by Technical Committee on Forensic Geotechnical Engineering (TC 302) of International Society for Soil Mechanics and Foundation Engineering (ISSMGE). The edited volume contains a few selected contributions on the topics relevant to forensic Geotechnical Engineering viz, (a) Evidence/data collection, (b) Distress characterization, (c) Use of diagnostic tests (laboratory and field tests), (d) Back analysis, (e) Failure hypothesis formulation, (f) Role of instrumentation and sensor based technologies, (g) Risk analysis and (h) Technical shortcomings.

The first two chapters are general and cover Framework for Forensic Foundation Engineering, by Prof. H.G. Poulos, Forensic Geotechnical Engineering—Theory and Practice by Dr. Suzanne Lacasse- Høeg and Dr. V.V.S. Rao presents general guidelines for Forensic Investigation of Geotechnical Failures. Caisson Failure Induced by Wave Action is addressed by Prof. Alonso and his co-workers.

Data Collection in Forensic Geotechnical Engineering Investigation is addressed by Dr. Peter Day, Diagnosis of Geotechnical Failure Causes Using Bayesian Networks is presented by Yao Xu and his colleagues. The role of Back Analyses in Geotechnical Engineering is addressed by G.L. Sivakumar Babu, Prof. Popescu and Dr. Richard Hwang. The role of instrumentation and monitoring for Forensic Geotechnical Engineering is presented by Prof. Yoshi Iwasaki.

Analysis of failures such as causes of undue movements of a retaining wall forming a deep basement in sandstone and consequent damage to an adjacent canal, failure of anchored retaining wall are studied by Jessep, Kazuya and their co-workers. The role of backwards problems in geotechnical earthquake engineering is examined by Dr. S. Iai and reanalysis of failure of underground construction is presented by Dr. B. Hsiung. Failure Analysis of slope slide as well as characterisation of failure at a large landslide are studied by Prof. Lee and Starr and others. Dr. Md. Mizanur Rahman presents a case study of Post-liquefaction Data Collection and Analyses for Earthquakes in New Zealand Prof. Leung presented some case studies in Forensics from Singapore.

A case study of failure due to Piping in Sand under a Dam and its Back Analysis and another failure in Construction of an Underground Station are presented by P. Sembenelli and Dr. Hwang. Dr. N. Santhosh Rao discussed the Influence of Vibrations on Installation of Bored Piles.

A case study of floor heaving of a condominium RC-Building with seven stories by Prof. Yoshi Iwasaki and his co-workers and failure Case Study of Tieback Wall in Urban Area, Korea by Prof. Jeong and Y.H. Kim are included.

Forensics of pile foundations subjected to earthquake-induced liquefaction and lateral spreading by Prof. Gopal Madabhushi and his co-workers, piping failure of a metro tunnel construction by Prof. Lee and his co-workers, forensic analysis of failure of retaining wall by Sivakumar Babu provide useful information.

Prof. K.K. Phoon and his co-workers discuss the role of reliability in forensic geotechnical engineering and Prof. Gilbert highlights role using case studies. Dhirendra Saxena discusses the technical, ethical and legal Issues in forensic geotechnical engineering using case studies. R.A. Jessep and his co-authors presented details of technical shortcomings causing geotechnical failures. The annexure contains the composition of the Technical Committee 302 during 2005–2013.

We wish to place on record our sincere thanks and appreciation to all the members of TC committee on Forensic Geotechnical Engineering and experts who have contributed to the activities of the Committee and only a select contributions were included in this volume. We wish to thank Prof. Pedro Seco e Pinto, President of ISSMFE during 2009–2013 for his encouragement and for writing the foreword for this contribution. We also wish to thank Mr. M. Pradeep, Editor in Chief of Master Builder for his help in organizing the programs and Springer for printing this important contribution in the area of Forensic Geotechnical Engineering.

Dr. V.V.S. Rao
G.L. Sivakumar Babu

Contents

About the Editors

V.V.S. Rao is an internationally acclaimed Geotechnical Consulting Engineer and is the founder of M/S Nagadi Consultants Pvt. Ltd. a leading Geotechnical Consultancy firm in India. His major field of specialization is soil dynamics including machine foundations and earthquake geotechnical engineering. He graduated in 1958 from University of Mysore and obtained Masters Degree from University of Roorkee in 1960, After conducting research at University of Rookee and at Technical University West Berlin, obtained the Ph.D. degree from the Institute of Technology, Delhi in 1968. He joined the faculty of Civil Engineering of Indian Institute of Technology Delhi in 1966. During 1966–1971 he headed the Soil Engineering Section of the Indian Institute of Technology Delhi and guided a number of research projects, in addition to conducting postgraduate programmes under USAID, and consultancy jobs. Since 1971 he has been practicing as a Geotechnical Consultant in his firm and has been involved in over 20,000 major to medium-sized projects. He was an advisor to NTPC and also to UNDP. He has been an active member for over 40 years in the Geotechnical Committees of the Bureau of Indian Standards.

During 1993–1997 he was the secretary and during 1997–2001 the Chair of ISSMGE Technical Committee, TC20 on Professional Practice. In 2001 he introduced the subject of Forensic Geotechnical Engineering at Indian Institute of Technology, Kanpur, India. His first paper on the same subject was presented in the International Workshop on Foundation Design Codes held at Kamakura in Japan in April 2002. He became the Chair of TC on Forensic Geotechnical Engineering, TC 40 during 2005–2009. This committee was designated as TC 302 in 2009. He continued to chair this TC until 2013.

He is a fellow member of a number of professional societies in India. He has been the national President of the Association of Consulting Civil Engineers (India) during 1989–2003. He is the Founder Chairman of IGS Chennai Chapter. As Joint Secretary of Association of Consulting Engineers (India), he has been involved in the initial drafting of the Engineers Bill.

Under his leadership a number of International Seminars/Workshops on Forensic Geotechnical Engineering have been organized between 2005 and 2013. He has chaired and delivered special lectures in more than eight international conferences and workshops. He also has over 40 published papers in international journals and conferences. He has received seven national awards in India, including special awards from Association of Consulting Civil Engineers, and from M/S Torsteel Research Foundation. He has received the prestigious Kueckelman Award from Indian Geotechnical Society for his significant contribution in Geotechnical Engineering.

G.L. Sivakumar Babu is well known for his contributions to research, teaching and consultancy. He completed Ph.D. (Geotechnical Engineering) in 1991 from Indian Institute of Science, Bangalore, India, Master's Degree (Soil Mechanics Foundation Engineering) in 1987 from Anna University, Madras and B.Tech. (Civil Engineering) in 1983 from Sri Venkateswara University, Tirupati. His teaching and research activities include courses on geomechanics, soil investigations, risk and reliability applications in Civil Engineering, pavement engineering, geosynthetics and ground improvement.

Prof. Babu served as Secretary of the Forensic Geotechnical Engineering Committee of ISSMGE for two terms (2005–2009 and 2009–2013) and currently serving as its Chairman. He organized a number of technical activities and progams in the area.

Dr. Babu is Editor-in-chief, Indian Geotechnical Journal and is Editorial Board Member of (1) American Society of Civil Engineers (ASCE), Journal of Hazardous, Toxic, and Radioactive Waste (2) International Journal of Georisk—Assessment and Management of Risk for Engineered Systems and Geohazards, (3) Ground Improvement, Journal of Institution of Civil Engineers, UK during 2006–2013 and (4) International Journal of Systems Assurance Engineering and Management. He is a Life Member of Indian Roads Congress, Fellow of Institution of Engineers India, Fellow of Indian Geotechnical Society, Fellow of Association of Consulting Engineers (ACCE), Member of American Society of Civil Engineers (ASCE), Member of International Geosynthetics Society (IGS).

He has written a book on soil reinforcement and geosynthetics, edited six books and proceedings, and has over 200 publications in journals and conferences to his credit. He received several awards and honours for his work for his work. He received the best Ph.D. thesis award in geotechnical engineering in India by the Indian Geotechnical Society in 1992. He is a Humboldt Fellow and a DST Boyscast Fellow and a recipient of several best paper awards for the papers published in ASCE journals and conferences of Indian Geotechnical Society.

Chapter 1
A Framework for Forensic Foundation Engineering

H.G. Poulos

Abstract The paper sets out a framework for investigating the possible causes of foundation failures in a systematic manner. Aspects relating to foundation failure such as conditions for failure, factors affecting geotechnical foundation resistance, structural foundation resistance, factors affecting applied loads and structural actions, development and testing of credible hypotheses are discussed. Attention is confined to foundation failures (ultimate failure conditions), although similar principles can be applied to cases involving excessive deformations of the foundation system (serviceability failures). An example is then presented to illustrate the application of the approach developed to a case involving failure of a piled foundation.

Keywords Foundation failure · Structural failure · Resistances · Loads · Failure hypothesis

1.1 Introduction

Forensic geotechnical engineering is gaining increasing importance in many countries where foundation failures may lead to litigation and even criminal action. Day (1999) has described various issues related to this emerging area of geotechnical practice while guidelines have been provided by Carper (1986).

This paper sets out a relatively simple framework for investigating the possible causes of foundation failures in a systematic manner. Attention will be confined to foundation failures (ultimate failure conditions), although similar principles can be applied to cases involving excessive deformations of the foundation system (serviceability failures). An example is then presented to illustrate the application of the approach developed to a case involving failure of a piled foundation.

H.G. Poulos (✉)
Coffey Geotechnics Pty. Ltd., University of Sydney, Sydney, Australia
e-mail: Harry.Poulos@coffey.com

V.V.S. Rao and G.L. Sivakumar Babu (eds.), *Forensic Geotechnical Engineering*,
Developments in Geotechnical Engineering, DOI 10.1007/978-81-322-2377-1_1

1.2 Conditions for Failure

Foundation failures can occur either because of failure of the supporting soil (geotechnical failure) or failure of the material of which the foundation is made (structural failure). For geotechnical failure to occur, the following condition must be satisfied:

$$R_g < SP \tag{1.1}$$

where
R_g ultimate geotechnical foundation resistance (bearing capacity)
SP sum of loads acting on foundation.

Similarly, for structural failure of the foundation to occur, the following condition will be satisfied:

$$R_s < SS \tag{1.2}$$

where
R_s ultimate structural foundation resistance
SS sum of loads and other structural action (such as bending moments) acting on the foundation.

Thus, in carrying out a forensic investigation of foundation failure, there are four broad possibilities:

1. The geotechnical resistance R_g may be inadequate to resist the applied loads;
2. The structural resistance R_s may be inadequate to resist the applied loads and structural actions;
3. The applied loads SP may be larger than anticipated in design, or may contain some components not accounted for in design;
4. The applied loads and structural actions SS may be larger than anticipated in design, or may contain some components not accounted for in design.

The assessment of each of the above possibilities will be considered below, with emphasis being placed on the geotechnical aspects. However, when engaging in forensic geotechnical investigations, it is unwise to ignore the possibility that the failure could be related to the structural aspects.

1.3 Factors Affecting Geotechnical Foundation Resistance

The geotechnical strength of a foundation (i.e. the bearing capacity) will depend on the following factors:

- The geotechnical profile below the foundation;
- The strength of the soils within the depth of influence;

- The nature of the foundation, e.g. whether it is a shallow or deep foundation;
- The directions of applied loading (e.g. the presence of lateral and moment loadings as well as vertical loads).

In carrying out a forensic investigation, each of these factors should be examined to assess whether they were relevant to the failure.

1.3.1 Geotechnical Profile

The geotechnical profile is generally assessed by some form of site investigation, consisting of drilling, in situ probing (for example via cone penetration testing) and/or geophysical methods. Clearly, if the profile is not appropriately characterized, there is a potential for the foundation bearing capacity to be mis-assessed. Among the questions which may need to be asked during a forensic investigation are the following:

- Has the site been characterized appropriately?
- Is the geotechnical characterization of the site consistent with the geological history of the site?
- Have any significant variations in the subsoil profile characteristics across the site been properly taken into account?
- Has the subsurface layering of the subsoil profile been considered and characterized appropriately?
- Have the groundwater and hydrogeological conditions been properly represented?

1.3.2 Soil Strength

There are a number of key questions with respect to the soil strength that need to be asked:

- What are the strength parameters of the various layers within the zone of influence of the foundation?
- Are short-term (undrained) or long-term (drained) conditions relevant?
- What were the pore pressures within the soil when the foundation strength was originally assessed?
- What were the pore pressures within the soil at the time of failure?
- To what extent is strain-softening likely to affect soil strength?
- If relevant, has the cyclic or repeated nature of loading influenced the soil strength?

- In particular, when dealing with sand layers, is there the potential for liquefaction to have occurred under the cyclic loadings applied to the foundation?

The question of undrained versus drained strength is a fundamental issue in soil mechanics. For sandy soils, this is generally not a concern, as the drained condition will generally be relevant unless earthquake or impact loading is to be considered. For clay soils, both the possibility of short-term failure (using undrained strength parameters) and long-term failure (using drained strength parameters) should be considered. Unfortunately, many geotechnical engineers ignore the possibility of long-term conditions being critical because of the emphasis on undrained bearing capacity in clays generally being the critical case. However, in problems involving earth retaining structures or excavations, the long-term drained condition may be critical. An example of the failure of a retaining structure in which the long-term condition was ignored has been described by Daniel and Olsen (1982).

1.3.3 Nature of the Foundation

The questions that may need to be explored in relation to the nature of the foundation include:

- What are the foundation dimensions? These may often be unknown when dealing with old foundations for which there are no records extant. In such cases, it may be necessary to carry out drilling or geophysical investigations to estimate these dimensions.
- Is the as-built foundation the same as the as-designed foundation? For example, if the base of a bored pile has not been properly cleaned, and debris is left between the pile tip and the underlying soil, the actual base resistance may be less than that assumed in design. Another example is in relation to high-rise buildings in Hong Kong in the 1990's, where some bored piles were constructed as much as 20 m shorter than designed, and thus were not founded on sound rock but on completely weathered rock.

1.3.4 Directions of Loading

It is unusual for the applied loading to be purely vertical, as is sometimes assumed in foundation bearing capacity calculations. In general, there will be some components of horizontal and moment loading, both of which may reduce the bearing capacity of a foundation (especially a shallow foundation).

1.4 Factors Affecting Structural Foundation Resistance

Many foundations consist primarily of steel or concrete members. Among the possible areas of concern and relevant questions to be answered are the following:

- Is the steel, or the concrete and its reinforcement, adequate to resist the structural actions within the foundation (axial stress (compression and tension), bending moment, shear, torsion)?
- Is there a possibility that some damage may have occurred to the foundation during installation (for example, damage to a driven pile because of over-driving or from tensile stresses induced in the pile when driving through a stiff layer into a softer layer)?
- Is there a possibility of problems with durability of the steel or concrete in relation to the foundation environment? For example, in highly acidic soils, there is a distinct danger that some corrosion may occur in the upper parts of the foundation which are exposed both to the soil and to air.
- Is there a possibility that loading may have caused the concrete to crack and allow ingress of water to the steel, with subsequent corrosion?

To answer such questions, the prudent geotechnical engineer will engage the services of a structural engineer who can investigate and advise on the likelihood that the failure has been caused by deficiencies in the strength or durability of the structure itself.

1.5 Factors Affecting Applied Loads

There are several forms of loading that may need to be considered in the original design and in the operation of a foundation system, including:

1. Dead loads;
2. Live loads due to occupation of the building:
3. Wind loading, which usually results in lateral, moment and torsional components of load, and which are often dynamic in nature;
4. Dynamic loads due to impact or machinery;
5. Repeated or cyclic loads, due for example to wind or wave loading;
6. Loads induced in the foundation by ground movements, which may be vertical and/or lateral;
7. Loads induced by seismic and earthquake events.

In a forensic investigation, the following questions may need to be answered:

- Were the loads acting on the foundation at the time of failure in excess of the design loads?
- Were all the relevant components of load acting on the foundation taken into account in the design?

- If applicable, was the dynamic nature of the loading taken into account in the design?
- If applicable, were the possible effects of repeated or cyclic loading (which may lead to a reduction in the strength of the supporting soil) taken into account in the design?
- Were there any circumstances under which soil movements could have been generated which would have then impacted on the foundation, and if so, were the resulting loads taken into account in the design?
- In considering the possible effects of earthquakes, was account taken of both the inertial effects (the structural loads generated by the structure) and the kinematic effects on the foundation (i.e. the loads induced by the ground movements generated by the earthquake)?

1.6 Factors Affecting Structural Actions

Structural actions include axial forces, lateral forces and shears, bending moments and torsional moments induced in the foundation, arising from the applied loadings and moments from the structure. The assessment of such structural actions will depend on a number of factors, including:

1. The method of calculation used;
2. The loads and moments assumed to act on the foundation from the structure;
3. The strength of the soil supporting the foundation.

In assessing the appropriateness or otherwise of the structural actions for which the foundation was designed, a number of questions may be asked, including:

- What was the basis of calculating the actions? It is customary for many structural engineers to make simplifying assumptions about the soil reactions acting on the foundation. For example, the soil can be represented by a Winkler or subgrade reaction model, in which the foundation-soil pressure is linearly related to the foundation deflection. Such a simplifying assumption may well lead to inaccurate (and underestimated) bending moments in a shallow footing.
- Were axial forces due to vertical ground movements taken into account in calculating axial stresses in the foundation (especially a pile or pier foundation)?
- Were shear forces and bending moments due to lateral ground movements taken into account in calculating bending moments and shears in the foundation (especially for pile and pier foundations)?
- Was the possibility of a reduction in strength of the supporting soil (e.g. due to potential liquefaction) taken into account?

1.7 Development and Testing of Credible Hypotheses

When the various factors mentioned above have been considered, it is necessary to formulate hypotheses on the most likely or credible factors which may have contributed to the failure. When this has been done, each of the credible hypotheses should be tested by carrying out calculations or some form of laboratory or field testing to assess whether the observed failure can be demonstrated to be consistent with the calculations or tests carried out. Allowance should be made for the likely variations in ground characteristics in carrying out such calculations. It is possible that more than one hypothesis will be found to be consistent, and in that case, further hypothesis testing will need to be carried out to try and assess which of the hypotheses is most likely to have occurred.

1.8 Example of Forensic Investigation

1.8.1 The Project

A commercial project located on the island of Java, Indonesia, involved the construction of three buildings, an office block, a hotel, and a shopping centre. Early in 1991, significant settlements started to develop in the 9-storey office block, which had been largely completed. Because of the continued increase in settlement, and the resulting tilting of the building, the decision was made to demolish the structure. Subsequently, investigations were carried out to assess the cause of the tilting, and these involved a consideration of a number of issues related to soil-structure interaction. A detailed description of this work is given by Poulos (1997), and a summary of some of the main features is presented below.

Figure 1.1 shows a plan of the project site and the locations of the boreholes and cone penetration tests carried out prior to and after construction. Figure 1.2 shows the details derived from the boreholes. A layer of soft to very soft silt extended to a depth of about 11 m, underlain by stiffer silt to about 18 m. This was in turn underlain by

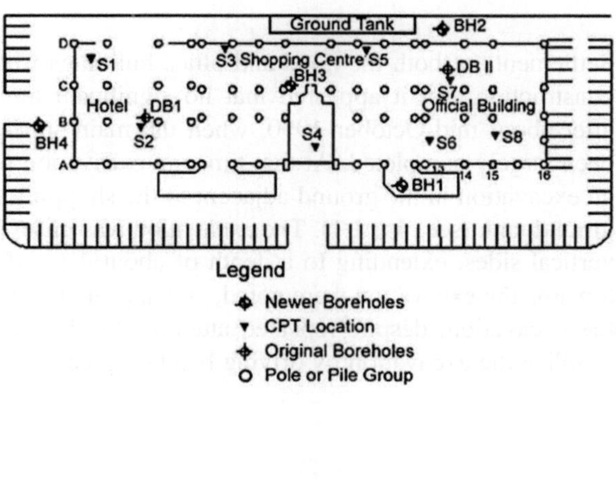

Fig. 1.1 Plan of project

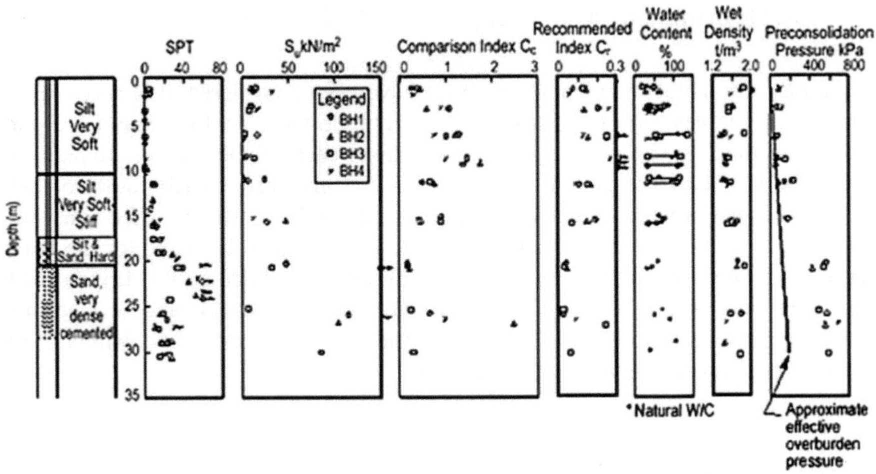

Fig. 1.2 Summary of engineering properties

very stiff and cemented layers of silt, silty sand and sand, in which the SPT values exceeded 60. The borelogs and SPT data obtained from the second investigation confirmed the earlier indications that the site was relatively uniform laterally. Laboratory testing was carried out to measure the undrained shear strength profile, and the one-dimensional compressibility characteristics. In the soft upper layer, the undrained shear strength was typically 20 kPa or less, and the compression index for the normally consolidated state was between about 0.6 and 1.7.

Figure 1.3 shows the plan of the foundation piles, which consisted of 0.5 m diameter driven cast-in situ piles, extending to a depth of about 20 m, and founded in the cemented hard layers of silt, sand and silty sand. A total of 90 piles were used for the office building.

1.8.2 Building Settlements

Settlements of both the hotel and office buildings were monitored during and after construction, but it appeared that no significant movements were observed until after about mid-October 1990, when the main superstructure of the building had been largely completed. At that time, an excavation sub-contractor started to make an excavation in the ground adjacent to the shopping centre for the installation of a ground tank (see Fig. 1.1). The excavation was unsupported and was meant to have vertical sides, extending to a depth of about 4 m. Movement of the soft silty soil towards the excavation were noted, and appears that it was not possible to complete the excavation, despite repeated attempts by the sub-contractor. He attempted to stabilize the excavation by driving bamboo poles, and then steel I-beams, but to no

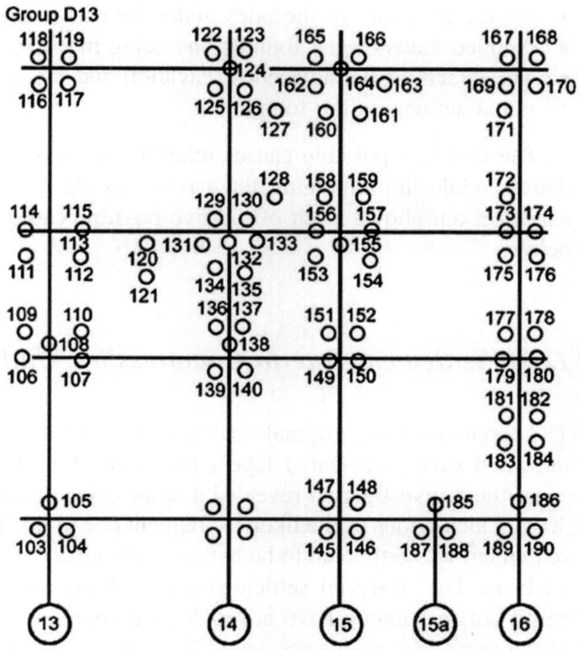

Fig. 1.3 Layout of piles for office

avail. It was also noted that some of the steel I-beams located near the attempted excavation had moved more than 1 m towards the excavation. It was clear then that the uncontrolled excavation had caused gross ground movements.

By mid-February, the average tilt of the office building floor was about 1 in 29, and the horizontal movement of the top of the building was about 1.2 m. Detailed settlement measurements were not available for the hotel building, but it was reported that no additional settlements had been recorded. In addition, there was no evidence of significant settlement of an apartment block on an adjoining site, although there were signs of some horizontal and vertical movements of the soil around this block towards the excavation, extending back about 7 m from the closest side of the excavation.

1.8.3 Investigation of Possible Causes of Failure of the Office Building

The following possible causes of the observed settlements and subsequent failure of the office building were examined:

- Settlements due to compressible underlying layers;
- Geotechnical failure of the piles;

- Structural failure of the piles under the design loading;
- Applied loads on the foundations being in excess of the design values;
- The effects of the adjacent excavation and the consequent imposition of additional actions on the foundation.

The first two possible causes relate to the geotechnical resistance of the foundation, while the remaining three relate to the remaining three areas of possible cause set out above. Each of the five possible causes was examined, as described below.

1.8.4 Settlement Due to Compressible Underlying Layers

The boreholes in the original investigation were taken to a depth of about 25 m and indicated strong cemented layers below the founding level of the piles, but the subsequent investigation revealed a relatively compressible layer from about 26 to 30 m. Calculations of the likely settlement due to this layer were carried out, and the settlement was estimated to be between 16 and 20 mm, depending on the method of analysis. The observed settlements were however of the order of 900 mm, and could not reasonably have arisen from compression of the underlying clay layer. Moreover, such settlements would demonstrate the characteristic slow development due to consolidation, and not the relatively sudden settlement which was observed in this case. It was therefore concluded that the observed settlement and failure of the office building was not due to compression of the underlying clay layers.

1.8.5 Geotechnical Failure of the Piles

Two modes of failure were considered when assessing the possibility of geotechnical failure of the piles:

- Individual failure of the piles in a group;
- Failure of pile group D13, at the corner of the office building, as a block.

In all cases, calculations were based on laboratory-measured undrained shear strength and static cone penetration test data, using the total stress approach. For a single pile, the calculated geotechnical capacity was about 3110 kN, which was somewhat greater than the values from load tests (about 2500 kN). At the time of failure, the estimated average load per pile was only about 900 kN (70 % of the design load). Thus, it was considered unlikely that "individual" pile failure would have occurred.

For group D13, the computed block capacity was 7400 kN, which was about 2.1 times the load acting on the pile group at the time of failure. Thus it was considered unlikely that block failure of pile groups below the building would have occurred.

1.8.6 Structural Failure of the Piles

Axial and lateral loading tests carried out indicated that the performance of the piles under the design loading appeared to be quite satisfactory. Under the design loads, the observed settlements and lateral displacements were relatively small, and there was no indication that structural distress had occurred. Design calculations indicated that the ultimate axial and moment capacity were well in excess of the maximum design values of axial load and bending moment.

It was also recognized that, even if one of the piles in a group was defective (e.g. because of construction problems), there would be the scope for re-distribution of load to the remaining piles and a relatively small increase in movements, but failure would not occur.

In view of the calculations carried out, and the experience of the contractor in constructing this type of pile, it was considered that structural failure of the piles, under the design loads, was not the cause of failure.

1.8.7 Applied Loads in Excess of the Design Loads

It was estimated that the loads on the structure at the end of construction were about 70 % of the design values. There appeared to be no evidence of any overloading of the structure, nor of any occurrence of earthquakes, which could induce additional loads in the foundation piles, at the time of failure.

It was concluded that overloading of the structure did not occur and was therefore not a cause of the failure.

1.8.8 Additional Actions on Piles Due to Ground Movements

The attempted excavation for the ground tank unquestionably caused significant horizontal and vertical movements of the surrounding ground. Such movements may have had the following effects on the existing pile foundations:

- Imposition of lateral soil pressures on the piles, and hence additional bending moments and shears;
- Imposition of additional axial shear stresses due to vertical soil movements, and a consequent increase in axial force in the pile due to negative skin friction.

It was significant to note that, prior to the excavation, there were no unusual settlements of either the office building, the hotel building, or the piles installed for the shopping centre. After the excavation, there were gross movements of the line of piles closest to the excavation. The lateral movements were of the order of 1 m, and the vertical movements were about 200 mm. Some time after the excavation commenced, settlement of the office building commenced, this settlement being most marked nearest the excavation. The hotel building, which was considerably further away from the excavation, did not appear to have suffered noticeable settlement when settlement of the office building commenced, and continued monitoring revealed no significant movement.

Thus, it was concluded that the most likely cause of failure was the additional structural actions (bending moments) caused by the lateral ground movements generated by the uncontrolled excavation. A more detailed examination of this hypothesis was the undertaken, as described below.

1.8.9 Examination of Most Likely Hypothesis

The effects of lateral soil movements were analyzed using the program ERCAP (CPI 1991), while the effects of vertical soil movements were analyzed using the program PIES (Poulos 1989). In estimating the vertical and lateral soil movements, use was made of the work of Mana and Clough (1981) and Clough et al. (1989). The assumed profile of movement (both axial and lateral) with depth is shown in Fig. 1.4. A depth of excavation of 4 m was assumed, although it was unlikely that this depth was reached in all sections of the excavation. The vertical and lateral movements appear to become relatively small at about 2–3 times the excavation depth behind the face of the excavation, and this characteristic appeared to be borne out on site. The second line of piles driven for the shopping centre (about 8–10 m away) did not appear to have suffered significant movements, at least initially. The ultimate skin friction, end bearing and lateral pressures were assessed on the basis of the laboratory measured undrained shear strength (su) values. Values of Young's modulus were estimated

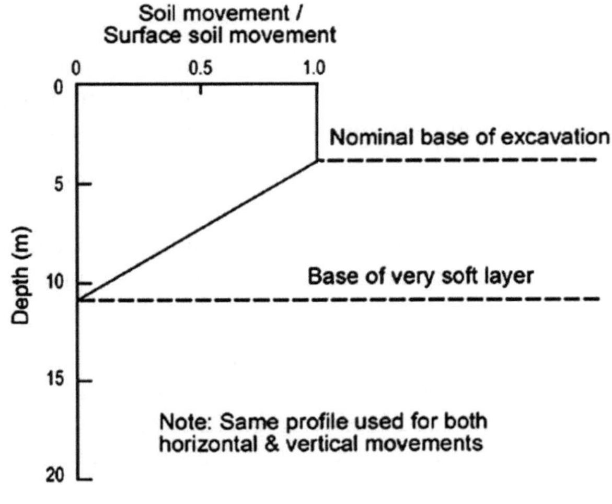

Fig. 1.4 Assumed soil movement profile

from the su values, and also from values backfigured from axial and lateral pile load test data. These values were found to be about 1000–1500 times su.

The piles in the group D13 nearest to the uncontrolled excavation were analyzed to assess the possible influence of soil movements. The estimated axial dead load on the piles was about 1000 kN per pile at the time of the excavation, and it was assumed that the pile heads were restrained against both rotation and translation, because of their fixity to the structure via the pile cap. From the analysis of pile response to vertical soil movements, soil movements of the order of 30–40 mm or more were found to develop an additional downdrag force of about 150 kN, with the additional pile settlement being about 5–6 mm.

In the analysis of pile response to lateral soil movements, it was found that a soil movement of 80 mm or more would cause the soil to flow past the piles, and would develop a bending moment of about 245 kNm in each pile, if the pile was assumed to be elastic. The corresponding head shear force would be 134 kN per pile, giving a total lateral force at the pile cap (for the four piles) of 536 kN.

Figure 1.5 compares the maximum applied bending moment and axial load with the structural capacity of the pile section. It is clear that the ultimate pile capacity was exceeded by a substantial margin. The analysis indicates that the yield moment of the pile section of 160 kNm would be reached at a lateral soil movement of only about 10 mm. Since movements in excess of 100 mm were measured at a nearby pile group of the shopping centre, it was likely that the soil movements occurring at D13 would have been enough to cause the pile to yield at the pile head. The additional restraining force and bending moment developed in the piles could not have reasonably been anticipated by the foundation designers, and were not allowed for in the structural design of the piles.

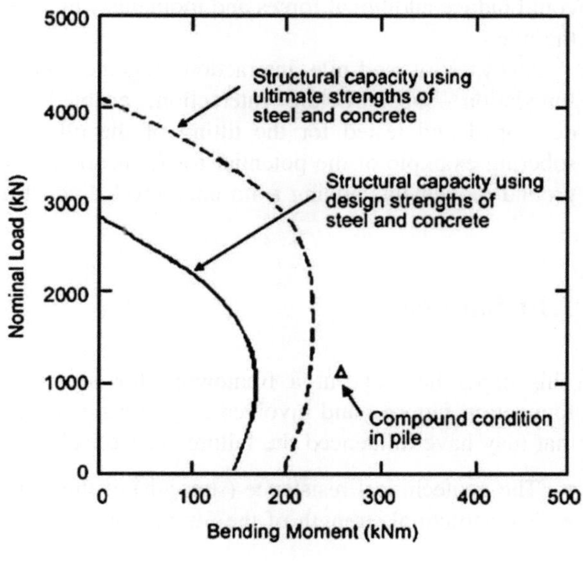

Fig. 1.5 Structural capacity of office building piles

It was therefore postulated that the tilting of the building was triggered by such structural failure of the piles, caused by ground movements resulting from the uncontrolled excavation. Deformation of the corner piles caused a transfer of structural load to adjacent columns of the building, and induced additional bending moments and tensile forces in the floor slab and the beams. Consequently, cracking of the floor slab and beams occurred, resulting in a further redistribution of structural loads to adjacent columns and panels, and hence further cracking. Because of the rigidity of the structure, it tilted subsequently. Tilting caused an increased moment due to the eccentricity of the weight of the building, and the tilting was exacerbated. Thus, the initial failure of the piles due to the ground movements was assessed to have triggered a progressive failure of both the structure and the foundations, extending over a period of 2–3 months.

Some confirmation for the postulated mechanism of tilting was provided by two observations:

- Reports from security personnel in early January 1991 of three loud "cracking" noises at about two-hourly intervals during the night. In the morning, the corner of the office building closest to the excavation had settled about 140 mm.
- Substantial cracking was found in the floor slab in both directions near the corner of the building supported by the pile group D13.

1.9 Conclusion

The analyses of soil-pile interaction for the foundation piles supporting the corner of the office building indicated clearly that excavation-induced ground movements could induce additional forces and moments which could cause structural failure of the pile.

Analyses of soil-pile interaction, together with qualitative consideration of foundation—superstructure interaction, enabled a credible mechanism to be developed and tested for the tilting of the office building. This case is also a sobering example of the potential for foundation damage which can be caused by ground movements arising from uncontrolled construction activities.

1.10 Summary

This paper has set out a framework for forensic geotechnical assessments of foundation failures, and involves a systematic consideration of the various factors that may have influenced the failure. These include:

- The geotechnical resistance (strength) of the foundation;
- The structural strength of the foundation;

- The magnitude and nature of the applied loads acting on the foundation;
- The structural actions developed within the foundation.

In each case, a series of questions which need to be addressed is set out. Having addressed these questions, it is then necessary to formulate one or more credible hypotheses and to test these for consistency with the observed behaviour.

An example illustrating the application of this approach is described. In that case, it has been deduced that the key feature of the failure was the bending moments induced in the piles by ground movements arising from an uncontrolled excavation nearby, i.e. additional actions induced in the foundation which were not considered in the design.

Acknowledgments The author is grateful to Dr. Ian Johnston, Mr. Patrick Wong and Ms. Frances Badelow for their constructive comments on the paper.

References

Carper KL (1986) Forensic engineering—learning from failures. ASCE, New York

Clough GW, Smith EM, Sweeney BP (1989) Movement control of excavation support systems by iterative design. Found. Eng.—Curr. Principles Practices ASCE 2:869–884

CPI (1991) ERCAP—users' manual. Coffey Partners International, Sydney

Daniel DE, Olsen RE (1982) Failure of an anchored bulkhead. Jnl Geot Eng ASCE 108 (GT10):1318–1327

Day RW (1999) Forensic geotechnical and foundation engineering. McGraw Hill, New York

Mana AI, Clough GW (1981) Prediction of movements for braced cuts in clay. J Geotech Div ASCE 107(GT6):759–777

Poulos HG (1989) PIES—users' manual. Centre for Geotechnical Research, University of Sydney, Australia

Poulos HG (1997) Failure of a building supported on piles. International Conference Foundation Failures, IES Singapore, pp. 53–66

Chapter 2
Forensic Geotechnical Engineering Theory and Practice

Suzanne Lacasse

Abstract Geotechnical engineers working with forensic evaluations must apply science and engineering within the rules and practice of the legal system, in order to be effective in representing reality and resolving conflicts. Such rules and practice will vary from country to country. The geotechnical work required for the documentation of forensic cases, however, should observe the same standards of quality in all countries. To provide the required assistance in the settlement of disputes, the engineer needs to combine high quality forensic investigations consistent with good science and engineering with an ability to clearly present the matters being disputed. This keynote lecture reviews the basic requirements of forensic geotechnical engineering. The technical forensic investigation requires collection of data, problem characterization, development of failure hypotheses, a realistic back-analysis, observations in situ and in some cases performance monitoring, and most importantly quality control of not only the formal but also the technical aspects of the work. Two case histories of landslides are presented. The role of the geotechnical engineer as a forensic expert is highlighted, in particular in investigating damage and failure, evaluating the hazards and consequences, developing repair recommendations and preparing reports.

Keywords Forensic · Risk · Expert witness · Standard of care

2.1 Introduction

The practice of forensic geotechnical engineering is the application of geotechnical engineering to answer questions pertaining to a conflict in the legal system. The word forensic comes from Latin, where forensic means "of" or "before the forum". In Roman times, a criminal charge required presenting the case in a forum before a

S. Lacasse (✉)
Norwegian Geotechnical Institute (NGI), Oslo, Norway
e-mail: suzanne.lacasse@ngi.no

V.V.S. Rao and G.L. Sivakumar Babu (eds.), *Forensic Geotechnical Engineering*, Developments in Geotechnical Engineering, DOI 10.1007/978-81-322-2377-1_2

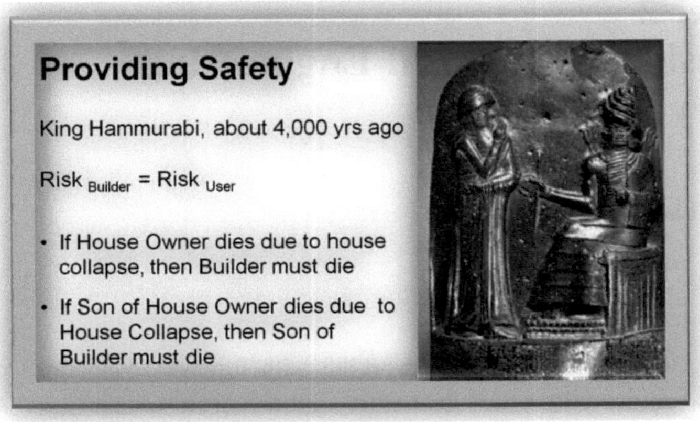

Fig. 2.1 Sanctions in construction code in old Babylonia

group of public individuals. The individual(s) with the best argument and delivery would determine the outcome of the case. The modern usages of the word forensic are a form of legal evidence and a category of public presentation (Lucia 2012).

The basic book on forensic engineering, "Forensic Geotechnical and Foundation Engineering" by Day (2011), interestingly, begins with the example of the sanctions in the legal code of construction of the great Babylonian King Hammurabi.[1] Two sanctions are illustrated in Fig. 2.1. Fortunately for our profession, the times and the legal codes have changed! Even more impressive, the code makers understood about risk, where Risk includes the Hazard and the Consequence components of an event.

Geotechnical engineers must apply science and engineering within the rules and practice of the legal system in order for their work to be effective in representing reality and resolving conflicts. Such rules and practice will vary from state to state and from country to country. However, the geotechnical work required for the documentation of forensic cases should observe the same standards of quality in all states and countries. To provide the required assistance in the settlement of disputes, the geotechnical engineer needs to combine high quality forensic investigations consistent with good science and engineering with an ability to clearly present the matters being disputed. The technical forensic investigation requires collection of data, problem characterization, development of failure hypotheses, a realistic back-analysis, field observations and in some cases performance monitoring, and most importantly quality control of not only the formal but also the technical aspects of the work.

[1]The Code of Hammurabi is a Babylonian law code dating back to about 1772 BC, and is one of the oldest writings in the world. The sixth Babylonian king, Hammurabi, enacted the code. The Code consists of 282 laws, with scaled punishments, adjusting "an eye for an eye, a tooth for a tooth" depending on social status (slave versus free man). Nearly one-half of the Code deals with matters of contract, establishing for example the wages to be paid to an ox driver or a surgeon. Other provisions set the terms of a transaction, for example, the liability of a builder for a house that collapses, or property that is damaged while left in the care of another.

Issues that have arisen requiring a forensic geotechnical analysis include for example (Day 2011): expansive soils, collapsible soils, settlement of shallow and deep compacted fills, moisture intrusion, corrosion, exposure to sulphates, runoff and drainage, pavement failures, slope instability, foundation failures, excavation failures, "differing" site conditions and underground pipeline failures.

The keynote lecture reviews the basic principles of forensic geotechnical engineering. Two case histories of slope instability, illustrating the importance of thorough and extensive geotechnical investigations and analyses, are presented. The first case study exemplifies how information from geological, geophysical and geotechnical investigations was integrated to establish a realistic model of the slope instability and to establish the trigger of the slide. This case ended up in court. The second case study tells an abridged story of the Storegga slide, one of the largest known underwater slides on earth. The role of the geotechnical engineer as a forensic expert is also highlighted, especially in preparing evidence and explaining the evidence in a clear manner.

2.1.1 The Practice of Forensic Engineering

The principles presented herein are based on the excellent review prepared by Lucia (2012) as part of the State-of-the-Art and State-of-Practice in Geotechnical Engineering GeoCongress in Oakland California USA.

In litigation, forensic engineering contains components of legal evidence and public presentation. The ability to present complicated technical facts to a layperson is the key to the outcome of a conflict. Unfortunately, facts are only a part of the resolution of a conflict. A poor presentation of the facts, in some case a distorted presentation of the facts, can determine the outcome in a direction that does not agree with the engineering state-of-the-art.

The evaluation of a failure in the case of litigation has to incorporate the laws of science, the practice of engineering and the rules of evidence within a court system.

The forensic evaluation leads to an opinion on the cause of a failure and then to the responsibility for that cause. The technical evaluation leads to an assignment of responsibility to a party.

Engineers are typically ill-equipped to deal with the forensic process. Lawyers on the other hand are skilled at the presentation of facts and the resolution of conflicts.

2.1.2 Standard of Care

The engineer's compliance to the "standard of care" is usually one of the main issues in the case of litigation. ASFE (1993) stated that the standard of care is "that level of skill and competence ordinarily and contemporaneously demonstrated by professionals of the same discipline practising in the same locale and faced with the same or similar facts and circumstances". Parties in litigation therefore look for factors that contributed to the failure that are not representative of the standards met by other engineers (including sampling, in situ testing, laboratory testing, type of analyses, assumptions made, recommendations and performance monitoring). The courts do recognize that the standards and practice of engineering can vary over time.

Because there are never complete data to describe a site, the geotechnical engineer must interpolate between the limited data, apply assumptions to an analysis, make recommendations based on the interpolations and assumptions, and possibly include a monitoring programme. All of these steps require "substantial judgment" (Lucia 2012). While extreme cases do happen where causes and effects are obvious, most of the time the evaluation of the compliance with the Standard of Care by engineers fall into a grey area which very often results in a difference of opinions between experts. The standard of care is never written down in a manner that has gained universal acceptance, and is therefore always subject to debate. The combination of science and empiricism is considered the definition of the Standard of Care (Lucia 2012).

2.1.3 Expert Evidence

The forensic evaluation is intended to come to an opinion as to the factors that led to the failure and were ultimately responsible for the failure. Lucia (2012) reported that the USA Supreme Court ruled that the judge must ensure that any and all scientific testimony or evidence admitted is both relevant and reliable. The rules of admissibility are based on four criteria:

- When a scientific theory or technique is used in the development of an opinion, has the theory or opinion been tested?
- Has the scientific theory or theory been subjected to peer review and publication?
- Are there standards to control the application of the scientific theory or technique, and is there a known or potential error rate?
- Has the scientific theory or technique gained acceptance within the relevant scientific community?

Knowledge needs to be more than subjective belief or speculation, and must apply to a body of data or facts. Expert testimony does not need to be accurate with 100 % certainty. It is recognized that uncertainties exist in science and engineering.

2.1.4 Forensic Investigation and Litigation Process

The process begins when the parties realize that something different than assumed by the engineer has occurred. The question is, for example, whether the situation could have been foreseen, was knowable, was the result of an error in calculations, an omission, a negligent act on the part of the engineer, or a defect in construction caused by the contractor.

The burden of proof is on the plaintiff to demonstrate that the geotechnical engineer breeched the Standard of Care.

After forensic investigation has been completed by both sides of the litigation, the process of financially resolving the conflict begins. The decision of a settlement or a court case is generally a business decision, including the consideration of the costs yet to be incurred and the risk of losing. Lucia (2012) reported that 90 % of all the litigation cases he has been involved in have been settled prior to trial.

2.1.5 Case Studies Presented by Lucia *(2012)*

Lucia (2012) presented two examples of how the quality of the expert testimony can determine the litigation outcome. Central facts are summarized in Table 2.1.

The one lesson learned from these two case studies is that convincing a jury or a judge in a case involving technical matters depends on presenting the best facts supporting the case in such a way that a layperson can understand. No matter how correct a theory can be, a layperson will not be convinced unless he/she can understand its effects. Often an analogy, e.g. with a simple physical demonstration, will end up in being the most effective tool.

Table 2.1 Two examples of expert testimonies (case studies from Lucia 2012)

Case no./description	Outcome/comments	Ruling	Expert testimony
Ground settlement after trenching for utility installation. Owner demanded that the contractor remedy to the defect. Mediation failed, and the case ended in court	Contractor did place the fill according to contract specification, but the contract specification was inadequate. Should the engineer have reviewed the specification before start of the work?	Jury ruled in favour of contractor	Good facts, good presentation (simple demonstration of wet and dry densities) The technical issues were narrow and were made easily understandable to jury
Shallow failures (landslides) in open spaces in large housing complex. Homeowner association sued developer and contractor. Repair costs estimated by opposing experts to USD 2.5 and 25 M	Technical issues debated: – Extent and residual strength of colluvium; – Assumptions on water level; – Effect of earthquake loading; – Appropriate factor of safety (1.25, 1.5 or >1.5?)	Jury ruled USD 6.5 M compensation to be paid by developer and contractor to homeowners	Good facts, bad presentation by experts on both sides Jury members told afterwards that they did not understand what the experts on either side were talking about, and did not know who had the best facts

2.2 Case Study I

2.2.1 Finneidfjord Nearshore Shallow Landslide

2.2.1.1 Description of Case Study

In June 1996, a landslide of over 1 million m^3 occurred just off the shoreline near the village of Finneidfjord in Northern Norway. During the Holocene, the marine deposits were exposed to freshwater flow and the leaching of salts resulted in very sensitive clays, called quick clays. The initial failure was believed to have occurred on the steepest slope of the foreshore (Janbu 1996; Longva et al. 2003; Gregersen 1999). Although 90 % of the slide was below sea level, its retrogressive nature meant that while encroaching 100–150 m inland, it destroyed 250–300 m of the Norwegian E6 highway and three residential houses (Figs. 2.2 and 2.3). Tragically four persons were killed, one in a car on the highway and three persons in one of the houses.

Geotechnical investigations prior and subsequent to the slide revealed large volumes of quick clay in the area. The Finneidfjord landslide developed along a well-defined plane. Based on seismic reflection and core data, the slip plane was identified as a stratigraphically weak, laminated, clay-rich bed (L'Heureux et al. 2012a, b; Steiner et al. 2012). The combined geophysical and geotechnical investigations indicated a complex, multi-stage failure.

A fill, from the excavated material for a nearby tunnelling project, was placed on the foreshore of the embankment just before the failure (Fig. 2.2). Blasting in the area had also taken place prior to the slide. In addition, intense rainfall had occurred in the days prior to the failure.

2.2.1.2 Post Failure Investigation

High-resolution swath bathymetry 2D reflection seismic profiles, a decimetre-resolution 3D seismic volume, numerous short cores, two long cores, and Free Fall cone penetrometer (FF-CPTu) profiles were used. The overlapping multi-beam bathymetric data sets were collected in the area between 2003 and 2009 (Vardy et al. 2012).

The geophysical data were complemented by geotechnical tests. Multi-sensor core logging (MSCL), detailed sedimentological description and X-ray imagery were used to describe structure, stratification and composition. The tests yielded magnetic susceptibility; density; P-wave velocity; porosity and water content; and grain size distribution. In situ geotechnical data were also acquired using free-fall CPTu. A total of 38 individual drops were done (Steiner et al. 2012).

Swath bathymetry images revealed evidence of mass wasting at several locations in the area. The landslide scars are 2–3 m high and devoid of debris, with the smooth surface interpreted as exposed slip planes. The planes correlate to a well-defined and high-amplitude reflection in high-resolution sub-bottom profiles (Fig. 2.4). This high-amplitude reflection was mapped throughout the fjord basin.

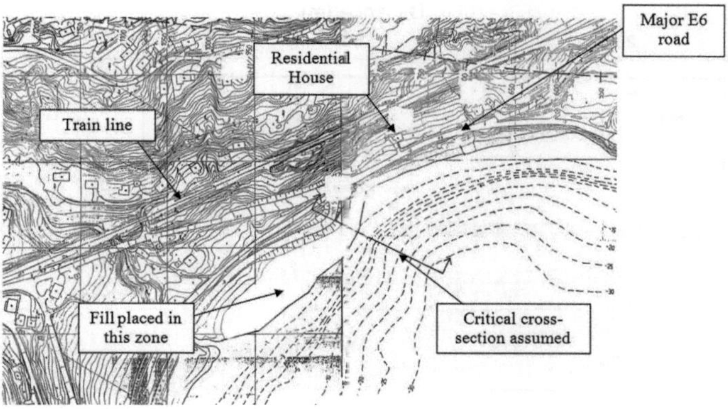

Fig. 2.2 Area prior to the Finneidfjord submarine slide (Gregersen 1999)

The bathymetry and seismic data and sediment cores showed that many of the underwater slides in the area were initiated along these “weak” beds (L’Heureux et al. 2012a). This includes the landslide of June 1996 (Fig. 2.3). Stage numbering in Fig. 2.3 refers to the phases of landslide development identified by Longva et al. (2003).

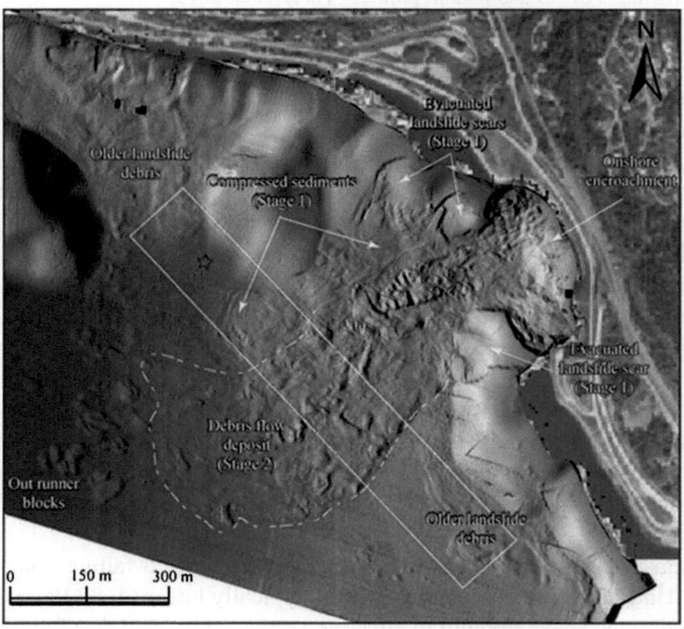

Fig. 2.3 Surface morphology for 1996 landslide imaged using high-resolution swath bathymetry (Vardy et al. 2012)

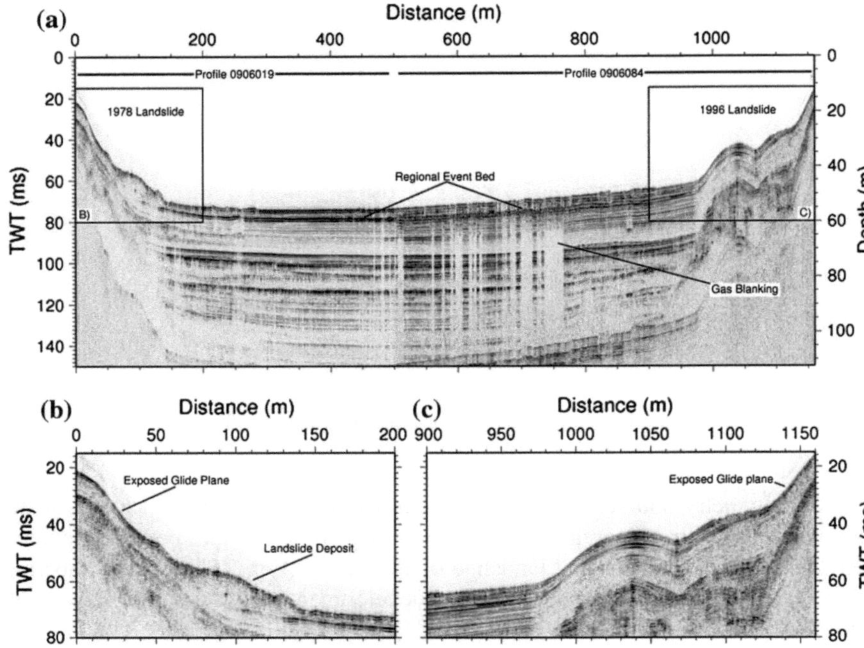

Fig. 2.4 Crossing TOPAS sub-bottom profiles (Vardy et al. 2012)

A two-stage mechanism describes the failure: the initial phase was a transitional movement of the foreshore slope; the second stage involved blocky debris flow deposition as the head wall retrogressed to the shoreline and beyond.

Figure 2.5 shows the results of seismic and geotechnical investigations. The upper part of the figure shows seismic evidence for the slip plane (weaker layer) as a composite seismic reflection event that is partly eroded underneath landslide debris. The lower part of Fig. 2.5 shows the results from piston core, laboratory analyses and free-fall CPTu tests (L'Heureux et al. 2012a).

The soil recovered in piston cores pushed adjacent to the slide deposit contained essentially homogeneous silty clays with shell fragments. The lithology changed suddenly at 2.9 m depth where a 45 cm thick bed, consisting of a 5 cm thick sand layer sandwiched between two distinct grey clay layers, 20 cm thick, was observed, with lows in both magnetic susceptibility and gamma density, and a very sharp peak in magnetic susceptibility and gamma density for the sandy layer. These features were confirmed by the free-fall CPTu tests. The water content for the laminated silty clay averaged 35 %.

In contrast, the water content was higher in the inferred failure zone, typically 45–65 %. Undrained shear strength (s_u) were typically lower (4–8 kPa), whereas the ratio of undrained shear strength to effective vertical stress (s_u/σ'_{vo}) fell between 0.2 and 0.3, which is a reasonable value for a normally consolidated clay. In contrast, the ratios s_u/σ'_{vo} for the background silt exceeded 0.3.

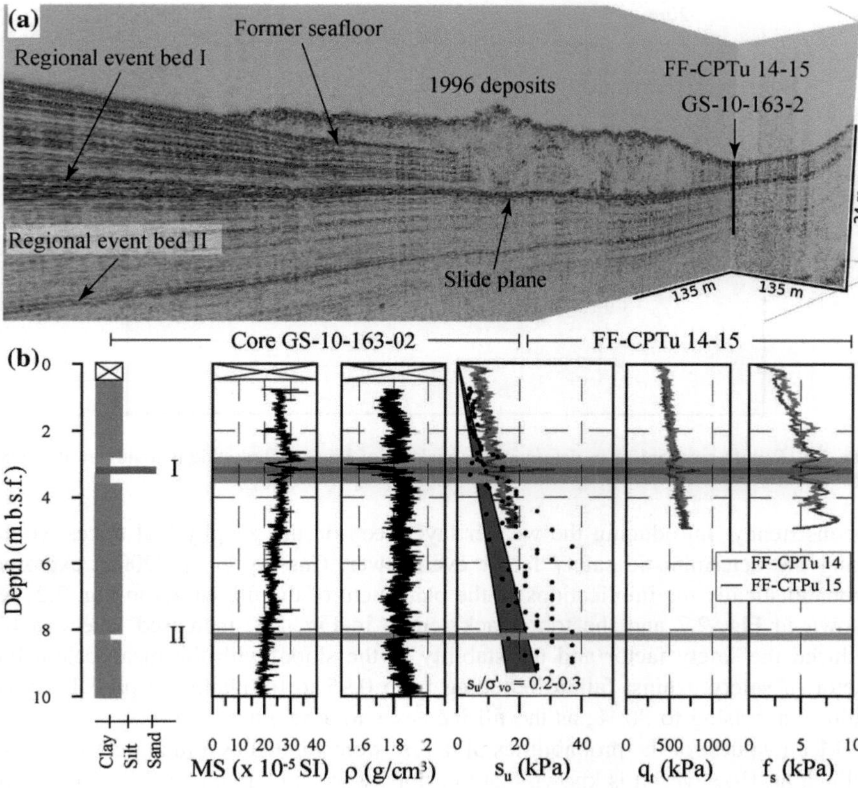

Fig. 2.5 Magnetic susceptibility (MS), gamma density (ρ), undrained shear strength (s_u), corrected cone resistance (q_t) and sleeve friction (f_s) in Finneidfjord (L'Heureux et al. 2012a)

2.2.1.3 Deterministic and Probabilistic Analysis of Slope Stability

An initial deterministic stability analysis was run along the steepest cross section in the region surrounding where a fill had been placed (Fig. 2.2). There was consensus that the slide was initially triggered in this north-west region (Janbu 1996; Gregersen 1999; Longva et al. 2003), based on eyewitness accounts of waves, bubbles and whirls moving away from the shore and sea bottom investigations. The stability analysis used the Morgenstern-Price method with the SLOPE/W software. The initial deterministic failure mechanism investigated is shown in Fig. 2.6. For the properties and geometry assumed, the slope was found unstable with a safety factor of 0.95.

The initial deterministic analysis predicted that the slope cannot hold even though it was actually standing before the fill was placed. Geometrically lower slope steepness, higher soil strengths or model bias can explain such apparent

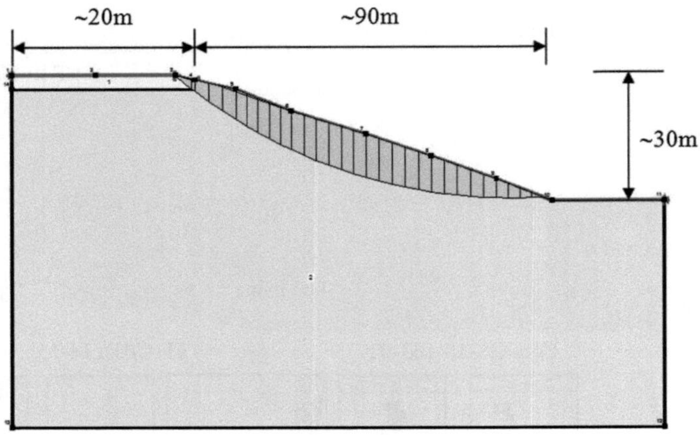

Fig. 2.6 Failure mechanism in deterministic analysis of Finneidfjord slide (Cassidy et al. 2008)

inconsistency. Introducing the weaker layer seen on the geophysical traces would bring the deterministic safety factor even lower. Cassidy et al. (2008) explored probabilistically the implications of the placement of the fill shown in Fig. 2.2. As shown in Fig. 2.7 and the top (black) curve in Fig. 2.8, increased layers of fill reduced the safety factor and the stability of the slope, with the mean calculable factor of safety against failure reducing from 0.95 to 0.91 and the probability of failure increasing to 80 %, as the fill increases to a height of 2.5 m.

Mean values for the probabilities of failure were studied as a function of varying fill height. However, it is known that the slope was standing under the condition of "no fill". Under the assumption that the placement of a fill is the only triggering mechanism, the probability of failure was again calculated.

Fig. 2.7 Probability of exceeding factor of safety (FoS) of unity in probabilistic analysis for increased fill heights (Cassidy et al. 2008)

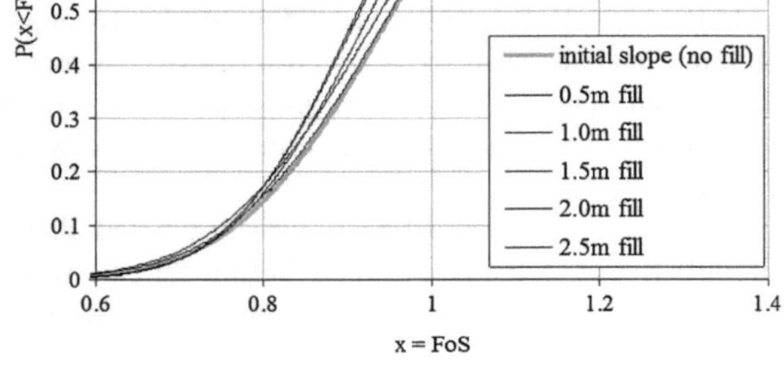

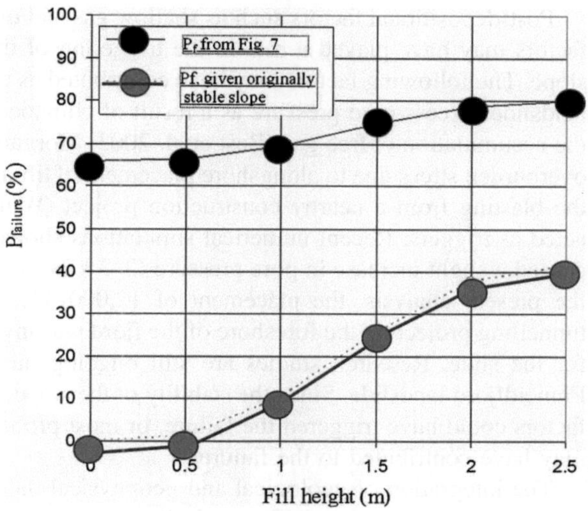

Fig. 2.8 Predicted probability of slope failure as a function of fill height (Lacasse et al. 2013)

The probability of slope failure given an initially stable slope was described as:

$$P(\text{FoS} < 1 | \text{FoS}_{\text{nofill}} \geq 1) = 1 - \frac{P(\text{FoS} \geq 1)}{P(\text{FoS}_{\text{nofill}} \geq 1)} \quad (2.1)$$

where FoS is the factor of safety. These results of the probabilistic analyses are shown in Fig. 2.8. The analysis suggests that by including the known fact that the slope was originally stable, the probability of failure increases from close to 0 to just over 40 % with a 2.5 m high fill (blue curve in Fig. 2.8). The increase is not linear, with initially very little effect from adding more fill, but then also tapering off between the 2 and 2.5 m scenario results. The safety was therefore marginal ($P_f \approx 40$ %) with the addition of a 2 m fill. Cassidy et al. (2008) also assessed the vulnerability and consequences of the Finneidfjord slide.

2.2.1.4 Possible Triggers for Finneidfjord Landslide

For the 1996 landslide near Finneidfjord, the identification of key reflections in the geophysical investigations indicated that the glide plane for the 1996 landslide may lie within the upper clay layer. This agreed with short cores higher up the foreshore slope that sample the exposed glide plane. In these cores, the sandy and lower clay layers were preserved. The upper clay was still present, but significantly thinner than observed in cores recovered adjacent to the slide (L'Heureux et al. 2012a). While the physical properties of the weaker material depended on the formation processes (i.e. rapid deposition of sensitive clay-rich material), it is unlikely that this alone made the layer weak enough to fail.

Post-depositional factors such as shallow gas and/or fluid flow and other external factors may have played a role in the triggering of the failure of the Finneidfjord slope. The following factors may have contributed as pre-conditioning factors to the landslide: excess pore pressure as a result of climatic and anthropogenic factors or the accumulation of free gas (Best et al. 2003; Morgan et al. 2009). The increase in overburden stress due to alongshore placement of fill material (Gregersen 1999) and the blasting from a nearby construction project (Woldeselassie 2012) could have acted as triggers. Recent numerical simulations showed that vibrations could have caused a slight increase in pore pressure (2–3 kPa) in the weaker silty clay layer. In the present analysis, the placement of 12,000–15,000 m^3 of fill from a nearby tunnelling project on the foreshore of the fjord was investigated as a possible trigger for the slide. Research studies are still ongoing on the possible triggers for the Finneidfjord landslide. Since the stability of the slope was initially marginal, several factors could have triggered the failure, or most probably, a combination of factors may have contributed to the failure.

The integration of geological and geophysical data was probably indispensable to construct a complete picture of all aspects influencing the conditions of the Finneidfjord slope, but also other instabilities in similar settings. The geophysics revealed the occurrence of a composite reflection of a thin sandy unit sandwiched between two clay units, which is new information. Similar identifications have been made in other fjords as the result of terrestrial quick-clay landslides in the catchment of the fjord, both in Norway and Canada (L'Heureux et al. 2012a). The stability analyses herein were done before all the geophysical information became available, and do not account for the presence of the effect of the weak layer. The weaker layer would change the shape of the slip plane, resulting in a lower safety factor and a higher probability of failure.

The jury involved in the litigation case concluded in disagreement with the placement of 12,000–15,000 m^3 of fill on the foreshore of the fjord being the trigger for the slide, and ruled that the cause of the failure could have been an increase in the pore pressure in the clay (at the time, the increase in pore pressure was still unexplained). The reason for this conclusion was the jury believing one of the experts more than the other one, probably because of a more forceful and more convincing presentation by this expert and the lawyer.

2.3 Case Study II

2.3.1 The Story of the Storegga Slide

The Storegga slide at the Ormen Lange site is one of the largest known submarine slides on earth. The head wall of the slide scar is 300 km long. About 3500 km^3 failed from the shelf edge, sliding out as far as 800 km in water depths as deep as 3000 m (Fig. 2.9). The failure started probably some 200 km downhill and crept

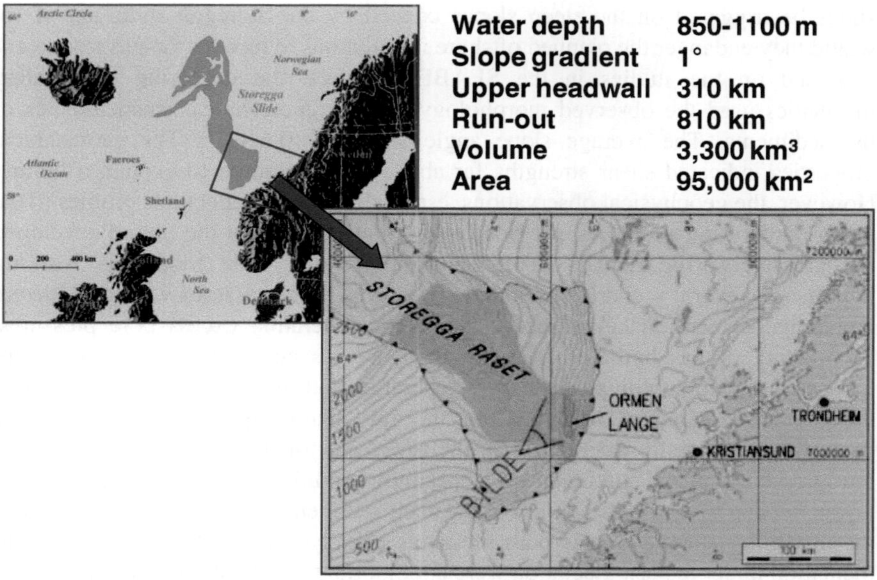

Fig. 2.9 The Storegga slide, 8200 years BP

rapidly upwards as the headwalls failed and slipped down towards the deep ocean floor. At the same time, the mass movement generated a huge tsunami that reached the shores of, among others, Norway, Scotland and the Shetland Islands. The sizable gas resources at Ormen Lange are located in the scar left by the giant underwater slide, beneath a relatively chaotic terrain created by the slide 8200 years ago.

The Storegga was not the subject of litigation, but it was the subject of a large, probably unprecedented, integrated study for the safe development of the deepwater gas field at the Ormen Lange site on the North Atlantic continental margin. In addition, the SEABED project was launched by the partners of the Ormen Lange field (Norsk Hydro ASA, A/S Norske Shell, Petoro AS, Statoil ASA, BP Norge AS and Esso Exploration and Production Norway AS) with the aim of improving the knowledge of the seafloor morphology, the shallow geology, and the potential hazards and risks associated with the area. The aim was to quantify the risks, reduce them as much as possible and ascertain any possible risk to third parties.

The project is an excellent example of the interweaving of research and practice and the cooperation of academia and industry. The reader should refer to Solheim et al. (2005a, b); Kvalstad et al. (2005a, b); Kvalstad (2007); Nadim et al. (2005) and the special issue of Marine and Petroleum Geology (2005) for a complete account of the slide and a summary of the studies by the parties involved.

The design questions that needed to be answered were: (1) Can a new large slide, capable of generating a tsunami, occur again, either due to natural processes or through the activities required for the exploitation of the field; and (2) Can smaller slides be triggered on the steep slopes created by the Storegga slide, and if so, would they endanger the planned offshore installations to recover the gas resources?

Based on the studies in the SEABED project, the triggering and sliding mechanics used the observed morphology and the geotechnical characteristics of the sediments. The average slope angle was only 0.6–0.7°. The geotechnical properties indicated shear strengths far above those required to explain a failure. However, the geophysical observations, especially seismic reflections profiles in the upper parts of the slide scar, provided strong indications that the failure developed retrogressively (Fig. 2.10). Using the retrogressive slide model as working hypothesis, several scenarios of sources of excess pore pressures were considered, including (1) earthquake-induced shear strain generating excess pore pressures, (2) melting of gas hydrates releasing methane gas and water, (3) shear strain-induced contraction with pore pressure generation and strain-softening and (4) rapid deposition. The studies concluded that the most likely trigger was an earthquake destabilizing a locally steep slope in the lower part of the present slide scar. The retrogressive process continued up-slope until conditions improved with stronger layers, related to the consolidation of the shelf sediments during glacial times. Once the instability started, excess pore pressures already generated during rapid sedimentation under the last glaciation were an important contribution to the large slope failure (Bryn et al. 2005).

Excess pore pressures still exist at the site, as demonstrated by in situ monitoring (Strout and Tjelta 2005). The excess pore pressures recorded in several locations and at several stratigraphic levels support the depositional role in the Storegga failure proposed by Bryn et al. (2005).

Seismic studies by Bungum et al. (2005) showed that strong, isostatically induced earthquakes along the mapped faults at the site and stress transfer-induced earthquakes had occurred earlier. They also suggested that multiple strong earthquakes with extended duration most likely occurred and could be the potential trigger for the Storegga slope instability.

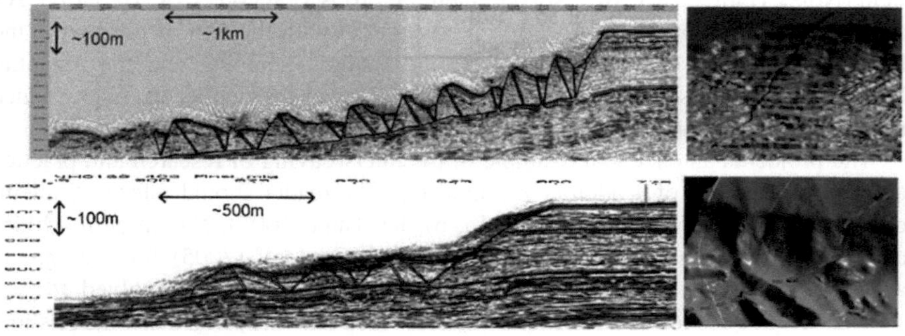

Fig. 2.10 Bathymetry and seismic profiles in the upper headwall at Ormen Lange and interpreted morphology of slide (Kvalstad et al. 2005a)

The tsunami generating potential of submarine slides is today widely recognized. Tsunami studies indicate that the field observations of tsunami run-up fitted will be the retrogressive slide model with velocity of 25–30 m/s, and short time lags of 15–20 s between individual slide blocks (Bondevik et al. 2005). The slide mass involved in the tsunami generation modelled was 2400 km^3.

Figure 2.11 illustrates the hemipelagic deposition of fine-grained sediments in the area of the Storegga slide. One of the pre-conditioning factors for sliding at Storegga was the presence of "weak layers". The stratigraphy and lateral extent of slide-prone deposits (i.e. the contourite drifts in Fig. 2.11) created "weaker" layers that increased the susceptibility of the slope to failure under earthquake loading. The contourites were controlled by the seabed topography and current direction. Deposition of soft marine clays is in fact still ongoing in the slide area.

Figure 2.12 presents a schematic illustration to explain the sedimentation process leading to failure, which supports the hypothesis that major slides have occurred in the Storegga area on a semi-regular basis, related to the glacial/interglacial cyclicity.

The bottom illustration in Fig. 2.12 (denoted 1) gives the last interglacial with deposition of soft marine clays. The middle illustration (denoted 2) presents the last glacial maximum (LGM) with the ice at the shelf edge and deposition of glacial sediments. The top illustration (denoted 3) presents the topography after the Storegga slide. Dating (BP, before present) is given for each illustration. The illustration denoted 3 also shows two older slide scars that were filled with marine clays. The slip planes were found in seismically stratified units of hemi pelagic

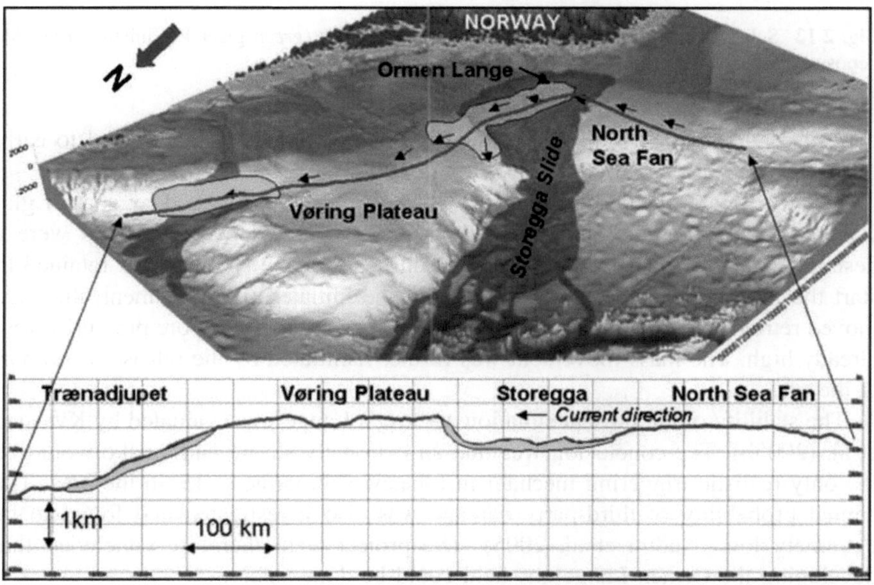

Fig. 2.11 Location of the main contouritic drifts (in *yellow*) (Bryn et al. 2005)

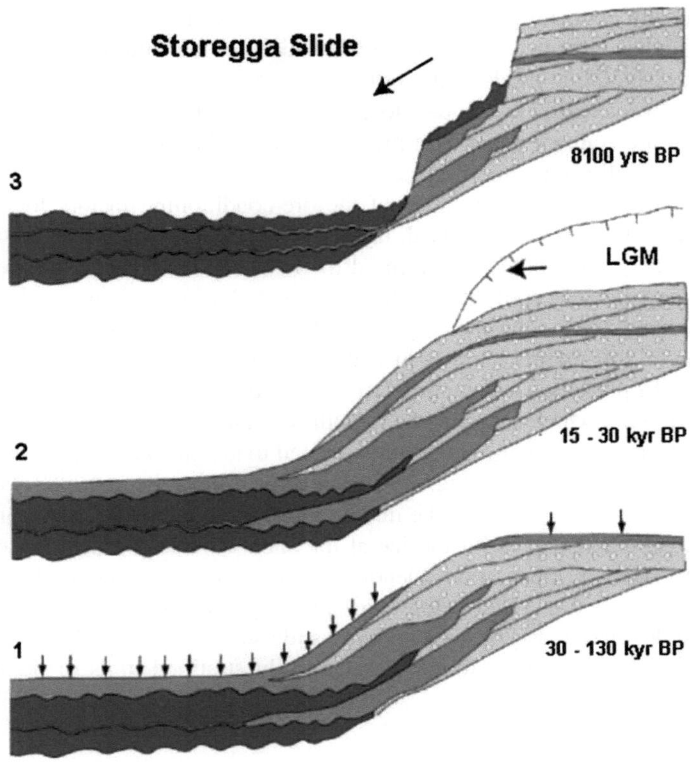

Fig. 2.12 Schematic of the deposition and sliding processes (*green* glacial sediments; *red* slide deposits; *blue* marine sediments) (Bryn et al. 2005)

deposits and the thick infill of stratified sediments indicate a late glacial to early interglacial occurrence of slides (Bryn et al. 2005).

The soft fine-grained hemipelagic deposits were rapidly loaded by coarser glacial deposits during the short glaciations period. Excess pore pressures were a destabilizing factor. The hypothesis of strong earthquake shaking was retained to start the underwater slide. After the earthquake initiated the movement, the slide moved retrogressively by back-stepping up the slope where the pore pressures were already high. The mass movement was further facilitated by the release of support at the toe.

The stability of the present situation at Ormen Lange was evaluated by Kvalstad et al. (2005b). The conclusion was that an extremely strong earthquake would be the only realistic triggering mechanism for new submarine slides in the area. The annual probability of third-party damage was also investigated and found to be extremely low (Nadim et al. 2005). The project team therefore concluded that developing the Ormen Lange gas field could be done safely.

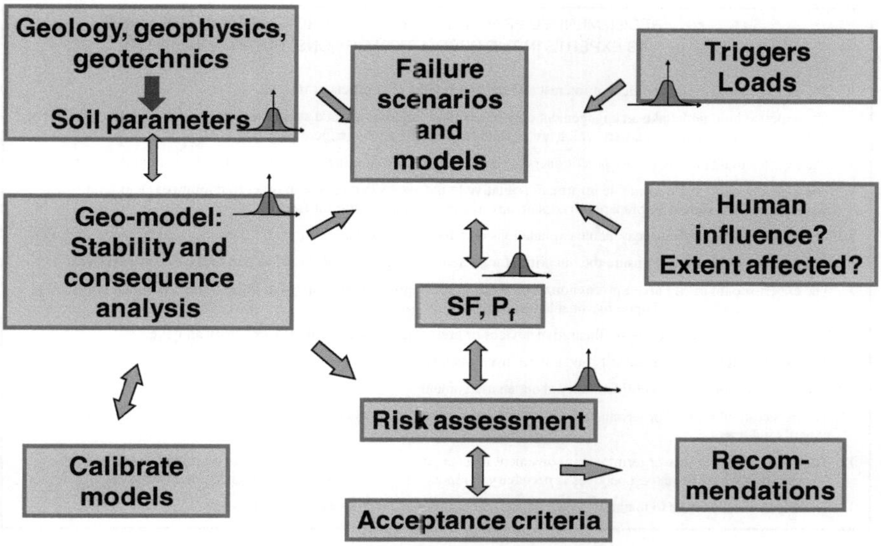

Fig. 2.13 Geohazards assessment methodology (*SF* factor of safety, P_f probability of failure)

In general, the geohazards assessment should include the components in Fig. 2.13, incorporating in each assessment the uncertainties in the parameters (represented in Fig. 2.13 by probability distribution functions in red).

2.3.1.1 The Role of the Expert in Forensic Engineering

Day (2011) presented a Recommended Practice for design professionals engaged as experts in the resolution of construction industry disputes prepared by the ASFE (1993). The 13-step guideline is summarized in Plate 2.1.

The author finds that the 13 guidelines are not specific to forensic engineering, but represents a good practice in all parts of geotechnical engineering. They speak of:

- avoiding conflicts of interest, placing integrity first;
- displaying professional demeanour, including respecting confidentiality;
- remaining within one's area of expertise;
- studying the facts and data in a thorough fashion and listening to other experts' opinions;
- striving for quality in all steps of the work;
- presenting findings in a concise and clear manner.

Any geotechnical engineering project, not only forensic engineering, needs to follow these guidelines.

RECOMMENDED PRACTICE FOR DESIGN PROFESSIONALS ENGAGED AS EXPERTS IN THE RESOLUTION OF CONSTRUCTION INDUSTRY

1. The expert should avoid conflicts of interest and the appearance of conflicts of interest.
2. The expert should undertake an engagement only when qualified to do so, and should rely upon other qualified parties for assistance for assistance in matters which are beyond the expert's area of expertise.
3. The expert should consider other practitioners' opinions relative to the principles associated with the matter at issue.
4. The expert should obtain available information relative to the events in question in order to minimize reliance on assumptions, and should be prepared to explain any assumptions to the trier of facts.
5. The expert should evaluate reasonable explanations of causes and effects.
6. The expert should strive to ensure the integrity of tests and investigations conducted as part of the expert's services.
7. The expert should testify about professional standards of care only with knowledge of those standards which prevailed at the time in question, based upon reasonable enquiry.
8. The expert should use only those illustrative devices or presentations which simplify or clarify an issue.
9. The expert should maintain custody and control over whatever materials are entrusted to the expert's care.
10. The expert should respect confidentiality about an assignment.
11. The expert should refuse or terminate involvement in an engagement when fee is used in an attempt to compromise the expert's judgment.
12. The expert should refuse or terminate involvement in an engagement when the expert is not permitted to perform the investigation which the expert believes is necessary to render an opinion with a reasonable degree of certainty.
13. The expert should strive to maintain a professional demeanor and be dispassionate at all times.

Plate 2.1 Recommended practice for design professionals engaged as experts in the resolution of construction industry disputes (ASFE 1993; Day 2011)

Nevertheless, forensic engineering is complex. Engineers typically work in collaboration where they challenge each other to arrive at the best engineered solution or design possible. In litigation, collaboration is replaced by criticism, sometimes of well-supported professional opinions.

Forensic engineering must therefore ally the best of science with the art of conflict resolution, all within a frame of important economical consequences for one of several parties. The requirements of forensic engineering are so demanding that not all geotechnical individuals can deliver good testimony. Specific qualities are required of forensic geotechnical engineers, including effective public speaking, a quick and logical mind and repartee, and exceptional pedagogical qualities, in addition to a thorough understanding of the subject of litigation.

An important responsibility is therefore to know when to say no to a request of expert opinion in the case of litigation.

2.4 Summary and Conclusions

The practice of forensic geotechnical engineering is the application of geotechnical engineering to answer questions of interest in the context of litigation, where the engineering and legal professions are brought together to resolve the conflict on responsibility for a failure. The geotechnical engineer must apply science and engineering within the rules of the legal system, in order to present effective arguments.

Practitioners are naturally drawn to failures as they provide opportunities to verify calculation procedures, and increasing judgment and understanding on how and why the application of existing knowledge failed to achieve the intended result. The practice of forensic engineering is therefore one of the most interesting and most challenging tasks for a geotechnical engineer.

Forensic engineering is however paradoxical. Engineers typically work in collaboration where they challenge each other to conclude with the best engineered solution possible. In litigation, collaboration is replaced by criticism, sometimes of well-grounded professional opinions. Forensic engineering must therefore ally the best of science with the art of conflict resolution, all within a frame of important economical consequences for one of several parties.

When an expert presents its opinion on the cause or responsibility for a failure, the opinion must be well founded, and most importantly, presented in a way that the judge or jury can understand the technical issues in dispute (Lucia 2012). Simple demonstrations have proven to be very effective to explain physical and/or mechanical behaviour. Lucia (2012), based on 25 years of forensic geotechnical engineering experience, stated that while jurors and judges do their best to sort out the issues, the results can often be confusing. He suggested that settlement of the dispute prior to proceeding to trial is almost always the preferable outcome. In any case, a thoughtful, high quality forensic technical investigation consistent with good science and engineering combined with an ability to clearly present the matters being disputed will always aid in the settlement of the dispute or the outcome of a trial.

The requirements of forensic engineering are so demanding that not all geotechnical individuals should go into forensic engineering, and that specific qualities are required of forensic geotechnical engineers, including effective public elocution, a quick and logical mind and repartee, and exceptional pedagogical qualities, in addition to a thorough understanding of the subject of litigation.

Acknowledgments The author is thankful to her many colleagues at NGI and from our research partners for their contribution to the case studies. Part of the work was funded by the project SEABED through the Norwegian Deepwater Programme and by NGI's Centre of Excellence "The International Centre for Geohazards". The author acknowledges in particular the key contributions made by Maarten Vanneste, Jean Sébastien L'Heureux, Carl-Fredrik Forsberg and Tore Kvalstad from NGI, Mark E. Vardy from Southampton University, Oddvar Longva and Shyam Chand from the Geological Survey of Norway, Haflidi Haflidason and Jo Brendryen from the University of Bergen and Alois Steiner from MARUM, in Bremerhaven.

References

ASFE (1993) Recommended practice for design professionals engaged as experts in the resolution of construction industry disputes. ASFE (association of engineering firms practicing in the geosciences). Silver Springs, MD, 8 pp

Best AI, Clayton CRI, Longva O, Szuman M (2003) The role of free gas in the activation of submarine slides in Finneidfjord. In: Locat J, Mienert J (eds) Submarine mass movements and their consequences. Kluwer Academic Publishers, Dordrecht, pp 491–498

Bondevik S, Løvholt F, Harbitz C, Mangerud J, Dawson A, Svendsen JI (2005) The Storegga slide tsunami—comparing field observations with numerical simulations. Mar Pet Geol 22(195):208
Bungum H, Lindholm C, Faleide JI (2005) Postglacial seismicity offshore mid-Norway with emphasis on spatio-temporal-magnitudal vatiations. Mar Pet Geol 22:137–148
Bryn P, Berg K, Solheim K, Kvalstad TJ, Forsberg CF (2005) Explaining the Storegga slide. Mar Pet Geol 22:11–19
Cassidy MJ, Uzielli M, Lacasse S (2008) Probability risk assessment of land slides: a case study at Finneidfjord. Can Geotech J 45:1250–1267
Day RW (2011) Forensic geotechical and foundation engineering, 2nd edn. McGraw Hill, New York. 508 p
Gregersen O (1999) Kvikkleireskredet i Finneidfjord 10. Juni 1996. NGI report 980005-1. Norwegian Geotechnical institute. Oslo, Norway
Janbu N (1996) Raset i Finneidfjord—20. Juni 1996. Unpublished expert`s report prepared for the County Sheriff of Nordland. Report number 1, Revision 1
Kvalstad TJ (2007) What is the current "best practice" in offshore Geohazard Investigations? A state-of-the-art review. OTC Paper 18545. Offshore technology conference, Houston, TX, USA
Kvalstad TJ, Andresen L, Forsberg CF, Berg K, Bryn P, Wangen M (2005a) The Storegga slide: evaluation of triggering sources and slide mechanics. Mar Pet Geol 22:245–256
Kvalstad TJ, Nadim F, Kaynia AM, Mokkelbost KH, Bryn P (2005b) Soil conditions and slope stability in the Ormen Lange area. Mar Pet Geol 22:299–310
Lacasse S, Nadim F, Vanneste M, L'Heureux JS, Forsberg CF, Kvalstad TJ (2013) Case studies of offshore slope stability. Keynote Lecture submitted to GeoCongress 2013. San Diego, CA, USA. March 2013
L'Heureux J-S, Longva O, Steiner A, Hansen L, Vardy ME, Vanneste M, Haflidason H, Brendryen J, Kvalstad TJ, Forsberg CF, Chand S, Kopf A (2012a) Identification of weak layers and their role for the stability of slopes at Finneidfjord, northern Norway. In: Yamada Y et al. (eds) Submarine mass movements and their consequences, advances in natural and technological hazards research, vol 31. Springer Science+Business Media
L'Heureux JS, Vanneste M, Rise L, Brendryen J, Forsberg CF, Nadim F, Longva O, Chand S, Kvalstad TJ, Haflidasen H (2012b) Stability, mobility and failure mechanism for landslides at the upper continental slope off Vesterålen, Norway. Subm. to Marine Geology
Longva O, Janbu N, Blikra LH, Boe R (2003) The 1996 Finneidfjord slide: seafloor failure and slide dynamics. In: Locat J, Mienert J (eds) Submarine mass movements and their consequences. Kluwer Academic Publishers, Dordrecht, pp 531–538
Lucia PC (2012) The practice of forensic engineering. Keynote Lecture. ACSE Geo-Institute Geo-Congress 2012. State of the art and practice in geotechnical engineering, vol 1. Oakland, CA, pp 765–785
Marine and Petroleum Geology (2005) Thematic set Ormen Lange. In: Solheim A, Bryn P, Berg K, Sejrup HP, Mienert J (eds) vol 22. pp 1–2. 318 p
Morgan E, Vanneste M, Longva O, Lecomte I, McAdoo B, Baise L (2009) Evaluating gas-generated pore pressure with seismic reflection data in a landslide-prone area: an example from Finneidfjord, Norway. In: Mosher DC et al (eds) Submarine mass movements and their consequences, advances in natural and technological hazards research, vol 28. Springer
Nadim F, Kvalstad TJ, Guttormsen TR (2005) Quantification of risks associated with seabed instability at Ormen Lange. Mar Pet Geol 22:311–318
Solheim A, Berg K, Forsberg CF, Bryn P (2005a) The Storegga slide complex: repetitive large scale sliding with similar cause and development. Mar Pet Geol 22:97–107
Solheim A, Bryn P, Sejrup HP, Mienert J, Berg K (2005b) Ormen Lange—an integrated study for safe development of a deep-water gas field within the Storegga slide complex, NE Atlantic continental margin. Mar Pet Geol 22:1–9
Steiner A, L'Heureux JS, Longva O, Lange M, Vanneste M, Haflidason H, Kopf A (2012) An in-situ free-fall piezocone penetrometer for characterizing soft and sensitive clays at Finneidfjord, northern Norway. In: Yamada et al (eds) Submarine mass movements and their consequences, advances in natural and technol. hazards research, vol 29. Springer, Dordrecht (NL)

Strout JM, Tjelta TI (2005) In situ pore pressures: what is their significance and how can they be reliably measures? Mar Pet Geol 22:275–286

Vardy ME, L'Heureux JS, Vanneste M, Longva O, Brendryen J, Steiner A, Forsberg CF, Haflidason H (2012) Multidisciplinary investigation of a shallow nearshore landslide, Finneidfjord, Norway. Near Surf Geophys 10(4):267–278. (Special Issue). In: Arthur et al (eds) Applied marine geophysics

Woldeselassie BH (2012) The effect of blansting in layered soils, example from Finneidfjord, Norway. MSc Thesis, Norwegian University of Science and Technology, Dept of Civil and Transport Engineering. June 2012. 114 p

Chapter 3
Guidelines for Forensic Investigation of Geotechnical Failures

V.V.S. Rao

Abstract When a failure occurs in a project, the first action is to immediately take necessary measures to ensure the stability of the remaining intact structure and protect the overall safety. The next action is to quickly assess the extent of the failure and undertake measures to reconstruct the failed portion and continue the repairs, etc. Only after these initial actions is the systematic analysis done to find the cause of the failure. Very often this analysis is omitted and lessons are not learned. The proper step would be to apply forensic methods to do the analysis of the failure using scientific and legalistic investigations to detect the causes of the failure which could be attributed to geotechnical origin. Such a critical analysis which should be undertaken simultaneously with the first action would provide answers to "what went wrong, when, where, why, how and by whom." This procedure gives strong inputs to improve future designs and also to identify the qualifications and expertise required for the staff. As the forensic analysis is basically a back analysis based on actual failure observations, normally adopted standard procedures of testing, analysis, design, and construction are not adequate in the majority of cases. This paper gives an overall view of the procedures to be adopted in forensic analysis.

Keywords Failure · Distress · Failure mechanisms · Data collection

3.1 Introduction

Forensic analysis in geotechnical engineering involves scientific and legalistic investigations and deductions to detect the causes as well as the process of distress in a structure, which are attributed to geotechnical origin. Cases of remedied installations, where the analysis and evaluation of adopted remedial measures with

V.V.S. Rao (✉)
Nagadi Consultants Pvt. Ltd., Geotechnical Consultant, Chennai, India
e-mail: drvvsr@gmail.com

V.V.S. Rao and G.L. Sivakumar Babu (eds.), *Forensic Geotechnical Engineering*,
Developments in Geotechnical Engineering, DOI 10.1007/978-81-322-2377-1_3

regard to their effectiveness and economy may be subjected to judicial scrutiny, also fall under this purview. The normally adopted standard procedures of testing, analysis, design, and construction are not adequate in the forensic analysis in the majority of cases. The test parameters and design assumptions will have to be representative of the actual conditions encountered at site. The forensic geotechnical engineer (who is different from the expert witness) should be able to justify the selection of these parameters in a court of law. Hence he has to be not only thorough in his field, but should also be familiar with legal procedures. This paper presents principles of planning and executing a forensic investigation.

3.2 Scope

While investigating any distress, the engineer should meticulously follow a well-planned program. The scope of work would broadly be under the following heads:

a. *Compulsory tasks*

   i. survey and documentation of the distress
   ii. scrutiny of all design documents
   iii. review results of original geotechnical investigations, their analysis
   iv. study the field reports of construction
   v. interview persons involved in planning, design, construction, and performance monitoring, etc.

b. *Optional tasks*

   i. perform additional investigations
   ii. develop and conduct special tests
   iii. nondestructive testing of structural element.

c. *Analyze all data and evaluate*

   i. the distress history
   ii. causes of distress
   iii. identify the shortcomings in the original investigation and analysis.

d. *Report*

   i. authority and scope
   ii. history
   iii. summary of original documents
   iv. data collected
   v. interviews

vi. meteorological information
vii. earthquake
viii. investigations performed, their methodology, and their results
ix. analysis
x. conclusions.

3.3 Types of Distress

The visual distress that commonly occurs in normal structures and the most probable causes are listed in Table 3.1.

Apart from the above causes, the most important causes would be inadequate and/or inappropriate soil investigation, selection of design parameters, and use of inappropriate theories.

3.4 Diagnostic Tests

After identifying the cause of distress, the following questions arise:

a. Has the distress fully occurred and if not, how much more can be expected? Quantify.
b. What were the precise causes for distress?

Table 3.1 Types of distress

Structure	Visual distress	Causes
Buildings and bridges	Cracks	Structural geotechnical: settlements heaving uplift vibrations
	Tilts	Differential settlements earthquakes
	Collapse	Excessive loading, soil erosion, repetitive loading, fatigue
Retaining structures	Lateral movement	Inadequate base resistance
	Tilting	Differential settlement between toe and heel
		Excessive surcharge on the backfill
		Excessive water pressure due to poor drainage
Slopes	Excessive settlements	Improper compaction
	Slope failure	Settlement of virgin strata
	Longitudinal cracks	Erosion due to water, rainfall inadequate drainage

c. Whether the soil underwent same stress-deformation history as was anticipated? If not, what was the actual history at site?
d. The effect and efficacy of remedial measures on the soil + structure behavior.

To answer these questions, detailed tests have to be conducted both in the field and in the laboratory. The choice of tests will normally be from among the following tests depending on the problem.

3.4.1 A. Field Tests

i. Borehole investigations including SPTs to a depth deeper than the influence zone.
ii. Cone penetration tests.
iii. Load tests.
iv. Special tests like pressure meter tests, vane shear tests, seismic, or dynamic tests.

3.4.2 B. Laboratory Tests

i. Triaxial shear tests; to simulate the actual field conditions, these tests should be done on stress increment basis on partially saturated sample. The effect of fluctuation in degree of saturation on the deformation behavior of the soil should also be investigated. In case of clays, the field stress history is also to be considered.
ii. Repeated cyclic shear tests, in cases like water towers, bridges, etc.
iii. Large deformation tests, to assess the residual strength and magnitude of final deformations in cases of slopes, etc.
iv. Compaction tests with different compaction energies.
v. Permeability tests.

In case of residual and wind deposited soils, the effect of change of soil structure due to loads, both static and moving ones. In all cases it is advisable to conduct regular borehole investigations and evaluate the subsoil profile.

3.5 Analysis

After collecting all data, detailed analysis can be done to evaluate the design parameters. Use of empirical relationships should be avoided, unless their validity in the particular site is established. The analysis should be based on:

- limit conditions
- partial factors of safety
- equilibrium state vis-à-vis flow state

- liquefaction potential
- critical void ratio in compacted fills.

With these design parameters the load-deformation history of the soil + structure combine can be reconstructed. This process will lead to identification of the causes of distress. A suitable and economically viable remedy can then follow.

3.6 Legal Issues

In the entire process of investigation, the forensic engineer should be careful to ensure that all the experimental and analytical procedures as well as the selected parameters for tests and analysis fully conform to the field conditions. The report should be comprehensive and intelligible to a legal person also. It is advisable to avoid "hi-tech" terminology and strong verbs like-should, must, etc. As far as possible, it is better to avoid too many details in the main text. At the same time, the report should have sufficient details for the client to give a comprehensive brief to the executing agency.

One should realize that the association of the engineer with the project is based on the principle of "contract of skill." Hence the consultant should ensure competent and reliable advice to the client. It is also imperative for the consultant to explicitly detail the risks that might be involved or expected in using the conclusions and recommendations.

The consultant should be aware of the importance and implications of the following Guidelines and literature (Leonards 1982; Green 1988; ASCE 1989; Day 1998):

Facts: These have to be true and should not be erroneous.

- exploration locations
- samples and cores available for inspection
- lithological descriptions of soils and rocks
- measured water tables
- test results

Interpretations: The skill of the engineer is judged here

- borelogs
- inferred stratigraphy between boreholes
- properties of various layers for use in the analysis
- seismic interpretations yielding velocity and layer depths

Opinion: May or may not be disputable

- assumptions
- judgment based on facts and interpretations

Negligence: Obviously, very serious

- performance of investigations
- description and analysis of information
- communication
- accuracy, in general.

Overall, it is emphasized that application of standard of skill and care is expected from a professional, irrespective of quantum of remuneration. However, the liability of the consultant is also limited as the owner pays for the "skill and service" and not for "insurance."

References

ASCE (1989) Guidelines for Failure Investigation. Task Committee on Guidelines for Failure Investigation

Day RW (1998) Forensic geotechnical and foundation engineering. McGraw Hill, New York.

Green DC (1988) Principles for providing geotechnical data in construction contracts. In: Conference on dams, Queenstown, Tasmania (also in Ancold Bulletin No. 81.)

Leonards GA (1982) Investigation of failures. J Geotech Eng Div ASCE GT2 108:187–246

Chapter 4
Caisson Failure Induced by Wave Action

E.E. Alonso, N.P. Pinyol and P. Fernández

Abstract In 2001 a storm induced the failure of four caissons recently built for the construction of a dyke in a new entrance to the Barcelona Harbor. Understanding the failure required the analysis of the construction history and the wave action. The caissons were founded on a thick layer of soft silty clays potentially liquefactable. The main properties of foundation soils are discussed based on laboratory and field data. A back-analysis of the failure is developed using an analytical and a numerical procedure. The analytical procedure is based on well-known solutions and concepts in Soil Mechanics. It involves the calculation of pore water pressure distributions on the foundation, its dissipation, the induced increase of the undrained strength, and the caisson stability during construction stages. An analysis of liquefaction, which explains the collapse observed, is also described. The problem is also solved numerically using a commercial finite difference code. This second analysis allowed a cross-checking of the analytical solution and the quantification of some simplifications introduced that can be properly accounted for in the numerical analysis. In the final part of the paper an alternative stable design of the caisson dyke is presented.

Keywords Caisson dyke · Failure · Liquefaction · Analytical solution · Upper bound · Consolidation · Undrained strength · Numerical analysis

E.E. Alonso (✉) · N.P. Pinyol · P. Fernández
Department of Geotechnical Engineering & Geosciences, C/Jordi Girona 1-3, Universitat Politècnica de Catalunya, Building D2, Barcelona 08034, Spain
e-mail: eduardo.alonso@upc.edu

N.P. Pinyol
Centre Internacional de Metodes Numerics en Enginyeria (CIMNE). C/Jordi Girona 1-3, Universitat Politècnica de Catalunya, Building C1, Barcelona 08034, Spain
e-mail: nuria.pinyol@upc.edu

V.V.S. Rao and G.L. Sivakumar Babu (eds.), *Forensic Geotechnical Engineering*, Developments in Geotechnical Engineering, DOI 10.1007/978-81-322-2377-1_4

4.1 Introduction

The design of a new entrance for the Barcelona Harbor involved opening a channel through the existing dyke and protection of the opening by means of a new dyke made of reinforced concrete caissons (Fig. 4.1).

Caissons (19.6 m wide, 19.5 m high and 33.75 m long) were built in a mobile platform and towed to their intended position, shown in Fig. 4.1. Caissons have a cellular structure. The inner vertical concrete walls allow filling the caisson in a controlled manner. In this way, the caisson may be precisely sunk (by controlled inundation of cells). Once "in situ," the total weight is increased by sand filling the caisson cells. Caisson foundation design should ensure stability against caisson weight and wave loading.

Foundation soils were deposited during the development of two overlapping deltas (Besós river delta toward the northeast and Llobregat river delta toward the southwest of the site). Soft silts and silty clays extend from the surface to substantial depths (tens of meters). A band close to the coastline is covered by a mantle of sand whose thickness decreases toward the sea.

The deep soft soils in the Barcelona Harbor area are a challenge for caisson stability. The favored design is to substitute part of the natural soils by a frictional fill extending on both sides of the breakwater (seaside and land side). Figure 4.2 shows a sketch of the foundation conditioning. A dredged trench is first excavated. Coarse granular soils are then extended and a final gravelly layer is leveled in preparation for caisson sinking.

Fig. 4.1 The new caisson breakwater (*background photograph* Google Earth)

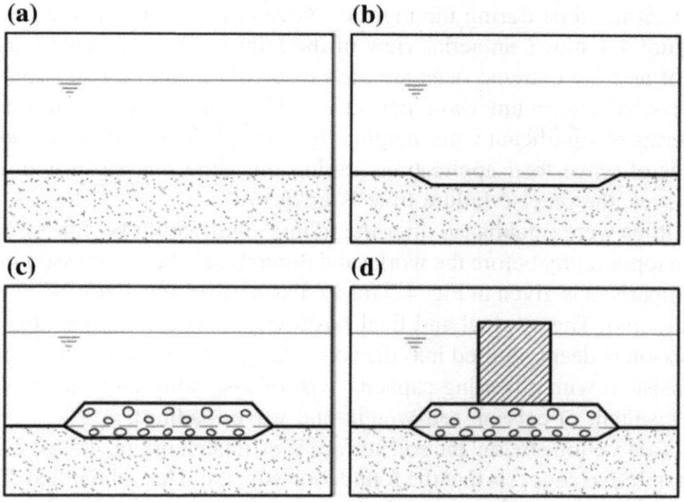

Fig. 4.2 Caisson construction sequence: **a** initial soil conditions; **b** trench excavation; **c** extension of granular embankment; **d** caisson sinking

Once sunk in place, caissons are finally capped with a concrete slab and a protective wall is built to avoid wave overrunning.

Dredging of the trench prepared to receive the coarse granular fill was finished on November 2000. Trench filling took the following 6 months. On May 10, 2001 the granular base was leveled and ready for the sinking of four caissons, in the position shown in Fig. 4.1 (caissons 1, 2, 3, and 4). Caisson sinking began, however, in the middle of October 2001. Cells were filled with sand some days later.

On November 10, 2001 an ENE storm with maximum significant wave heights of 4 m hit the coast. The time record of wave period and significant height is given

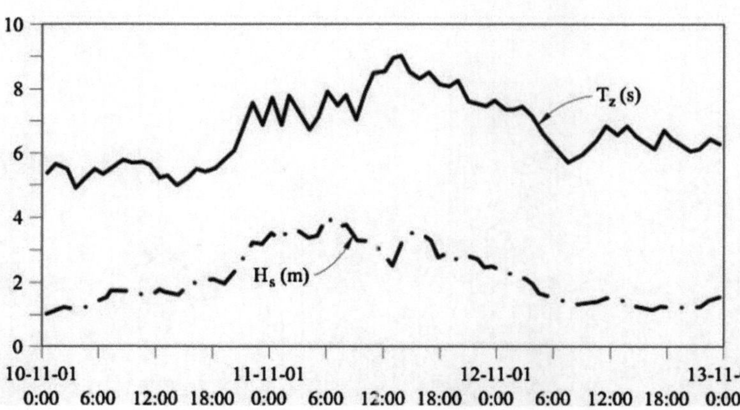

Fig. 4.3 Significant wave heights (H_s) and wave period (T_z) of storm

in Fig. 4.3. Some time during the night of November 10 and 11, the four caissons failed. Figure 4.4 shows an aerial view of the failure. The two central caissons are not in sight and the extreme ones are seen to be tilted and partially submerged.

The recorded maximum wave period was 9 s. The maximum intensity of the storm in terms of significant wave height ($H_s = 4$ m) lasted around 1 h and therefore the number of wave load applications during this time interval was around 250–350. However, the precise failure time is unknown.

Soil profiles were established after the failure. They could be compared with the sea bottom topography before the works and immediately before caisson installation. Such a comparison is given in Fig. 4.5 for a cross-section of caisson No. 3 (one of the central caissons). The original and final positions of the caisson are also plotted.

The caisson is deeply buried into the soil. The tilt of caisson top, toward the open sea, is consistent with a bearing capacity type of instability induced by an inclined load (the resultant of caisson self-weight and wave loading).

The caisson volume below the soil surface is estimated in Fig. 4.5 to be 240 m^3/m. The depth of burial suggests that the foundation soil could have liquefied. This aspect will be examined later. The internal caisson walls were severely damaged. Wall reinforcement was not intended to resist the efforts associated with a large tilt.

The four caissons involved in the failure were later covered by a conventional fill-type breakwater. However, the remaining caissons envisaged in the project were built after a revision of the foundation design. They provided settlement data, which

Fig. 4.4 Failed caissons

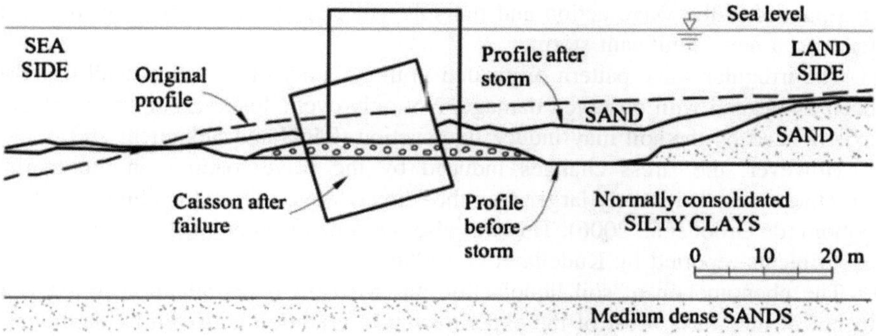

Fig. 4.5 Cross-section through caisson 3 before and after failure. Also shown in the original soil profile, the excavated profile, the granular berm, and the soil profile after the failure

helped to derive some foundation parameters (average stiffness and consolidation coefficient).

The profiles given in Fig. 4.5 indicate that the initial excavation in sands, in the land direction, was substantially filled again after the caisson failure. The calculated soil volume between the surface profiles before and after the storm is 220 m^3/m, which is a value similar to the buried caisson volume under the foundation level. It is then reasonable to accept that the caisson failure displaced the foundation soil toward the land side following a deep failure surface. It is also inferred that wave action after the caisson failure distributed the volume of soil initially displaced by the caisson failure over a wider area.

The failure discussed here is an additional case history to a caisson failures discussed in the literature. Vertical caisson failures were collected and discussed by Oumeraci (1994). The author describes the history of design and construction of vertical breakwaters and identifies the causes of failures. However, the field information in most cases is very limited, especially as far as soil conditions are concerned. He concludes that forces induced by irregular breaking waves are of special concern and are probably a main reason for the observed failures. Concerning the role of the foundation soil, he observes that scour and erosion at the toe of the structures has been often observed. But major failures are probably associated with pore pressure build-up during cyclic loading and eventually with a full or partial liquefaction of the foundation soils. Unlike earthquake loading, wave action is characterized by long time periods (low frequency of loading), a relatively large number of loading applications, and by a sequence of fast wave impacts if a broken wave hits the wall. Once the soil is liquefied or maintains a certain level of pore water overpressure, failure may occur by a critical slip failure mode.

It has also been found that the direct wave action on the sea bottom may induce the accumulation of excess pore pressures, a phenomenon which may lead to soil instability. Damage of some marine structures has been attributed to this phenomenon which has been reviewed by Jeng (1998, 2001) and Jeng and Lin (2000). Most studies refer to sand beds of high porosity. Excess pore pressures eventually dissipate after the wave action and the soil will consolidate, increasing the resistance to a new significant storm.

The irregular wave pattern originated in the vicinity of a vertical wall has also been associated with observed damage. Not only cyclic loading but stress rotation experienced by the soil may induce liquefaction (Sassa and Sekiguchi 2001).

However, the stress changes induced by the wave loading on the marine structures are significantly larger than the (direct) wave-induced loading on the sea bottom (de Groot et al. 2006). This was also the conclusion of large-scale laboratory experiments reported by Kudella et al. (2006).

The phenomenon of soil liquefaction has received much attention in the geotechnical literature. Most of the experimental information is driven by earthquake research. A comprehensive description of liquefaction behavior was provided by Ishihara (1993) and Youd and Idriss (2001).

Model testing has provided further insight into the mechanisms of caisson behavior under wave loading. Scaled model testing under normal gravitational conditions cannot reproduce the stress conditions prevailing at a prototype scale under the foundation and they can hardly reproduce liquefaction behavior. The type of failure observed in these tests (seaward motion of caisson top) is not supported by some field observations, including the failure reported here.

Centrifuge tests have been reported by Rowe and Craig (1976), Van der Poel and de Groot (1998) and by Zhang et al. (2009). They all report the behavior of caissons founded on sand beds. Mechanisms of caisson failure by tilting toward the sea direction have been found. Softening and erosion of the sand in the vicinity of the caisson heel, as well as large liquefied zones beneath the caisson edge, have been proposed by these authors. This is consistent with the observations and analysis reported in this study. Centrifuge tests as well as laboratory cyclic shear test on sands indicate that reversal and irregular shear loading enhances soil liquefaction (if compared with non-reversal regular cyclic loading).

This paper presents a back-analysis of the failure. Two procedures are compared: a purely analytical one, based on well-known solutions in Soil Mechanics and a numerical elasto-plastic analysis. The case illustrates the capabilities of analytical methods and the role they still play to get considerable insight into the foundation behavior. The case also teaches a few practical lessons in the difficult problem of founding safely heavy concrete caissons in soft deposits capable of liquefying under repeated wave action.

4.2 Soil Conditions

Figure 4.6 shows a simplified representative stratigraphic profile under caissons. An upper layer of loose silty sand, 10 m thick, overlies a deposit of clayey silts and silty clays, 20 m thick. Below, a level of medium to dense sands was found. The upper 9 m of sand was removed by dredging. It appears that the thickness of the coarse granular fill below the caisson was rather small (around 2 m). The figure also shows the estimated lateral extent of the coarse granular embankment in the land direction. A detailed stratigraphic record with additional information on soil parameters derived from a few undisturbed samples tested is given in Fig. 4.7. The figure also includes SPT N values.

The silty clay is a soft deposit as revealed by the low N values. It has a moderate plasticity (w_L = 30–32.6 %) and the plasticity index is particularly low (4–10 %). These deltaic deposits classify as ML, CL-ML, or CL. The void ratio is as high as 0.92–0.96.

The calculated virgin compression coefficients (C_c = 0.22–0.26) in oedometer tests performed on undisturbed samples are high.

The coefficient of consolidation determined in small specimens is of limited reliability. The settlement records of the caissons built later were analyzed to determine this parameter and to estimate the permeability "in situ." The lower silty sands and clean sands are markedly stiffer. This is reflected in the high SPT values (N = 15–46), in the lower void ratio (e = 0.7) and in the small compressibility index, C_c = 0.06, measured in an oedometer test on a recovered sample.

Unconfined compression strengths, measured in samples (12–19 kPa) remain below the minimum accepted values for a normally consolidated low plasticity deposit, a result that may be explained by sample disturbance.

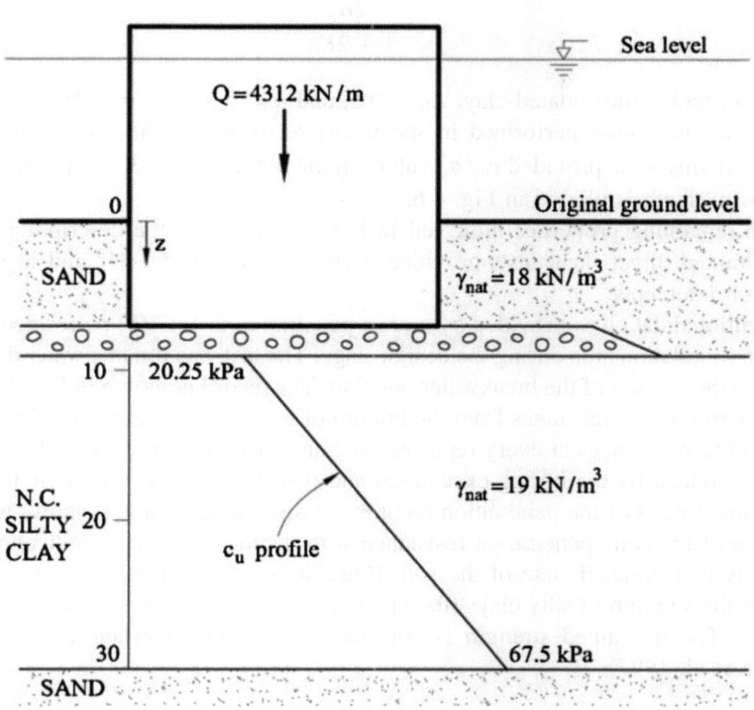

Fig. 4.6 Simplified soil profile under caissons

The undrained strength of normally consolidated soils increases linearly with the confining effective stress. A useful relationship is

$$c_u = a\sigma'_v \tag{4.1}$$

where σ'_v is the vertical effective stress and a is a coefficient that takes values in the range $a = 0.25$–0.30. Expressions have also been found for c_u in terms of the mean effective stress σ'_m, which are also useful in applications:

$$c_u = \bar{a}\sigma'_m \tag{4.2}$$

The expression for coefficients a and $\bar{a}$ can be derived following a theoretical procedure. For instance, Wood (1990) and Potts and Zdravkovic (1999) presented such derivations for a Cam Clay elasto-plastic model. If the coefficient of earth pressure at rest, K_0, is known, the mean effective stress is given as

$$\sigma'_m = \frac{1 + 2K_0}{3}\sigma'_v \tag{4.3}$$

and, therefore,

$$\bar{a} = \frac{3a}{1 + 2K_0} \tag{4.4}$$

For a normally consolidated clay, $K_0 = 0.5$, and $\bar{a} = 0.38$ if $a = 0.25$.

Simple shear tests performed in specimens recovered in the same deltaic formation of silty soils provided c_u/σ'_v values in the range 0.25–0.30. The distribution of c_u with depth is plotted in Fig. 4.6.

The remaining properties indicated in Fig. 4.7 complete the description of the soil. Drained direct shear tests provided friction angles of 25°–31° and negligible cohesion intercepts.

Additional data was provided by a cone penetration test (CPT) performed in the caisson foundation area during the design stage. The test was run at a water depth of 24 m on the seaside of the breakwater position. The record is shown in Fig. 4.8. The test was run in several stages from the bottom of an advancing borehole. The initial penetration resistances at every repeated pushing operation are affected by a stress release induced by the boring excavation and possibly by some soil remolding. If these initial parts of the penetration records are disregarded, the test shows a linear increase of the cone penetration resistance with depth, which is an indication of a normally consolidated state of the soil. Being at a water depth of 24 m the cone records the strength of silty clays (the upper sand layer is not present at these water depths). The undrained strength is correlated with point resistance, q_c, through (Lunne et al. 1997):

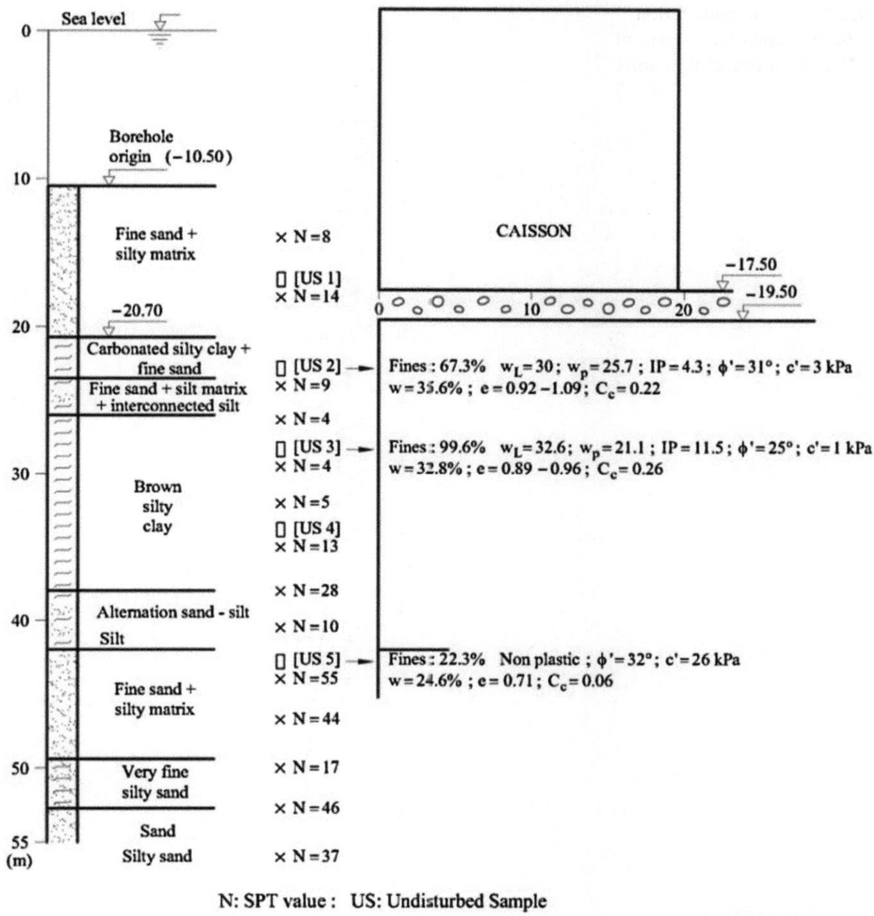

Fig. 4.7 Detailed soil profile under caissons

$$c_u = \frac{q_c - \sigma_v}{N_k} \tag{4.5}$$

where σ_v is the total vertical stress at the current location and N_k is a "bearing capacity" factor, which takes values in the range 10–20. For the Barcelona Harbor soils, a value $N_k = 15$ provides c_u values consistent with undrained simple shear data. The peak resistance values indicated in Fig. 4.8 correspond to more resistant and dilatant sand layers. The silty clay strength corresponds to the minimum envelope of the q_c record, leaving aside the peaks and the disturbed initial parts of successive records.

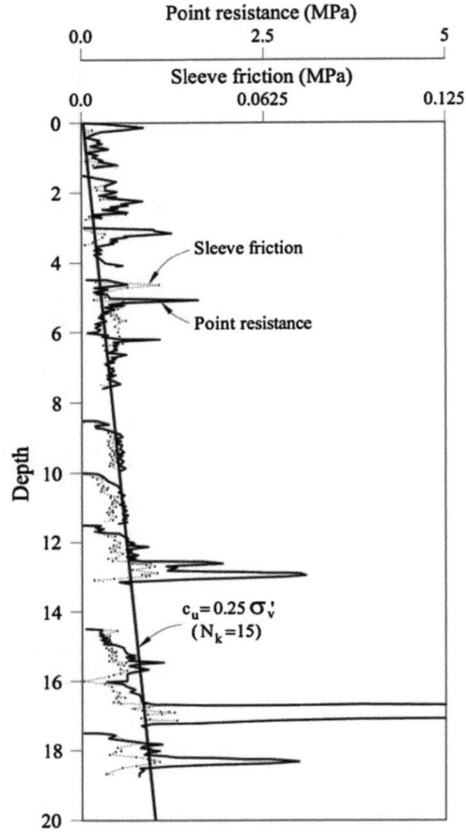

Fig. 4.8 Cone penetration resistance and sleeve ratio of CPT test on foundation soils

4.3 Liquefaction

The susceptibility to liquefaction of the low plasticity silty soil of the caisson foundation may be evaluated by performing undrained cyclic simple shear tests on undisturbed silt specimens, from the same geological formation, recovered in other Barcelona Harbor emplacements. The cyclic shear excitation is defined in Fig. 4.9.

The periodic shear stress signal is described by an average value, τ_{ave}, a purely cyclic component, τ_{cycl}, and a time period T. The results of tests performed are represented in a two-dimensional plot relating τ_{cycl}/σ'_v and τ_{ave}/σ'_v (Fig. 4.10). Each of the points in this plot indicates a combination of the pair (τ_{cycl}/σ'_v, τ_{ave}/σ'_v), which leads to failure of the specimen. The number associated with each point is the number of cycles applied. Increasing τ_{cycl}/σ'_v and/or τ_{ave}/σ'_v leads to a progressively smaller number of applied cycles necessary to induce failure. Failure was accepted when the shear strain reached 10 %. Also indicated in the plot are the results for Norwegian Drammen clay for normally consolidated conditions.

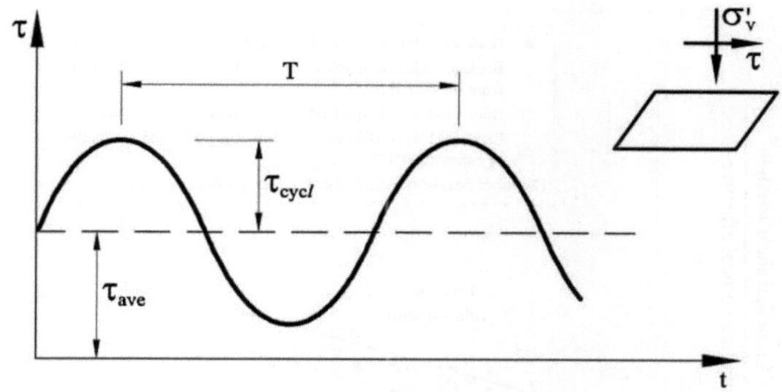

Fig. 4.9 Definition of variables in cyclic shear loading

Fig. 4.10 Results of cyclic undrained simple shear tests. Also indicated in the figure is data from Drammen clay (NGI 2002)

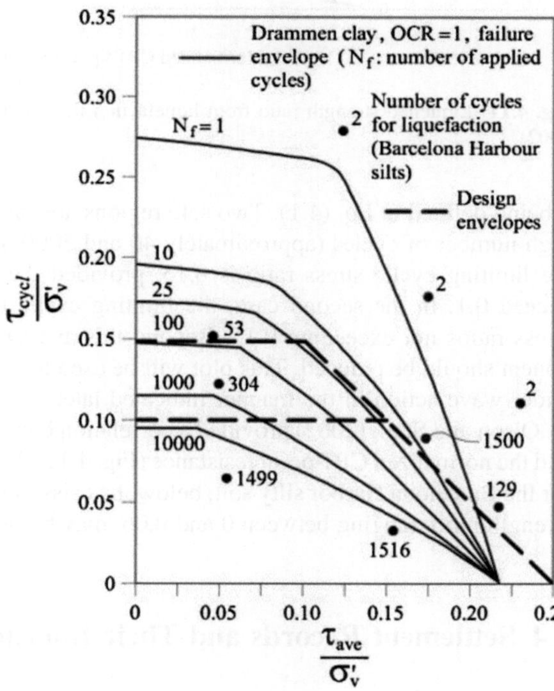

The information given in the plot was used to isolate a safe region for a given number of stress cycles applied. Safe regions are limited by the line

$$\frac{\tau_{\text{cycl}}}{\sigma'_{\text{v}}} + \frac{\tau_{\text{ave}}}{\sigma'_{\text{v}}} = \frac{c_{\text{u}}}{\sigma'_{\text{v}}} = a \tag{4.6}$$

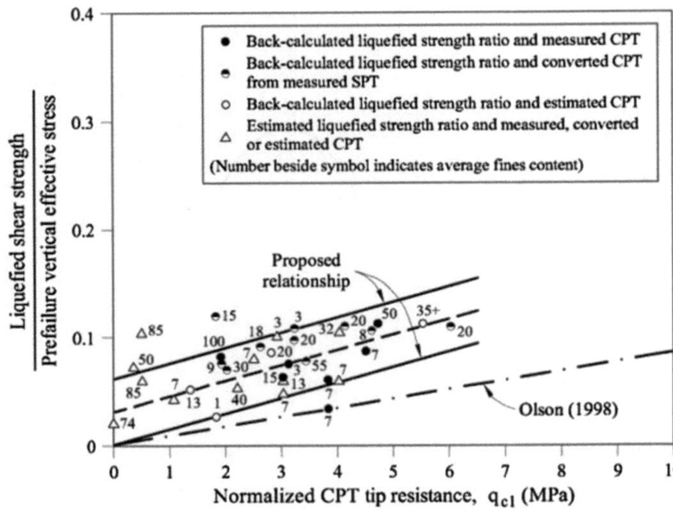

Fig. 4.11 Liquefied strength ratio from liquefaction flow failure case histories (Olson and Stark 2002)

a being defined in Eq. (4.1). Two safe regions are shown in the plot for a low and a high number of cycles (approximately 40 and 5000, respectively). In the first case, the limiting cyclic stress ratio is 0.15, provided the average stress ratio does not exceed 0.1. In the second case, the limiting cyclic stress ratio is 0.1 for average stress ratios not exceeding 0.15. Beyond this average stress ratio, the cyclic component should be reduced. This plot will be used to estimate liquefaction conditions under wave action in the manner indicated later.

Olson and Stark (2002) provided a correlation between the liquefied strength ratio and the normalized CPT point resistance (Fig. 4.11). For the q_c values reported before for the Barcelona Harbor silty soil, below the caisson foundation, a post-liquefaction strength ratio, ranging between 0 and 0.06, may be operative in this case.

4.4 Settlement Records and Their Interpretation

New caissons, built after the failure, were monitored and settlement records for an extended period of time were obtained (Fig. 4.12). They were interpreted to derive average values of the foundation soil coefficient consolidation and stiffness. Soil stiffness was determined in oedometer tests but the integrated field value provided by caisson settlement records is more reliable.

All the settlement records of caissons located in the vicinity of the failed ones were similar. They could be used to derive a field relationship between degree of consolidation, *U*, and time.

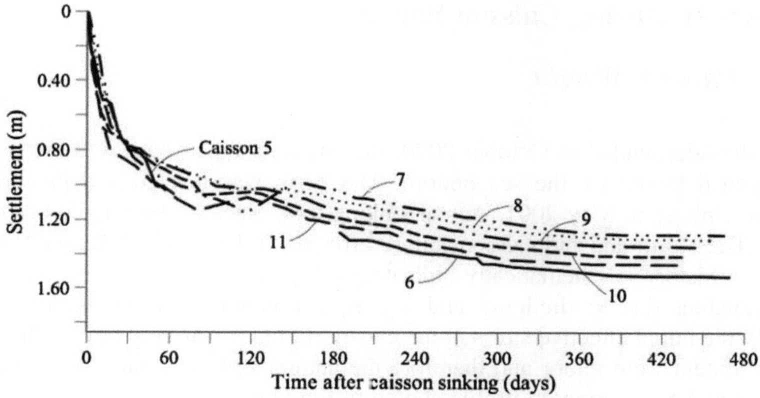

Fig. 4.12 Settlement records for caissons *5, 6, 7, 8, 9, 10,* and *11* built after the failure of the first four caissons. Settlement records are plotted with a common time origin

Davis and Poulos (1972) published the solution for consolidation of a strip loading, which is useful to interpret the settlements of a caisson. They made assumptions equivalent to the classical one-dimensional Terzaghi consolidation equation (constant coefficient of compressibility and permeability, soil deforms only in vertical direction). The solution for a permeable top and base was used. For every pair of (U, t) values, an estimation of c_v was found. Table 4.1 provides c_v values for the first 3 months of the consolidation process.

As expected, c_v decreases with time probably because of the reduction in permeability as the soil void ratio decreases. In order to estimate the soil permeability, it is necessary to know the soil confined (elastic) stiffness, E_m, which is found in terms of the compressibility index and the effective vertical stress.

Below the caisson base, at a depth equal to caisson half-width, the vertical stress in the soil is estimated to be 300 kPa. Therefore, $E_m \approx 5250$ kPa. Then, $k = 1.5 \times 10^{-8}$ m/s for $c_v = 0.7$ m^2/day. For the first stages of consolidation, which are the relevant ones in our case, the foundation soil reacts with a c_v value in the vicinity of 0.75 m^2/day.

The result, even if it is only approximate, indicates that the foundation soil is rather impervious. It will react in an undrained manner when subjected to relatively rapid loading (wave action or caisson sinking) and failure will be also undrained. The relevant strength property will be the undrained strength.

Table 4.1 Coefficient of consolidation from time-settlement records

t (days)	15	30	45	60	90
c_v (m^2/day)	0.748	0.702	0.60	0.53	0.46
c_v (cm^2/s)	0.086	0.081	0.07	0.061	0.053

4.5 Safety During Caisson Sinking

4.5.1 Caisson Weight

Once dredging ended in October 2000, the coarse granular base was extended on the opened trench on the sea bottom. The berm was leveled and prepared for caisson sinking in May 2001. Actual sinking took place in the middle of October 2001. The storm and the caisson failure arrived 20 days later. (The caisson construction history is schematically indicated in Fig. 4.13.)

Unloading due to dredging and subsequent granular filling did not restore exactly the initial effective stress in the clayey silt but it was very close. Changes in water content were minor and therefore the natural soil essentially maintained the original undrained strength profile shown in Fig. 4.6.

Caissons cell volume amounts to 55 % of the total volume. When cells are filled with water, the caisson sinks. The granular base was leveled at elevation −7.50 m and, therefore, after sinking, the upper 2 m of caissons remained above sea level. The net stress of the caisson against the foundation was calculated as $q_w = Q/b = 134\,\text{kPa}$, where Q is the total load (per meter) on the caisson base of width b. When filled with submerged sand the caisson effective weight against its foundation, per unit longitudinal length of caisson, is $q_s = 220$ kPa.

4.5.2 Bearing Capacity

Davis and Booker (1973) found the exact solution for the bearing capacity of a strip footing when the undrained strength increases linearly with depth according to the relationship.

$$c_u = c_{u0} + \rho z \tag{4.7}$$

where ρ is a constant. The upper granular layer (Fig. 4.6) introduces some additional bearing capacity, but it is probably very small. In fact, the lateral extent of the

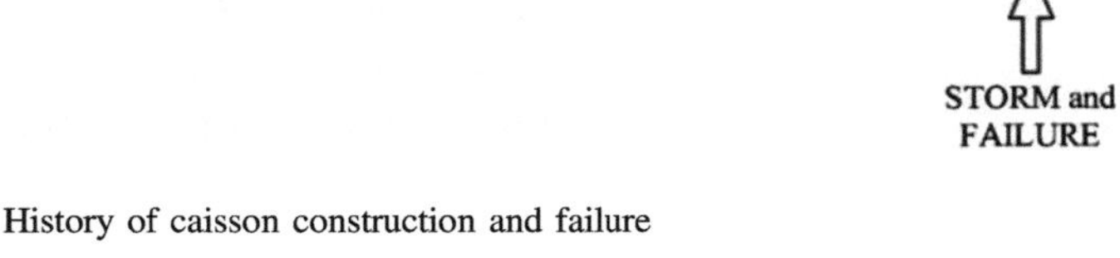

Fig. 4.13 History of caisson construction and failure

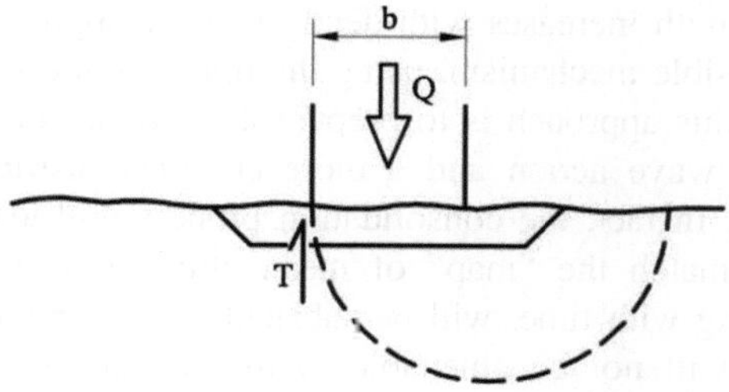

Fig. 4.14 Sketch to illustrate the frictional resistance to failure offered by the granular berm

berm is small. If a simple failure mechanism is considered, the sole effect of the granular layer is to provide a frictional resistance T (Fig. 4.14). Being at the surface and having only a thickness of around 2 m, the confining stress in the granular berm is very small compared with the contribution of the natural soil. It will be accepted, for simplicity, that only the natural soil contributed to the bearing capacity.

The theoretical bearing capacity was written by Davis and Booker (1973)

$$Q/b = F[(\pi + 2)c_{u0} + \rho b/4] \tag{4.8}$$

where F is a correction factor that depends on the ratio $\rho b/c_{u0}$.

Equation (4.8), provides $Q/b = 155\,\text{kPa}$ which is lower than the applied caisson net stress if filled with sand (220 kPa).

Caissons, however, were initially filled with water before replacing it with sand. The net stress on the foundation of a water filled caisson is 134 kPa. Therefore, the safety factor when the caissons were sunk with water ballast can be calculated as $\text{SF} = 155/134 = 1.16$.

The theoretical expression (4.8) predicts conditions very close to failure at the time of caisson sinking. Three-dimensional effects, due to the finite rectangular shape of the caissons base, leads to an increase in the bearing capacity and this effect may explain that the caissons remained stable. Also, the limited thickness of the upper granular term provided some additional bearing capacity. On the other hand, if the undrained strength was actually somewhat higher (Eq. (4.1) with $a = 0.30$, for instance), the safety factor against failure would also increase. Safety factor increases linearly with c_u and a value of $a = 0.30$ would result in $\text{SF} = 186/134 = 1.4$.

It is difficult to be more precise, but the fact is that the caissons did not fail during sinking. However, the estimation made points toward a small safety factor above one. Beyond this moment, the consolidation of the soil under the caisson's weight will increase the available shear strength in the natural soil. The caisson was then loaded with its definite weight (when filled with saturated sand), it consolidated for a few weeks, and eventually received storm-wave loading.

Before analyzing these processes, it is worth investigating if the theoretical undrained bearing capacity obtained by Davis and Booker for a strip footing on a clay soil whose strength increases with depth, may be approximated by a simpler cinematically admissible mechanism, using the upper bound theorem of plasticity. The reason behind this approach is to prepare the ground for subsequent calculations involving the wave action and a more complex distribution of undrained strength with depth. In fact, the consolidation process will lead to a "map" of c_u values which will match the "map" of mean effective stresses Eq. (4.2). This distribution, changing with time, will be substantially different from a linear variation of c_u and it will not be amenable to theoretical solutions. However, the plasticity theorems still provide an approximation. It then seems wise to try to establish some confidence on the assumed failure mechanism (based on the upper bound theorem). The way to do it is by comparing the exact solution provided by Davis and Booker (1973) with the upper bound approximation.

4.5.3 An Upper Bound Solution for a Strip Footing Founded on Clay with a Linearly Increasing Strength with Depth

Figure 4.15 shows the critical velocity field found by Davis and Booker (1973) in their exact solution to the bearing capacity problem (infinite, rigid smooth footing). The mechanism is symmetric with respect to the axis of the footing.

Even if the rough footing is a more realistic case, this mechanism suggests the simplified symmetric mechanism, based on rigid triangular wedges, indicated in Fig. 4.16.

Consider one of the two symmetric mechanisms which receive half the external load ($Q/2$). This mechanism will be optimized with respect to the angle α shown in Fig. 4.16. The motion of this mechanism is characterized by the virtual displacement rate vector δ_1, which defines the motion of wedge $A'B'E'$ sliding on the straight segment $A'B'$. The undrained strength will be defined by the linear function $c_u(z) = c_{u0} + \rho z$.

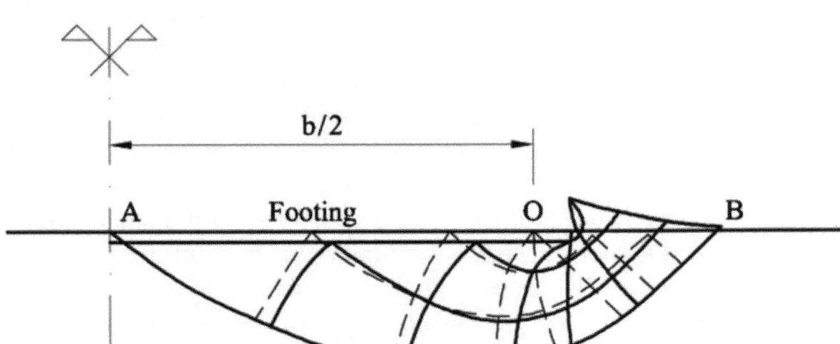

Fig. 4.15 Velocity field for smooth footing (after Davis and Booker 1973)

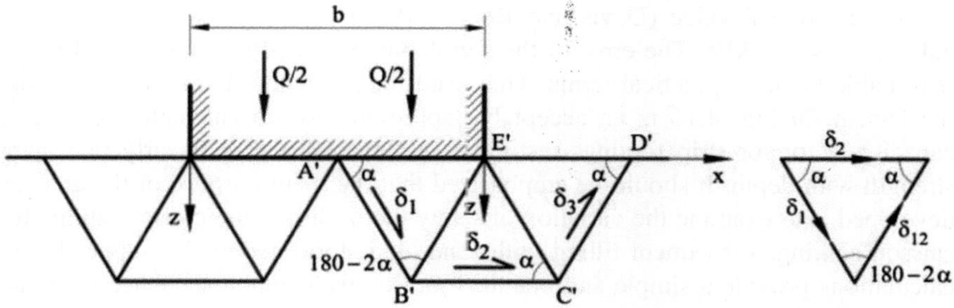

Fig. 4.16 Symmetric failure mechanism for upper bound analysis of strip footing under vertical load

The dissipation along $A'B'$ is calculated as follows:

$$W_{A'B'} = \delta_1 \frac{b}{4\cos\alpha}\left(c_{u0} + \frac{\rho b}{8}\tan\alpha\right) \tag{4.9}$$

Also,

$$W_{B'C'} = \left(c_{u0} b\cos\alpha + \rho\frac{b^2}{4}\sin\alpha\right)\delta_1 \tag{4.10}$$

Considering the hypothesis of rough footing, dissipation along the line $A'E'$ is

$$W_{A'E'} = c_{u0} b\delta_1 \cos\alpha \tag{4.11}$$

Dissipation along lines $A'B'$, $B'E'$, $E'C'$, and $C'D'$ are equal and therefore the total internal dissipation work on the mechanism will be

$$W_{\text{int}} = 4W_{A'B'} + W_{B'C'} + W_{A'E'} \tag{4.12}$$

The external work performed by $Q/2$ is calculated as

$$W_{\text{ext}} = \frac{Q}{2}\delta_1 \sin\alpha \tag{4.13}$$

Making $W_{\text{int}} = W_{\text{ext}}$, Q is isolated in terms of α, c_{u0}, ρ, b:

$$Q = \frac{1}{4}\frac{b(8c_{u0} + \rho b\tan\alpha + 12c_{u0}\cos^2\alpha + 2\rho b\sin\alpha\cos\alpha)}{\sin\alpha\cos\alpha} \tag{4.14}$$

The best upper bound solution for Q is its minimum value with respect to α. This minimization calculation was performed with the help of the "solve" function including in Excel. For $b = 19.6$ m, $c_{u0} = 20.25$ kPa, and $\rho = 2.25$ kN/m^3 a minimum value of $Q/b = 182$ kPa is obtained for a critical angle α equal to 50.8°.

The theoretical value (Davis and Booker 1973) for rough footing provides a value $Q/b = 155$ kPa. The error of the simple upper bound mechanism is 17 %, a reasonable value in practical terms. This result indicates that the triangular wedge mechanism in Fig. 4.16 is an acceptable approximation to calculate the bearing capacity factor for strip footings resting on a clay soil with a linearly increasing strength with depth. It should be emphasized that the main purpose of the analysis developed is to examine the variation of safety factor during the consolidation after caisson sinking, subsequent filling with sand, and storm action. The upper bound calculations provide a simple and practical tool to evaluate the sequence of events leading to caisson failure.

However, if a horizontal load is also acting on the caisson, because of wave action, a symmetric mechanism cannot possibly occur and a more likely mechanism is indicated in Fig. 4.17. We call it a nonsymmetric failure mechanism. A similar upper bound calculation provides:

$$Q = \frac{b}{2}\frac{(4c_{u0}\cos\alpha + \rho b\sin\alpha)}{\sin\alpha\cos^2\alpha} + \frac{(4c_{u0}\cos^3\alpha + 2\rho b\sin\alpha\cos^2\alpha)}{\sin\alpha\cos^2\alpha} \tag{4.15}$$

which is different from the Q value calculated for the symmetric mechanism.

The minimization of Q with respect to α was also performed on an Excel sheet. For similar parameters ($b = 19.6$ m; $c_{u0} = 20.25$ kPa; $\rho = 2.25$ kN/m^3), a failure unit load $Q/b = 209$ kPa was calculated (for a critical angle α equal to 44.5°) which is 15 % higher than the load calculated for the symmetric mechanism. This nonsymmetric mechanism will be used when wave action is considered.

Caisson consolidation resulted in increasing c_u values and increasing safety factor against bearing capacity failure.

Two weeks after the caisson first sinking, caisson cells were filled with sand. This increase in net weight will be also analyzed below in order to determine the associated safety factor. In the days which followed the sand filling, caisson

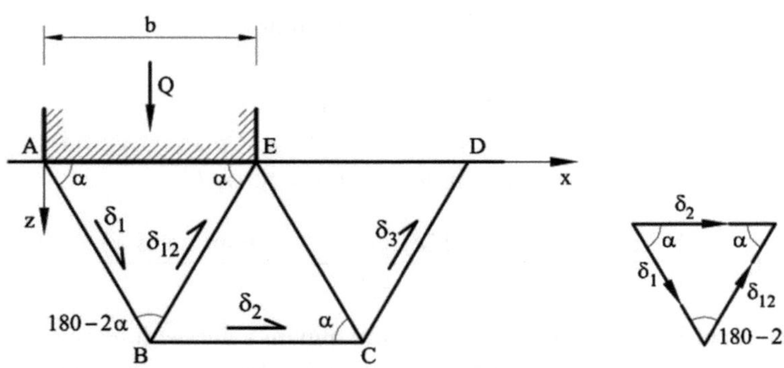

Fig. 4.17 Nonsymmetric failure mechanism for upper bound analysis of strip footing under vertical load

consolidation continued and the soil undrained strength had to be estimated before analyzing the storm effect on caissons.

Therefore, the next step in the analysis was to analyze the caisson consolidation and the increase in soil strength.

4.6 Caisson Consolidation. Increase in Soil Strength

Actual settlement records of caissons built after the failure, shown in Fig. 4.12, indicate that a significant consolidation may be achieved in a few weeks. Points of the soft foundation soil close to the upper pervious granular layer would experience a rapid consolidation under the full caisson loading. Clay levels located close to the lower pervious sandy boundary will also consolidate fast but the stress increments reaching the lower sand levels will be significantly lower. Caisson consolidation leads to a progressive increase in effective stress and therefore to an increase in undrained strength. For the reasons mentioned, however, the new distribution of undrained strength values will be nonhomogeneous and also far from the initial linear distribution with depth. This will be especially the case of the foundation soil directly under the caissons.

The increments of undrained strength will be simply calculated as a fraction of the increment in effective mean stress through Eq. (4.2). Therefore, the objective now is to calculate the distribution of effective mean stress under the caisson loading, taking into account the consolidation process. The calculation will be split into two parts:

- Stress increments under a strip footing and determination of excess pore pressures.
- Dissipation of the induced excess pore pressures.

4.7 Stress Increments Under a Strip Footing and Determination of Excess Pore Pressures

This analysis will be guided by the subsequent use of the calculated undrained strengths. In fact, the ultimate objective is to determine the failure load and to compare it with the actual caisson loading. Failure conditions will be calculated by means of the upper bound theorem of plasticity, through the mechanisms already examined (Figs. 4.16 and 4.17). Consider in Fig. 4.18 the two alternative mechanisms proposed here: a symmetric one (already identified as an adequate solution for vertical loading only) and the nonsymmetric one which will be employed when including wave action.

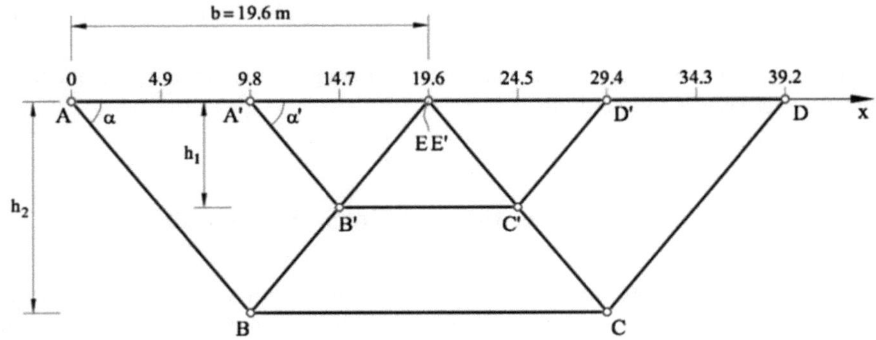

Fig. 4.18 Geometry of the symmetric and nonsymmetric failure mechanisms. Position of vertical profiles for the calculation of undrained soil strength

Upper bound calculation will require the determination of plastic work dissipation on segments AB, $A'B'$, etc. Since nonlinear strength variations will be the rule, a minimum of three control points are proposed to estimate by a simple numerical integration, the average strength on each of the sliding surfaces of the two mechanisms shown in the figure. Those points (marked as open circles) define a number of vertical profiles characterized by the horizontal coordinate distances to the left caisson foundation corner indicated in Fig. 4.18.

Pore pressure dissipation will be dominated by the vertical flow toward the upper and lower drained boundaries. A hypothesis of vertical consolidation, which is close to real conditions, as justified in Eq. (4.4), helps to perform the consolidation analysis. The following sequence of steps will be considered (the starting point −time = 0—will be the initial sinking of caissons):

- Stress increments in the foundations are determined. Elastic solutions for strip loading are used. Stress calculations are performed in vertical profiles located on the horizontal coordinates shown in Fig. 4.18 ($x = 0$, 4.9, 9.8 m, … etc.). The reason for this choice has already been given.
- Mean total stress will be computed in points located on the vertical profiles. Excess pore pressures (over hydrostatic values) will be made equal to the increments of mean stress. This is a reasonable and sufficiently accurate assumption.
- Excess pore pressures will be dissipated vertically toward the upper and lower drainage boundaries in a one-dimensional process. A time period of 14 days will provide the state of the foundation before the sand filling of the caissons.
- Mean effective stress will be calculated as a difference between the calculated total stresses and pore water pressures. An updated distribution of c_u values will be calculated through Eq. (4.2).
- Caisson failure loads will be determined through the upper bound theorem. The safety factor will be determined.

4.8 Stress Increments and Excess Pore Pressures

The actual sequence of caisson loading is shown in Fig. 4.19. The time of sinking is $t = 0$ for the remainder of the analysis. At $t = 14$ days caisson cells were filled with sand and the net stress on foundation increased to 220 kPa. The storm arrived at $t = 21$ days.

Mean stresses calculated for a uniformly loaded strip footing lead to the pore pressure profiles shown in Fig. 4.20 for a few vertical profiles and $t = 0$. The figure shows two kinds of shapes for the distribution of the initial excess pore pressure due to caisson loading (solid line). Under the caisson ($x = 0^+$ to $x = 19.6$ m) the mean stress reaches a maximum at the caisson-soil contact and a minimum at the bottom of the clay stratum (at a depth of 21 m under the caisson's base where the pervious sand layer is encountered). On both sides of the caisson ($x > 19.6$ m) the stress increment at the surface is zero. It increases to reach a maximum at some intermediate depth and decreases again. The intensity of the mean stress increments decreases as the distance to the caisson base increases.

A dissipation process of excess pore pressures will immediately start toward the upper and lower drainage boundaries and, in parallel, mean effective stresses will increase. These are steps (c) and (d) of the description of the process leading to an increase in soil strength.

The initial excess pore pressures plotted in Fig. 4.20 do not follow simple linear laws. Probably the simplest curve fitting all the profiles shown in the figure to a reasonable approximation is a parabola:

$$f(z) = A + Bz + Cz^2 \tag{4.16}$$

where $f(z)$ is the initial excess pore pressure, which is a function of the vertical coordinate. A, B, C can be determined by curve fitting procedures or simply by selecting three points on the curve at different z to be fitted.

The result of this fitting process is also given in Fig. 4.20. The parabolic fit is very good under the caisson and at a certain distance from it. The most difficult fit is for vertical profiles outside the caisson but in its immediate vicinity. However, increasing the degree of the polynomial approximation was probably not a

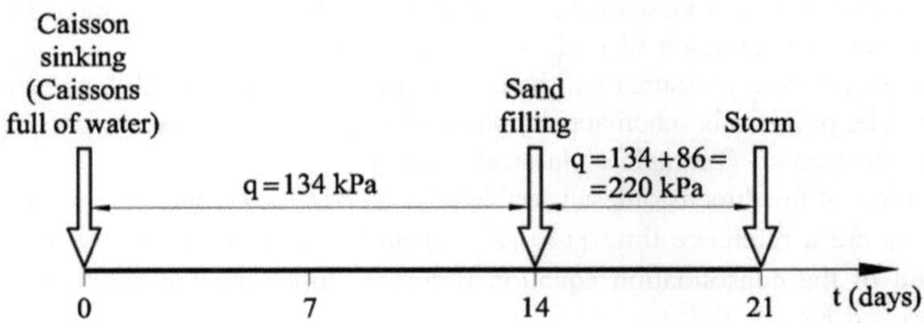

Fig. 4.19 Caisson loading sequence

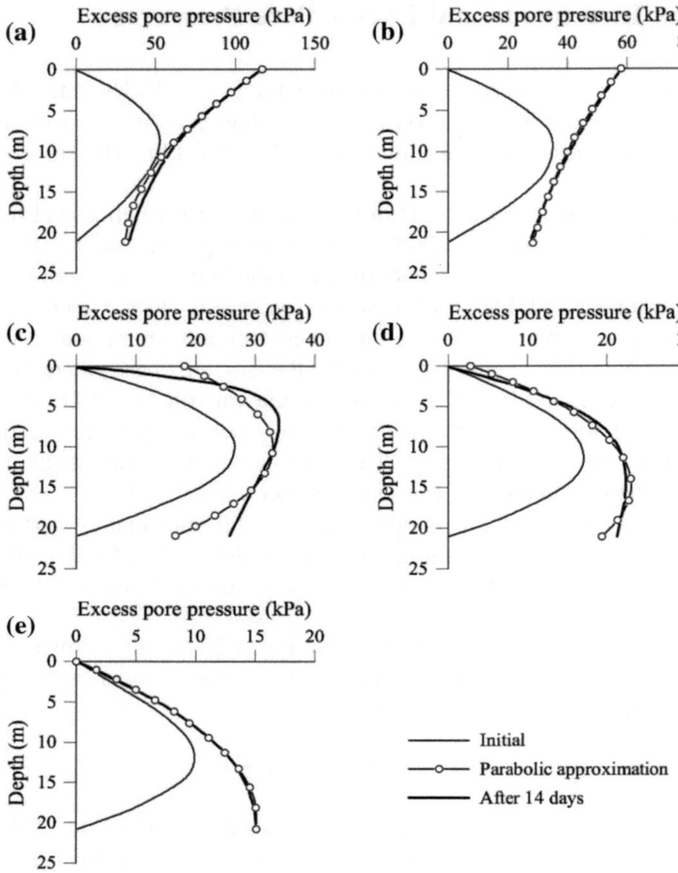

Fig. 4.20 Excess pore pressures at the vertical profiles defined by $x = 9.8$, 19.6, 22.05, 26.95, and 34.30 m in Fig. 4.18. Each of the plots provides the calculated initial excess pore pressure, its parabolic approximation, and the excess pore pressure profile calculated after 14 days of consolidation

reasonable decision in view of the associated complexity and the limited influence of an exact representation of initial excess pore pressures.

The excess pore pressures will be assumed to dissipate vertically, as mentioned before. The problem is schematically shown in Fig. 4.21. The excess pore pressure $u(z,t)$, must satisfy Terzaghi's classical equation.

In terms of the dimensionless variables $Z = z/H$; $T = t/\tau$ and $W = u/u_0$, where τ and u_0 are a reference time $\left(\tau = H^2/c_v\right)$ and a reference pressure, the general solution of the consolidation equation becomes (following the solution given in Alonso and Krizek 1975):

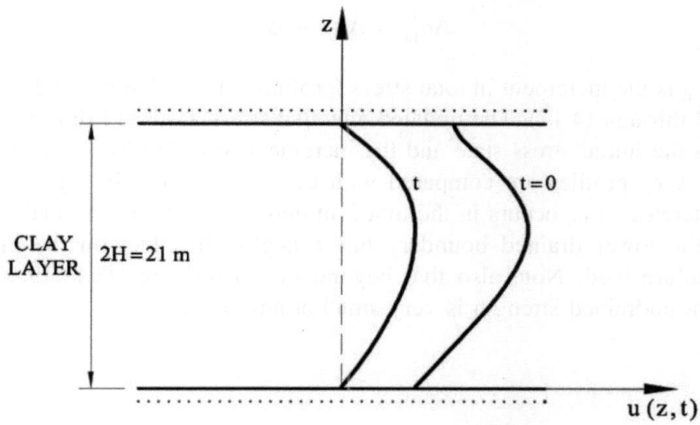

Fig. 4.21 Double-drained consolidation induced by caisson sinking at $t = 0$ and excess pore pressures

$$W(Z,T) = \int_0^2 \left(\sum_{n=1}^{\infty} \sin\left(\frac{n\pi Z_0}{2}\right) \sin\left(\frac{n\pi Z}{e}\right) \exp\left(-\frac{n^2\pi^2 T}{4}\right)\right)$$

$$(\bar{A} + \bar{B}Z_0 + \bar{C}Z_0^2)\mathrm{d}Z_0 = \sum_{n=1}^{\infty} \sin\left(\frac{n\pi Z}{2}\right) \exp\left(-\frac{n^2\pi^2 T}{4}\right)$$

$$\int_0^2 \left[\bar{A} \sin\left(\frac{n\pi Z_0}{2}\right) + \bar{B}Z_0 \sin\left(\frac{n\pi Z_0}{2}\right) + \bar{C}Z_0^2 \sin\left(\frac{n\pi Z_0}{2}\right)\right] \mathrm{d}Z_0$$

$$= \sum_{n=1}^{\infty} \frac{2}{n\pi} \sin\left(\frac{n\pi Z}{e}\right) \exp\left(-\frac{n^2\pi^2 T}{4}\right)$$

$$\left\{\bar{A}\left[(1)^{n+1}+1\right] + 2\bar{B}(-1)^{n+1} + 4\bar{C}\left[\frac{2}{n^2\pi^2}[(-1)^n-1] - (-1)^n\right]\right\} \tag{4.17}$$

4.9 Effective Stresses and Updated Undrained Strength

Equation (4.17) was used to calculate the excess pore pressure 14 days after caisson sinking (when full of water). Adding the first five terms in Eq. (4.17) provides almost the exact solution. The calculated pore water pressures are also plotted in the vertical profiles given in Fig. 4.20. Fourteen days after caisson sinking, the excess pore pressure in all profiles has a similar shape with maximum values at depths ranging between 7 and 13 m. The increase in mean effective stress is calculated as

$$\Delta\sigma'_{\mathrm{m}} = \Delta\sigma_{\mathrm{m}} - \Delta u \tag{4.18}$$

where $\Delta\sigma_{\mathrm{m}}$ is the increment in total stress (profiles for $t = 0$ in Fig. 4.20) and Δu is calculated through (4.17). The updated effective stress at $t = 14$ days is calculated by adding the initial stress state and the increment provided by Eq. (4.18).

The new c_{u} profiles are compared with the initial values in Fig. 4.22. A significant increase in c_{u} occurs in the first 5 m under the caisson. A smaller increase close to the lower drained boundary has a negligible effect on the increase in caisson failure load. Note also that beyond the loaded area ($x > 19.60$ m), the increase in undrained strength is very small at any depth.

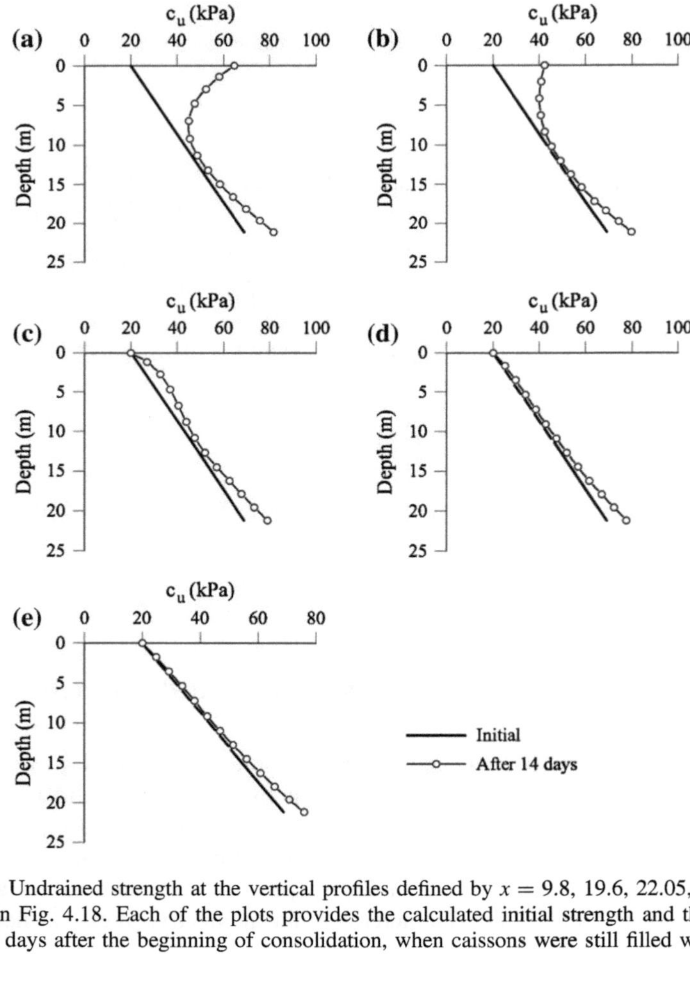

Fig. 4.22 Undrained strength at the vertical profiles defined by $x = 9.8$, 19.6, 22.05, 26.95, and 34.30 m in Fig. 4.18. Each of the plots provides the calculated initial strength and the resulting profile 14 days after the beginning of consolidation, when caissons were still filled with water

The new distribution of c_u values under the caisson is now fundamentally different compared with the initial linear increase with depth. Now c_u reaches a relative maximum directly under the caisson and decreases with depth until reaching the initial distribution of undrained strength which increases linearly with depth. This change will have interesting effects on the critical failure mechanism, as explained later.

The next step in caisson construction was increasing caisson loading to its full design value (sand filling). The safety factor against failure will now be calculated as well as the subsequent consolidation process under the new load. This will lead to an updated distribution of undrained strength, which will be operating at the time of storm arrival.

4.10 Caisson Full Weight. Safety Factor Against Failure and Additional Consolidation

Consider again in Fig. 4.16 the symmetric, shallow, failure mechanism (half width of the caisson). The dissipation work on the edges of the triangular wedges will be approximated by three values (two nodes and the midpoint). For instance the average c_u value on segment $A'B'$ will be

$$c_u^{A'B'} = (c_u^{A'} + c_u^{1} + c_u^{B'})/3 \tag{4.19}$$

where c_u^1 is the strength of the intermediate point between A' and B'. The calculation of the internal dissipation work follows the procedure developed before:

$$\begin{aligned} W_{\text{int}} &= \left[\left(c_u^{A'B'} + c_u^{B'E'} + c_u^{E'C'} + c_u^{C'D'}\right)\right]\frac{b}{4\cos\alpha}\delta_1 + c_u^{B'C'}\frac{b}{2}2\cos\alpha\delta_1 + c_u^{A'E'}\frac{b}{2}2\cos\alpha\delta_1 \\ &= \left[\left(c_u^{A'B'} + c_u^{B'E'} + c_u^{E'C'} + c_u^{C'D'}\right)\frac{b}{4\cos\alpha} + \left(c_u^{B'C'} + c_u^{A'E'}\right)b\cos\alpha\right]\delta_1 \end{aligned} \tag{4.20}$$

The external work, W_{ext}, was given in Eq. (4.13) Making $W_{ext} = W_{int}$ and isolating the collapse load Q,

$$Q = \frac{2}{\sin\alpha}\left(\frac{W_{\text{int}}}{\delta_1}\right) \tag{4.21}$$

The preceding Q value should be minimized with respect to α. This collapse load could be compared with the net caisson stress applied when cells are filled with saturated sand ($Q/b = 220$ kN/m).

An alternative failure mechanism is the nonsymmetric zone sketched in Fig. 4.18. Mean c_u values on segments limiting the moving triangular wedges are calculated following previous results. The internal dissipation is now given as

$$W_{\text{int}} = \frac{b}{2\cos\alpha}\left(c_u^{AB} + c_u^{BE} + c_u^{CD} + c_u^{EC}\right) \\ \delta_1 + 2b\cos\alpha c_u^{BC}\delta_1 \tag{4.22}$$

and

$$W_{\text{ext}} = Q\sin\alpha\delta_1 \tag{4.23}$$

Therefore, since $W_{\text{int}} = W_{\text{ext}}$

$$Q = \frac{1}{\sin\alpha}\left(\frac{W_{\text{int}}}{\delta_1}\right) \tag{4.24}$$

Again, a minimization process with respect to angle α should be performed. The two cases (symmetric and nonsymmetric mechanisms) have been solved in an Excel sheet following the preceding methodology and equations. The following results are obtained for $t = 14$ days: Symmetric mechanism

$$Q/b = 264\,\text{kN/m};\ \text{SF} = \frac{264}{220} = 1.20;\ \alpha = 56^\circ$$

Nonsymmetric mechanism

$$Q/b = 244\,\text{kN/m}\ ;\ \text{SF} = \frac{244}{220} = 1.11;\ \alpha = 47^\circ$$

Application of the upper bound theorem, in the manner indicated above, leads to variation of the failure load with angle α shown in Fig. 4.23 for the two mechanisms. The critical failure mechanism is now the nonsymmetric one. The reason is that a

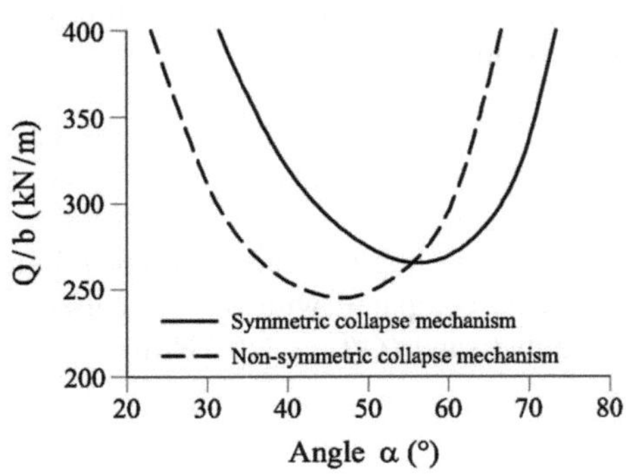

Fig. 4.23 Variation of upper bound collapse load with angle α of the failure mechanism. Caisson partially consolidated under full weight

deeper nonsymmetric mechanism goes in search of lower strength values compared to the shallower symmetric one. This is an effect of the increase in undrained strength, which is maximum at the upper dissipation boundary (where mean stresses are maximum and the drainage more effective) and decreases with depth.

Though the calculated minimum safety factor is small (1.1), the fact is that caissons survived also the sand filling and a new consolidation process started under the added load.

4.11 Caissons Under Full Weight

The best way to analyze the consolidation in the days that followed sand filling is to imagine the process divided into two parts. The first part is simply the continuation of the consolidation induced by caisson sinking (caisson filled with water). The second is a new process whose starting time is the application of the new load. It has been calculated before that filling the caissons with sand implied an increase in uniform load amounting to $\Delta Q/b = 86$ kPa on the base of the caissons.

The previous analysis is repeated having in mind that the objective is now to determinate the undrained strength distribution at the time of the storm arrival. Therefore, the distribution of mean stresses will now be obtained for $q = Q/b = 220$ kN/m. The pore pressures are the result of two contributions:

- A consolidation process induced by excess pore pressures calculated for $\Delta q = \Delta Q/b = 134$ kN/m, lasting for $t = 21$ days.
- A consolidation process induced by excess pore pressures calculated for $\Delta q = Q/b = 86$ kN/m, lasting for $t = 7$ days.

The analysis follows steps previously described and is not repeated here. The resulting distribution of undrained strength is calculated as follows:

$$\begin{aligned} c_u(21 \text{ days}) = c_u(\text{initial}) + \Delta c_u(t = 21 \text{ days; water filled caissons}) \\ + \Delta c_u(t = 7 \text{ days; sand filled caissons}) \end{aligned} \tag{4.25}$$

The result of this calculating process is given in Fig. 4.24. The added load (sand filling) had a limited time to be transformed into effective stresses and, accordingly, into available undrained strength but nevertheless the upper levels of the foundation soil under the caisson significantly increased the undrained strength. The strength profile shows a maximum at the caisson-foundation contact. Strength decreases rather fast with depth to meet the initial values which increase linearly with depth. This distribution is plotted in a two-dimensional cross-section in Fig. 4.25. It shows that caisson consolidation was able to build a "strong" soil nucleus in the upper 7 m although strength decreased continuously with depth. Beyond a depth of approximately 10 m, the initial undrained strength is recovered. A small increase in strength is calculated in the vicinity of the lower drainage layer.

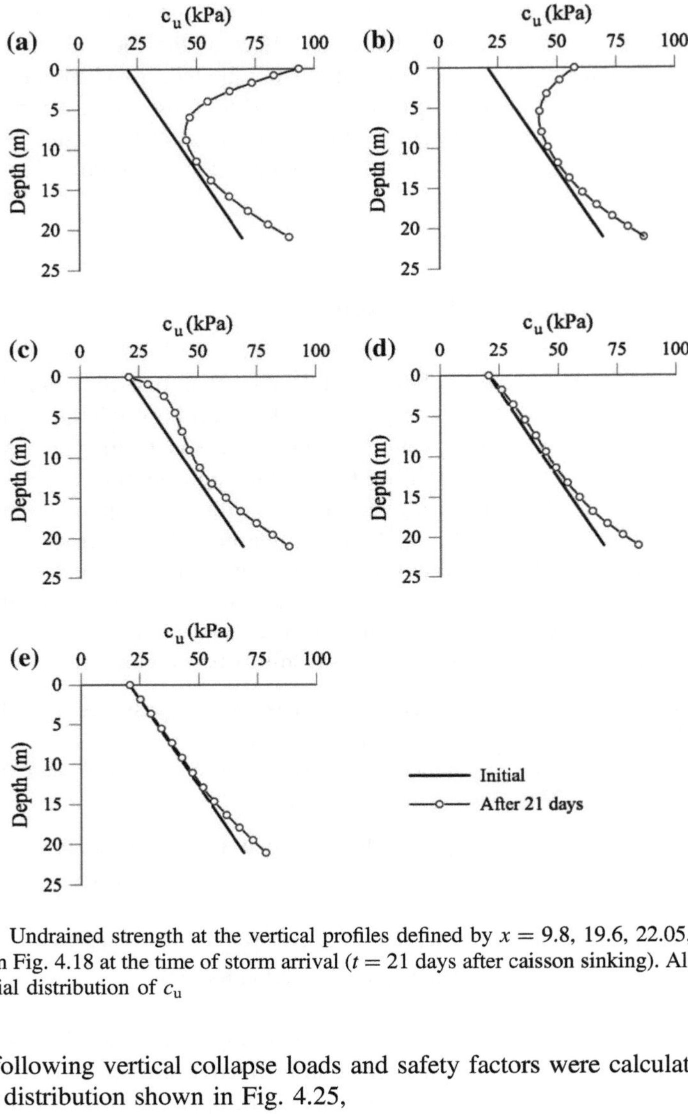

Fig. 4.24 Undrained strength at the vertical profiles defined by $x = 9.8$, 19.6, 22.05, 26.95, and 34.30 m in Fig. 4.18 at the time of storm arrival ($t = 21$ days after caisson sinking). Also indicated is the initial distribution of c_u

The following vertical collapse loads and safety factors were calculated for the strength distribution shown in Fig. 4.25,

- Symmetric mechanism: SF = 1.42 for $\alpha = 58°$
- Nonsymmetric mechanism: SF = 1.22 for $\alpha = 44°$

This was the situation when the eastern storm of November 11, 2001 hit the caissons.

Note that the critical mechanism is now clearly the deep (nonsymmetric) one. The lower failure surface reaches in this case a depth of 9.46 m, against a maximum depth of 6.72 m for the symmetric, shallower mechanism. This is a consequence of the "inverted" profile of undrained strength created by the caisson weight.

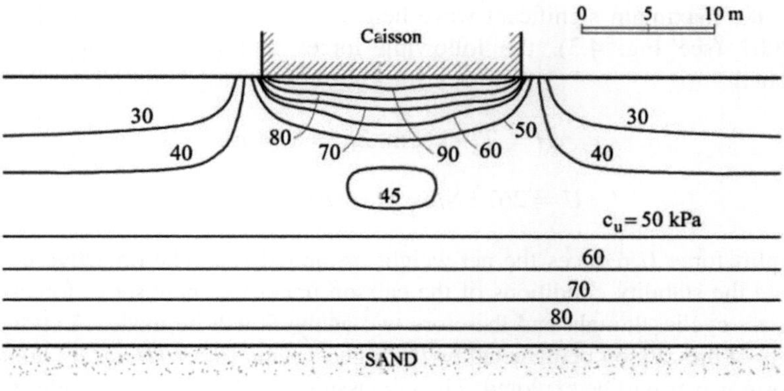

Fig. 4.25 Calculated distribution of available undrained strength at the time of storm arrival (November 11, 2001)

4.12 Caissons Under Storm Loading

4.12.1 Wave Forces on Caissons

A well-known calculation method to find wave forces on vertical dykes was proposed by Goda (1985). The wave load has two components: an excess pressure on the exposed wall and an uplift pressure on the caisson base. The assumed distribution of these pressures is shown in the sketch of Fig. 4.26.

The pressure distributions are equivalent to the concentrated forces:

- Horizontal force H at an elevation Z_H.
- Uplift force U at a horizontal coordinate $X_U = 1/3b$.

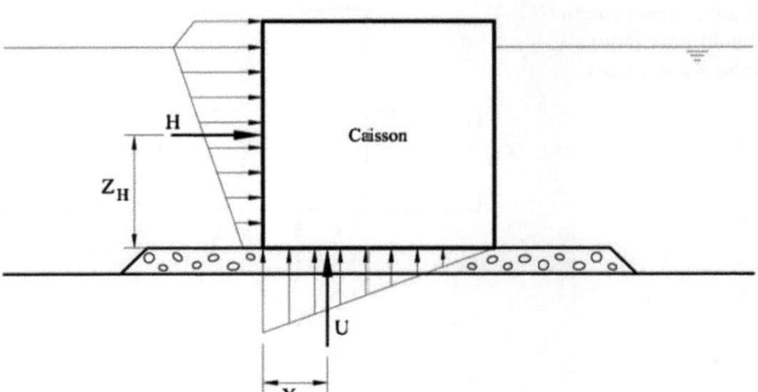

Fig. 4.26 Goda's wave pressures

For the maximum significant wave height recorded in the storm of November 11, 2001 (see Fig. 4.3), the following forces and points of application are calculated:

$$H = 779 \text{ kN/m}, Z_H = 8.94 \text{ m},$$

$$U = 267 \text{ kN/m}, X_U = b/3 = 6.5 \text{ m}$$

The uplift force U reduces the net weight of the caisson. The objective now is to estimate the stability conditions of the caisson under this new set of forces. Wave forces are cyclic, though, and therefore two analyses will be made. First, the static stability of the caisson being acted by the highest static forces (H and V) induced by the storm waves will be explored. Then the issue of soil liquefaction under repeated wave loading will be examined.

4.12.2 Static Analysis. Safety Factor

The problem is represented in Fig. 4.27. The weight Q is now reduced by the calculated uplift force U and a horizontal wave loading H is added. The assumed failure mechanism for the application of the upper bound theorem is also shown. Under virtual displacement rate δ_1 of the wedge under the caisson, only the vertical force Q_r and the horizontal force H perform external work. Note that in this case, the symmetric mechanism is no longer possible and only the nonsymmetric "deep" mechanism will be analyzed. The external work is written as

$$W_{ext} = (Q - U)\delta_1 \sin \alpha + H\delta_1 \cos \alpha \tag{4.26}$$

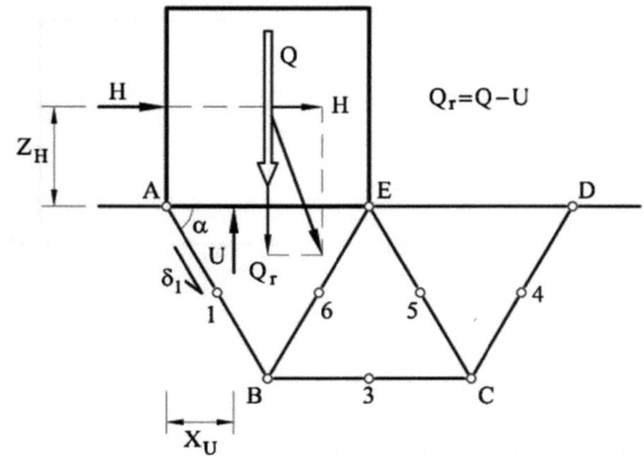

Fig. 4.27 Failure mechanism under combined gravitational load and static wave forces

The internal dissipation work follows previous developments. Since the soil has a marked nonuniform distribution of undrained strength, the calculations are based on the c_u values estimated at the corners of the mechanism (*A*, *B*, *C*, *D*, *E*) and at the midpoints (1, 3, 4, 5, 6). Note also that the caisson and the wedge *ABE* displace as a solid body and therefore no shear dissipation occurs on *AE*.

Now the determination of the safety factor against failure requires more careful consideration. So far the safety factor has been defined as the ratio between the limiting vertical load (determined through the upper bound theorem) and the actual load induced by the caisson defined by its own weight. But now three different forces are acting on the caisson: weight *Q*, horizontal load *H*, and uplift *U*.

We will distinguish the real forces from the forces actually inducing the failure (Q^{real} vs. $Q^{failure}$, etc.). A single safety factor SF is defined as follows:

$$\mathrm{SF} = \frac{Q^{\mathrm{failure}}}{Q^{\mathrm{real}}} = \frac{H^{\mathrm{failure}}}{H^{\mathrm{real}}} = \frac{U^{\mathrm{failure}}}{U^{\mathrm{real}}} \tag{4.27}$$

The upper bound equilibrium equation ($W_{ext} = W_{int}$) which defines the failure loads may be written as follows:

$$\mathrm{SF}\left[\left(Q^{\mathrm{real}} - U^{\mathrm{real}}\right)\sin\alpha + H^{\mathrm{real}}\cos\alpha\right] = \frac{W_{\mathrm{int}}}{\delta_1} = W^*_{\mathrm{int}} \tag{4.28}$$

This equation defines the safety factor as

$$\mathrm{SF} = \frac{W^*_{\mathrm{int}}}{\left(Q^{\mathrm{real}} - U^{\mathrm{real}}\right)\sin\alpha + H^{\mathrm{real}}\cos\alpha} \tag{4.29}$$

W^*_{int} is a function of α and the minimization of Eq. (4.29) with respect to α will provide the best upper bound approximation to SF.

Consider now the structure of W^*_{int} given in Eq. (4.22). It is, in fact, a weighted sum of c_u values taken at different positions (c_{uj}) (the nodes of the mechanism and the auxiliary intermediate points):

$$W^*_{\mathrm{int}} = \sum_{j=1}^{N} \beta_j c_{uj} \tag{4.30}$$

where β_j are coefficients that depend on the geometry of the mechanism. Therefore, Eq. (4.28), which establishes the balance between external work and internal dissipation, can be written as

$$\mathrm{SF}\left(Q^{\mathrm{real}} - U^{\mathrm{real}}\right)\sin\alpha + H^{\mathrm{real}}\cos\alpha = \sum \beta_j c_{uj} \tag{4.31}$$

or

$$(Q^{\rm real} - U^{\rm real}) \sin \alpha + H^{\rm real} \cos \alpha = \sum \beta_j \frac{c_{uj}}{\rm SF} = \sum \beta_j c^*_{uj} \tag{4.32}$$

where

$$c^*_{uj} = \frac{c_{uj}}{\rm SF} \tag{4.33}$$

is the set of mobilized undrained shear strengths necessary to satisfy the balance implied by the upper bound mechanism in terms of the actual real loads.

But this definition is essentially similar to the safety factor concept used in limit equilibrium analysis, i.e., the reduction factor which has to be applied to the strength parameters in order to achieve strict equilibrium under the actual real loads.

The fact that the safety factor defined in (4.27) as a ratio of loads becomes a strength reduction factor in (4.33) is simply a consequence of the linear relationship between failure loads and undrained strength. This is the case in undrained stability analysis but it is far from being so in drained analysis because failure loads and the ($\tan \phi'$) drained strength parameter are not linearly related.

The expression in (4.33) has to be seen as a sound "geotechnical" definition of safety factor. It is equivalent to loading ratios (Eq. 4.27) only under conditions of undrained analysis.

But the safety factor is not a theoretical or particularly well-defined measure of uncertainty. Other alternatives may suit the needs of the designer. For instance, in the case discussed now, it may be argued that wave loading is rather uncertain compared with caisson weight $Q^{\rm real}$. It may be also accepted that the soil undrained strength is well known. Then, a safety factor aimed at judging the risk of failure induced by wave loading may be defined as

$${\rm SF}^* = \frac{H^{\rm failure}}{H^{\rm real}} = \frac{U^{\rm failure}}{U^{\rm real}} \tag{4.34}$$

Then, following the previous steps, SF* is obtained as

$${\rm SF}^* = \frac{W^*_{\rm int} - Q^{\rm real} \sin \alpha}{H^{\rm real} \cos \alpha - U^{\rm real} \sin \alpha} \tag{4.35}$$

This function of α should be now minimized, following the procedures associated with the upper bound theorem.

Safety factors SF and SF* have been calculated for the following set of forces: $Q^{\rm real} = 4312$ kN/m; $H^{\rm real} = 779$ kN/m; $U^{\rm real} = 267$ kN/m, and for the calculated distribution of c_u values at the time of the storm arrival ($t = 21$ days after first sinking).

The calculated safety factors and the angle α that defines the mechanism are

$$SF = 1.10(\alpha = 41°), SF^* = 1.77\,(\alpha = 41°).$$

The calculation indicates that the static wave force was not enough to induce a generalized failure of caissons but it was quite close. However, the deep failure of caissons was also an indication that the foundation soil had experienced additional reduction in strength most likely associated with soil liquefaction. The next section deals with the liquefaction of foundation soil.

4.13 Analysis of Liquefaction

Stability of caissons subjected to horizontal loading is checked in practice by examining a few failure possibilities: bearing capacity (or overall stability), sliding on its base and overturning. In the preceding section the bearing capacity of caissons against its own weight, combined or not with a static estimation of wave loading, has been discussed. The survey after the failure did not provide any support for sliding or overturning modes of failures.

The deep sinking of caissons after the failure (Fig. 4.5) and the type of failure (tilting of caisson top in the seaward direction) suggest that soil liquefaction played a significant role in the failure. The cyclic interaction diagram (Fig. 4.10) shows that liquefaction conditions are defined by the static or average shear stress ratio, the cyclic ratio, and the number of applied cycles.

Instead of performing a comprehensive dynamic analysis, the following simplified approach will be followed here:

- Shear stresses on horizontal planes (τ_{xz}) will be computed on the foundation for the following two states:
- Caisson weight,
- Wave action.

The theory of elasticity will be used in these calculations.

- Stress ratios (shear stress over vertical effective stress) will then be calculated. The distribution of vertical effective stresses (σ_z') corresponds to the consolidation time at the time of storm arrival (t = 21 days after caisson sinking).
- The calculated stress ratios will be compared with the information provided by the cyclic interaction diagram (Fig. 4.10). For the estimated number of loading cycles applied by the storm, points in the foundation soil may either fall in a stable zone or in an unstable (liquefied) domain. Points "satisfying" the liquefaction condition will define an area where undrained soil strength will decrease to a post-liquefaction state.

- A new stability analysis following the upper bound methodology will be carried out. The spatial distribution of undrained strength will now be a consequence of the initial state (linear increase of c_u with depth), the previous consolidation history under caissons weight, and the c_u reduced values on the liquefied areas.

4.14 Shear Stresses on Horizontal Planes

Consider the stress distribution under the caisson induced by its own weight (Fig. 4.28).

Note the point-symmetric distribution of shear stresses, and the existence of a maximum/minimum at some distance from caisson axis. Close to the foundation surface, a sharp peak is calculated at the transition between the loaded and unloaded areas. Away from that zone of intense shearing, the absolute value of the shear stress is small. At depth, the peak "widens" and the shear stresses distributes more evenly. The caisson weight induces significant shear stresses at depth. Even at a depth of 20 m, where the sandy stiff layer marks the lower boundary of the silty foundation soil, the maximum shear stress is close to 40 kPa.

The fact that a shear stress is positive or negative is not of particular significance here because they are equally capable of inducing limiting conditions (isotropic soil properties are assumed).

Consider now the wave action in Fig. 4.29. The resultant horizontal force H acting at a height Z_H over the caisson base is made equivalent to a shear force

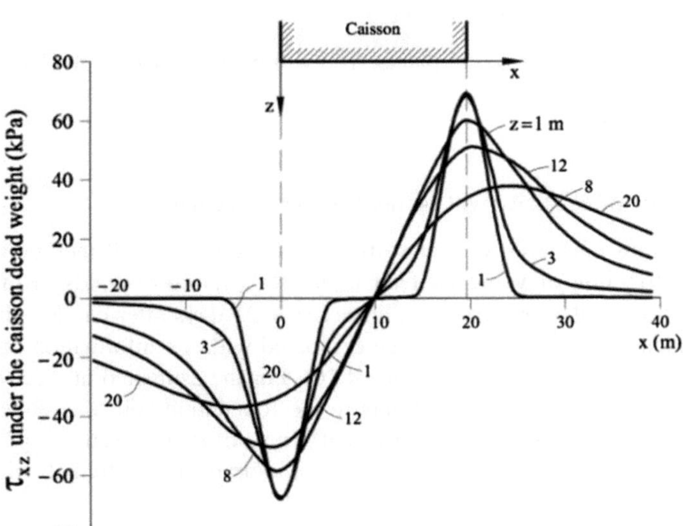

Fig. 4.28 Calculated τ_{xz} values under the caisson dead weight at depths $z = 1, 3, 8, 12$, and 20 m

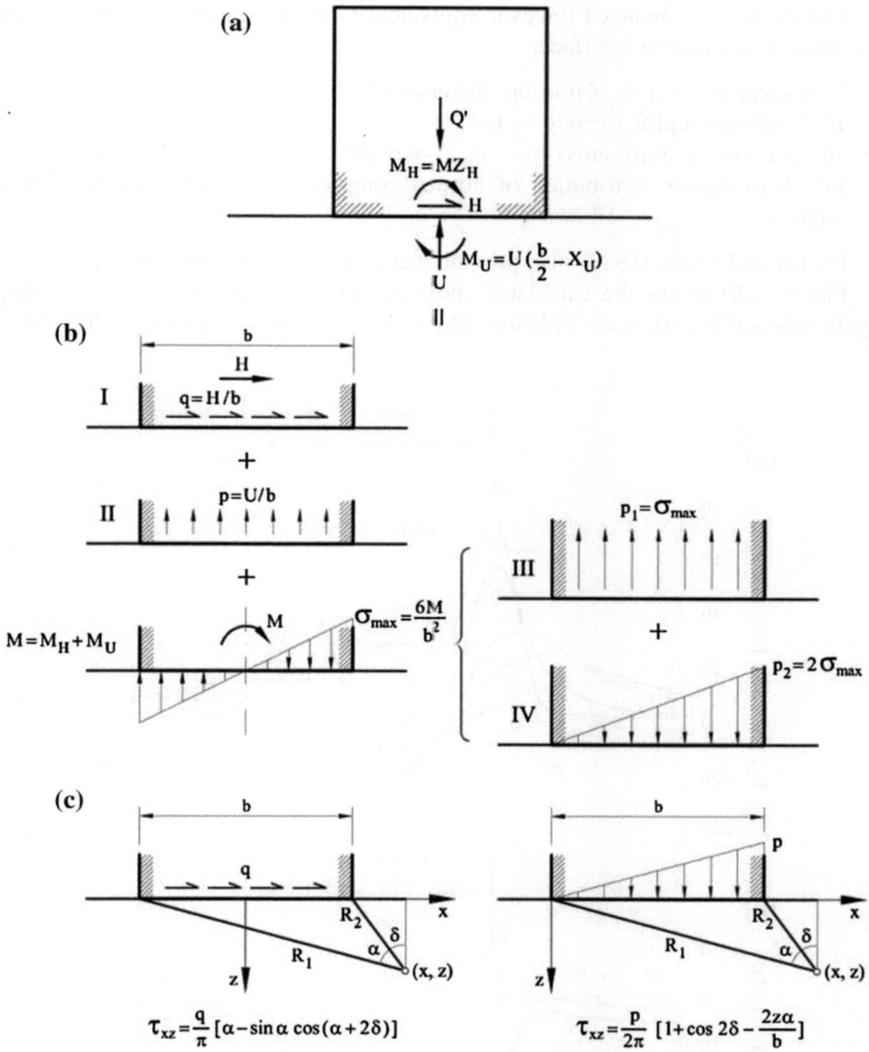

Fig. 4.29 **a** Forces in caisson in the presence of wave loading; **b** stresses on the caisson-foundation interface; **c** shear stresses, τ_{xz}, in the subsoil induced by uniform distribution of boundary shear forces and a triangular normal loading

H at the foundation base and a moment $M_H = H \times Z_H$. A triangular distribution of water overpressures at the caisson-soil interface is also acting, following Goda (1985). It is equivalent to a uniform uplift load *U* and a moment M_U which adds to M_H. The value of these forces for the storm of November 11, 2001 is calculated in the Appendix.

The set of wave-induced forces is equivalent to the following set of stresses on the caisson-foundation interface:

- I: A shear stress $q = H/b$ in the direction of H.
- II: A uniform uplift stress $p = U/b$.
- III: A uniform uplift stress $p_1 = \sigma_{max} = 6\ M/b^2$ where $M = M_H + M_U$.
- IV: A triangular distribution of normal compression stresses with maximum value $p_2 = 2\sigma_{max} = 12\ M/b^2$.

Poulos and Davis (1973) also provide elastic solutions for all these cases.

Figure 4.30 shows the calculated shear and normal stresses for the following cyclic forces (Gouda): $H = 779$ kN/m; $M_H = H, Z_H = 7242$ kN m/m, $U = 267$ kN/m;

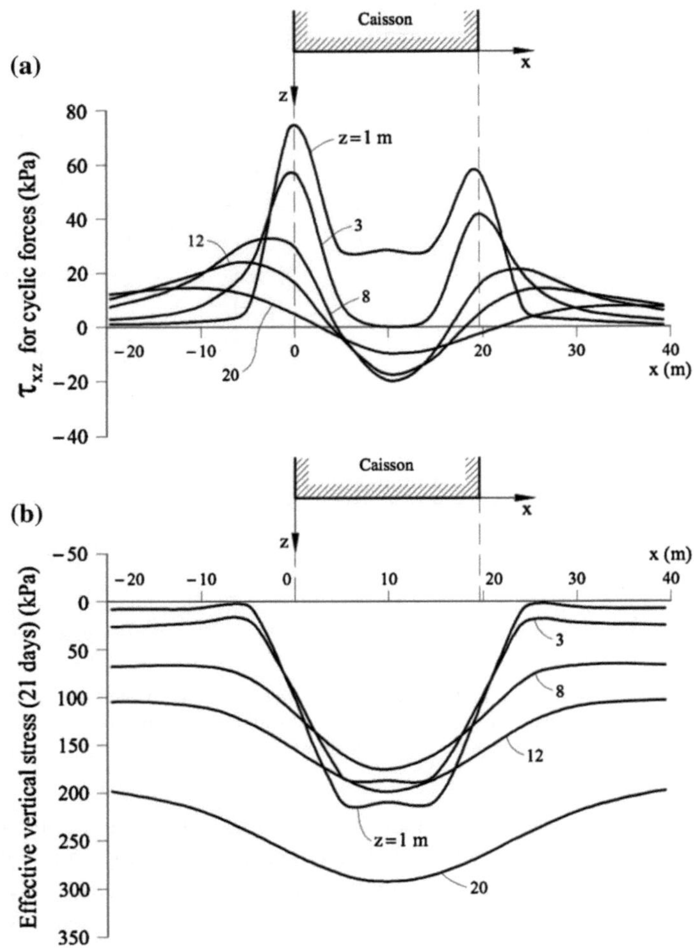

Fig. 4.30 Calculated **a** τ_{xz} and **b** σ'_v values for cyclic forces at depths z = 1, 3, 8, 12, and 20 m

$M_U = 869$ kN m/m. Now the point symmetry of shear stress distribution is lost. However, the peak values for shallow depths concentrate again in the vicinity of the caisson edges. In addition, the shearing force applied at the interface results in significant shear stresses at shallow depths, under the caisson base. Cyclic shear stresses are of the same order of magnitude as static shear stresses.

4.15 Stress Ratios

Normal effective stresses on horizontal planes were calculated previously. They are required to normalize shear stresses.

The ratios τ_{ave}/σ'_v and τ_{cycl}/σ'_v are now calculated. They are plotted for the same depths in Fig. 4.31a, b. The two figures show that stress ratios reach very high

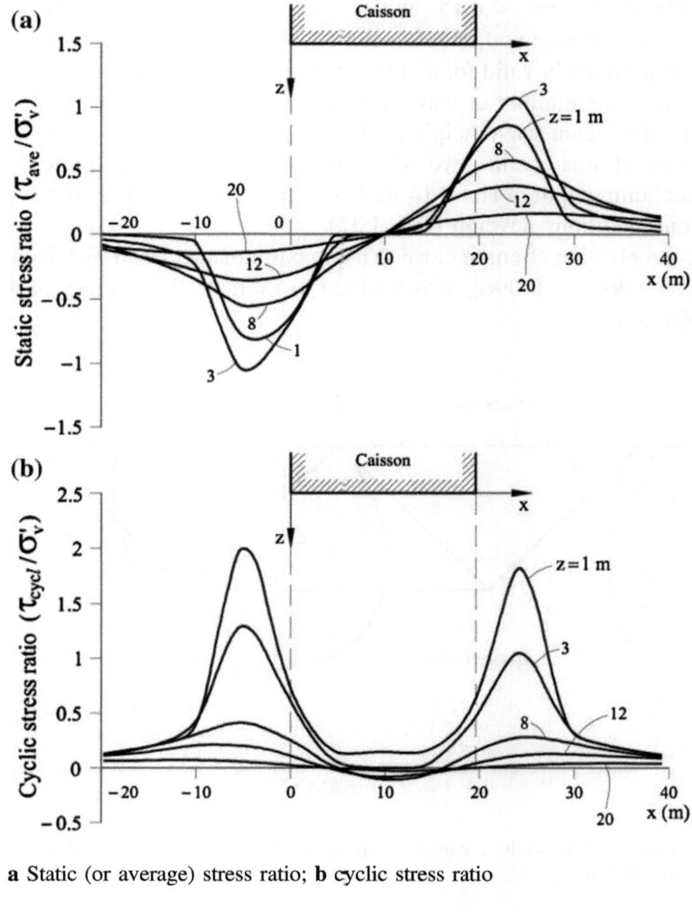

Fig. 4.31 **a** Static (or average) stress ratio; **b** cyclic stress ratio

values, at shallow depths, in the transition zone between loaded and unloaded areas. Of course, these high stress ratios cannot be resisted by the soil and stress redistributions will occur. The figure indicates, however, that these shallow zones at the edge of caissons are plastified and, in addition, they are particularly critical against cyclic loading. Under the caisson, the stress ratio decreases rapidly due to high confining stress. At increasing depths the transition between loaded–unloaded surface areas remains in critical conditions. This is better appreciated if the stress ratios in Fig. 4.31 are compared with the cyclic interaction diagram in Fig. 4.10.

4.16 Liquefied Zones

Calculated stress ratios have been compared with the cyclic interaction diagram given in Fig. 4.10. Stress ratios falling on the unstable or liquefied zone have been indicated in Fig. 4.32. The "stable" zone is defined in a simplified manner by the two conditions $\tau_{\text{cycl}}/\sigma'_{\text{v}} < 0.15$ and $\tau_{\text{ave}}/\sigma'_{\text{v}} < 0.25$. Outside this rectangular domain, the soil is assumed to reach liquefaction. The limiting cyclic stress ratio (0.15) is approximately valid for a number of cyclic stress applications of 100–200. This is close to the number of wave impacts when the storm reaches its maximum intensity (a significant wave height of 4 m).

The liquefied areas define two wide bulbs on both edges of the caisson. They reach a maximum depth of about 14 m. Under the caisson central area, however, no liquefaction conditions develop (Fig. 4.32).

A post-liquefaction strength close to the maximum suggested by Olson and Stark (2002) $c_u = 0.06\sigma'_v = 0.09\sigma'_m$ was adopted to estimate the stability in the case of soil liquefaction.

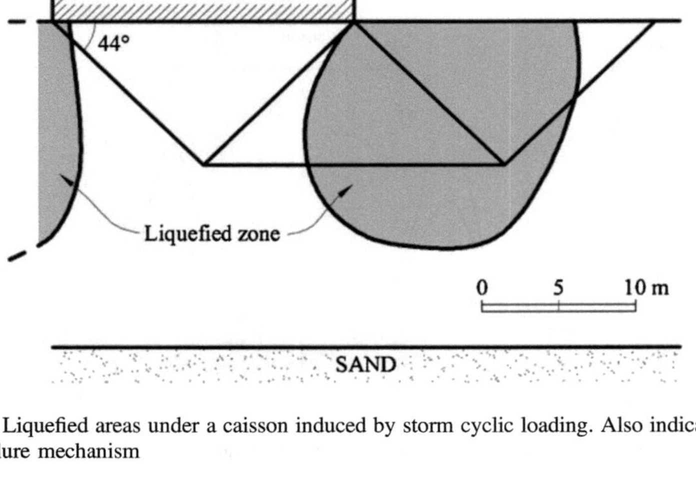

Fig. 4.32 Liquefied areas under a caisson induced by storm cyclic loading. Also indicated is the critical failure mechanism

4.17 Stability of Caissons

The final step of the analysis is to calculate the failure conditions when part of the foundation soil is liquefied. The failure mechanism crosses the liquefied zone located on the leeward side of the caisson foundation (Fig. 4.32). In the non-liquefied areas, the distribution of undrained strength was reported before. Within the liquefied area $c_u = 0.09\,\sigma'_m$.

The calculated safety factor, defined as in Eq. (4.29), drops to 0.56 for a critical mechanism defined by an angle $\alpha = 44°$. Even if the liquefaction zone is characterized by a relatively high value of the residual undrained strength ($0.09\,\sigma'_m$), the drop in safety factor is very significant.

If a more substantial reduction of strength is specified, the safety factor drops to very low values. This result helps to explain the deep burial of caissons as a consequence of the failure.

4.18 Discussion. Analytical Results

The upper bound theorem of plasticity provides an unsafe estimate of bearing capacity. A comparison against the available analytical solution for increasing undrained soil strength shows that errors may be of the order of 17 % for the failure mechanism selected here. The mechanism is, however, simple and it may accommodate, at a limited effort, complicated spatial distributions of undrained shear strength. Figure 4.33 shows the variation in time of the safety factor of the caissons,

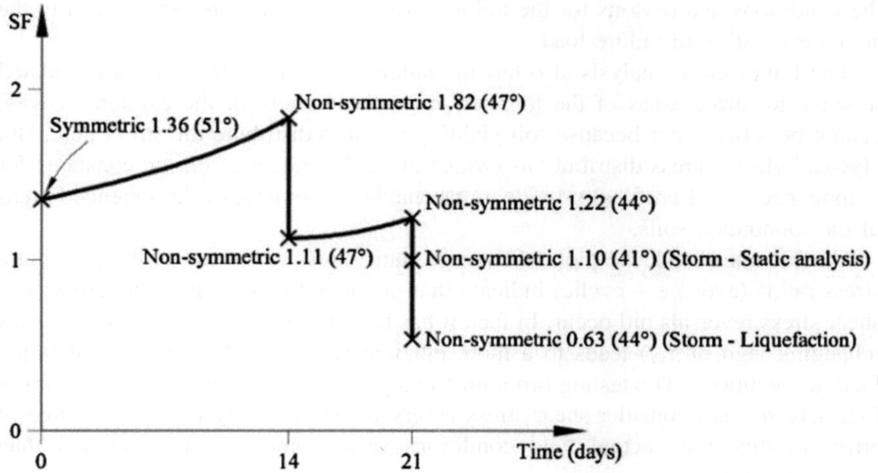

Fig. 4.33 Evolution of caisson safety factor against bearing capacity failure. The value in parenthesis indicates the critical α angle of the failure mechanism

starting at the time of caisson sinking ($t = 0$) and ending at the failure time, 21 days later. Symmetric (shallower) and nonsymmetric (deeper) mechanisms are considered when appropriate. The plotted variation of safety factor with time follows the minimum values.

The calculated safety factors, taking the upper bound solution as a correct estimate, should be regarded in sequence, paying attention to the changes in SF but not to its precise values. Its evolution in time is the most valuable information. The fact that caissons did not fail during the first sinking provides a reference condition (SF > 1) for $t = 0$, irrespective of calculations. Presumably, in view of the results obtained, the actual value of SF was in the vicinity of 1.3. Then the calculation process described, which maintains the failure mechanism, results in an SF value, at $t = 21$ days and for a static equivalent loading to wave action, which is greater but close to one.

Therefore, it is concluded that the failure and the deep sinking of caissons should be explained by a strength-loss mechanism associated with the cyclic wave loading. A soil liquefaction phenomenon seems a good candidate to explain the strength loss, especially if one considers the deep burial of failed caissons into the foundation soil. Of course, the silty deltaic deposits seem prone to liquefy in view of the accumulated experience synthesized in Fig. 4.10.

The consideration of liquefaction, which has been analyzed by a procedure inspired in the cyclic mobility diagram of the natural soil, indicates that extensive zones of the caisson foundation soils reached critical conditions under the storm of November 11, 2001. A substantial reduction in safety factor is calculated, even if the loss of undrained strength is moderate. In other words, irrespective of the safety factor at $t = 21$ days under pseudostatic wave loading, the liquefaction of the foundation soil is capable of inducing a catastrophic failure. These comments help to accept the upper bound calculation as a useful and simple procedure to analyze the conditions and reasons for the failure, even if some error is experienced in the actual estimation of failure load.

The liquefaction analysis also has limitations. The large stress ratios calculated in some localized areas of the foundation (in the vicinity of the caisson's edges) cannot possibly occur because soil yielding would redistribute the stress field. But classical elastic stress distributions (which are independent of elastic constants for homogeneous soil profiles) provide a reasonable approximation for extended areas of the foundation soils.

On the other hand, the plots of average static stress ratios and of the maximum stress ratios (average + cyclic) indicate that in some zones of the foundations soil shear stress reversals did occur. In fact, it has been found that shear stress reversals (changing sign of τ_{xz}) leads to a more efficient and rapid development of liquefaction conditions. The testing program leading to the cyclic interaction diagram of Fig. 4.10 did not consider shear stress reversals and this may be also a source of error. In this case, actual field conditions seem somewhat more critical than experimental ones.

The four failed caissons were essentially isolated and surrounded by open sea conditions. Therefore, only wave forces in the landward direction should be considered. If caissons protect a water body, the wave trough results in an unbalanced hydrostatic load in the seaward direction. This load, which enhances stress reversals, was unlikely in this case.

4.19 Simplified Numerical Analysis

This analysis was performed as a cross-check of the analytical solution presented before and also as a procedure to quantify some simplifications which were necessarily introduced in the analytical work.

The numerical model was developed in FLAC 3D 3.10. Figure 4.34 shows the mesh used in all cases reported below (8000 cells).

The liquefaction is modeled following the same procedure used in the analytical analysis. A reduced undrained strength equal to 0.9 σ'_m is assigned to the calculation points in which the liquefaction criterion is reached.

The comparison of some results (pore water pressure and undrained strength distributions, cyclic shear stress ratios and liquefaction volumes) was remarkably similar in both analyses. The safety factor in the case of FLAC was derived by a (c, ϕ) reduction technique. Figure 4.35 shows the critical symmetric and nonsymmetric mechanisms described previously.

Safety factors are compared in Table 4.2.

The differences in SF are most probably a consequence of the errors in the upper bound theorem of plasticity. Note that numerical analysis predicts lower SF, as

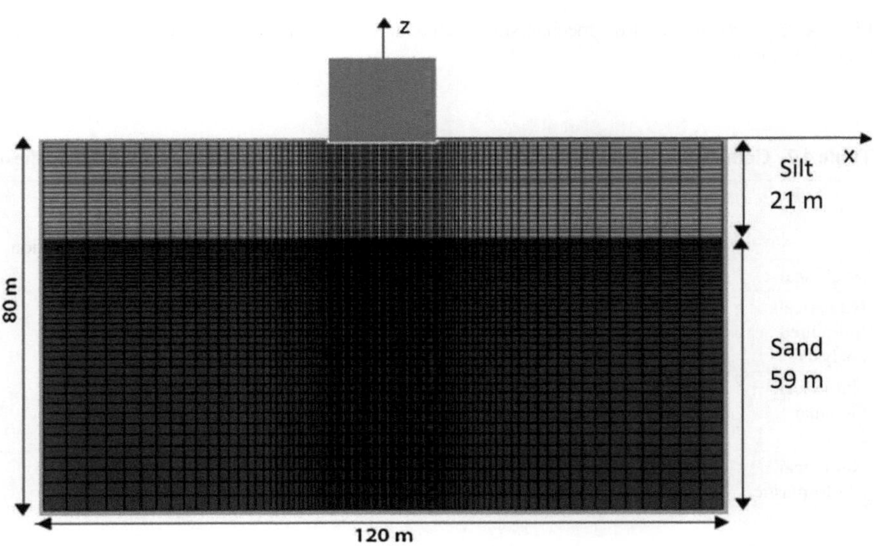

Fig. 4.34 Finite difference mesh used for FLAC calculations

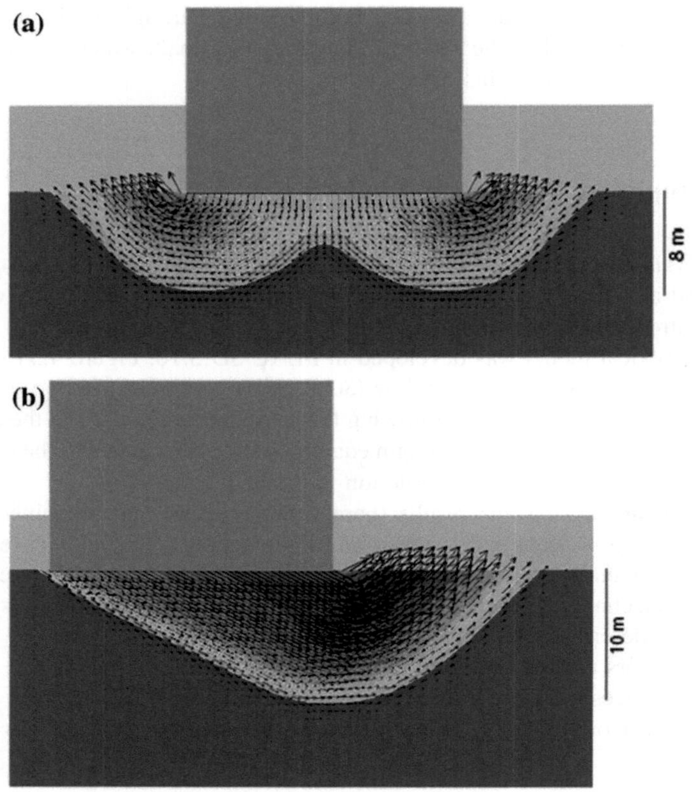

Fig. 4.35 Critical failure mechanisms calculated numerically (FLAC). **a** Symmetric; **b** nonsymmetric

Table 4.2 Comparison of safety factors. Analytical and numerical analyses. Simplified hypothesis

	Phase 1. Sunk	Phase 2. Consolidation t = 14 days	Phase 3. Consolidation t = 21 days	Phase 2. Consolidation t = 21 days	Phase 5. Storm	Phase 6. Storm + liquefaction
Analytical	1.36	1.82	1.11	1.22	1.1	0.63
Numerical. Simplified analysis	1.24	1.56	0.94	1.02	0.86	0.38
Numerical. Granular layer	1.46	1.76	1.06	1.14	0.92	0.42
Numerical. Elasto-plastic	1.46	1.84	1.18	1.26	1.08	0.54

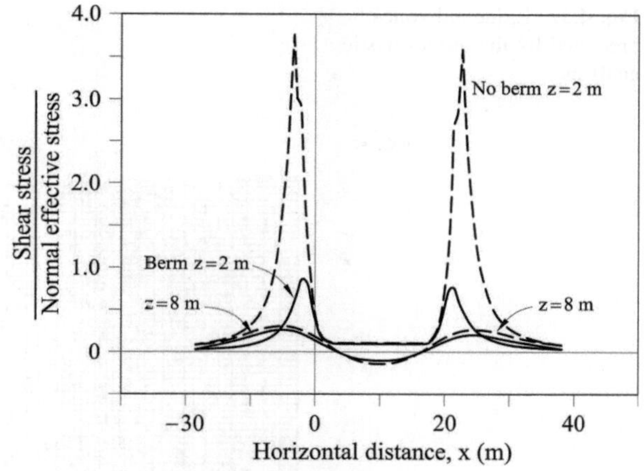

Fig. 4.36 Cyclic shear stress ratio analyses. Effect of upper granular layer

expected. The numerical analysis predicts the failure in Phase 3. Since it did not happen the effect of simplifications introduced should be investigated.

The most significant aspect is probably the effect of the granular foundation layer on the development of shear stresses in the soft silty soil. A 2 m thick upper granular layer has been simulated and characterized with a Mohr-Coulomb model ($E = 5.25$ MPa, $\nu = 0.3$, $c' = 0$ and $\phi' = 35°$). The stiffness of the upper granular layer and its thickness contributes to reducing shear stresses on the silt layer surface. This is shown in Fig. 4.36 for the cyclic shear stress ratio. The high peak stress ratios in the vicinity of the caisson edges are largely reduced.

This result explains the improvement in SF achieved when the granular layer is introduced (Table 4.2).

Note that the new calculation predicts an SF > 1 in Phase 2, which is consistent with actual caisson behavior.

There was also an interest in introducing the plastic character of the soil because it would imply a further reduction of the maximum shear stress ratios.

4.20 Elasto-Plastic Numerical Analysis

Given the nature of the silty layer, a soft reasonably homogeneous normally consolidated material, a Cam clay model was selected to proceed with further analysis of the case. The following material parameters were derived from soil data already presented: $\lambda = 0.109$; $\kappa = 0.027$; $M = 1\ 11$. Some anisotropy in permeability was also introduced because of the presence of some fine sand lenses in the soil. Following the criterion given by Jamiolkowsky et al. (1985) a ratio Kh/Kv = 3 was accepted. In addition, a rigid caisson was also adopted. If compared with a flexible foundation, a rigid base will enhance shear stressing under caisson edges. The upper granular layer was also introduced.

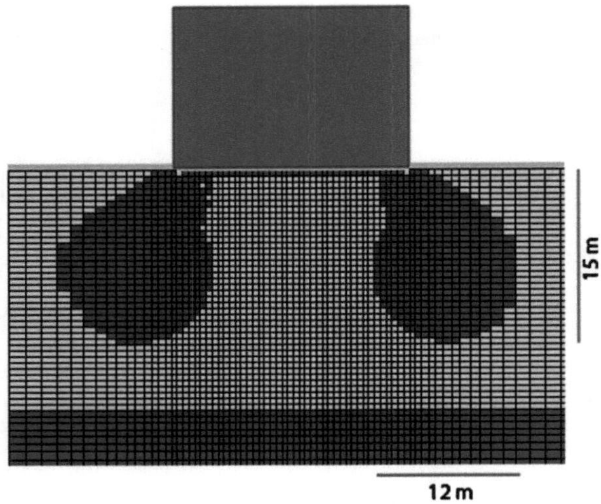

Fig. 4.37 Liquefied zones predicted by the elasto-plastic analysis

The liquefied area is given in Fig. 4.37. If compared with the analytical solution, the intensity of liquefaction is significantly reduced in the upper part of the silt layer. The depth of the liquefied bulb is similar in both approaches. Safety factors were calculated with the help of the (c, ϕ) reduction procedure and are compared in Table 4.2.

The more accurate elasto-plastic analysis predicts again that no failure occurred in Phase 3 (caissons filled with sand). It also predicts that the static forces derived from the storm were not able to sink the caissons. However, the liquefaction strength reduction resulted in a major drop of available foundation bearing capacity. This drop is of such a magnitude that possible errors introduced during the first part of the analysis lose practical significance. This is also true for the analytical results.

Figure 4.38 provides a graphical representation of the life of the caissons in terms of the evolution of safety factors. Two cases are compared: the simple numerical solution and the elasto-plastic analysis, closer to real conditions. The elasto-plastic analysis only predicts failure if liquefaction conditions are introduced. It is consistent with observations and offers the opportunity to discuss further the design of a stable caisson in the next section.

4.21 Designing a Stable Caisson Dyke

The increase in undrained strength with time under the weight of the granular berm and the caisson itself is a key aspect of the stability of caissons. If the basic design (caissons founded on a granular layer extended on the bottom of the sea) is maintained the safe design should essentially define the thickness and the lateral extent of the granular foundation blanket. There are, of course, other possibilities: improving the foundation soils by alternative procedures such as dynamic consolidation, piles, gravel columns, or by changing caisson dimensions.

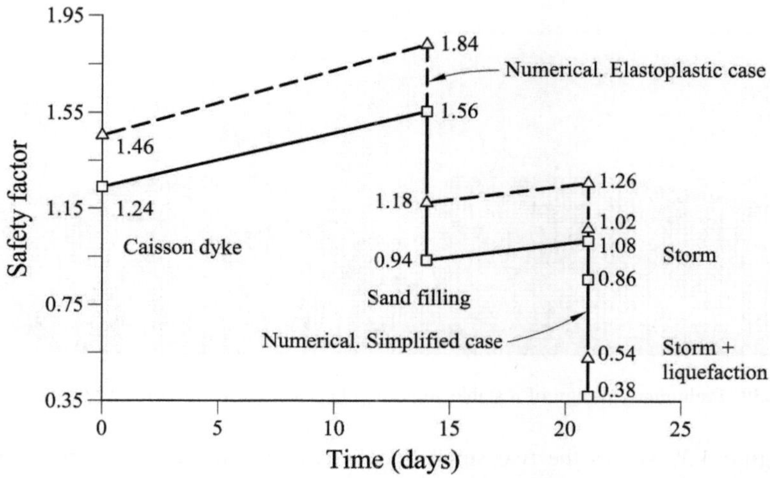

Fig. 4.38 Safety factors calculated numerically for the simplified and the elasto-plastic analysis at different construction stages

The following discussion assumes that caisson dimensions are maintained and a conventional design—a granular berm—is maintained.

Two conditions have been defined to illustrate the analysis: Short term and long term. In both cases the liquefaction of the foundation is accepted as the determinant criterion in view of the previous analysis.

For the short term the condition imposed is to require an acceptable SF, including the case of liquefaction, 3 weeks after caisson sinking, for the storm intensity which led to the failure of caissons (wave significant height: $H_s = 4.5$ m).

For the long term a stronger storm, corresponding to a return period of 475 years has been defined. Actual wave data recorded by an instrumented buoy outside Barcelona Harbor at a distance of 5.1 km from the site was used to define the long-term storm. The storm was defined by a characteristic wave height $H_s = 7$ m and a period $T = 11.6$ s. The design criterion is again to ensure an SF against foundation failure 6 months after the caisson sinking. The sequence of operations an initial water filling following by a sand filling 2 weeks later has been maintained.

Soil properties have already been defined. Goda's formulas are used to transform wave action into forces acting on caissons.

The analysis started by calculating the final SF, for short- and long-term conditions, for increasing thickness and width of the granular berm.

The symmetric design is shown in Fig. 4.39. A granular layer 8 m thick and two lateral "wings" 65 m long) led to the following safety factors:

- Short term: SF = 1.18
- Long term: SF = 1.04

This design requires 840 m^3 of granular material per meter on dyke length. The short term SF may be acceptable but the long term one is too low.

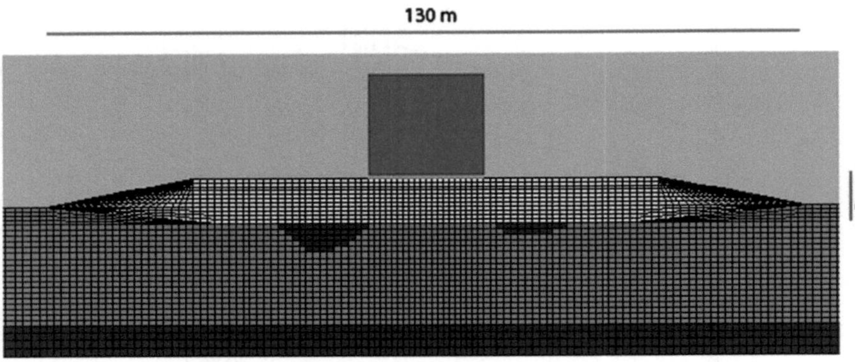

Fig. 4.39 Preliminary design of a stable solution. Open sea on the left of the figure

Figure 4.39 shows the two small liquefied zones calculated for the long-term storm. A symmetric design does not seem to be the best option because foundation failure will go in the land direction. Therefore, part of the berm extended on the seaside is probably not required. On the other hand, eliminating the possibility of liquefaction under the caisson edges is also a desirable situation. The objective is to optimize the geometry of the granular berm, possibly maintaining the overall volume of added material so that acceptable Safety Factors are calculated and, at the same time, liquefaction is totally avoided.

Decreasing the length of the berm on the seaside, increasing its thickness under the caisson base, increasing the length of the berm on the land side, and reducing somewhat the berm thickness on the land side results in the design shown in Fig. 4.40.

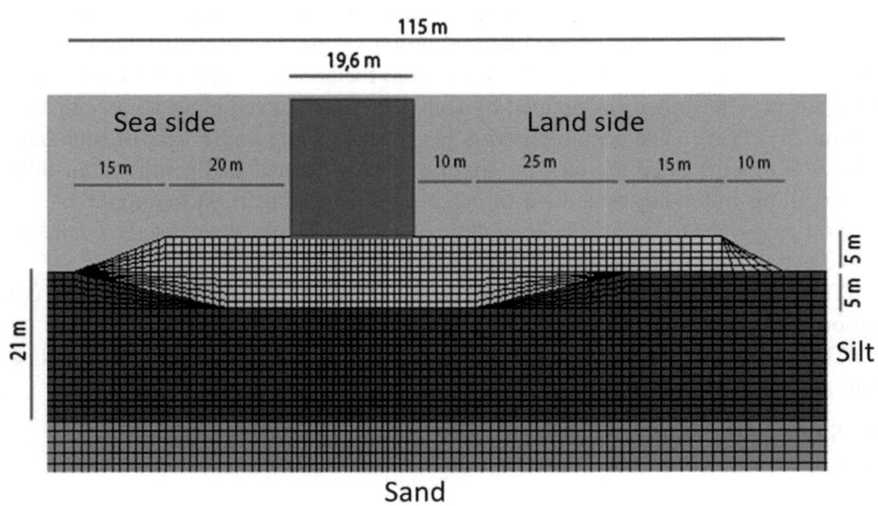

Fig. 4.40 Proposed design of a stable solution. Open sea on the *left* of the figure

Table 4.3 Calculated safety factors for the proposed design (long term involves total dissipation of excess pore water pressure)

	Static (fill with sand)	Static long term	Storm after 21 days	Storm long term
SF	1.28	1.5	1.14	1.12

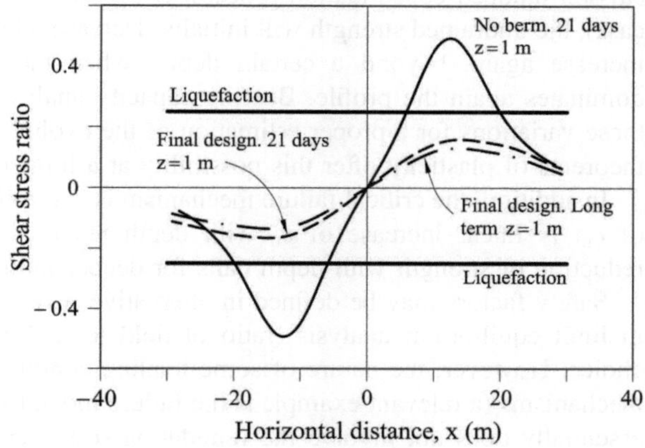

Fig. 4.41 Cyclic stress ratios 1 m below the contact between the granular berm and the foundation silty soil

The solution given in Fig. 4.40 implies a similar volume of granular materials (837.5 m^3/m). However, the calculated SFs, given in Table 4.3, are now acceptable. "Static" safety factors in the absence of dynamic wave loading are 1.28–1.5. Under wave action safety factors are larger than 1.1, a criterion that can be accepted. Liquefaction is avoided, and this, as indicated in Fig. 4.41, is that the calculated cyclic stress ratios close to the silt-berm interphase remain in a stable zone.

4.22 Concluding Remarks

The design of safe foundations for caissons dykes in an open sea environment sitting on normally consolidated soft silty and silty clay deposits of low plasticity poses a significant geotechnical challenge.

The following aspects should be accounted for in an analysis of global or bearing capacity type of failure:

- The protocol of caisson sinking and subsequent weight increase.
- The consolidation process under the history of caisson increase in weight.
- The strength increase induced by consolidation.
- The forces induced by wave action under the design storm.
- The risk of soil liquefaction under repeated loading (if foundation soils are prone to liquefaction).

This paper describes first the capabilities of analytic solutions to follow the sequence of events. Later a numerical procedure is also used. Differences remain moderate provided the main phenomena are properly addressed.

In low permeability soft soils, the most critical limiting condition is an undrained failure. The initial profile of undrained strength of a normally consolidated deep layer of soil (a linear increase of c_u with depth) is substantially modified during the caisson-induced consolidation process. If upward drainage is allowed (a common case), the undrained strength will initially decrease with depth and eventually it will increase again, beyond a certain depth, when the geostatic stress distribution dominates again the profile. Bearing capacity analyses should be able to include these variations for a proper estimation of the evolution of safety with time. Limit theorems of plasticity offer this possibility at a limited calculation effort.

In addition, the critical failure mechanism is controlled by the spatial distribution of c_u. A linear increase of c_u with depth results in shallower mechanisms. A reduction of strength with depth calls for deeper failure mechanisms.

Safety factors may be defined in alternative ways. The classical definition used in limit equilibrium analysis (ratio of field to mobilized strength) is a practical choice. However, the nature of some loading conditions or, indeed, of the failure mechanisms (a relevant example is the failure mode by overturning of walls, which essentially does not involve the foundation soil strength) may require a different formulation. A natural choice is to compare failure loads and actual loads. One example has been given in the section when examining the caisson safety subjected to wave loading and caisson weight. Defining soil liquefaction condition.

Liquefaction conditions were defined by means of a cyclic interaction diagram. It combines, in a purely experimental criterion, the static, maintained, or averaged stress ratio, the cyclic stress ratio, and the number of applied stress cycles which marks the onset of liquefaction conditions. Such a diagram was available for the deltaic silty foundation soils of the caissons.

The cyclic interaction diagram has inspired a simplified procedure to analyze liquefaction under field conditions. It was found that liquefaction takes place on both sides of the caisson, under and beyond caisson edges, and at depths which reach the caisson width.

The flexibility and capabilities of the upper bound procedure was demonstrated when the liquefied zones were integrated into a failure analysis of the caisson once liquefaction had occurred. A substantial reduction of failure load was calculated for a reasonable estimation of post-liquefaction undrained strength.

After the failure caissons were found tilted, their cap displaced in the seaward direction. The failure mechanism implies that both the caisson weight and the wave forces performed a positive work, balanced by the internal dissipation in the soil. No evidence of overturning induced by wave forces was found.

The numerical elasto-plastic analysis performed was capable of following more precisely the calculation stages. Once the model was validated against actual observations it allowed a designing process for a stable caisson dyke under strong wave action. Avoiding soil liquefaction was the main criterion. It was found that a non-symmetric granular berm foundation over the soft silty soils was the optimized solution.

References

Alonso EE, Krizek RJ (1975) Consolidation of randomly heterogeneous clay strata. Transp Res Rec 548:30–47

Davis EH, Booker JR (1973) The effect of increasing strength with depth on the bearing capacity of clays. Géotechnique 23(4):551–563

Davis EH, Poulos HG (1972) Rate of settlement under two- and three-dimensional conditions. Géotechnique 22(1):95–114

Goda Y (1985) Random seas and design of maritime structures. University of Tokyo Press, Tokyo, 323 pp

de Groot MB, Bolton MD, Foray P, Meijers P, Palmer AC, Sandven R, Sawicki A, The TC (2006) Physics of liquefaction phenomena around marine structures. J Waterw Port Coast Ocean Eng 132(4):227–243

Ishihara K (1993) Liquefaction and flow failure during earthquakes. Géotechnique 43(3):351–415

Jamiolkowski M, Ladd CC, Germaine JT, Lancellotta R (1985) New developments in field and laboratory testing of soils. In: Proceedings, 11th international conference on Soil mechanics and foundation engineering, San Francisco, vol 1, pp 57–154

Jeng DS (1998) Wave-induced seabed response in a cross anisotropic seabed in front of a breakwater: an analytical solution. Ocean Eng 25(1):49–67

Jeng DS (2001) Mechanism of the wave-induced seabed instability in the vicinity of a breakwater: a review. Ocean Eng 28:537–570

Jeng DS, Lin YS (2000) Poroelastic analysis of the wave-sea interaction problem. Comput Geotech 26:43–64

Kudella M, Oumeraci H, de Groot MB, Meijers P (2006) Large-scale experiments on pore pressure generation underneath a caisson breakwater. J Waterw Port Coast Ocean Eng 132(4):310–324

Lunne T, Robertson PK, Powell JJM (1997) Cone penetration testing in geotechnical practice. Blackie Academic/Chapman and Hall, E&FN Spon, London

NGI (2002) Report on DSS tests. Port Authority, Barcelona

Olson SM, Stark TD (2002) Liquefied strength ratio from liquefaction flow failure case histories. Can Geotech J 39:629–647

Oumeraci H (1994) Review and analysis of vertical breakwater failures-lessons learned. Coast Eng 22:3–29

Potts D, Zdravkovic L (1999) Finite element analysis in geotechnical engineering, vol I, theory. Telford Publishing, London

Poulos HG, Davis EH (1973) Elastic solutions for soil and rock mechanics. Wiley, New York

Rowe PW, Craig WH (1976) Studies of offshore caissons founded on Ostercheelde sand. Design and construction of offshore structures. ICE, London, pp 49–60

Sassa S, Sekiguchi H (2001) Analysis of wave-induced liquefaction of sand beds. Géotechnique 51(2):115–126

Van der Poel JT, de Groot MB (1998) Cyclic load tests on a caisson breakwater placed on sand. Proc. Int. Conf. Centrifuge 98 1:403–408

Wood DM (1990) Soil behaviour and critical state soil mechanics. Cambridge University Press, Cambridge

Youd TL, Idriss IM (2001) Liquefaction resistance of soils: summary report from the 1996 NCEER and 1998 NCEER/NSF workshops on evaluation of liquefaction resistance of soils. J Geotech Geo-environ Eng ASCE 127(10):297–313

Zhang XY, Lee FH, Leung CF (2009) Géotechnique 59(1):3–16

Chapter 5
Forensic Geotechnical Engineering Investigations: Data Collection

Peter Day

Abstract This paper describes the initial stage of a forensic geotechnical investigation, namely gathering of data on the site, the works, and the failure for further detailed analysis. It has been written as part of the effort by the ISSMGE Technical Committee TC40 to prepare a handbook on forensic geotechnical engineering. The paper provides guidance to the investigator on the objectives of the investigation, the nature of the data required, sources of information to be considered and the recording and storage of data.

Keywords Failure · Distress · Failure mechanisms · Data collection

5.1 Background

Forensic geotechnical engineering deals with the investigation of failures of geotechnical origin, not only from a technical viewpoint but also with the possibility of legal proceedings in mind (Rao 2005).

Forensic investigations differ from conventional geotechnical investigations in that they are retrospective. They seek to explain what has happened rather than to predict future performance. A further distinguishing factor is that, following a failure, there is an urgency to clean up the site and rebuild or repair the works. This limits the time available for investigation and makes it essential that all relevant data is recorded before the evidence is removed.

The ideal outcome of the data collection stage of a forensic investigation would be to have a body of information that is (a) as complete as reasonable possible, (b) accepted by all parties as an accurate record of the facts and events, and (c) is stored in an accessible and readily understood way. This paper describes procedures aimed at achieving this outcome.

P. Day (✉)
Jones & Wagener Consulting Engineers, Rivonia 2128, South Africa
e-mail: day@jaws.co.za

V.V.S. Rao and G.L. Sivakumar Babu (eds.), *Forensic Geotechnical Engineering*, Developments in Geotechnical Engineering, DOI 10.1007/978-81-322-2377-1_5

5.2 Scope of Investigation

Although some geotechnical failures such as landslides occur in the absence of any human intervention, most geotechnical failures involve both the ground (soil, rock, and groundwater) and the works (some man made structure or intervention). The works may be a structure that imposes loads on the ground, the alteration of the surface geometry (cuts or fills), alteration of drainage patterns or the creation of underground openings. Thus, the forensic geotechnical investigation must include a study of the event or failure which gave rise to the investigation, the site on which the failure occurred, and the nature of the works. Only after these three aspects have been investigated and recorded can post-failure diagnostic testing and back-analyses commence.

5.2.1 The Failure

In any forensic investigation, it is essential that the circumstances and events surrounding the failure are investigated and recorded as soon as possible, before any evidence is removed.

Details of this stage of the investigation will vary from case to case. Nevertheless, there are common aspects that apply to all failure investigations. These include the conditions that prevailed immediately prior to the failure, the sequence of events, and the condition of the works and surrounding areas following the failure.

5.2.1.1 Circumstances Prior to Failure

Before one can establish the cause of the failure, it is essential to investigate and record the condition of the works immediately prior to the failure. Typical factors to be recorded include:

- the stage of completion of the works at the time (see Sect. 5.2.3.3),
- the occurrence of accidental actions (impact, explosion, earthquake, flooding or water leakage, etc.) or abnormal loading,
- abnormal meteorological conditions (wind, snow, rainfall, temperature, etc.),
- the results of any monitoring (pore pressures, deformations, settlements, anchor loads, etc.), and
- any early warnings of incipient failure (cracking of ground or structure, falls of ground, changes in anchor loads, etc.).

5.2.1.2 Sequence of Events

Obtaining an accurate record of the sequence of events that lead to the failure will assist greatly in determining the failure mechanism and, in many cases identifying the trigger. The information to be recorded will vary from site to site. However, an attempt should be made to obtain as much information as possible from the time when the first signs of distress were noted and, if possible, to draw up a time line from which the sequence of events and the speed of progression can be ascertained.

5.2.1.3 Resulting Distress

The two subsections above deal with the "before" and "during" situations. This subsection deals with the "after" or post-failure state of the works and surrounding areas. An accurate description of the distress caused by the failure may be pivotal in determining the value of any claim for compensation or damages which may follow.

Unlike the "before" and "during" situations which must be investigated by reference to historical records, the "after" situation can be directly observed, photographed, and recorded. Note that the recording of the condition of the works after the occurrence as described in this paper is distinct from the detailed diagnostic tests that may be required to provide parameters for back-analysis of the failure.

Typical information to be recorded may include:

- the extent and severity of distress,
- the magnitude of deformations and trajectory of movement,
- any signs of where rupture may have occurred (slickensided shear zones, yielded construction elements, etc.),
- indications of abnormal surface or subsurface water conditions (seepage, high water marks, etc.),
- condition of any exposed rupture surfaces whether on the structure or in the ground,
- any deviations from expected ground conditions (paleo channels, intrusives, faults, adverse jointing, seepage, etc.), and
- damage or changes in surrounding areas (physical damage, settlement, lateral movement, drop in groundwater levels, rupture of services, etc.).

5.2.1.4 Sources of Information

When determining the condition of the works prior to the failure, reference should be made to any available reports (geotechnical, structural, etc.), to construction records, progress payment certificates, photographs, as-built drawings and any information that can be provided by site staff. Depending on the circumstances, it may be necessary to obtain additional information from outside the site such as water meter readings, flood levels, metrological data, etc. Recent aerial photos or satellite images may also be of assistance.

Determining the sequence of events normally relies on eye-witness accounts and photographs taken immediately prior to, during and immediately after the event. Care should be taken when interpreting eye-witness accounts as vested interests may be involved. For this reason, it is preferable to obtain information from as many site personnel as possible and preferably not only from personnel employed by either the engineer, the owner or the contractor. Any discrepancies in the information provided by witnesses should be revisited and clarified at the time. Written records should be kept of eye-witness interviews and these should preferably be signed by both the investigator and the witness. Where appropriate, verbatim transcripts or recordings of eye-witness evidence should be kept.

Information of the condition of the works after the occurrence is generally obtained by direct observation. This will typically include photography, sketches, written descriptions, post-failure survey drawings, etc.

It is preferable that the post-failure information is jointly recorded by the parties involved (owner, engineer, contractor, insurer etc.) to minimize disputes at a later stage.

5.2.1.5 Potential Failure Mechanisms

The scientific method relies on a process of postulation and verification. Many scientific endeavors have floundered as a result of failing to consider alternative postulates and gathering only information which supports a particular point of view. In any scientific investigation, it is as important to record both supporting data and data which is inconsistent with various postulates.

The purpose of this first phase of the investigation is to gather data. Back analyses and the identification of the most likely failure mechanism(s) will follow at a later stage. In fact, forming a fixed opinion on the cause or mechanism of failure too early in the process may result in certain essential information relating to possible alternative causes being overlooked. Nevertheless, it is recommended that investigators should seek to identify all the potential triggers, sequences of events, and failure mechanisms. Simple logic is likely to eliminate many of these from the start. An attempt should then be made to obtain the data which will enable the likelihood of all plausible failure mechanisms to be assessed at a later stage.

5.2.2 *The Site*

In order to carry out a competent analysis into the causes of the failure, information is required on the site on which the works were undertaken. Much of this information should already be available in the form of existing reports and other documents.

5.2.2.1 Essential Data

The essential data required will vary from site to site. In most instances, it will include:

- the location and extent of the site,
- surface topography and alterations thereto,
- surrounding services and development,
- site description including vegetation, drainage, climate, previous land use, existing development, etc.,
- geological setting, site geology, regional geological structures (faults, folding, etc.), and seismicity,
- site stratigraphy including the identification of typical soil profiles for various areas of the site,
- detailed soil and rock profiles from boreholes, pits, or other exposures at particular locations on the site,
- information on groundwater including depth of water table(s), seasonal fluctuations, gradients, flow characteristics, etc., and
- results of field and laboratory tests.

5.2.2.2 Sources of Information

The main source of geotechnical information on the site should be the geotechnical and geological reports prepared for the project. The absence of such reports may, in itself, be a contributor to the failure.

The next most likely source of geotechnical information on the site is the construction records. Much valuable information may be gleaned from site instructions, recorded founding depths, tunnel face maps, and other similar information.

On some projects, borehole core or samples from the original investigation may still be available. This creates the possibility of re-inspecting the core or retesting of samples.

5.2.3 The Works

In considering the works, there are three main aspects which require consideration. These are (a) the works as designed, (b) the works as constructed, and (c) if appropriate, the state of completion of the works at the time of failure.

5.2.3.1 Works as Designed

Information on the works as designed is generally contained in the construction drawings and project specifications. These documents specify the work to be carried out by the contractor and provide a basis from which any deviations may be assessed.

Of equal importance are the calculations on which the design of the works was based. A review of the design calculations will enable the investigator to ascertain whether the conditions on site are consistent with those assumed by the designer during the design process.

5.2.3.2 Works as Constructed

The works as constructed may differ from the works as described on the construction drawings and project specifications for a number of reasons such as on-site design modifications, substandard materials or workmanship, geometric deviations (both construction tolerance and setting out errors), and concessions granted to remedy noncompliance.

Identification of differences between the "as designed" and "as constructed" works will generally require physical measurement, survey, inspection and testing on site. Supplementary information can be obtained from site documents such as nonconformance reports, design modification reports, site instructions, minutes of site meetings, etc.

5.2.3.3 State of Completion

A significant proportion of all geotechnical failures occur during construction. Where this is the case, it is essential to determine:

- the state of completion of the works at the time,
- applied loading (both self weight and imposed loads),
- changes in loading or construction activities immediately prior to failure and
- strength of materials that show time dependency at the time of failure (e.g., clays undergoing consolidation, freshly cast concrete or grout, etc.).

5.3 Recording of Data

5.3.1 Attention to Detail

The forensic investigator should strive to ensure that all information obtained during the course of the investigation is complete in every respect. Not only does this require that the scope of the investigation is adequate, that all plausible failure mechanisms are considered and that all relevant documentation is examined, it also requires meticulous attention to detail.

In this regard, it is recommended that written records be kept of all discussions and inspections and that these records be dated and signed. All photographs should be uniquely referenced and the date, time, location and orientation of the photographs should be recorded. Any samples taken should be photographed in situ (prior to sampling), provided with a unique sample number and records kept of the tests undertaken and the results obtained.

5.3.2 *Agreement Between Parties*

Forensic geotechnical investigations may either be carried out by an independent investigator or by different investigators employed by the various parties (employer, contractor, engineer, insurer, etc.).

Wherever possible, all parties should be afforded the opportunity to witness critical stages of the investigation such as exhuming of foundations, taking of samples, removing collapsed structures, etc. and observations made during such crucial stages of the investigation should be shared with all parties. This is to avoid the situation where crucial evidence is ruled to be inadmissible during the legal process as the accuracy of the observations, location of the samples or photographs, and other such details can only be vouched for by one party.

In investigations of this nature, it is preferable that as much agreement as possible be obtained during the investigation stage as it is likely that much of the evidence will have been removed when the time arrives for resolving disputes or legal argument.

5.3.3 *Reporting and Data Storage*

It goes without saying that all data obtained during the course of the investigation must be adequately documented and stored in a well referenced and easily retrievable format. Even the simplest of failure investigations can drag on for a number of years. Given the mobility in the job market, it is likely that the data will be analyzed by a different team of engineers to those responsible for its collection.

In the case of small investigations, it is preferable that a report be compiled which details the extent and findings of the investigation and appends any relevant supporting documents.

In the case of larger investigations, the report is more likely to be a summarized version of events and an index of the various supporting documents. On major projects, the number of documents may run into the hundreds and possibly thousands. Under these circumstances, electronic storage of documents should be considered due to the ease of reproduction and portability of electronic records.

5.4 Concluding Remarks

The gathering of data for a forensic geotechnical investigation should be conducted with an open mind. Collection of selective data intended to support a particular hypothesis may be counterproductive in that the data so collected will be inadequate to test the veracity of alternative theories. It is as important to collect supporting evidence as it is to note evidence that is inconsistent with postulated failure mechanisms.

The investigator should be mindful of the fact that the data gathered may be subject to scrutiny in subsequent legal proceedings. Obvious bias in the collection and reporting of data will discredit the findings of the investigation.

Where the investigator has access to all parties (e.g. in the case of a joint appointment), consultations should be held with all parties involved. More often than not, follow-up consultations will be required as information obtained from one party may be queried or refuted by another. Wherever possible, an attempt should be made to obtain agreement between the parties on important issues at the time as this reduces the amount of evidence to be lead in any future legal proceedings.

Reference

Rao VVS (2005) TC40 terms of reference. Report submitted to ISSMGE, Dec 2005

Chapter 6
Diagnosis of Geotechnical Failure Causes Using Bayesian Networks

Y. Xu and L. Zhang

Abstract Failure of a complex geotechnical system may be caused by many possible mechanisms. This paper introduces a Bayesian network-based method for the diagnosis of geotechnical distress mechanisms considering uncertainties in these mechanisms and interrelationships among these mechanisms. The methodology is illustrated through the diagnosis of a distressed dam with seepage problems. First, common patterns and causes of dam distresses are identified based on the information in a database of 993 distressed in-service embankment dams in China and used as the prior information. The interrelations among the dam distresses and their causes are quantified using conditional probabilities determined based on historical frequencies from the database. The observed leakage rates, seepage exit locations, and boundary conditions of a specific distressed dam are used as project-specific evidences. Bayesian networks are used to diagnose the distressed dam by combining the prior information based on the database and the project-specific evidence in a systematic way. Based on results of the diagnosis, key distress factors for the dam can be identified and suitable remedial measures can be suggested.

Keywords Diagnosis · Bayesian networks · Distresses · Embankment dams

6.1 Introduction

Three principal approaches have widely been used in dam safety studies, including failure modes and effects analysis, event tree analysis, and fault tree analysis. Hartford and Baecher (2004) made an excellent summary of the concepts and applications of these three approaches. When the modes and causes of dam failures

Y. Xu (✉)
China Institute of Water Resources and Hydropower Research, Beijing, China
e-mail: xuyao@iwhr.com

Y. Xu · L. Zhang
The Hong Kong University of Science and Technology, Hksar, China
e-mail: cezhangl@ust.hk

V.V.S. Rao and G.L. Sivakumar Babu (eds.), *Forensic Geotechnical Engineering*, Developments in Geotechnical Engineering, DOI 10.1007/978-81-322-2377-1_6

are multiple and interrelated, however, these traditional approaches may not be effectively used to diagnose dam distress mechanisms. Often one triggering cause might not truly have resulted in failure had the dam not had a secondary weakness. Such a special feature requires a global consideration of a distressed dam by putting into perspective all the components and their interrelationships. Bayesian network analysis (Jensen 1996) has recently drawn increasing interest as an alternative for dam safety studies. Some disadvantages of the traditional methods, which become apparent in dam safety applications, can be potentially overcome by applying the Bayesian network technique. Therefore, Bayesian networks are well suited for studying a complex system with multiple elements and their interactive influences in a systematic way.

The main objective of this paper is to develop a robust probability-based tool by using Bayesian networks for the diagnosis of dam distresses. Bayesian network analysis can tackle not only the multiplicity of dam distresses and causes but also the complex interrelations within them. In this paper, general characteristics of embankment dam distresses are first studied using Bayesian networks based on a database of 993 in-service distressed embankment dams in China. The common patterns and causes of embankment dam distresses are identified; the global-level Bayesian networks are also constructed. Then, project-specific evidence is input into the existing global-level Bayesian networks to update the diagnosing results for a specific distressed dam. A case study is worked out to illustrate the methodology.

6.2 Fundamentals of Bayesian Networks

A Bayesian network is a directed acyclic graph formed by the variables (nodes) together with the directed edges, attached by tables of conditional probabilities of each variable on all its parents (Jensen 1996). Each variable has a probability distribution, which may be continuous or discrete. In this study, each variable is assumed to have a finite set of mutually exclusive states in which each state is associated with a probability measure. The nodes that have arrows directed into them are called child nodes and the nodes that have arrows directed from them are called parent nodes. Each child has a conditional probability table, given the state of the parent nodes. Note that the terms node and variable are used interchangeably in this paper.

Figure 6.1 shows three typical simple Bayesian networks, involving a serial connection, a diverging connection, and a converging connection, respectively. In practical applications, a Bayesian network often consists of all the above three types of connections. The joint probability function of random variables in a Bayesian network can be expressed in a general form as

$$p(x_1, \ldots, x_n) = \prod_{i=1}^{n} p(x_i | \mathrm{pa}(X_i)) \tag{6.1}$$

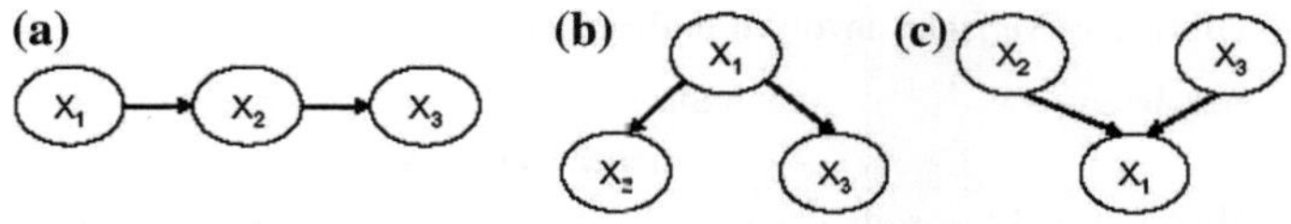

Fig. 6.1 Illustration of three typical simple Bayesian networks: **a** serial connection, **b** diverging connection, and **c** converging connection

in which pa(X_i) is the parent set of X_i. Note that if X_i has no parents, then the function reduces to the unconditional probability of $p(x_i)$.

It is indicated from the concepts that Bayesian networks are a combination of graph theory and probability theory. More specifically, two primary issues are involved in a Bayesian network: (1) establishing the cause and effect relationships among the variables, and (2) determining the conditional probabilities associated with these relationships (Smith 2006).

6.3 Global-Level Bayesian Networks

The general information of 993 in-service distressed embankment dams in China and details of their distresses are compiled into a database in this paper. Among these distressed embankment dams, a total of 610 cases (591 homogeneous earthfill dams and 19 composite earthfill dams) are grouped into Dataset A due to their similar structure and distress characteristics. Similarly, a total of 383 cases (317 earthfill dams with vertical clay cores, 59 earthfill dams with inclined clay cores, three rockfill dams with vertical clay cores, and four rockfill dams with inclined clay cores) are grouped into Dataset B. In the latter distress diagnosis, separate analyses will be conducted for homogeneous/composite dams and clay-core dams based on Dataset A and Dataset B, respectively.

Most of the past embankment dam failures were caused by either overtopping or seepage erosion/piping (Xu and Zhang 2009). Therefore, this paper focuses on the distress mechanisms that may cause overtopping or seepage erosion/piping. In this paper, a distress is regarded as a component of a dam system that cannot satisfy the performance requirements for which it was intended. In other words, a distress may be treated as nonsatisfaction of a performance requirement. Table 6.1 shows the common performance variables and causes of distresses associated with the embankment dams in Dataset A and Dataset B. Each variable is taken as a discrete variable with two to four states. A state indicates the degree the performance requirements are violated.

Considering the two primary types of dams in Dataset A and Dataset B and the two major failure modes, four Bayesian networks are constructed for diagnosing distresses of (1) homogeneous/composite dams resulting in overtopping, (2) homogeneous/composite dams resulting in seepage erosion/piping, (3) clay-core dams resulting in overtopping, and (4) clay-core dams resulting in seepage erosion/piping.

Table 6.1 Performance variables involved in diagnosing distressed dams

Code of node	Variable	State	
		Number	Description
ARS	Abutment rocks or soils	2	Satisfactory, unsatisfactory
ASS	Abutment seepage situation	4	Normal, unstable, moderate leakage, serious leakage
CF	Cutoff at foundation	2	Satisfactory, unsatisfactory
CMG	Clay-core materials and geometries	2	Satisfactory, unsatisfactory
DEH	Designed embankment height	2	Satisfactory, unsatisfactory
DEW	Designed embankment width	2	Satisfactory, unsatisfactory
EBI	Embankment-abutment interface	2	Satisfactory, unsatisfactory
EC	Embankment cracking	2	Occurred, non-occured
ECE	Embankment crest elevation	2	Satisfactory, unsatisfactory
EM	Embankment material	2	Satisfactory, unsatisfactory
ES	Embankment settlement	2	Normal, excessive
ESP	Emergency spillway	3	Satisfactory, moderately unsatisfactory, seriously unsatisfactory
ESS	Embankment seepage situation	4	Normal, unstable, moderate leakage, serious leakage
FD	Filtered drainage	2	Satisfactory, unsatisfactory
FDC	Rood design criteria	2	Satisfactory, unsatisfactory
FRC	Flood releasing capacity	2	Satisfactory, unsatisfactory
FRT	Rood releasing tunnel	3	Satisfactory, moderately unsatisfactory, seriously unsatisfactory
FSS	Foundation seepage situation	4	Normal, unstable, moderate leakage, serious leakage
GLD	Gales and lifting device	3	Satisfactory, moderately unsatisfactory, seriously unsatisfactory
OT	Overtopping	2	Yes, no
PS	Primary spillway	4	Satisfactory, moderately unsatisfactory, seriously unsatisfactory, no spillway
PW	Parapet wall	2	Satisfactory, unsatisfactory
RBSS	Reservoir bank slope stability	2	Satisfactory, unsatisfactory
RSS	Reservoir sedimentation situation	2	Satisfactory, unsatisfactory
SCF	Sludge cleaning at foundation	2	Satisfactory, unsatisfactory
SEP	Seepage erosion or piping	2	Yes, no
SSC	Seepage situation around embedded culvert	4	Normal, unstable, moderate leakage, serious leakage
TB	Termite burrows	2	Occurred, non-occured
WRC	Water retaining capacity	2	Satisfactory, unsatisfactory

To develop a Bayesian network for diagnosing dam distresses, the first step is to identify all possible distress mechanisms and display them in a causal network. Based on the dam distress database, an inventory of dam distresses and corresponding causes has been constructed. Each distress mechanism is described as a causal chain consisting of several variables, which represents the distress process from root causes to the final failure in a transparent way. Table 6.1 shows the variables involved in diagnosing distressed embankment dams. An example is shown in Fig. 6.2 for studying distresses associated with seepage erosion/piping of homogeneous/composite dams.

The probability tables for the relationships in the causal network in Fig. 6.2 are constructed based on the results of the historical frequencies from the dam distress database. The constructed probability tables reflect the statistics of the data currently available. Further case information may be gathered if necessary to update these probability tables.

Based on the causal network and the corresponding probability tables constructed, Bayesian network analysis is conducted using program Hugin Lite developed by Hugin Expert A/S (2004) in this paper. Still take the analysis of distresses of homogeneous/composite dams resulting in seepage erosion/piping (SEP) as an example. The calculation results are P(SEP = Yes) = 0.0377 and P(SEP = No) = 0.9623. In other words, the distressed homogeneous/composite dams have an average failure frequency of 3.77 % associated with seepage erosion/piping over their operation life, and have an average annual seepage erosion/piping failure probability of 2.37×10^{-4} (the dam operation time is assumed as $T = 50$ years). Table 6.2 summarizes the calculated failure frequencies over the operation time and the calculated average annual failure probabilities for homogeneous/composite dams and clay-core dams in China. The historical average annual failure probabilities for dams in China (Li et al. 2006) are also included in Table 6.2. It is found that the calculated average annual failure probabilities using Bayesian networks are close to the historical average annual failure probabilities, which shows that the outcome from the Bayesian network analysis is reasonable at the global level.

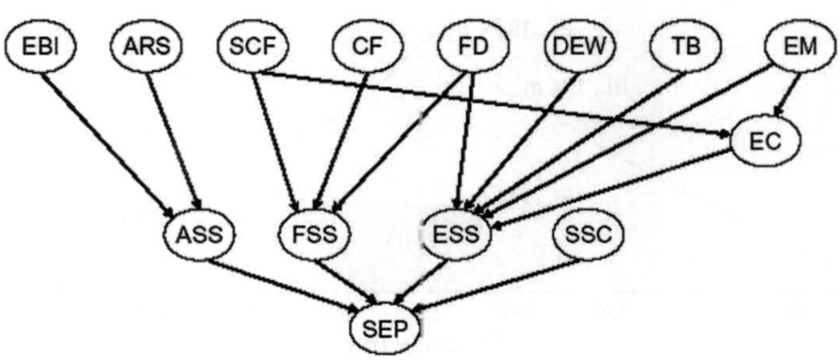

Fig. 6.2 Causal network for diagnosing distresses associated with seepage erosion/piping of homogeneous/composite dams. The symbols are defined in Table 6.1

Table 6.2 Calculated failure frequencies over the operation time and average annual failure probabilities

	Homogeneous/composite dams		Clay-core dams	
	Seepage erosion piping	Overtopping	Seepage erosion/piping	Overtopping
Calculated failure frequency over the operation life (%)	3.77	6.75	2.72	6.99
Calculated average annual failure probability	2.37×10^{-4}	4.24×10^{-4}	1.71×10^{-4}	4.39×10^{-4}
Historical average annual failure probability	2.19×10^{-4}	4.39×10^{-4}	2.19×10^{-4}	4.39×10^{-4}

6.4 Bayesian Network Analysis of a Specific Distressed Dam

Bayesian networks are as a dynamic tool in the sense that new evidence may be continuously utilized to update the knowledge. Based on the above Bayesian networks developed for diagnosing dam distresses at the global level, a Bayesian network for diagnosing a specific distressed dam may be developed considering local-level evidence. The network can incorporate project-specific variables, which reflect local-level performance records and/or physical analysis results. In this paper, a case study on the Chenbihe dam is conducted to illustrate the methodology.

The Chenbihe dam was located on the Chenbihe river in Guangxi, China (Niu 2002). It is an earthfill dam with a height of 70.4 m. The crest elevation and the base elevation of the dam are 190.4 and 120.0 m, respectively. Figure 6.3 shows a cross section of the dam. The dam was originally designed as an earthfill dam with a clay core. The dam construction was divided into three stages from December 1958 to October 1961. In the first stage, both the quality of the fill and clay materials and the

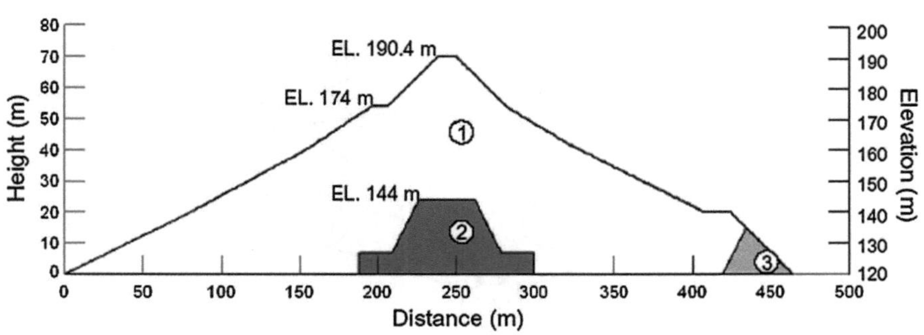

Fig. 6.3 Cross-section of Chenbihe Dam. *1* Earthfill, *2* clay core, *3* drainage

compaction effort were controlled rigorously to the design requirements. The thick clay core under elevation 144.0 m was constructed well according to the required quality standard. In contrast, the quality control became loose during the second and third stages due to the short construction period allowed and the shortage of proper clayey soils for the core. The clay core constructed in the second and third stages became thinner than required and contained a large proportion of coarse soils. Therefore, the dam above elevation 144.0 m can be treated as a homogeneous earthfill dam. After the initial filling of the reservoir in 1961, the downstream wetted area and the leakage rate increased as the reservoir water level rose. When the water level reached elevation 181.5 m in August 1971, there were five areas with serious leakage on the downstream slope at elevations of 144.0–170.0 m. The total leakage area was 4315 m^2 with a length of about 165 m along the dam axis. The total seepage rate in these areas was measured as 200 L/min.

In the Chenbihe dam case, the total seepage rates, the seepage exit locations, and the boundary conditions of the embankment are known. Therefore, back analyses of the seepage through the embankment may be performed to improve the knowledge on the embankment materials, e.g., the coefficients of permeability of relevant materials: k_1 of the earth fill and k_3 of the drainage. Considering the back analysis results on k_1 and k_3, nodes K_1 and K_3 are input into the existing global-level Bayesian network, resulting in a new Bayesian network in Fig. 6.4. Based on the new Bayesian network, the diagnosis result is summarized in the posterior probability column in Table 6.3. These posterior probability values for the nodes reflect the distress characteristics at the local level by considering the evidence on the seepage rate through the embankment.

For comparison purposes, the analysis was also conducted without considering the knowledge on K_1 and K_3 deduced from the project-specific evidence on the measured seepage rate (see Fig. 6.2). The results are summarized in the

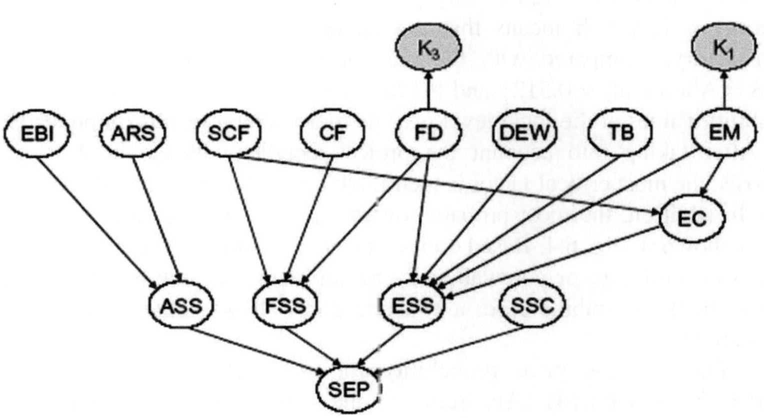

Fig. 6.4 Causal network for diagnosing the distressed Chenbihe Dam

Table 6.3 Probability table for the variables relevant to embankment seepage erosion/piping for the distressed Chenbihe Dam

Variable	State	Probability	
		Prior	Posterior
Clay-core materials and geometries (CMG)	Satisfactory	–	–
	Unsatisfactory	–	–
Designed embankment width (DEW)	Satisfactory	0.9300	0.9300
	Unsatisfactory	0.0700	0.0700
Embankment cracking (EC)	Satisfactory	0.9640	0.9220
	Unsatisfactory	0.0360	0.0780
Embankment materials (EM)	Satisfactory	0.5390	0.0000
	Unsatisfactory	0.4610	1.0000
Embankment seepage situation (ESS)	Normal	0.4879	0.1324
	Abnormal	0.5121	0.8676
Filtered drainage (FD)	Satisfactory	0.8432	1.0000
	Unsatisfactory	0.1568	0.0000
Seepage erosion or piping (SEP)	Yes	0.0141	0.0223
	No	0.9859	0.9777

prior-probability column in Table 6.3. Note that these prior-probability values for the variables reflect the distress characteristics at the global level.

It is found in Table 6.3 that the prior probabilities for node ESS at the global level are P(ESS = Normal) = 0.4879 and P(ESS = Abnormal) = 0.5121, which means that the seepage condition in the embankment has about one half of chance of being abnormal. In contrast, the posterior probabilities for node ESS at the local level are P(ESS = Normal | Posterior) = 0.1324 and P(ESS = Abnormal | Posterior) = 0.8676, which means that the seepage in the embankment has a great chance of being abnormal. Similarly, the posterior probabilities for node EM at the local level are P(EM = Satisfactory | Posterior) = 0 and P(EM = Unsatisfactory | Posterior) = 1, which means that the embankment materials are certain to be unsatisfactory. Compared with the prior-probability values at the global level, P(ESS = Abnormal) = 0.5121 and P(EM = Unsatisfactory) = 0.4610, the posterior probability values at the local level give us more certain results on nodes ESS and EM. After taking into account the project-specific evidence in the local-level diagnosis, the most critical factor is identified: unsatisfactory embankment materials (EM). In addition, the most probable distress process is also identified: EM-ESS-SEP, as shown in Fig. 6.4. Based on the diagnosis results, remedial actions, such as slurry wall, grouting, or corewall, may be suggested for cutting off embankment seepage. In the Chenbihe case, a concrete corewall was constructed later in the embankment.

In Table 6.3, the prior probability for node SEP at the global level is P(SEP = Yes) = 0.0141. According to the classification criteria for hydraulic structures issued by MWR (1994), Chenbihe Dam belongs to an important structure in the first grade, and the corresponding design service life is 100 years. Therefore, the prior average annual probability of seepage erosion/piping failure is Pf = 0.0141/100 = 1.41×10^{-4}. The posterior-I probability for node SEP at the local level is P(SEP = Yes | Posterior) = 0.0223. Therefore, the posterior average annual probability of seepage erosion/piping failure is Pf = 0.0223/100 = 2.23×10^{-4}. It is found that the prior average annual probability of seepage erosion/piping failure, 1.41×10^{-4}, is underestimated by 37 % compared with the posterior value at the local level of 2.23×10^{-4}. Dam failures are regarded as one of "low-probability, high-loss" events. From such a standpoint of view, the potential failure losses may be underestimated if the prior global-level value is used in the dam risk management.

6.5 Conclusions

Bayesian networks allow an effective global consideration of a distressed embankment dam by putting into perspective all of its components. The interrelations among dam distresses and causes in the networks are measured by probability tables, which have been constructed in this paper based on the historical frequencies from a dam distress database. The calculated average annual failure probabilities for embankment dams in China by using Bayesian networks agree well with the historical records. Based on the prior knowledge of common characteristics of dam distresses extracted from the database, Bayesian networks are further used to diagnose a specific distressed dam by combining global-level knowledge and local-level evidence in the same probabilistic model systematically and to identify key influence factors for a specific dam. The case study on Chenbihe Dam with seepage problems proves the effectiveness of Bayesian networks in the diagnosis of dam distresses.

Acknowledgments This research is substantially supported by the Research Grants Council of Hong Kong (Grant No. 622207).

References

Hartford DND, Baecher GB (2004) Risk and uncertainty in dam safety. Thomas Telford, London

Hugin Expert A/S. 2004. Hugin Lite. Online: http://www.hugin.com

Jensen FV (1996) An introduction to Bayesian networks. Springer, New York

Li L, Wang RZ, Sheng JB et al (2006) Risk assessment and risk management for dam. Water Power Press, Beijing (in Chinese)

MWR (1994) Flood control standard, GB50201-94. Ministry of Water Resources of the People's Republic of China (MWR), Beijing (in Chinese)

Niu YG (2002) Case studies on rehabilitation of dangerous reservoirs. Water Power Press, Beijing (in Chinese)

Smith M (2006) Dam risk analysis using bayesian networks. In: Nadim F, Pöttler R, Einstein H, Klapperich H, Kramer S (eds) Proceedings of the 2006 ECI conference on geohazards, ECI symposium Series No. 7, engineering conferences international (ECI), New York. http://services.bepress.com/eci/geohazards/10 Accessed 10 Mar 2009

Xu Y, Zhang LM (2009) Breaching parameters for earth and rockfill dams. J Geotech Geoenviron Eng ASCE 135(12):1957–1970

Chapter 7
Back Analyses in Geotechnical Engineering

G.L. Sivakumar Babu and Vikas Pratap Singh

Abstract The paper highlights the use of back analyses as a means to analyze the failures that frequently occurs in geotechnical engineering. The importance of in situ conditions, investigations, techniques, inputs complexity of the problem in terms of three-dimensional effects, etc., in the field of forensic studies in geotechnical engineering is highlighted.

Keywords Failure · Back analysis · In situ conditions · Boundary conditions

7.1 Introduction

Back analysis is an approach commonly used in geotechnical engineering to estimate operable material parameters in situ (Deschamps and Yankey 2006). Further, back analyses are required to provide technical evidences to prove or to disprove the hypotheses made on the cause of failures and to establish scenarios of failure (Hwang 2008). The approach of back analyses is popular because there are significant limitations in the use of laboratory and in situ test results to accurately characterize a soil profile. Back analyses have been commonly used to study the causes of failures in geotechnical engineering applications such as slope stability, landslides, earth retaining structures, dams, highways, and foundations. Back analyses are being used extensively in the geotechnical engineering practice, several studies including Leroueil and Tavenas (1981), Leonards (1982), Stark and Eid (1998), and Tang et al. (1999), describe various applications and the limitations of back analyses.

G.L. Sivakumar Babu (✉) · V.P. Singh
Department of Civil Engineering, Indian Institute of Science,
Bangalore 560012, India
e-mail: gls@civil.iisc.ernet.in; glsivakumar@gmail.com

V.V.S. Rao and G.L. Sivakumar Babu (eds.), *Forensic Geotechnical Engineering*,
Developments in Geotechnical Engineering, DOI 10.1007/978-81-322-2377-1_7

7.2 Basic Considerations

In accordance with Hwang (2008), following points shall be considered while performing back analyses for the given problem.

1. Back analyses should be performed for "as-built" conditions because many of the assumptions made in design are either nonexisting or different from reality.
2. Reconnaissance of the site must be done preferably jointly with all the parties involved so as to sort out differences in opinions, if any. Site observations may be captured in the form of photographs/videos that plays a vital role in decision making and provides support to the conclusions drawn.
3. To ensure that the results of the analyses are reliable, the data available must be carefully verified. This includes an appropriate appraisal of local geology of the site of interest to help in understanding actual ground conditions and related historical events.
4. Design drawings and calculations, if available, must be checked to ensure that the works have been executed in accordance with appropriate design.
5. Depending upon the complexity of the problem, analyses can be performed using one or more of the following: (a) rules of thumb that includes indices such as stability number and overload factor, (b) empirical relationships, (c) closed-from solutions, (d) simple numerical models, and (e) sophisticated numerical models.
6. While using numerical methods (generally using commercial software packages), an understanding of the suitability, capability, and the limitations of the method for the particular problem must be developed and the judicious interpretation of the output shall be made by the expert analyst. The software package adopted for analyses must have sufficient technical backup.
7. Depending on the algorithm and the numerical scheme adopted, different software packages may lead to drastically different results. Therefore, it is necessary to conduct a few test runs so as to calibrate the parameters to be adopted by comparing the results of analyses with observations or with known solutions.
8. Following important source of errors in the numerical analyses shall be given due importance while drawings conclusions and final recommendations based on the interpretations from the computed output: (a) implications of modeling of a 3D system as a 2D model, (b) modeling time dependency of soil behavior (e.g., rate of dissipation of pore water pressures), and (c) modeling of the nonlinearity of the soil behavior (i.e., choice of an appropriate constitutive soil model).

7.3 Scope

Following are some the areas that can be identified to have scope for back analyses in geotechnical engineering applications:

1. Back analyses may be used to study the settlement response of different types of foundations and structures, classification of cracking damage, backfill

settlement, to identify causes of settlements such as limestone cavities or sinkholes, consolidation of soft and/or organic soils, underground mines and tunnels, extraction of ground water or oil, landfills, and decomposition of organic matter.

2. Back analyses may used to study failures/behavior of structures founded on expansive soil by analyzing factors such as lateral and vertical movements of expansive soils, special considerations in design of foundations and pavements on expansive soils and treatment measures.
3. Back analyses may be used to study potential causes of lateral movement in applications such as rock falls, surficial slope failures, landslides, slope softening and creep, and dam failures.
4. Back analyses may be used to earthquakes induced phenomenons such as surface faulting and ground rupture, liquefaction, slope movement and settlement, and foundation behavior.
5. Back analyses may be used to study bearing capacity failure of buildings, roads, retaining walls, and historical structures.
6. Back analyses may be used to study problems in structures such as slopes and pavement due to the ground water and presence of moisture.

7.4 Examples of Potential Errors

Deschamps and Yankey (2006) provided a few project examples to illustrate the challenge and potential errors that can be present in back analysis for material strengths, rupture (slip) surface, pore pressures, and three-dimensional or "End Effects". A brief overview of the examples related to material strength and 3D effects from the study of Deschamps and Yankey (2006) is presented below.

7.4.1 Material Strengths

This example was drawn from a case history related to the Grandview Lake Dam located in Bartholomew County, central Indiana to illustrate the dependence that the back-calculated strength along a weak seam has on assumptions of strength in other zones. It was desired to perform back analysis for estimating the strength along the planer slip surface. An assumed rupture surface was developed from inclinometer measurements. The dam was constructed primarily of glacial till and residual soils weathered from claystones. The first challenge was to select the operable strength of the dam materials. There was no distinct zonation of materials in the dam; and therefore, no basis for subdividing the dam into discrete materials.

Table 7.1 Back-calculated strength (Deschamps and Yankey 2006)

Embankment strength	Back-calculated friction angle (°)
Lower bound	22–24
Upper bound (high friction angle)	16
Upper bound (high cohesion)	11
Average	18

To select the operable strength of the dam materials, results of the consolidated undrained (CU) triaxial tests were used. Four different characterizations of embankment strength were considered for back analyses purposes to characterize weak seam strength. A summary of the back-calculated friction angles in the weak seam is shown in Table 7.1.

The strength along the weak seam was characterized as having a zero cohesive intercept because it was rationalized that this material was at or near its residual strength because the deformations along the very thin seam were significant, at least several inches. Moreover, there is little tendency for volume change at residual conditions such that shear-induced pore pressure changes were considered negligible.

Table 7.1 illustrates the range in back-calculated friction angles is from 11° to 24°, with the average strength providing 18°. Although this range can be viewed as extreme, it is apparent that even if a narrower range of strengths were used to characterize the dam, the back-calculated strength would still vary over an appreciable range. A conservative (low) estimate of embankment strength leads to a relatively high interpretation of strength along the weak seam. Note also the significant difference in back-calculated strengths for the two upper bound cases, wherein the case with primarily cohesive strength leads to a much lower back-calculated friction angle for the geometry considered because of the higher shear strength of the compacted materials.

7.4.2 Three-Dimensional or "End Effects"

This example was drawn from a case history related to the 'Lock and Dam 10' on the Kentucky River to demonstrate the importance of understanding of three-dimensional effects when back-calculating strengths. 'Lock and Dam 10' is a relatively small concrete gravity dam owned by the Commonwealth of Kentucky and built circa 1905. The dam is a spillway over its complete length of 240 ft; it has a height of 34 ft; a width of 32 ft; and is made up of ten monoliths 24 ft long. Following are key features of the study:

Problem Stability analyses are conventionally performed on idealized two-dimensional cross sections, which are based on plane strain conditions. At 'Lock and Dam 10,' coring through the dam indicated that the construction joints between concrete monoliths were essentially rubble, and could not be relied upon as shear connections between monoliths. Although it was considered imprudent to rely on the shear resistance between monoliths as a design consideration, it was recognized that some resistance was likely to be present.

Back Analyses An attempt was made to estimate the magnitude of this resistance in order to understand the inconsistency between required strength and interpreted strength. Accordingly, numerical modeling with the program *FLAC* was used to model the entire dam as a beam. The assumption was made that only a nominal frictional resistance (35°) was available between monoliths (no tensile or cohesive strength) and that the ends were fixed at the ends. Figure 7.1 illustrates the idealized model of the dam taken in plan view. The dam is attached to an abutment and training wall on the left, and the lock river wall on the right, both assumed to be stable. The distributed load on the beam was progressively increased to represent increasing the net hydrostatic pressures from higher pools.

Interpretation The modeling effort produced a surprising result in which a zone within the dam formed a compressive arch that developed significant flexural resistance. Based on this analysis, the compressive arch that develops has sufficient capacity to carry the complete hydraulic load acting on the dam during the maximum design flood, independent of any frictional resistance at the base, and with only frictional resistance between monoliths. This example clearly illustrates how difficult it would be to back-calculate strengths if there is a significant, but uncertain, three-dimensional influence. Although, the three-dimensional effects are extreme in the present case, influences of 5–30 % are generally believed to be expected.

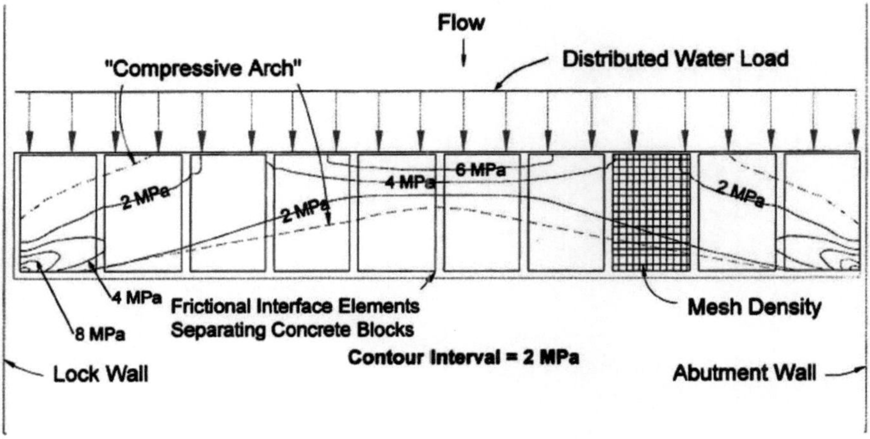

Fig. 7.1 Plan view of kentucky river dam No. 10, modeled as a beam (Deschamps and Yankey 2006)

7.5 Concluding Remarks

Back analyses provide a means to analyze the failures that frequently occurs in various technical fields including geotechnical engineering. It is of prime importance that back analyses must represent the in situ conditions to the extent possible. Choice of back analysis methodology must based on the technical inputs and data available from the failure site, detailed laboratory investigations, and other factors such as complexity of problem, availability of experts, and the cost of analyses. Back analyses also form the key step in the field of forensic studies, and provides basis for the techno-legal aspects in geotechnical engineering.

References

Deschamps R, Yankey G (2006) Limitations in the back-analysis of strength from failures. ASCE J Geotech Geoenviron Eng 132(4):532–536

Hwang RN (2008) Back analyses in forensic geotechnical engineering. ISSMGE/TC40—Forensic Geotechnical Engineering. (Draft: 2008/10/20), 1–8

Leonards GA (1982) Investigation of failures. ASCE J Geotech Eng Div 108(2):185–246

Leroueil S, Tavenas F (1981) Pitfalls of back-analysis. In: Proceedings of the 10th international conference on soil mechanics and foundation engineering, Balkema, Rotterdam 1:185–190

Stark TD, Eid HT (1998) Performance of three-dimensional slope stability methods in practice. ASCE J Geotech Geoenviron Eng 124(11):1049–1060

Tang WH, Stark TD, Angulo M (1999) Reliability in back analysis of slope failures. Soils Found 39(5):73–80

Chapter 8
Back Analysis of Slope Failures to Design Landslide Stabilizing Piles

M.E. Popescu and V.R. Schaefer

Abstract It is generally accepted that shear strength parameters obtained by back analysis of slope failures ensure more reliability than those obtained by laboratory or in situ testing when used to design remedial measures. In many cases, back analysis is an effective tool, and sometimes the only tool, for investigating the strength features of a soil deposit. The fundamental problem involved is always the data quality and consequently the back analysis approach must be applied with care and the results interpreted with caution. Procedures to determine the magnitude of both shear strength parameters (c' and ϕ') or the relationship between them by considering the position of the actual slip surface within the failed slope are discussed. Using the concept of limit equilibrium the effect of any remedial measure (drainage, modification of slope geometry, restraining structures) can easily be evaluated by considering the intercepts of the $c'-\tan\phi'$ lines for the failed slope (c'_0, $\tan\phi'_0$) and for the same slope after installing some remedial works (c'_{nec}, $\tan\phi'_{\mathrm{nec}}$), respectively. This procedure is illustrated to design piles to stabilize landslides taking into account both driving and resisting force acting on each pile in a row as a function of the nondimensional pile interval ratio B/D. The accurate estimation of the lateral force on pile is an important parameter for the stability analysis because its effects on both the pile- and slope stability are conflicting. That is, safe assumptions for the stability of slope are unsafe assumptions for the pile stability, and vice versa. Consequently, in order to obtain an economic and safe design it is necessary to avoid excessive safety factors.

Keywords Slope failures · Limit equilibrium methods · Failure envelope · Lateral piles

---

M.E. Popescu (✉)
Parsons Brinckerhoff Americas Inc./Illinois Institute of Technology,
Chicago Illinois, USA
e-mail: mihail.e.popescu@gmail.com

V.R. Schaefer
Iowa State University, Iowa Ames, USA

V.V.S. Rao and G.L. Sivakumar Babu (eds.), *Forensic Geotechnical Engineering*,
Developments in Geotechnical Engineering, DOI 10.1007/978-81-322-2377-1_8

8.1 Introduction

Correction of an existing landslide or the prevention of a pending landslide is a function of a reduction in the driving forces or an increase in the available resisting forces. Any remedial measure used must involve one or both of the above parameters. The IUGS Working Group on Landslides (Popescu 2001) has prepared a short checklist of landslide remedial measures arranged in four practical groups, namely: modification of slope geometry, drainage, retaining structures, and internal slope reinforcement, as shown in Table 8.1 As many of the geological features, such as sheared discontinuities, are not well known in advance, it is better to put remedial measures in hand on a "design as you go basis". That is the design has to be flexible enough for changes during or subsequent construction of remedial works.

Although slope instability processes are generally seen to be "engineering problems" requiring "engineering solutions" involving correction by the use of structural techniques, nonstructural solutions including classical methods such as drainage and modification of slope geometry, as well as some novel methods such as lime/cement stabilization, grouting, or soil nailing, are increasingly being used (Popescu 1996). The cost of nonstructural remedial measures is considerably lower when compared with the cost of structural solutions.

Terzaghi (1950) stated that, "if a slope has started to move, the means for stopping movement must be adapted to the processes which started the slide." For example, if erosion is a causal process of the slide, an efficient remediation technique would involve armoring the slope against erosion, or removing the source of erosion. An erosive spring can be made nonerosive by either blanketing with filter materials or drying up the spring with horizontal drains, etc.

Morgenstern (1992) followed this theme when he noted that post-failure analyses can be used to provide a consistent explanation for landslide causal events. The back analyses can then be used as a basis for designing the stabilizing measures if engineering works are required. This approach has the added appeal that the remedial design is normalized in terms of the post-failure analytical model.

Most landslides must usually be dealt with sooner or later. How they are handled depends on the processes that prepared and precipitated the movement, the landslide type, the kinds of materials involved, the size and location of the landslide, the place or components affected by or the situation created as a result of the landslide, available resources, etc. The technical solution must be in harmony with the natural system, otherwise the remedial work will be either short lived or excessively expensive. In fact, landslides are so varied in type and size, and in most instances, so dependent upon special local circumstances, that for a given landslide problem there is more than one method of prevention or correction that can be successfully applied. The success of each measure depends, to a large extent, on the degree to which the specific soil and groundwater conditions are prudently recognized in an investigation and incorporated in design.

In this paper a methodology involving back analysis of the slope failure and the use of piles to remediate the landslide are presented.

Table 8.1 A brief list of landslide remedial measures

1. Modification of slope geometry
1.1. Removing material from the area driving the landslide (with possible substitution by lightweight fill)
1.2. Adding material to the area maintaining stability (counterweight berm or fill)
1.3. Reducing general slope angle
2. Drainage
2.1. Surface drains to divert water from flowing onto the slide area (collecting ditches and pipes)
2.2. Shallow or deep trench drains filled with free-draining geomaterials (coarse granular fills and geosynthetics)
2.3. Buttress counterforts of coarse-grained materials (hydrological effect)
2.4. Vertical (small diameter) boreholes with pumping or self draining
2.5. Vertical (large diameter) wells with gravity draining
2.6. Subhorizontal or subvertical boreholes
2.7. Drainage tunnels, galleries, or adits
2.8. Vacuum dewatering
2.9. Drainage by siphoning
2.10. Electroosmotic dewatering
2.11. Vegetation planting (hydrological effect)
3. Retaining structures
3.1. Gravity retaining walls
3.2. Crib-block walls
3.3. Gabion walls
3.4. Passive piles, piers, and caissons
3.5. Cast-in situ reinforced concrete walls
3.6. Reinforced earth retaining structures with strip/sheet—polymer/metallic reinforcement elements
3.7. Buttress counterforts of coarse-grained material (mechanical effect)
3.8. Retention nets for rock slope faces
3.9. Rockfall attenuation or stopping systems (rocktrap ditches, benches, fences, and walls)
3.10. Protective rock/concrete blocks against erosion
4. Internal slope reinforcement
4.1. Rock bolts
4.2. Micropiles
4.3. Soil nailing
4.4. Anchors (prestressed or not)
4.5. Grouting
4.6. Stone or lime/cement columns
4.7. Heat treatment
4.8. Freezing
4.9. Electroosmotic anchors
4.10. Vegetation planting (root strength mechanical effect)

8.2 Back Analysis of Failed Slopes to Design Remedial Measures

8.2.1 Failure Envelope Parameters

A slope failure can reasonably be considered as a full scale shear test capable to give a measure of the strength mobilized at failure along the slip surface. The back calculated shear strength parameters, which are intended to be closely matched with the observed real-life performance of the slope, can then be used in further limit equilibrium analyses to design remedial works. The limit equilibrium methods forming the framework of slope stability/instability analysis generally accept the Mohr–Coulomb failure criterion:

$$\tau_f = c' + \sigma' \tan \phi' \tag{8.1}$$

where τ_f and σ' are the shear stress and effective normal stress, respectively, on the failure surface and c' and ϕ' are parameters assumed approximately constant for a particular soil.

A significant limitation in the use of this criterion is that the constant of proportionality is not really a constant when wide range of stress is under consideration. There is now considerable experimental evidence to show that the Mohr failure envelope exhibits significant curvature for many different types of soil and compacted rockfill. Therefore, if the assumption of a linear failure envelope is adopted, it is important to know what range of stress is appropriate to a particular slope instability problem. To avoid this difficulty a curved failure envelope can be approximated by the following power law equation:

$$\tau_f = A\,(\sigma')^b \tag{8.2}$$

which was initially suggested by De Mello (1977) for compacted rockfills and subsequently found appropriate for soils (Atkinson and Farrar 1985).

8.2.2 Procedures for Back Analysis of Slope Failures

Shear strength parameters obtained by back analysis ensure more reliability than those obtained by laboratory or in situ testing when used to design remedial measures. In many cases, back analysis is an effective tool, and sometimes the only tool, for investigating the strength features of a soil deposit (Duncan 1999). However, one has to be aware of the many pitfalls of the back analysis approach that involves a number of basic assumptions regarding soil homogeneity, slope and slip surface geometry and pore pressure conditions along the failure surface (e.g., Leroueil and Tavenas 1981) A position of total confidence in all these assumptions is rarely if ever achieved.

While the topographical profile can generally be determined with enough accuracy, the slip surface is almost always known in only few points and interpolations with a considerable degree of subjectivity are necessary. Errors in the position of the slip surface result in errors in back calculated shear strength parameters. If the slip surface used in back analysis is deeper than the actual one, c' is overestimated and ϕ' is underestimated and vice versa.

The data concerning the pore pressure on the slip surface are generally few and imprecise. More exactly, the pore pressure at failure is almost always unknown. If the assumed pore pressures are higher than the actual ones, the shear strength is overestimated. As a consequence, a conservative assessment of the shear strength is obtainable only by underestimating the pore pressures.

Procedures to determine the magnitude of both shear strength parameters or the relationship between them by considering the position of the actual slip surface within a slope are discussed by Popescu and Yamagami (1994). The two unknowns —i.e., the shear strength parameters c' and ϕ'—can be simultaneously determined from the following two requirements:

(a) $F = 1$ for the given failure surface. That means the back calculated strength parameters have to satisfy the c'–tan ϕ' limit equilibrium relationship;
(b) F = minimum for the given failure surface and the slope under consideration. That means the factors of safety for slip surfaces slightly inside and slightly outside the actual slip surface should be greater than one (Fig. 8.1a).

Based on the above-mentioned requirements, Saito (1980) developed a semi-graphical procedure using trial and error to determine unique values of c' and tan ϕ' by back analysis (Fig. 8.1b). An envelope of the limit equilibrium lines c'−tan ϕ', corresponding to different trial sliding surfaces, is drawn and the unique values c' and tan ϕ' are found as the coordinates of the contact point held in common by the envelope and the limit equilibrium line corresponding to the actual failure surface. A more systematic procedure to find the very narrow range of back calculated shear strength parameters based on the same requirements is illustrated in Fig. 8.1c.

The procedures discussed above to back calculate the linear strength envelope parameters, c' and ϕ' in Eq. (8.1) can be equally applied to back calculate the nonlinear strength envelope parameters, A and b in Eq. (8.2) (Popescu et al. 1995).

The fundamental problem involved is always the data quality and consequently the back analysis approach must be applied with care and the results interpreted with caution. Back analysis is of use only if the soil conditions at failure are unaffected by the failure. For example, back calculated parameters for a first-time slide in a stiff overconsolidated clay could not be used to predict subsequent stability of the sliding mass, since the shear strength parameters will have been reduced to their residual values by the failure. In such cases an assumption of $c' = 0$ and the use of a residual friction angle, ϕ_r are warranted (Bromhead 1992). If the three-dimensional geometrical effects are important for the failed slope under consideration and a two-dimensional back analysis is performed, the back calculated shear strength will be too high and thus unsafe.

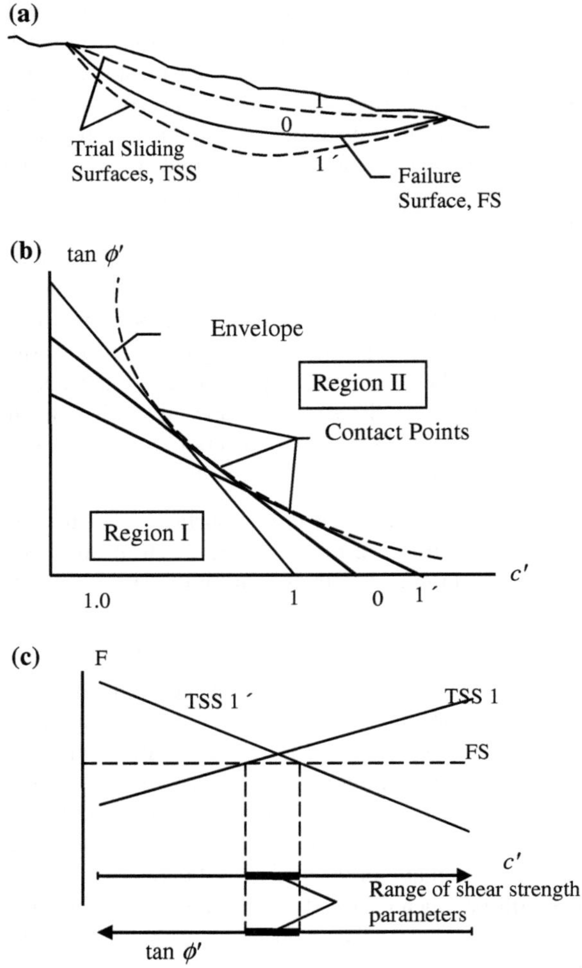

Fig. 8.1 Shear strength back analysis methods

8.2.3 Design of Remedial Measures Based on Back Analysis Results

In order to avoid the questionable problem of the representativeness of the back calculated unique set of shear strength parameters a method for designing remedial works based on the limit equilibrium relationship $c'-\phi'$ rather than a unique set of shear strength parameters can be used (Popescu 1991).

The method principle is shown in Fig. 8.2. It is considered that a slope failure provides a single piece of information which results in a linear limit equilibrium relationship between shear strength parameters. That piece of information is that the factor of safety is equal to unity ($F = 1$) or the horizontal force at the slope toe is

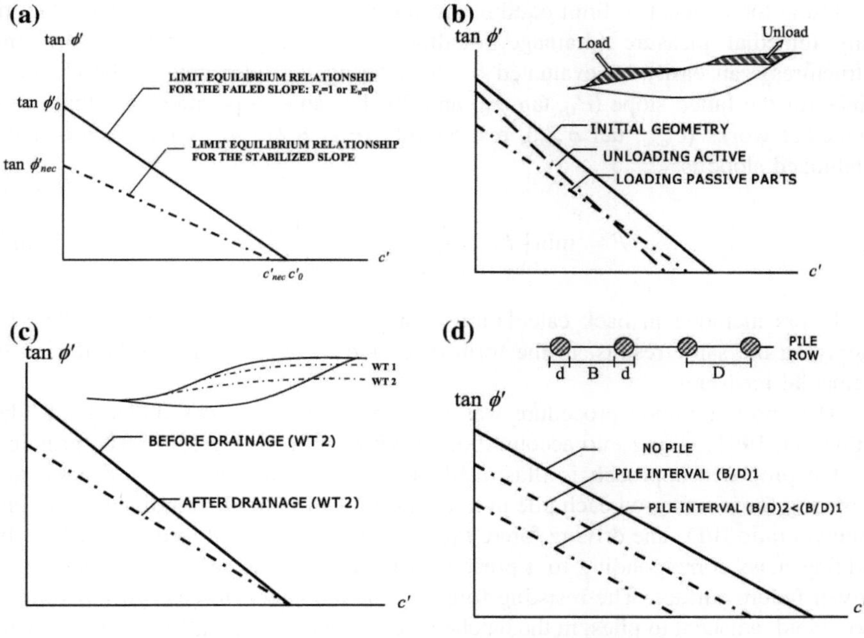

Fig. 8.2 Limit equilibrium relationship and design of slope remedial measures

equal to zero ($E = 0$) for the conditions prevailing at failure. Each of the two conditions ($F = 1$ or $E = 0$) results in the same relationship c'–tan ϕ' which for any practical purpose might be considered linear.

The linear relationship c'–tan ϕ' can be obtained using standard computer software for slope stability limit equilibrium analysis by manipulations of trial values of c' and tan ϕ' and corresponding factor of safety value. It is simple to show that in an analysis using arbitrary ϕ' alone ($c' = 0$) to yield a nonunity factor of safety, F_{ϕ}^{*}, the intercept of the c'–tan ϕ' line (corresponding to $F = 1$) on the tan ϕ' axis results as:

$$\tan \phi_0' = \tan \phi' \big/ F_{\phi}^{*} \tag{8.3}$$

Similarly, the intercept of the c'–tan ϕ' line (corresponding to $F = 1$) on the c' axis can be found assuming $\phi' = 0$ and an arbitrary c' value which yield to a nonunity factor of safety, F_c^{*}:

$$c_0' = c' / F_c^{*} \tag{8.4}$$

Using the concept of limit equilibrium linear relationship c'–tan ϕ', the effect of any remedial measure (drainage, modification of slope geometry, restraining structures) can easily be evaluated by considering the intercepts of the c'–tan ϕ' lines for the failed slope (c'_0, tan ϕ'_0) and for the same slope after installing some remedial works (c'_{nec}, tan ϕ'_{nec}), respectively (Fig. 8.2). The safety factor of the stabilized slope is:

$$F = \min\left(F_c = \frac{c'_0}{c'_{nec}}, F_\phi = \frac{\tan \phi'_0}{\tan \phi'_{nec}}\right) \quad (8.5)$$

Errors included in back calculation of a given slope failure will be offset by applying the same results, in the form of c′−tan ϕ' relationship, to the design of remedial measures.

The above outlined procedure was used to design piles to stabilize landslides (Popescu 1991) taking into account both driving and resisting force. The principle of the proposed approach is illustrated in Fig. 8.3 which gives the driving and resisting force acting on each pile in a row as a function of the nondimensional pile interval ratio B/D. The driving force, F_D, is the total horizontal force exerted by the sliding mass corresponding to a prescribed increase in the safety factor along the given failure surface. The resisting force, F_R, is the lateral force corresponding to soil yield, adjacent to piles, in the hatched area shown in Fig. 8.3. F_D increases with the pile interval while F_R decreases with the same interval. The intersection point of the two curves which represent the two forces gives the pile interval ratio satisfying the equality between driving and resisting force.

The accurate estimation of the lateral force on pile is an important parameter for the stability analysis because its effects on both the pile—and slope stability are conflicting. That is, safe assumptions for the stability of slope are unsafe assumptions for the pile stability, and vice versa. Consequently, in order to obtain an economic and safe design it is necessary to avoid excessive safety factors.

The problem is clearly three-dimensional and some simplification must be accepted in order to develop a two-dimensional analysis method based on the principles outlined above. However, the only simplicity to be accepted and trusted is the simplicity that lies beyond the problem complexity and makes all details and difficulties simple by a sound and profound understanding.

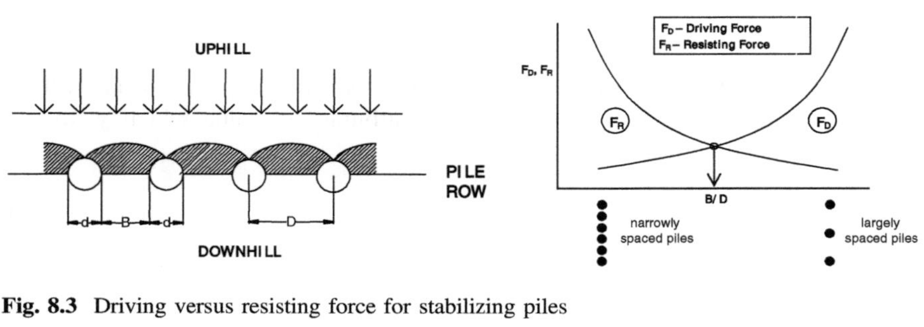

Fig. 8.3 Driving versus resisting force for stabilizing piles

8.3 Application

8.3.1 Site Conditions

The described methodology is applied to a landslide in Ohio in the United States. The site is located along the Ohio River in south-central Ohio. A replacement bridge was proposed at the site and site preparations reactivated an ancient slide. The cross section is shown in Fig. 8.4. The slope consists of shale bedrock overlain by shale weathered to a residual clay. Overlying the residual clay is alluvial silts and clays. Construction activities at the site led to the reactivation of an ancient slide. The slip plane discerned from surface scarps and inclinometer data is shown in Fig. 8.4. It can be seen that the failure surface is planar in nature and occurs just above the shale bedrock in the weathered residual clay.

To accommodate the new bridge a fill was proposed on the existing slope, which was now moving and would have exacerbated the instability. Hence the use of piles to stabilize the slope was proposed.

8.3.2 Back Analysis

Back analyses were conducted for the slope failure using limit equilibrium techniques as described previously (Popescu 1991). The back analyzed relationship between friction angle and cohesion for the residual clay and the failure surface are shown in Fig. 8.5. The resulting strength parameters vary depending upon the water level in the Ohio River. The relationship between the friction angle and the cohesion are shown in Fig. 8.5 for cases of no pile and pile B/D ratios of 0.75 and 0.5. The back analyzed friction angle of about 13° for the no pile case compares favorably with residual shear test results.

The numerical results for the B/D ratios in Fig. 8.5 were obtained using the methodology proposed by Liang (2002) and coded into an Excel spreadsheet.

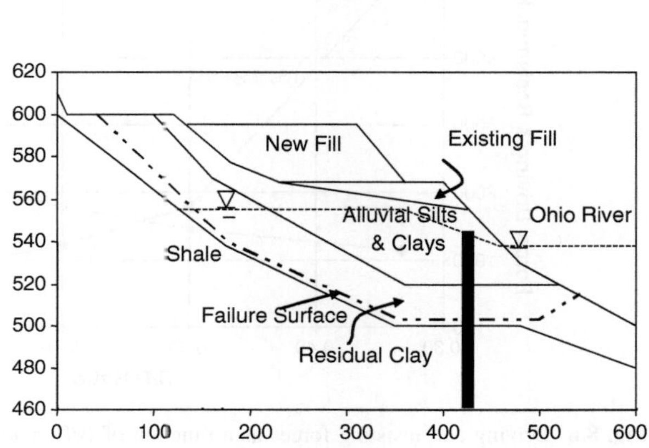

Fig. 8.4 Cross section of slope

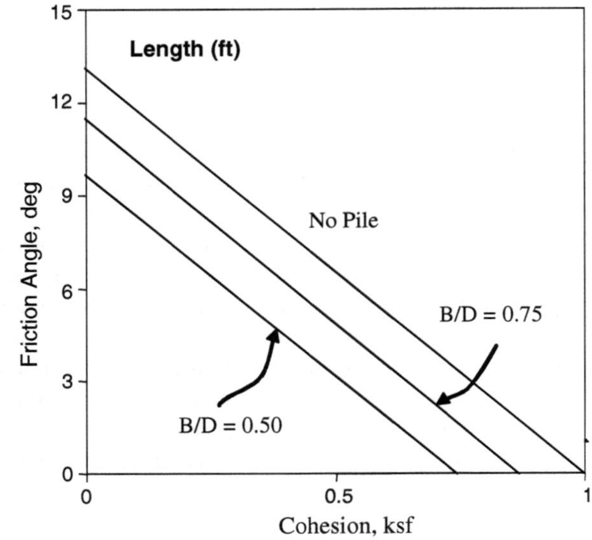

Fig. 8.5 Back analyzed relationship between friction angle and cohesion

8.3.3 Driving and Resisting Forces

The driving forces were determined using limit equilibrium analyses utilizing the program XSTABL (Interactive Software Designs, Inc. 1994) and spreadsheet analyses. The driving forces are shown in Fig. 8.6 for various B/D ratios. The resisting forces were determined using Ito and Matsui (1975) method as outlined by Popescu (1995). The resisting forces are shown in Fig. 8.6 for various B/D ratios.

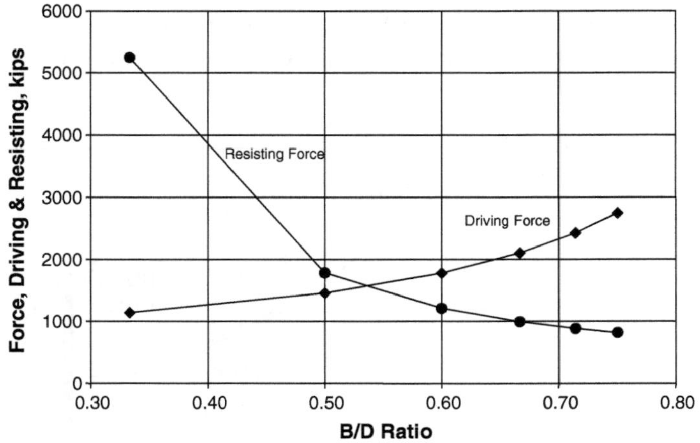

Fig. 8.6 Driving and resisting forces as a function of B/D ratio

From the results in Fig. 8.6 it can be seen that the resisting force and driving force cross at a B/D ratio slightly larger than 0.5 with a required resisting force of about 1800 kips. A shear force of this magnitude could be obtained using six-foot diameter shafts; however, eight-foot diameter shafts were selected to provide a margin of safety for the drilled shafts.

8.4 Conclusions

This paper has outlined an approach to back analyzing the strength parameters in a slope failure and determining the force required to stabilize a slope using piles considering the back analysis results. The use of the technique has been demonstrated through application to a case history.

References

Atkinson JH, Farrar DM (1985) Stress path tests to measure soils strength parameters for shallow landslips. In: Proceedings of the 11th international conference soil mechanics foundation engineering, San Francisco, vol 2. pp 983–986

Bromhead EN (1992) Slope stability, 2nd edn. Blackie Academic and Professional, London, 411 pp

De Mello VFB (1977) Reflections on design decisions of practical significance to embankment dams. Géotechnique 27(3):281–354

Duncan JM (1999) The use of back analysis to reduce slope failure risk. J Boston Soc Civil Eng 14 (1):75–91

Interactive Software Designs, Inc. 1994. XSTABL An Integrated Slope Stability Analysis Program for Personal Computers. Reference Manual, Version 5, Moscow, ID

Ito T, Matsui T (1975) Methods to estimate landslide forces acting on stabilizing piles. Soils Found 15(4):43–59

Leroueil S, Tavenas F (1981) Pitfalls of back-analyses. In: Proceedings of the 10th international conference. Soil Mech Found Eng 1:185–190

Liang RY (2002) Drilled shaft foundations for noise barrier walls and slope stabilization. University of Akron, Akron

Morgenstern NR (1992) Keynote paper: the role of analysis in the evaluation of slope stability. In: Proceedings of the 6th international symposium on landslides. Christchurch 3:1615–1629

Popescu ME (1991) Landslide control by means of a row of piles. Keynote paper. In: Proceedings of the international conference on slope stability engineering, Isle of Wight, Thomas Telford, pp 389–394

Popescu ME (1995) Keynote lecture: back analysis of slope failures to design stabilizing piles. In: 2nd turkish symposium on landslides, Oct 15–26, Adapazari, Turkey

Popescu ME (1996) From landslide causes to landslide remediation, special lecture. In: Proceedings of the 7th international symposium on landslides, Trondheim 1:75–96

Popescu ME (2001) A suggested method for reporting landslide remedial measures. IAEG Bulletin 60(1):69–74

Popescu ME, Yamagami T (1994) Back analysis of slope failures—a possibility or a challenge? In: Proceedings of the 7th congress int assoc. Eng geology Lisbon 6:4737–4744

Popescu ME, Yamagami T, Stefanescu S (1995) Nonlinear strength envelope parameters from slope failures. In: Proceedings of the 11th ECSMFE. Copenhagen 1:211–216

Saito M (1980) Reverse calculation method to obtain c and ϕ on a slip surface. In: Proceedings of the International symposium landslides, New Delhi, vol 1. pp 281–284.

Terzaghi K (1950) Mechanisms of landslides. Geological Society of America, Berkley, pp 83–123

Chapter 9
Back Analyses in Forensic Geotechnical Engineering

Richard N. Hwang

Abstract Back analyses are required to provide technical evidences to prove or to disprove the hypotheses made on the causes of failures and to establish scenarios of the failures. Analyses must be performed by experienced engineers who are familiar with the analytical tools to be adopted for analyses. The analytical tools adopted must be suitable for the cases to be investigated and the constitutive laws to be used in the analyses must be representative of the materials to be simulated. It is important to realize the fact that there are limitations associated with analyses and the results obtained must be interpreted by experienced engineers who have sufficient practical experience.

Keywords Forensic · Geotechnical · Failure · Analyses

9.1 Introduction

Failures seldom occur for a single reason and for cases in which litigation is involved, the causes of failures are inevitably difficult to ascertain. It is necessary to make various assumptions regarding why and how failures happen and perform analyses to prove or disprove these assumptions. Concession among the parties involved is often required in reaching conclusions. Sometimes, it has to be left to moderators to make final judgments.

Discussed herein are the general principles of back analyses. A case history is presented to illustrate the complexity of back analyses and the difficulties associated with the interpretation of the results obtained.

R.N. Hwang (✉)
Moh and Associates, Inc., Oriental Technopolis Building A, 22 FL,
No. 112 Xintai Wu Road, Sec. 1, Xizhi District, 22102 New Taipei City, Taiwan
e-mail: richard.hwang@maaconsultants.com

V.V.S. Rao and G.L. Sivakumar Babu (eds.), *Forensic Geotechnical Engineering*,
Developments in Geotechnical Engineering, DOI 10.1007/978-81-322-2377-1_9

9.2 Common Causes of Failures

Defective design of temporary works will likely lead to failures during construction and such failures usually occur rather suddenly. The collapse of Nicoll Highway during the construction of the Circle Line of the Singapore Mass Rapid Transit System in April 2004 is a good example. The retaining system of a cut-and-cover tunnel collapsed suddenly while the excavation was about to reach the formation level and a section of the six-lane expressway fell into the sinkhole. Failure started as the waling on the northern wall buckled at 9 am on 20 April 2004 and by 3 pm all the struts in a 100 m section totally failed. The incident led to the death of three construction workers and one supervisory staff. The expressway was closed for 7.5 months. A Committee of Inquiry was immediately appointed by the Ministry of Manpower to investigate the causes of failure. It was concluded that the failure was mainly caused by the inadequate design of the retaining system (COI 2005; Moh and Hwang 2007; Yong and Lee 2007). Effective stress parameters were adopted for marine clay which is essentially an undrained material, resulting in much overestimates of soil strengths and underestimates of wall deflections and bending moments. However, the failure was triggered by buckling of the stiffeners of walings and propagated to all other members of the retaining system.

Defective design of permanent works, on the other hand, will lead to problems which may last for a very long time. The leaning Tower of Pisa is a notable example. The construction of the tower started in 1173. The incline of the tower was already apparent when the construction was halted in 1178. The tower was only 10.6 m tall and consisted of three stories then. The tower, 58 m in height, was completed in 1319 and the bell-chamber was not finally added till 1372. Geographical surveys show that the tower is founded on loose sediments accumulated in a river course. The tilt was obviously caused by differential settlements. There have been several attempts made for preventing the tilting from worsening but resulted in only adverse consequences. The incline reached 5.5° and the government officially closed the tower in 1990. The tilt was finally halted by extracting soil beneath the foundation on the north side and the tower was re-opened at the end of 2001. It is necessary to study the history of the tower and the measures taken in all these attempts to establish the entire scenario.

Failures are not always related to geotechnical problems. If there are no apparent ground movements, failures are most likely due to defective design of structures. A 15-year-old building, Lian Yak Building (commonly referred to as "Hotel New World"), in Singapore collapsed in 1986 and claimed 33 lives. An intensive investigation was conducted by the government and failure of the foundation was one of the possible causes studied. Of 33 columns in the building, the foundations of 5 of them were selected for detailed examination and serious defects were observed in piles and in pile caps (Hulme et al. 1993). However, it was found that the basement walls and slabs were virtually undamaged and there was no evidence of differential settlements. Checking of design calculations revealed that the dead loads of the structures were either grossly underestimated, as in the case of brickwalls, or completely missed, as in some case of self-weight of slabs. It was thus concluded that the failure was due to under-design of the structures and was not related to geotechnical problems.

Many failures could be attributed to conditions, such as adverse soil conditions, abnormal changes of groundwater level, heavy storms, or unexpected loading, etc., which were not accounted for in designs. Some of these conditions are unforeseeable but most of them are merely unforeseen because of the limited data available. Groundwater rushed into an arrival shaft as a portal was made on diaphragm wall for receiving the shield machine during the construction of the Taipei Metro. As a result, a section of twin tunnels was seriously damaged and 39 rings in the up-track and 34 rings in the down-track tunnels had to be replaced (Ju et al. 1998). A PVC pipe was found at the invert of the up-track tunnel after the shaft was drained and the shield machine exposed. It could be one time used for pumping water from the underlying water-bearing gravelly layer for irrigation or fish farming. The soils surrounding the tunnel had been solidified by jet grouting previously. A large piece of timber was found next to the PVC pipe and it is suspected that the treated ground was much disturbed, resulting in cracks which became water paths, as the shield machine forced its way out. Groundwater was then able to rush from the water-bearing gravelly layer into the shaft from the abandoned PVC pipe. Drift woods were frequently encountered during constructions in the Taipei Basin; however, there is no way to ascertain their locations beforehand. It was also a common practice to pumping groundwater for irrigation and fish farming in the area; however, pumping wells are equally difficult to locate.

Failures due to defective constructions are not uncommon. Of the five disastrous events in metro constructions in Asia Pacific in the period of 2001–2007, one was caused by defective ground freezing work, one was caused by defective diaphragm wall and two were caused by defective ground treatment by jet grouting (Moh and Hwang 2007). In all these four cases, water was the major source of the problem. In fact, water was responsible for a majority of failures in underground works carried out in soft ground.

9.3 Back Analyses

A timeline showing all the events chronologically will be very helpful in understanding what has happened. It will even be better if all major events can be presented in a Gantt chart with their durations clearly identified.

Analyses should be performed for "as-built" conditions because many of the assumptions made in design are either nonexisting or different from reality. Therefore, reconnaissance of the site is a must and may have to be conducted for more than once. It is preferably to be conducted jointly with all the parties involved in the case so differences in opinions can be sorted out at the site. Photos and videos will be very useful in helping one's memory and will provide vital supports to the conclusions to be made.

In many cases, failures involve soil-structural interaction and advices from structural engineers are desirable. For cases involving failure of structures, analyses should be performed jointly with structural engineers.

9.3.1 Data Collection and Verification

A checklist should be prepared to document the data available in hand and the data which are still missing. The data needed will depend on the modes of failures. Although the cause of failure may appear to be obvious right at the beginning, it is still necessary to investigate all the potential modes of failure before they can be eliminated.

To ensure that the results of analyses are reliable, the data available must be carefully verified. For example, ground conditions may be misinterpreted, instrument readings may be erroneous, construction activities may be wrongly logged, etc. Some of the data may be misleading without being realized, and some of data may be contradictory and have to be sorted out. It is natural for patties involved to hind facts, intentionally or unintentionally, which are not in their favor; therefore, data must be critically reviewed and judgments must be applied whenever data appear to be dubious.

An appropriate appraisal of local geology sometimes is helpful in understanding ground conditions and historical events related to the site of interest should be investigated.

9.3.2 Checking Design and Calculations

If design drawings are available, it is necessary to check whether or not the works were constructed as designed. Design calculations, if available, should be first checked to see if they are appropriate. However, inconformity with specifications and/or codes of practice will not necessarily lead to failures and the redundancy demanded by specifications and codes may not be required at the time of failure.

It should be noted that ground conditions might have been altered once a failure occurs and it is necessary to figure out what they were prior to the event. This may not always be possible and guesswork may sometimes be required.

9.3.3 Analytical Tools

Depending on the complexity of the problem, analyses can be performed by using:

(a) rules of thumb using indices such as stability number, overload factor, etc.
(b) empirical relationship

(c) closed-form solutions
(d) simple numerical models
(e) sophisticated numerical models

Simple models are available for many failure modes and many of them can be solved by hand calculations. Complicated problems can be solved by numerical analyses. In any case, the tools adopted for analyses must be suitable for the type of problem to be solved and must have sufficient technical backup.

The analytical tools adopted for analyses must have been validated in accordance with stringent quality assurance program and well documented. For this reason, commercial software packages are preferred to in-house programs because the former are usually well tested and improved based on the feedback from users.

Numerical methods have their limitations and the results cannot be trusted blindly. Depending on the algorithm and the numerical scheme adopted, different software packages many lead to drastically different results. Therefore, a few test runs are necessary to calibrate the parameters to be adopted by comparing the results of analyses with observations or with known solutions.

9.3.4 Limitations

With the rapid advancement of computer technology, finite element method and finite difference method have become important tools for design and they are very useful tools for back analyses as well. They have become so user-friendly that even fresh graduates from colleges can perform the analyses with little guidance. This, however, leads to the danger of mishandling of problems and misinterpretation of the results of analyses. After all, geotechnical engineering is an art rather than science. The results of analyses provide important evidences for judgments to be based on and experience should prevail at the end.

It is important to realize the limitations associated with analyses so the results can be correctly interpreted. First of all, although tools are available, three-dimensional analyses are extremely labor intensive and time consuming. Therefore, unless the cases are critical enough and resources are available, one-dimensional and two-dimensional analyses are performed and the results are usually appropriate for practical purposes. However, simulation of a 3D system by a 2D model inevitably alters the nature of the problem and introduces errors in the results of analyses.

Second, soils are usually classified as either drained materials, i.e., pure sand, or undrained materials, i.e., pure clay, in analyses. However, in reality, most of soils are neither pure sand nor pure clay and are mixtures of sand and clay. Their behavior will depend on how fast pore water pressures dissipate. In other words, problems are time-dependent. The rate of dissipation of pore water pressure will not only depend on the rate of loading, but will also depend on the permeability of soil and the length of drainage path. The reliability of the results of analyses will depend on how well this time-dependency of soil behavior is handled.

Third, it is a well-known fact that soil behavior is highly nonlinear and there are many soil models to handle the nonlinearity. The results of analyses may vary considerably if different constitutive laws, which describe the stress–strain relationship of soils, are selected.

9.4 Case Study

The collapse of Nicoll Highway, refer to Fig. 9.1, is an ideal case for illustrating the points mentioned above. The site is located in a piece of land reclaimed in the 80s. As shown in Fig. 9.2, the subsoils at this site contain mainly two thick layers of marine deposits (namely, the upper marine clay and the lower marine clay) and are underlain by the Old Alluvium which is a competent base stratum. Figure 9.3 is a plot of the results of a cone penetration test carried out in the vicinity of the site. The excavation was supposed to be carried out to a depth of 33.5 m and diaphragm walls with a thickness of 800 mm (locally, 1000 mm) were used.

9.4.1 The Incident

Excavation for constructing the cut-and-cover tunnels was carried out by using the bottom-up method of construction. The collapse occurred in Section M3 on 20 April 2004 while the tenth dig was completed and excavation reached a depth of

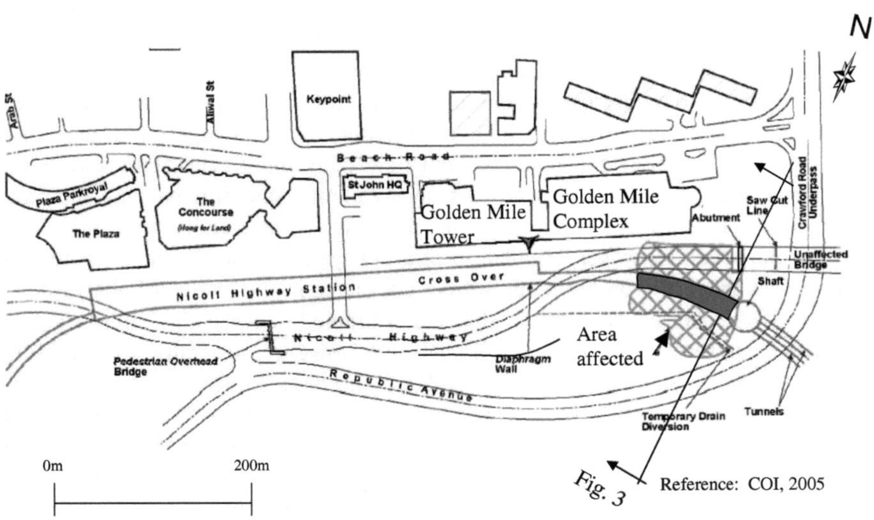

Fig. 9.1 Collapse of Nicoll highway

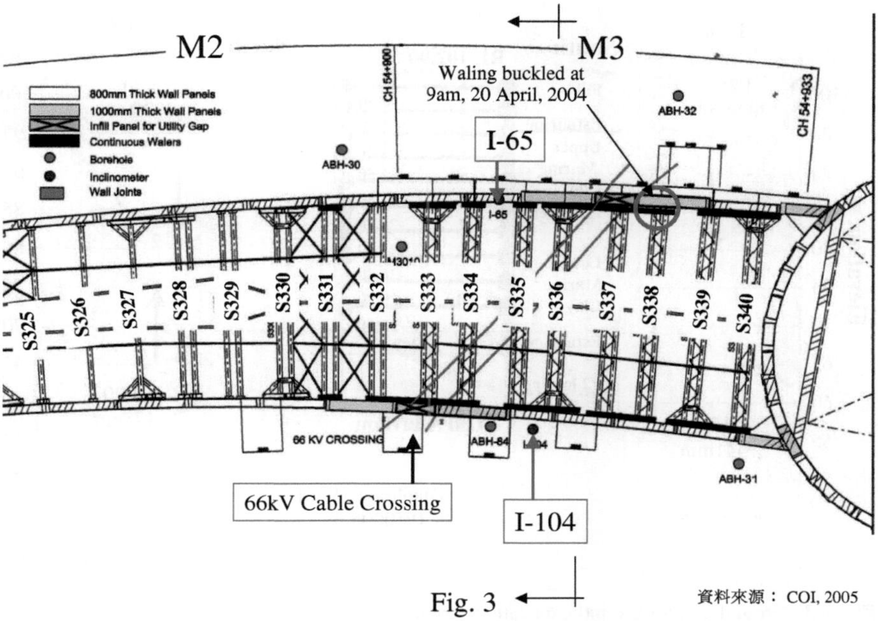

Fig. 9.2 Retaining system in section M3

30.5 m 4 days earlier. There were two inclinometers, i.e., I-65 and I-104, available for monitoring wall deflections. Figure 9.4 shows the wall deflection paths, which are the plots of maximum wall deflections versus depth of excavation in a log-log scale, for these two inclinometers (Hwang et al. 2007). Wall deflections on the two sides of the excavation were about the same till 9 March, 2004, when excavation reached a depth of 25 m, and deflections of 198 and 215 mm were recorded by Inclinometers I-65 and I-104, respectively. Subsequently, there was a period in which I-104 was not read because it was damaged. When monitoring resumed on 26 March, the deflection of the southern wall was found to have increased by 67 to 282 mm while the readings for Inclinometer I65 on the north were fairly steady in this period. The readings for Inclinometer I-104 kept on increasing while those for I-65 remained to be steady subsequently, presumably, because of the asymmetry of ground conditions. In fact, I-65 appeared to move outward by 27 mm and the maximum deflection reduced from 202 mm on 26 March to 175 mm on 20 April, as depicted in Figs. 9.2 and 9.4. On the other hand, Inclinometer I-104 moved inward by 90 to 441 mm in the 3-day period from 17 April to 20 April as shown in Fig. 9.5.

Failure started as the waling at the northern end of Strut S338 at the ninth level, refer to Fig. 9.2, buckled at 9 am on 20 April and failure propagated to other struts. By 3 pm of the day, all the struts for a 100 m section totally failed. As Nicoll Highway sank, gas, water, and electricity cables ruptured, causing power to go out for about 15,000 people and 700 businesses in the Marina and Suntec City area.

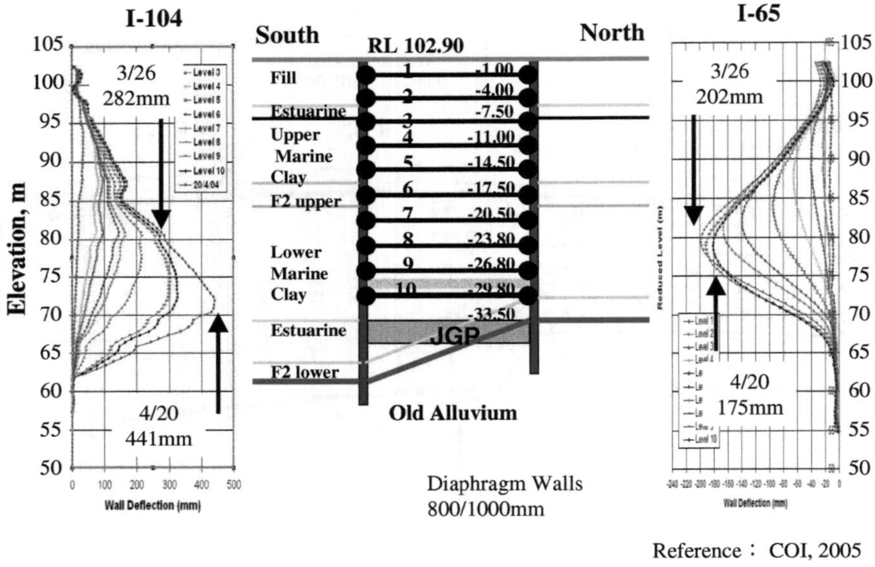

Fig. 9.3 Ground conditions and excavation scheme in section M3

Fig. 9.4 Results of a cone penetration test

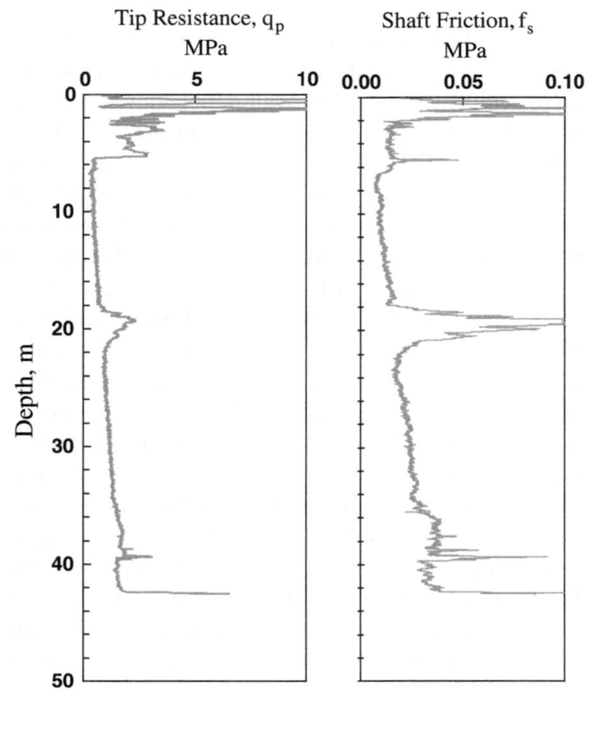

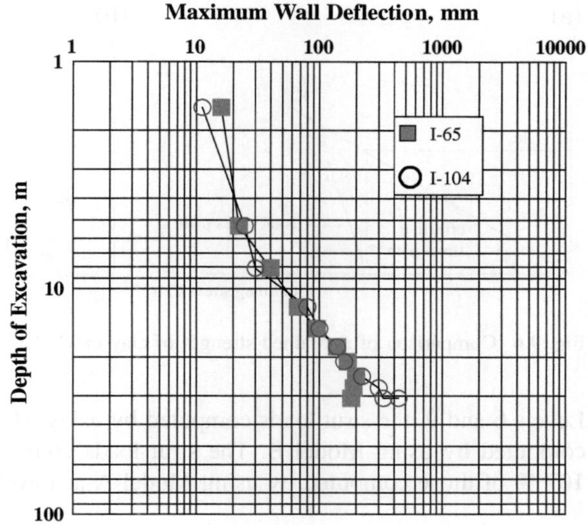

Fig. 9.5 Deflection paths for diaphragm walls

Tremors were felt at Golden Mile Complex. Tenants and residents in the building were also evacuated.

9.4.2 The Design

The contract was awarded as a design-and-build contract and temporary works were designed by the contractor. The computer program PLAXIS, which was first developed in 1987 at the Technical University of Delft and made commercially available by PLAXIS bv of Delft, Netherlands, was adopted in the design of retaining system. PLAXIS allows users to select different material types (drained, undrained, and nonporous) and different material modes (Mohr-Coulomb, soft soil, etc.) Mohr-Coulomb Model was adopted for defining failure of soils for the case of interest. When using the Mohr-Coulomb model, it is possible to input either effective stress parameters (c' and ϕ') or the undrained strength parameters ($c' = c_u$ and $\phi' = 0$). The use of a Mohr-Coulomb soil model with effective strength parameter has been referred to as Method A in the report by the Committee of Inquiry (COI 2005). Method B refers to the use of Mohr-Coulomb soil model with undrained strength parameters in combination with undrained material type. Method A was adopted in the analyses and effective strength parameters, c' and ϕ', instead of the undrained strength parameters, $c' = c_u$ and $\phi' = 0$, were used for marine clays.

This, as depicted in Fig. 9.6, drastically over-estimated the undrained strength of clays and analyses indicated that wall deflections and the associated bending moments were grossly underestimated by a factor of, roughly, 2. However, except

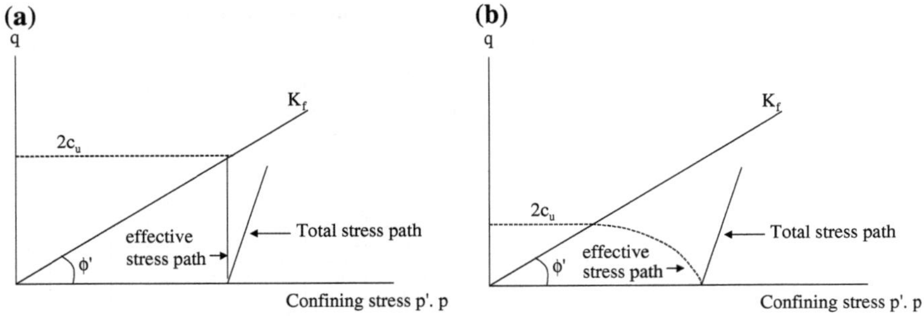

Fig. 9.6 Comparison of undrained strength of clay in Models A and B. **a** Model A. **b** Model B

Levels 6 and 9, the strut loads computed by using Model A were larger than those computed by using Model B. The strut loads computed by using Model B were 107 % of those computed by using Model A at Level 6 and 110 % at Level 9.

9.4.3 Back Analyses

Back analyses were carried out by reducing the stiffness of soils and JGP slabs to match the performance of the wall when the reading of I-65 reached 159 mm, exceeding the alert level of 105 mm, as the excavation reached a depth of 18 m in the sixth stage of excavation on February 23, 2004. The results of analyses predicted a maximum wall deflection of 253 mm in the subsequent stages of excavation. On April 1, the reading of I-104 reached 302.9 mm, back analyses were again carried out by further reducing the stiffness of materials and the maximum deflection was revised to 359 mm. Model A was continuously adopted in these back analyses.

9.4.4 Forensic Studies

Subsequent to the occurrence of failure, the Singapore Government immediately formed an independent Committee of Inquiry (COI), headed by a Senior District Judge, to look into the incident. After thorough investigations, in which 173 witnesses were interviewed and 20 experts offered their professional opinions, an Interim Report was released on 13 September 2004 and a very comprehensive Final Report was made available to the public on 13 May 2005 (COI 2005).

The Committee identified critical design and construction errors, particularly the design of stiffeners on the walings at the connections between the diaphragm walls and the struts, that led to the failure of the earth retaining system. The Committee also found deficiencies in the project management that perpetuated and aggravated the design errors, including inadequate instrumentation and monitoring of works, improper management of instrumentation data, and lack of competency of persons carrying out specialized work.

At the request of the Land Transport Authority, the Engineering Advisory Panel (EAP) conducted an extensive study to review the case and the design of other parts of the Circle Line. Three-dimensional analyses were carried out using the commercial software ABAQUS to analyze the performance of the retaining system. Each of the first nine layers of excavation was modeled as a uniform lift with all the elements removed at the same time. The tenth excavation step was simulated in three substeps, starting with the removal of the elements at the eastern end of the excavation and proceeding westwards (Yong and Lee 2007).

9.4.5 Causes of Failure

As mentioned in Sect. 9.4.2, wrong strength parameters were adopted for marine clay, leading to under-design of wall deflections and bending moments. Although back analyses were performed during the course of construction, they were performed by engineers who did not have experience on the program PLAXIS and the results were not reviewed by experienced engineers. In this regard, however, regarding soil parameters, the experts did differ in opinions in the Inquiry. It is amusing to note the statement made by the representative of the Contractor in the Inquiry that:

> ….. despite more than 6 months of intensive work by the six teams of experts who have been reviewing the collapse at M3, the experts still can not reach any agreement on the correct input parameters to be adopted in a back analysis. Very significant difference still remain, particularly respect to the parameters to be adopted for JGP. It is also noted that because of the stiffness characteristics of the ground, diaphragm wall and JGP are all highly non linear, it is virtually impossible to obtain agreement between the monitored strut loads and wall displacements throughout all the stages of excavation sequence using a single set of linear elastic stiffness value, as adopted in Plaxis analyses.

Although these statements may be exaggerating, they nevertheless illustrate the difficulty associated with analyses.

Model A was adopted by the contractor in analyses for all the cut-and-cover sections in this contract. Although considerable efforts have been made to deal with excessive wall deflections, the excavation in other sections was able to be completed. This indicates that the under-design of diaphragm walls is not solely responsible for the incident. The failure must be due to a combination of adverse factors.

As depicted in Fig. 9.3, the soft deposits on the two sides of the excavation differed in thickness while analyses were performed for only the southern half of the mesh on the assumption that the results would be conservative. This assumption may not be valid reality because the deflections of the southern wall tended to be larger due to the imbalance of earthpressures on the two sides of the excavation.

The 3D finite element analyses conducted by EAP also indicated the significance of the curvature of the alignment. The wall panels tended to split at the joints as the wall moved inward as the excavation proceeded. This is particularly true for the south wall where wall deflections tend to widen the gaps.

The failure in fact was triggered by the buckling of walings. As mentioned in Sect. 9.4.2, except Levels 6 and 9, the strut loads computed by using Model A were larger than those computed by using Model B. Even for Levels 6 and 9, the strut loads computed by using Model B were only 7–10 % larger. Therefore, the buckling was not a result of the use of wrong material model for clay.

The Inquiry revealed that the loads at some of the connections between struts and walings were under-estimated either because a wrong type of joint was installed or because slays were omitted. Second, around February 2004, several instances of stiffener plate buckling as well as the buckling of a waling were reported at the Nicoll Highway Station, the contractor proposed replacing the stiffener plates by C-channels in an attempt to improve the performance of the connection. The replacement of double stiffener plates with C-channels, refer to Fig. 9.7, provided only minor improvement in terms of axial load bearing capacity for the waling connections, but this came at the expense of ductility. The change rendered it more susceptible to the brittle "sway" failure mode. This is clearly seen in results of the post collapse finite element analyses and physical tests. The poor detailing is exacerbated by the discontinuity of walings, as pointed out in the EAP study, which is particularly important at the curved south wall where deflection due to excavation caused separation of the joint between diaphragm wall panels.

The instrumentation and monitoring system was not properly executed and some of the crucial readings were found erroneous and misleading. This made it impossible to correlate the strut loads with the performance of the retaining system.

In addition to all the above-mentioned shortcomings in the design, the performance of the two grouted slabs was also questioned and investigated in the Inquiry and in the EAP study. The removal of sacrificial JGP layer above the formation level was not followed with timely installation and preloading of struts to compensate for the loss of reaction load sustained in the JGP layer. As a result, it caused significant lateral movement to the temporary walls and increase in load to the struts

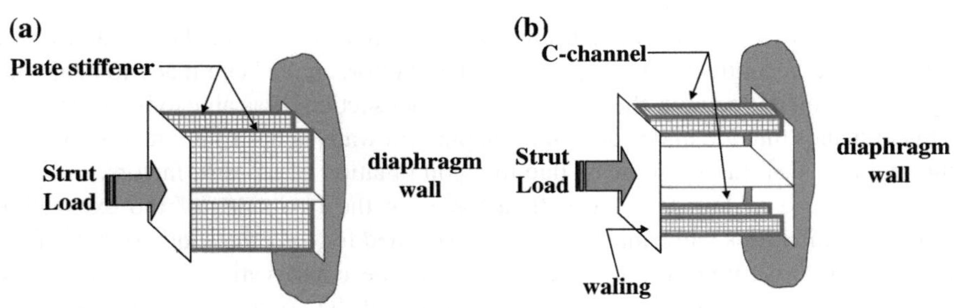

Fig. 9.7 Stiffeners of walings. **a** Original design. **b** Revised design

above. This triggered the local failure of connections at the joint between struts and walings at critical locations and contributed to the chain of collapse of diaphragm walls.

9.5 Conclusions

Failures frequently occur and back analyses enable their causes to be identified. The experience learned can then be passed on from generation to generation. For the results of analyses to be reliable,

(1) the analytical tool used must be suitable for the type of problem to be analyzed
(2) the analyses must be conducted by experienced engineers who have sufficient background on the algorithm of analyses
(3) the input parameters must be representative of the materials to be studied
(4) the construction sequence must be reliable
(5) instrument readings must be faithful
6) the results must be interpreted by experienced engineers who have sufficient practical experience on the performance of soils and structures.

It is however important to realize that there are limitations associated with analyses and the results obtained only play a role of providing supporting evidences and engineering judgments should always prevail.

References

COI (2005) Final report of the committee of inquiry into the incident of the MRT circle line worksite that led to the collapse of Nicoll highway on 20 April 2004, presented by committee of inquiry to minister for manpower on 10 May 2005, Singapore

Hulme TW, Parmar HS, Hou KH, Sripathy P (1993) The collapse of the hotel new world, Singapore, a technical inquiry. Struct Eng 71(6/16):91–98

Hwang RN, Moh ZC, Wong KS (2007) Reference envelopes for deflections of diaphragm walls in Singapore marine clay. In: Proceedings of the 16th southeast asian geotechnical conference, Kuala Lumpur, Malaysia, 8–11 May

Ju DH, Duann SW, Tsai HH (1998) Ground freezing for restoration of damaged tunnel. In: Proceedings of the 13th southeast asian geotechnical conference, Taipei, 16–20 Nov, pp 615–620

Moh ZC, Hwang RN (2007) Lessons learned from recent MRT construction failures in Asia *Pacific*, opening keynote address. In: Proceedings of the 16th southeast Asian geotechnical conference, 8–11 May, pp 3–20, also, Geotechnical Engineering, special issue, 38(3) December, Bangkok, Thailand, pp 121–137

Yong KY, Lee SL (2007) Collapse of Nicoll highway—a global failure at the curved section of a cut-and-cover tunnel construction, Chin Fun Kee Lecture. In: Proceedings of the 16th southeast Asian geotechnical conference 8–11 May, pp 23–36

Chapter 10
Instrumentation and Monitoring for Forensic Geotechnical Engineering

Y. Iwasaki

Abstract Instrumentation plays an important role in forensic geotechnical engineering. In soil mechanics, there are at least several reasons or mechanisms to explain a problematic phenomenon. The scientific facts that were recorded by instrumentation during the construction and/or post-construction give a direction to select one of several possible hypotheses. Two case histories are presented to show the effects of instrumentation upon the forensic investigation of the failure of geotechnical constructions. One is Teton Dam, Idaho in the U.S.A. that was earthen dam and failed during the first filling stage in 1976. Some conclusions were given by Independent Panel to review the failure; however, Teton Dam has been providing endless long debate because of not enough monitored data to rely on. Another one is Nicoll Highway Collapse that was caused by failure of retaining wall during subway construction in Singapore in 2004. In this second case, monitored records of the deflection of retaining wall gave very important information. From the seventh excavation step, deflection of wall at south side became much larger than at north side. One-sided deformation was due to the inclined dense gravel layer toward south below the soft layer and deeper soft soil that differed from the initial design of horizontal layer condition. After the failure, it was confirmed thicker soft soil layer in the south compared to north side by boring study. If the retaining wall had failed without knowing different performance between these walls, it might be much work needed to reach the right conclusion. Instruments and sensors in geotechnical engineering are briefly reviewed. In recent decades, new technologies were introduced to sensor and instrumentation in geotechnical engineering. Among these innovative instruments, two techniques are introduced. Carrier Phase Tracking GPS for measuring displacement between reference GPS receiver and target GPS receiver with a high accuracy of displacement of 2–5 mm. BOTDR (Brillouin Optical Fiber Reflectometer) is another cutting edge technology that provides strain and temperature. Technical basic knowledge of these instruments is explained to understand the principle that is useful not only to follow but also to take advantage of the next generation of instrumentation becoming to prevail at present and in near future.

Y. Iwasaki (✉)
Geo-Research Institute, Osaka, Japan
e-mail: dec17yoshi@gmail.com

V.V.S. Rao and G.L. Sivakumar Babu (eds.), *Forensic Geotechnical Engineering*, Developments in Geotechnical Engineering, DOI 10.1007/978-81-322-2377-1_10

Keywords Forensic · Teton dam · Nicoll highway collapse · Carrier phase tracking GPS · BOTDR

10.1 Introduction

In the preface to second edition of the textbook on Soil mechanics by Terzaghi and Peck in 1967, Prof. Ralph B. Peck added a new chapter on Performance Observation. Chapter 12 was intended to aid the engineer in the use of observational method which is at the very heart of successful application of soil mechanics. Among several good reasons to adapt observation, Dr. Peck indicates field observations for providing evidence in lawsuits. The textbook indicates that lawsuits frequently arise from conflicts between the owner and contractor because of the completed structures or between the contractor and neighbor because of damage to the latter's property.

First, two cases are presented where instrumentation was key factor in providing fact process of failure. Second, instrumentations and sensors are reviewed including some of the cutting edge technology that provides innovative instrumentation in geotechnical engineering at present and in near future.

10.2 Lessons in the Past

In recent cases, big disasters in geotechnical engineering show the need of the instrumentation and monitoring. Two examples are shown in the following section. One is Teton Dam that failed on June 5, 1976 and another is Nicoll Highway Collapse on April 21, 2004.

10.2.1 Teton Dam Failure

The Teton Dam was a federally built earthen dam on the Teton River in southeastern Idaho, USA which when filling for the first time suffered a catastrophic failure on June 5, 1976 as shown in Photo 10.1. The collapse of the dam resulted in the deaths of 11 people and 13,000 head of cattle. The dam cost about USD $100 million to build, and the federal government paid over $300 million in claims related to the dam failure. Total damage estimates have ranged up to $2 billion. The dam was never rebuilt (Wikipedia).

Not enough instrumentation was installed to monitor pore pressure in the dam.

Hydraulic fracture in the core was considered as most likely the key mechanism estimated by desk work with laboratory aids.

Photo 10.1 Teton Dam, Idaho, June 5, 1976 from http://en.wikipedia.org/wiki/Teton_Dam

Independent Panel to Review Cause of Teton Dam Failure (1976) reached a conclusion as follows:

The Panel had quickly identified piping as the most probable cause of the failure, and then focused its efforts on determining how the piping started. Two mechanisms were possible. The first was the flow of water under highly erodible and unprotected fill, through joints in unsealed rock beneath the grout cap, and development of an erosion tunnel. The second was "cracking caused by differential strains or hydraulic fracturing of the core material." The Panel was unable to determine whether one or the other mechanism occurred, or a combination.

If pore pressure sensors had been installed in the dam section, process of increasing pore pressures during filling water might have provided vital information of the failure process. Figure 10.1 shows the Geotechnical characteristics of section.

The Panel referred on monitoring as

> There were not enough instruments in the dam to provide adequate information about changing conditions of the embankment and abutments.

The debate on the cause of the failure still continues until recently (Solava and Delatte 2003) and at present. If pore pressure monitored data had been provided, we could have reached the valuable understanding.

10.2.2 Collapse of Nicoll Highway

The Collapse of Nicoll Highway (The Committee of Inquiry 2005) that took place on April 21, 2004 was caused by an accident of subway construction in Singapore.

The failed section in the subway construction near Nicoll Highway was a cut and cover excavation shown in Fig. 10.1.

(Number with underline is displacement of JGP slab).

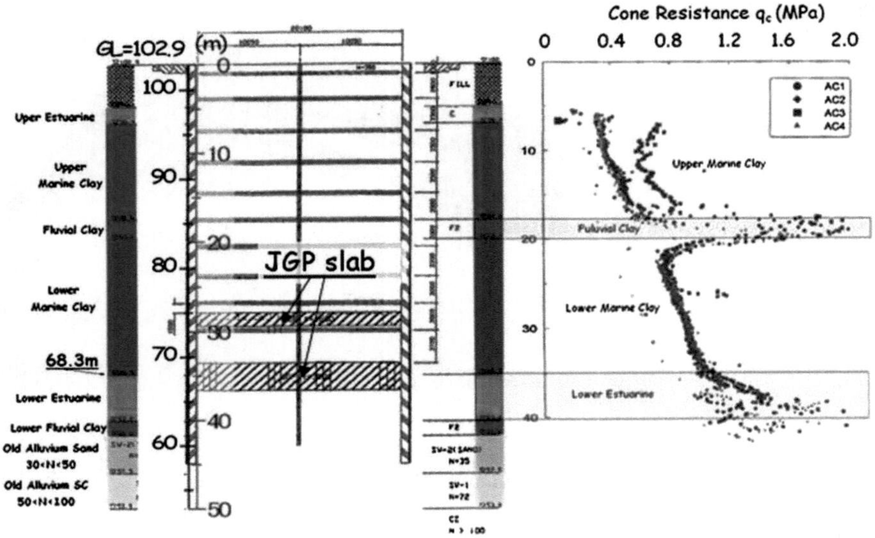

Fig. 10.1 Geotechnical condition for design of the failed section

The excavation is planned in a very deep soft marine clay layer and some special design of JGP (Jet Grouting Pile) was considered. The slab near the bottom of the excavation was to take strong horizontal load from the soft clay layer.

Among various instrumentations, the most important information was the inclination of the temporal retaining wall.

Though under emergency, the measurement was not made daily. However, the obtained data gives clear tendency of the deformation as shown in Fig. 10.2. Two things should be noted.

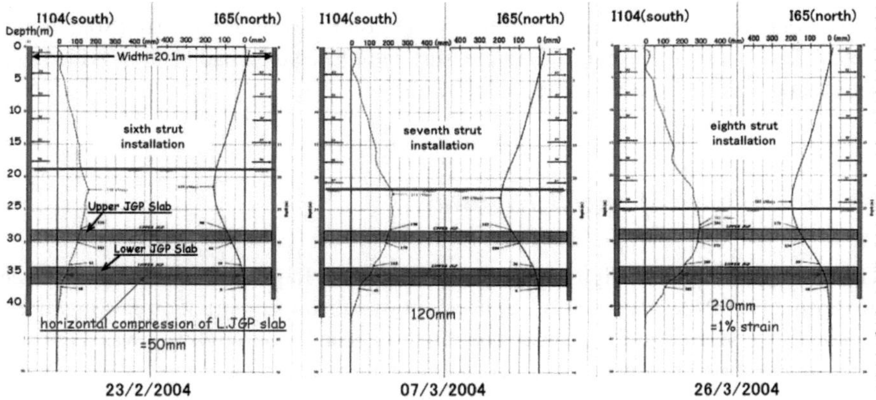

Fig. 10.2 Deflection of retaining wall of the failed section

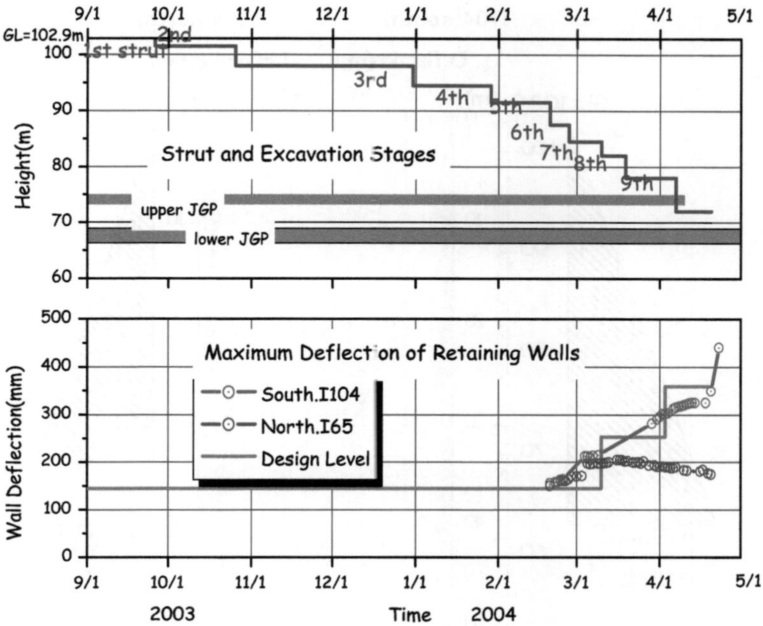

Fig. 10.3 Maximum deflection of walls of the south and north side

One is the deflection of retaining wall. The other is the horizontal compression of JGP slab. It should be noticed not only the increment of the maximum deflection itself but also the difference between south and north walls. As shown in Figs. 10.2 and 10.3, the maximum deflections at sixth and seventh excavation level show the same order of 160 and 200 mm for north and south sides.

However, after the seventh excavation the maximum deflection of the south side wall increased with the excavation depth. At the same time, the deflection of the north side was kept constant about 20 cm and even decreasing. However, this difference of deflection needs some reasons of either weaker soil condition or smaller stiffness of retaining wall in south side.

The original design condition was assumed as the same soil layers at both sides. Boring study after the failure have revealed that the dense sand gravel layer below the soft marine clay is inclined toward south and the soft marine layer is deposited deeper than the original design as shown in Fig. 10.4. If the site engineer could evaluate the different response of the walls of south and north, he might proceed to perform boring study and could prevent the failure. The site engineer just agreed with the proposal of modifying design parameter by the contractor and allowed to increase the allowable design level without any geotechnical consideration as shown in Fig. 10.3. The JGP slab is rather stiff elements compared to the soft soil at the site. The horizontal compression displacement allowable in the slab might be controlled by failure strain level under compression of the order less than possibly a

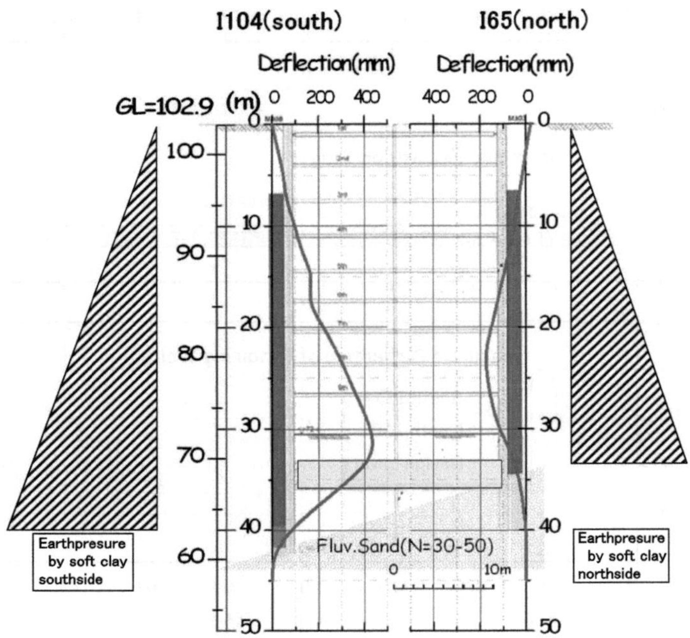

Fig. 10.4 Deflection of walls at the final step and the confirmed geotechnical condition by boring after the failure

few percents. Since the horizontal length of the JGP slab is about 20 m, 20 cm of the compression at 8th strut stage (Fig. 10.2) corresponds to 1 % of compression strain, which was the design allowable strain. The instrumentation and monitored data really had provided not only before the failure but also after the failure to give constrains of the possible mechanism discussed in forensic engineering process.

10.3 Instrumentation for Monitoring

Recent technologies of monitoring displacements or strains have much advanced within the past decade. Among these several technologies, only two methods of GPS (Global Positioning System) and BOTDR (Brillouin Optical fiber Time Domain Reflectometer) are introduced here.

However, conventional methods are still important and used as daily practice.

Due to the limitation of the number of paper, the author will review only a trend of instrumentation for monitoring in geotechnical engineering in some aspect.

Reader should refer an introductory book as well as containing professional level by Dunnicliff (1988).

10.3.1 Mechanical Sensor

Photo 10.2 shows a simple and yet accurate monitoring device to measure gap movement for foundation of Hudson–Athens Lighthouse (U.S. National Register of Historic Places) in the Hudson River in the state of New York in the United States.

The two halves of the device are rigidly attached to the foundation wall on opposite sides of the crack. The red reference lines are on the transparent half and the grid is on the opaque white half. Both vertical and horizontal movements can be monitored over time. In this picture, the crack can be seen to have widened by approximately 0.3 mm (and no vertical movement) since the gage was installed.

10.3.2 Strain Gauge Sensor

Electric sensor is also common for automatic monitoring. Photo 10.3 is strain gauge-based displacement sensor. The range of measurement is 5–10 mm with accuracy of 0.5–1 %/Full Scale.

Strain gauge for steel plate was also developed based upon vibrating wire sensor as shown Photo 10.4.

Photo 10.2 Mechanical type of gap sensor (http://en.wikipedia.org/wiki/Strain_gauge)

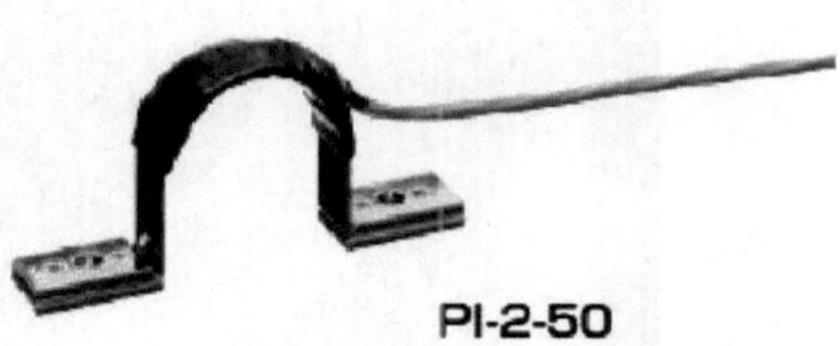

Photo 10.3 Strain gauge-based displacement sensor (Tokyo Sokki)

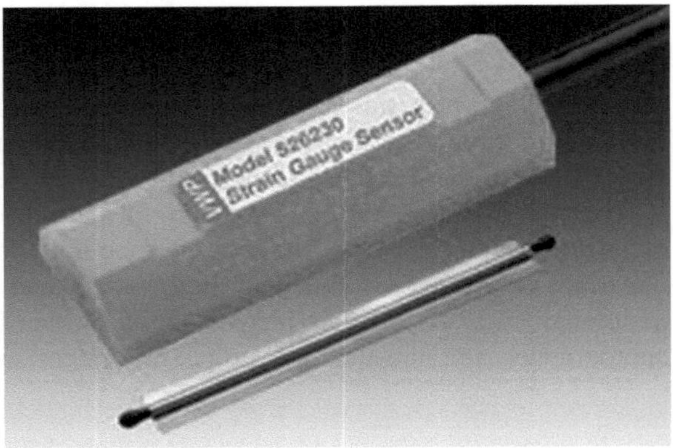

Photo 10.4 VW spot-welded strain Gauge (Durham Geo Slope indicator Co.)

Vibrating wire transducer measures change of natural frequency of a wire installed in a very small pipe on a steel plate. It is designed to set the sensor by spot welding to steel surface (Photos 10.5 and 10.6).

Photo 10.5 GPS receiver (http://www.furuno.co.jp/product/gps/terrain/dana2000.html)

Photo 10.6 Optical fiber (http://en.wikipedia.org/wiki/opticalfiber)

The accuracy of the vibrating wire sensor is about 0.05 % of full scale of 2500 μ strain.

10.3.3 Displacement, Load Cell, and Pore Pressure Sensor

Conventional strain gage and vibrating wire are basic transducer for not only strain but also other application of displacement, load, and pore pressure sensors.

Other types of transducers like LVDT (Linear Voltage Differential Transducer) are also used for various types of sensors. The accuracy of LVDT sensor is around 0.1–0.0.5 %/F.S.

10.3.4 Inclinometer

Inclinometer sensor system that had been developed in the 1970s was based upon force balance sensor. The force balance principle that is to measure how much force needed to get back the original zero position of the displacement sensor have resulted in the wider range to measure more than 30° with the accuracy of 0.01–0.02 %/FS.

Recently, Microelectromechanical Systems (MEMS) have realized much cost-effective with the same performance becomes available in the market.

10.4 Recent Development of Sensors

In the past decade, sensor technology had entered another era of innovative principles like MEMS (Microelectromechanical Systems).

In the chapter, among these cutting edge technologies, GPS and fiber optic sensors are introduced.

10.4.1 Carrier Phase Tracking of GPS

Carrier phase tracking signal of GPS has resulted in a revolution in land surveying. A line of sight along the ground is no longer necessary for precise positioning. Positions can be measured up to about 30 km from reference point without intermediate points. This use of GPS requires specially equipped carrier tracking receivers.

Displacements of multiple points may be monitored by Carrier Phase Tracking GPS method that provides relative displacements of target monitoring points to the reference point.

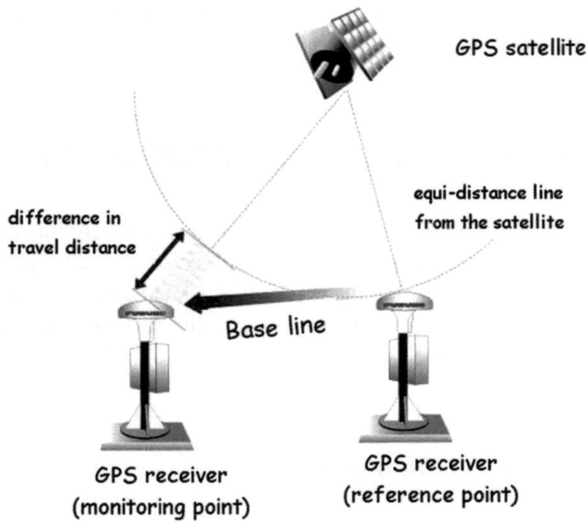

Fig. 10.5 Basic concept of the carrier phase tracking modified from http://www.shamen-net.com/word/img/ (Difference in travel distance = $n19 + \delta$ (cm))

As shown in Fig. 10.5, distance between the monitoring station and GPS is divided into two segments of equidistance and the difference.

The difference in travel distance is obtained as to detect the time difference between the receiving times of the same phase of the signal carrier signals. The $L1$ and/or $L2$ carrier signals are used in carrier phase surveying. The wavelength used for the system is about 19 cm of the $L1$ carrier band.

The difference in travel distance = $n19 + \delta$ (cm))
where n is an integer number and called integer bias.

With accuracy of 1 % of wavelength in detecting the leading edge, the error might be as low as 2 mm.

Errors are caused by several factors. One of the common factors is the variation of wave velocity near the surface among the monitoring points. Wave velocity may be affected by such space conditions like humidity, temperature, and air pressure. To avoid such errors, it is better to set monitoring position with relative position within some limited range of several hundred meters.

The most time consuming process is to estimate the wave number n. Satellite is moving over the receivers from $L1$ to $L2$ along the known orbit. Using four GPS satellite signals, the phase difference as δ is obtained for every satellite. As shown in Fig. 10.6, point P is the crossing point between ray path from satellite to the monitoring receiver and the equidistance circle (Circle 1 in Fig. 10.6). As the satellite moves from $L1$ to $L2$, the crossing point $P1$ moves to $P2$. Since the radius of the circle is very long compared with baseline, the arc from $P1$ and $P2$ to the reference receiver (RR) is considered as a straight line. The locus of the points $P1$ and $P2$ forms an elliptical orbit (Fig. 10.7).

The wave number n is obtained as to give the minimum error to show the orbit as elliptical shape in 3D space.

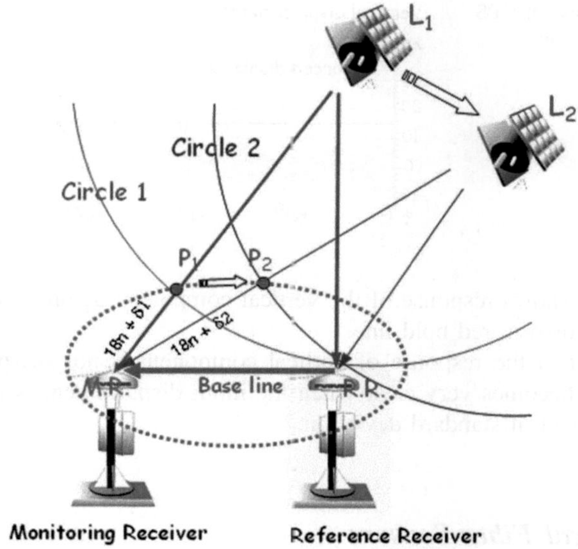

Fig. 10.6 Determining wave number of integer bias for carrier phase tracking of GPS method

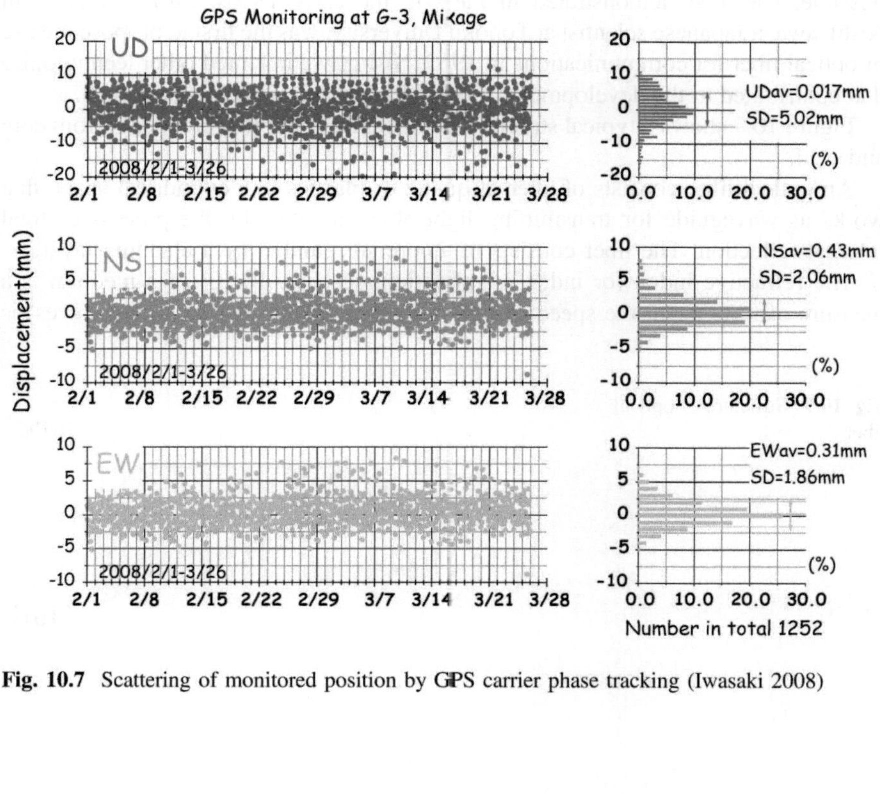

Fig. 10.7 Scattering of monitored position by GPS carrier phase tracking (Iwasaki 2008)

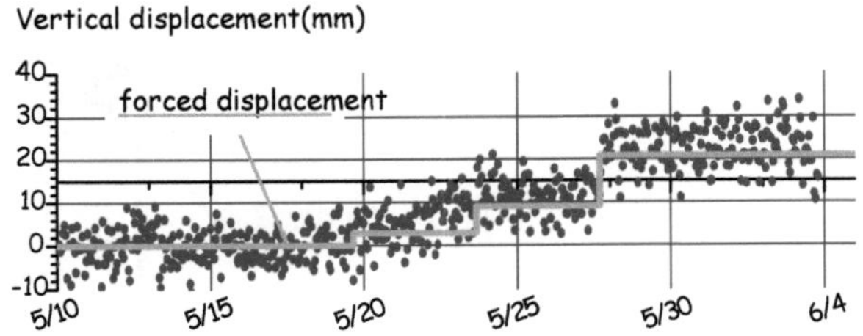

Fig. 10.8 Response of GPS in vertical component

Figure 10.8 shows response of the vertical component against forced displacement that is shown in red bold line.

It is found that the response of vertical component is not sharp to small displacement but becomes very clear when the input displacement is larger than the 5 mm of the level of standard deviation.

10.4.2 Optical Fiber Sensors

Fiber optics, though used extensively in the modern world, is a fairly simple and old technology. Guiding of light by refraction, the principle that makes fiber optics possible, was first demonstrated in Paris in the early 1840s. Professor Jun-ichi Nishizawa, a Japanese scientist at Tohoku University, was the first to propose the use of optical fiber for communications in 1963. Nishizawa invented other technologies that contributed to the development of optical fiber communications as well.

Figure 10.9 shows a typical structure of the optical fiber that composes from core and clad.

An optical fiber consists of fiber of quartz or plastics in a cylindrical shape that works as waveguide for transmitting light along its axis, by the process of total internal reflection. The fiber consists of a core surrounded by a cladding layer.

The refractive index (or index of refraction, see Fig. 10.10) of a medium is a measure of how much the speed of light (or other waves such as sound waves) is

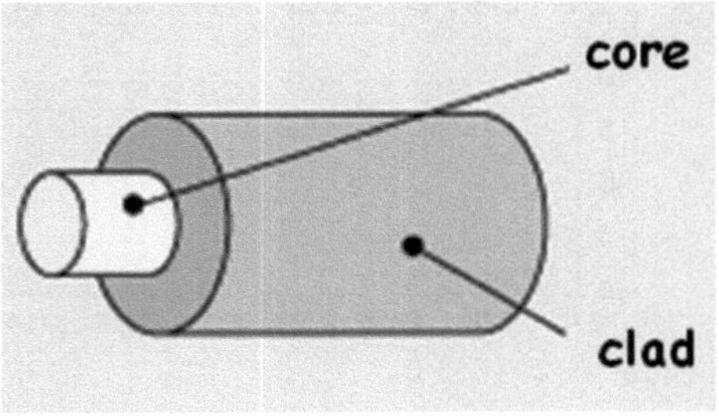

Fig. 10.9 Structure of optical fiber

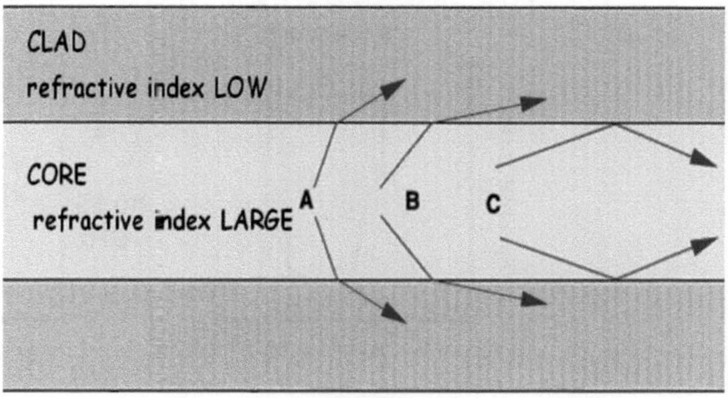

Fig. 10.10 Refractive Index (*A* and *B* refraction, *C* full reflection)

reduced in the medium compared to a reference medium. In treating light, the reference medium is usually vacuum air

The RI of 1.0 means light travel in the medium with the speed in vacuum air. The RI greater than 1.0 means light travel in the medium with lower speed than in vacuum air. To confine the optical signal in the core, the refractive index of the core must be greater than that of the cladding. If the wavelength of the light is longer than compared to diameter of the core cable, the light is confined within the core and can travel very long distance with little scattering. The boundary between the core and cladding may either be abrupt change of refractive index, in step-index fiber, or gradual, in graded-index fiber. Single mode optical cable (SM) is designed as abrupt change to show little loss with capability of traveling long distance of 10 km.

10.4.3 BOTDR (Brillouin Optical Fiber Time Domain Reflectometer)

When a sharp and strong optical pulse is sent into an optical fiber cable, some weak optical signal was recognized as returned to the entered gate. The signals are originated from the scattering of the light at the propagating front. Three different kinds of scattering are identified as Rayleigh, Raman, and Brillouin. The scattered signals of Brillouin are found to show special characteristics. The frequency of the scattered pulse is shifted about 10 GHz from that of the input pulse and changes proportional to strain and temperature in the fiber as shown in Fig. 10.11.

As shown in Fig. 10.12, when a signal is sent to an optical fiber, backward traveling scattered light continuously arrives at the front gate. The point where the scattered signal originated is estimated by the delayed arrival time from the time of sending pulse.

Since the relationship between the strain and Brillouin frequency shift is known, the analyzed frequency shift is easily transformed to strains or temperature. When temperature effect should be known independently, another optical fiber cable with free strain may be installed. The accuracy of the measurement depends upon accuracy to obtain the number of the frequency of Brillouin shift. At present, the

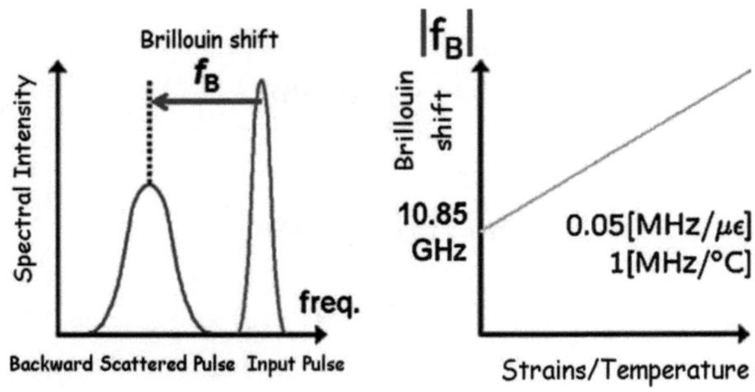

Fig. 10.11 Brillouin shift versus strains and temperature (http://park.itc.u-tokyo.ac.jp/hotalab/research/bocda-r)

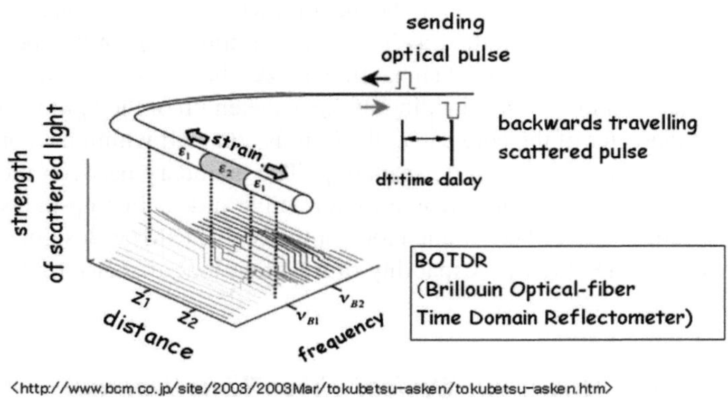

Fig. 10.12 Basic concept of BOTDR

measurement is usually given for 1 m segment of the cable. The error range of strain is about 10–20 μ strain and that of temperature is 0.5–1 °C.

10.4.4 BOTDR as Pipe Strain Gage

Nakano et al. (2003) reported a case history of the application of the BOTDR to identify sliding surface by inserting a pipe equipped with optical fiber in a slope as shown in Fig. 10.13.

An optical fiber was fixed along a plastic pipe. The fiber cable was given some pretension strain. The pipe strain gage of optical fiber gives continuous change of the strain compared to conventional pipe strain measurement that gives only

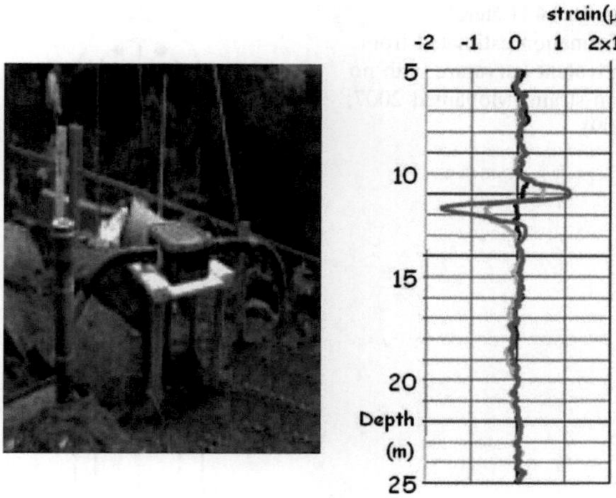

Fig. 10.13 Monitoring slope instability by vertical pipe with optic cable

discrete points. As shown in Fig. 10.13, the strain shows continuous change from compression to extension that corresponds to bending deformation at the depth of 10–11 m. This depth was considered as a sliding surface.

10.4.5 BOTDR as Inclinometer Equipped with Secant Pile Wall

Mohamad et al. (2007, 2009) show some BOTDR application to monitor axial strain of bored piles and lateral deflection of retaining wall of secant pile wall of a series of intersecting reinforced concrete piles.

The diameter of the secant wall was 450 mm and a pair of optical fiber cable was installed along the reinforced steel was installed along steel cage giving a pretension of 2000 μ strain. Another single optic fiber loose tubed cable was equipped along the monitored pile. Figure 10.14 shows how to evaluate lateral deformation from the edge strain along a pile. Since the edge strains are induced by mix modes of axial compression and bending deformation. To obtain bending strains, axial compression should be separated. These axial components should be subtracted from the apparent edge strains as shown in Fig. 10.14.

Temperature is another factor that affects upon the apparent strain values.

Figure 10.15 shows an example of the measured strains of BOTDR for retaining wall after excavation. The excavated level was about −4.5 m. Figure 10.15a shows two edge strains of outside and nearside of the pipe as well as axial strain of the averaged values.

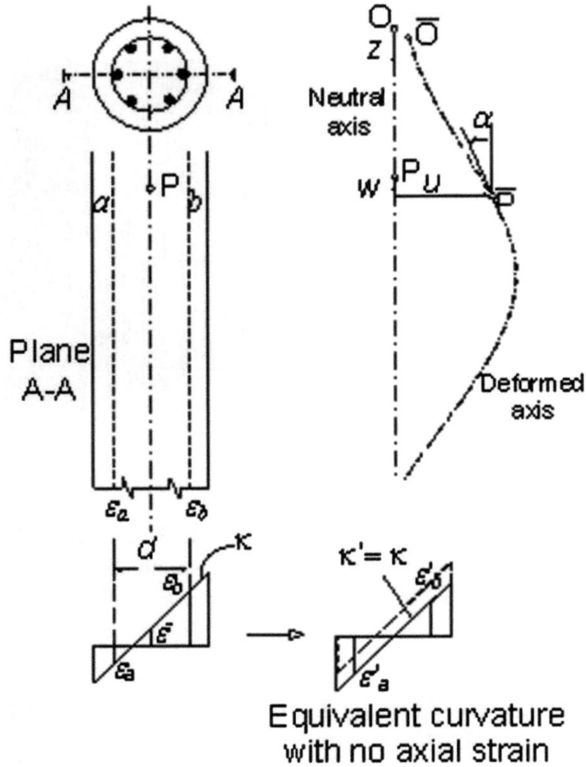

Fig. 10.14 Lateral deformation estimated from equivalent curvature with no axial strain (Mohamad 2007, 2009)

Figure 10.15b shows rather constant temperature in the soil up to a level of −6.5 m and gradually decreases to −13° at the top of the pipe.

The air temperature was lower than the in the ground, temperature decreases from near the ground surface to the top of the pile.

Figure 10.16 shows comparison between a lateral deformation curve obtained from BOTDR and two curves by conventional inclinometer measurements. It is noticed that the averaged curve of the two deflection curves from the conventional inclinometer corresponds well to the one from BOTDR.

Compare to the conventional inclinometer of manual handling, inclinometer system by BOTDR provides automatic and real-time evaluation of deformation of retaining wall.

Inclinometer measures inclination angle relative to gravity. BOTDR measures a pair of the edge strains under bending mode. It should be noticed that the distance

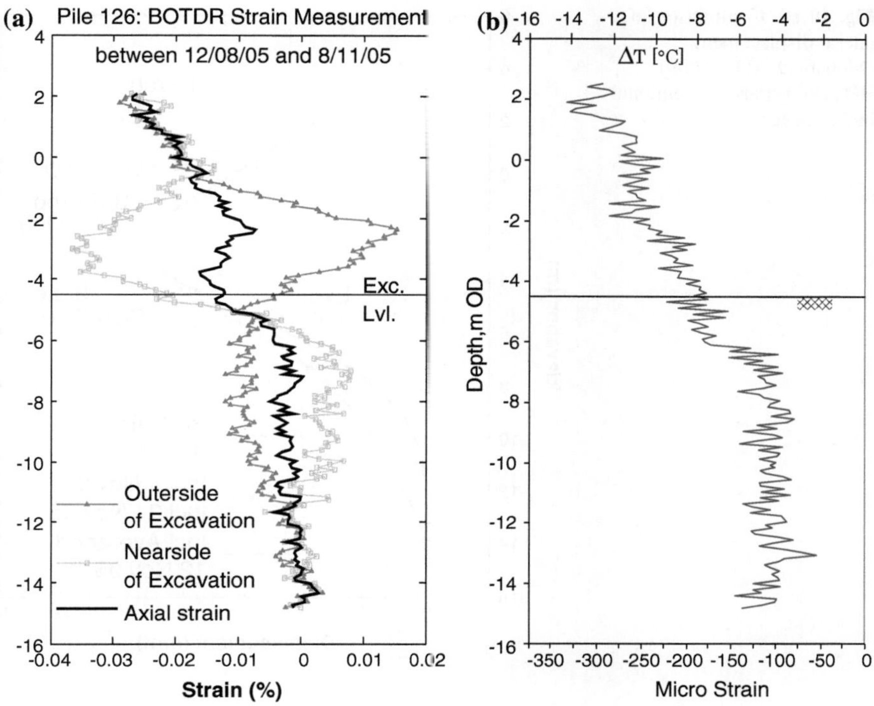

Fig. 10.15 Measured strains along edges and in a loosed tube, **a** strains at pipe edges, **b** strain in a loosed tube cable, Mohamad et al. (2009)

between two edges must be wide enough to keep acceptable accuracy of bending deformation (Table 10.1).

At present, the facility to send and receive the signal as well as analyzing the BOTDR is rather expensive. In the near future, the cost-effective system is expected to be available and become useful especially for big and deep excavation works.

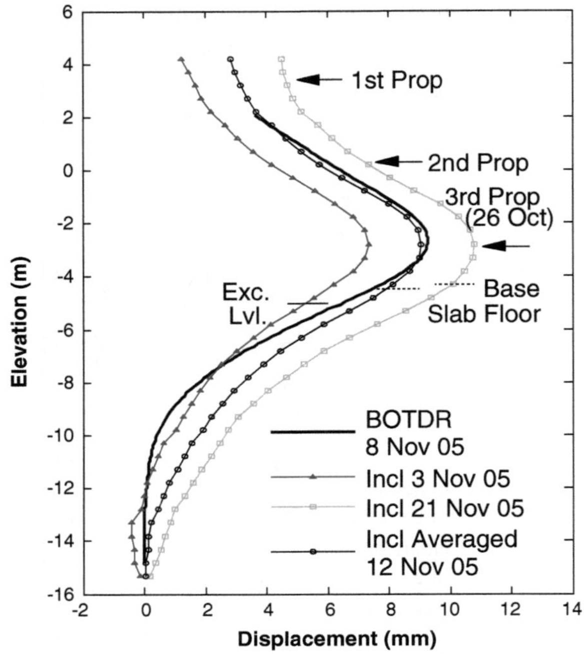

Fig. 10.16 Comparison of lateral displacement (Mohamad 2007, 2009), BOTDR versus conventional inclinometer

Table 10.1 Errors of displacement obtained by GPS

Component	Standard deviation (mm)
UD	5.02
NS	2.06
EW	1.86

10.5 Conclusion

In the first part, two comparative case studies are overviewed. Teton Dam that was failed by piping. No further detailed mechanism is clarified due to not enough instrumentation.

In the collapse of Nicoll Highway, though the data were not collected as it should be the available data of inclination of the subway construction nearby the Highway show clear fact of one-sided deformation of the wall from which the geotechnical condition for the design was anticipated different from the design condition. If the construction had been modified, the collapse might well be avoided.

After the failure, the results of instrumentation contributed to identify the process of the failure from the geotechnical point of view.

In the second part, review of sensors is made including some edge cutting technologies of two sensors of Carrier phase tracking of GPS method and BOTDR.

These sensing technologies in addition with other sensor like MEM (Micromachine Electronic Machine) sensor are still on a way of development.

Forensic engineers, especially in the field of geotechnical field where most of the phenomena takes place in dark underground, should use and take advantages of installation and monitored data to reach the most likely mechanism among several possibilities.

References

Dunnicliff J (1988) Geotechnical instrumentation for monitoring field performance. Wiley, New York

Independent Panel to Review Cause of Teton Dam Failure (1976) Report to the U.S. department of the interior and state of idaho on failure of Teton Dam. Idaho Falls, Idaho

Iwasaki Y (2008) Basic study of carrier-phase tracking of GPS method at Mikage site. GRI report

Mohamad H, Bennett PJ, Soga K, Klar A, Pellow A (2007) Distributed optical fibre strain sensing in a secant piled wall. In: Proceedings of the seventh international symposium on field measurements in geomechanics (FMGM2007), ASCE geotechnical special publication, no 175

Mohamad H, Soga K, Bennett PJ (2009) Fibre optic installation techniques for pile instrumentation. In: Proceedings of the 17th ICSMGE (in print)

Nakano M, Yamazaki H, Okuno M (2003) The earthquake disaster prevention monitoring system using optical fiber technology. JSCE J Earthq Eng 27:1–4

Solava S, Delatte N (2003) Lessons from the failure of the Teton Dam. In: Proceedings of the 3rd ASCE Forensics Congress, San Diego, California, 19–21 Oct 2003

The Committee of Inquiry (2005) Report on the incident at the MRT circleline worksite that led to the collapse of the Nicoll highway on 20 April 2004. Ministry of Manpower, Government of Singapore, Singapore

Chapter 11
Analysis of the Causes of Undue Movements of a Retaining Wall Used to Form a Deep Basement in Sandstone and Consequent Damage to an Adjacent Canal

R.A. Jessep

Abstract A large new building was to be constructed with a 15 m deep basement formed in sandstone using a contiguous bored pile retaining wall. The site was bounded on one side by a canal. The wall was designed assuming a construction sequence involving props near the top of the wall and temporary rock beams. In the event, the construction sequence was altered to essentially remove the top level of propping and the propping as constructed contained serious shortcomings. The wall was monitored and deflected up to around 30 mm resulting in significant water ingress into the excavation from the canal. To reduce the rate of ingress sprayed concrete was applied between the piles. However, the design of the propping system had considered the resultant water pressure. The wall deflection increased dramatically to around 100 mm in a short period of time until holes were cut in the sprayed concrete to relieve the pressure and avert a failure. An analysis of the available information was carried out, including a back analysis of the wall behaviour, so that others could appropriately allocate the cost of the incident.

Keywords Movements · Construction sequence · Retaining wall · Propping system

11.1 Introduction

The firm was appointed to investigate the causes of undue movement of a piled retaining wall during the construction of a new building next to a canal. The movements of the retaining wall had damaged the canal and resulted in significant inflows of water into the basement excavation.

R.A. Jessep (✉)
William J Marshall & Partners, 35 Westminster Palace Gardens Artillery Row, London SW1P 1RR, UK
e-mail: raj@williamjmarshall.co.uk

V.V.S. Rao and G.L. Sivakumar Babu (eds.), *Forensic Geotechnical Engineering*,
Developments in Geotechnical Engineering, DOI 10.1007/978-81-322-2377-1_11

The time available for our investigation was very limited (less than 2 weeks) because we were appointed by solicitors advising one of the parties in the context of a short, interim dispute resolution process introduced in the UK in May 1998, known as adjudication. For this reason, our views were based solely on a review of the available information. Our advice was given by Experts in accordance with English Civil Procedure Rules to assist the adjudicator in his decision so that the cost of the incident could be allocated appropriately.

Description of the building and site. The new multistorey building includes a 15 m deep basement with three floors. The building is formed of reinforced concrete with reinforced concrete floor slabs founded on pad foundations on the underlying rock. The basement is formed of a perimeter retaining wall constructed of contiguous bored cast in situ piles that were 17 m long. The piles are nominally 900 mm diameter spaced at 1050 mm diameter so there is usually a space between the piles. A deep capping beam ties the pile heads together. In the final state, reinforced sprayed concrete was applied on the face of the piles and a drained cavity system formed using a proprietary system behind blockwork walls.

The canal is about 2 m deep and has side walls formed of masonry, the nearest of which is a little over 1 m from the capping beam. The gap between the capping beam and canal was filled with mass concrete during construction. The base of the canal was originally lined with a thin layer of puddle clay but this has eroded in places.

The geology of the area of the site is a Sandstone Group. The site is underlain by less than 1 m of Made Ground over 2.5 m of dense silty Sand, which is a residual soil derived from the Sandstone. This is underlain by about 3 m of very weak Sandstone over about 6.5 m of very weak Mudstone, which is underlain by moderately weak Sandstone. The natural groundwater level is deep in the moderately weak Sandstone due to artificial lowering, but there is a perched water table on the Mudstone supplied in part by leakage from the canal. Figure 11.1 below shows a section through the sub-structure, canal and the subsoils.

Construction sequence. The bored piles for the contiguous piled retaining wall were installed around the full perimeter of the site and the capping beam excavated

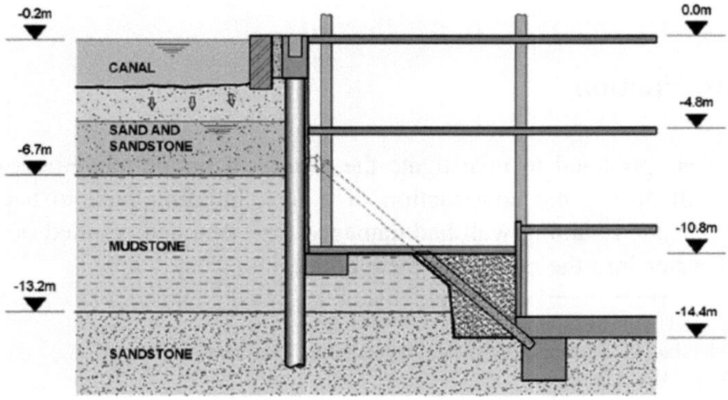

Fig. 11.1 Typical section through the basement

and then cast and backfilled with mass concrete in sections whilst propping the canal wall as necessary. In order to excavate and form the basement floors columns and walls, a sequence of construction was required that would limit deflection of the retaining wall and maintain an adequate factor of safety against wall instability in the temporary condition where the floors to prop the retaining wall had not yet been constructed.

The designer of the retaining wall, who was a piling contractor, envisaged a construction sequence comprising as follows:

- Excavate to the base of the capping beam and subsequently form a rock berm against the retaining walls: excavate in the centre of the site continuing to the lowest basement level.
- Installing of temporary raking steel props supporting the capping beam and founded within a bench in the rock berm formed 6 m below the top of the wall. The unfactored prop force was 100 kN/m wall.
- Excavate of the berm above the bench level and construct the basement bottom-up until the upper two basement floors supported the wall.
- Excavate next to the wall to 10 m below the top of the wall and subsequently complete the remaining basement construction.

The maximum wall movement with this construction sequence was expected to be about 5 mm. Figure 11.2 below is a section showing the key stage in this construction sequence.

The sub-contractor responsible for the excavation and construction of the basement engaged a specialist temporary works designer. These parties amended the construction sequence to the following:

- Excavate across the site leaving a rock berm 4 m wide at the crest, which was 6 m below the top of the wall, and 7 m wide at the base of the excavation. At this stage the retaining wall would effectively cantilever 6 m.

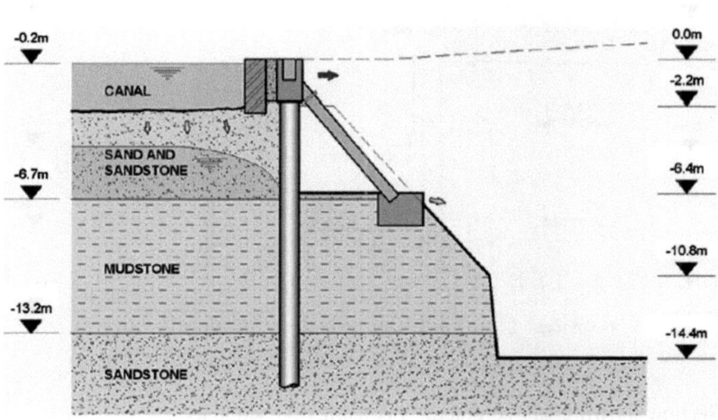

Fig. 11.2 Section showing piling contractor's construction sequence

- Generally install temporary raking props in slots cut in the berm supporting the wall at about 6 m below the top of the wall with a horizontal waling beam used to support all the piles by means of concrete infill between the waling beam and piles. These temporary props were spaced at 7.5 m centres, founded at the base of the basement excavation and were to be prestressed to 100 tonnes compression. Horizontal diagonal temporary steel props were positioned across the corners of the basement.
- Excavate the berm to 10 m below the top of the wall and construct the lower two basement floors.
- Construct the upper two basement floors and remove the temporary props.

No analysis to predict the wall movement for this construction sequence was carried out. Figure 11.3 below is a section showing the key stages in this construction sequence.

Wall movements. Having reviewed the sub-contractor's proposed construction sequence, the structural engineer responsible for the permanent works design save the pile retaining wall, specified a "red trigger level" of 15 mm on the permissible movement of the retaining wall, above which contingency measures were to be implemented. The structural engineer implemented a system of monitoring with surveying of the position of the top of the canal and retaining walls at approximately 7.5 m centres and inclinometers installed in eleven piles. The allowable damage category for the canal was classified as Category 0 after Burland (2001).

During construction, the movement of the retaining wall adjacent to the canal exceeded the limit set down in the specification and there was a significant increase in the rate of inflow of water through the retaining wall. The dispute. There was a dispute between the sub-contractor responsible for the basement construction and the developer over the causes of the undue movements of the retaining wall and the responsibility for the movements.

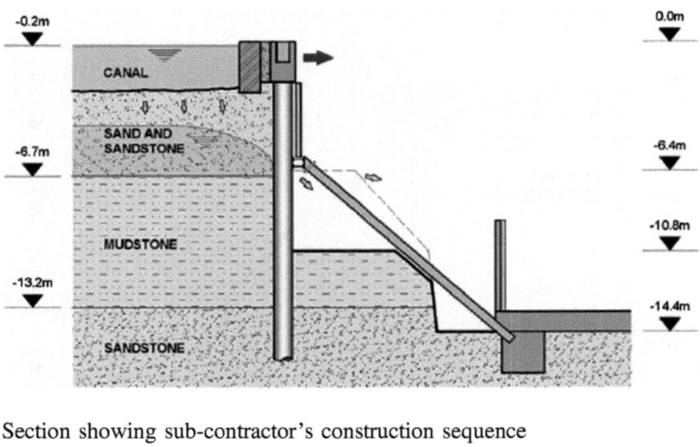

Fig. 11.3 Section showing sub-contractor's construction sequence

11.2 Key Events During Construction

The sub-contractor commenced excavation of the basement after the capping beam had cured sufficiently. The top of the berm formed against the retaining wall on the canal side of the basement was 6 m below the top of the wall as the sub-contractor had intended. The top of the berm was, however, only 3 m wide. At this stage, the inclinometers in the piles showed the deflection at the top of the retaining wall to be up to 19 mm, thereby exceeding the red trigger level for this wall. However, no action was taken to implement any contingency measures.

Photographs from the site reveal that the berm was then reduced so that the top of the berm was 10 m below the top of the retaining wall even though the installation of the temporary steel props had not been completed. Following this excavation, the deflection at the top of the retaining wall increased to up to 22 mm.

The sub-contractor proceeded to form further the excavations in the top of the berm to construct pad foundations at the second lowest basement level. In the period when these excavations were being formed, the retaining wall deflection increased to up to 33 mm at the top. The movement of the retaining wall caused deflection and settlement of the canal wall. In this period, there was a significant increase in the ingress of water through the retaining wall adjacent to the canal.

The parties agreed that the sprayed concrete works should be brought forward to reduce the water ingress. These works were implemented about 2 weeks after the water ingress increased significantly. Within about 3 days of the completion of the sprayed concrete works, the retaining wall deflection increased up to about 80 mm at the top and the wall moved upward about 6 mm.

Holes were cut in the shotcrete to relieve the water pressure but the deflection of the retaining wall continued to increase to up to about 100 mm at the top a few days later. Thereafter, there was essentially no further retaining wall deflection. The records for the maximum deflection of the top of the retaining wall are shown in Fig. 11.4.

An inspection of the wall by the temporary works designer revealed that there were gaps between the waling beam and 37 piles in the wall adjacent to the canal. The engineer also found that no jacks had been installed in two of the temporary raking props, and hence these props had not been prestressed. Evidence suggests that the remaining two props had only been prestressed to 100 kN (10 tonnes). The props were also not built in their design positions resulting in a 7.5 m cantilever on one part of the waling beam.

The slab at the second lowest basement level about 10 m below the top of the wall was cast up to the retaining wall adjacent to the canal about a week later. A further temporary raking prop supporting the cantilever section of waling beam was installed and prestressed around 10 days later.

Within another month, the second highest basement slab was cast and the temporary props were removed.

A subsequent inspection of the canal found significant open cracks in the base through which water was flowing at a substantial rate.

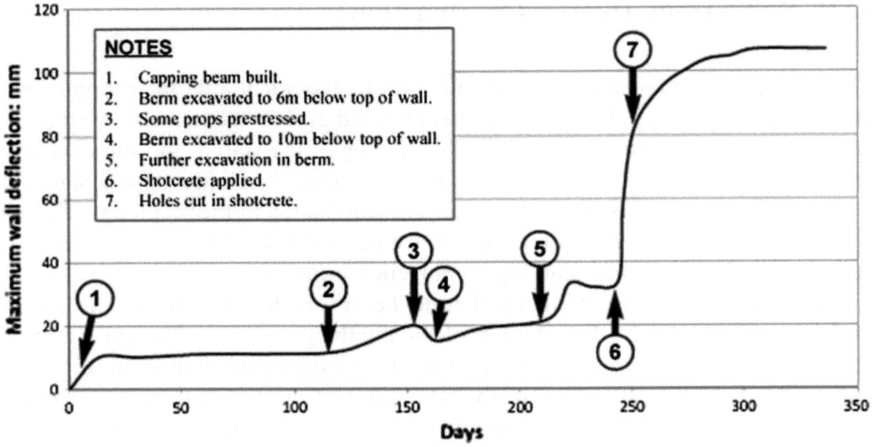

Fig. 11.4 Maximum deflection of the retaining wall during construction

11.3 Analysis—Causes of Undue Movements

Design concept. One way of limiting the inflow of water during construction would have been to install a secant piled retaining wall. However, the water pressure behind this form of wall could have been expected to be significantly greater than behind a contiguous piled retaining wall which allows water to drain between the piles. Hence with a secant piled wall, substantially increased temporary propping works would have required.

The parties involved in the project had previous experience of building basements on adjacent sites (also adjoining the canal) and had previously successfully constructed contiguous piled retaining walls and only encountered modest groundwater flows during the works: very limited recharge of the perched water above the Mudstone occurred with a substantial decrease in inflows observed after the initial excavation. Hence, experience in this area had shown that a carefully considered construction sequence with properly designed and constructed temporary works could be used to limit wall deflection and limit ingress of water into the excavation. For this reason, all parties agreed to proceed using a contiguous bored piled retaining wall on this development.

The flow of groundwater between the piles in the contiguous pile retaining wall would pose a risk of washout of material, and hence volume loss in the ground and consequent damage to the canal. However, because the Sand at the near surface was very dense and the Sandstone below was cemented the risk of loss of material was relatively low provided the flows of groundwater were limited.

If the project was to proceed using a contiguous bored pile retaining wall, the risk associated with the potential for substantial inflow of water would have to be reviewed and appropriate safeguards and/or mitigation procedures put in place to limit the wall deflections and hence ingress of water into the excavation.

On this project, the structural engineer advised that grouting of the Sand and Sandstone may be required if the ingress of water increased during construction and told the sub-contractor to provide contingency measures such as additional propping to limit the wall movements and adequately support the retaining wall. The sub-contractor was fully responsible for devising the construction sequence, including temporary works, and for carrying out the basement construction whilst limiting the movement of the retaining wall within specified limits and limiting the damage to the canal.

It seems the ingress of water was relatively modest during the initial stages of basement construction, whist the movements were within the limit specified, and the seepage was pumped to disposal without difficultly. This effectively confirmed the suitability of the construction method, provided deflections of the retaining wall were small and the water ingress was tolerable and/or managed.

Construction sequence and temporary works design. In original design, the piling contractor allowed only for 2 m depth of water in the canal but no other groundwater during construction. It would have been prudent to have also included the water pressure from the perched water table above the Mudstone in the temporary design stages. A simplified two-dimensional finite element analysis (Borin 2002) of the construction sequence devised by the piling contractor was carried out which predicted a maximum deflection of 6 mm at the top of the retaining wall.

The soil parameters devised following a site investigation for the building and used in the original design by the piling contractor were adopted for all the analysis. This was considered acceptable in the circumstances and because the deflections of the retaining walls on the other three sides of the basement predicted by the piling contractor were in close agreement with those observed on site.

The wall was analysed assuming plane strain conditions: this was considered appropriate because the restraint afforded by the capping beam spanning about 50 m, was assessed using methods in British Standard BS 8110, Part 2: 1985 (BSI 1985) and would not be significant in the mid-length of the retaining wall.

An analysis of the retaining wall with the construction sequence devised by the sub-contractor and temporary works designer predicted a wall deflection of 13 mm. For this sequence, it would be essential to prestress the props to limit the deflection.

Unlike the construction sequence envisaged by the piling contractor, which provided a degree of redundancy, the scheme devised and adopted by the sub-contractor relied on a single level of raking props. The deflection of the wall was significantly increased from the piling contractor's envisaged scheme because the wall cantilevered 6 m above the level of the raking props and the stiffness of the raking props was substantially less than the concrete floor slabs in the upper two basement levels. The load capacity of the temporary works was assessed to British Standard BS 5950, Part 1: 2000 (BSI 2001) and found to be structurally adequate.

Works as constructed. The analysis of the sub-contractor's scheme with the as-constructed prop positions indicated that if the prestress was not applied to the raking props the wall deflection to be expected would be a little over 20 mm with an allowance made for elastic movements in the foundations to the props, and hence would be unacceptable. This is in close agreement with the recorded deflection of the retaining wall of up to 22 mm when the top of the berm was reduced to 10 m below the top of the wall. This analysis supports the view that the lack of prestress in the raking props was probably the principal factor causing the undue movements of the retaining wall prior to the sprayed concrete being applied. Other contributing factors were probably the nonconforming prop spacing and reduced berm size.

The excavations in the top of the berm in front of the retaining wall to form pad foundations further reduced the berm size from that identified in the construction sequence and the wall deflection increased to over 30 mm. It was concluded that the primary cause of the increased deflection were shortcomings in the temporary works including lack of prestress in the props and the further reduced size of berm. At this time, the documents reveal there was a substantial increase in the rate of ingress of water into the basement.

Sprayed concrete was then applied to the piled wall. The increase in the deflection of the retaining wall to 100 mm following the application of the shotcrete between the piles is consistent with the movements being caused by an increase in water pressure acting on the retaining wall.

The uplift of the wall observed at this time, which was probably caused by buoyant effects following the increase in water pressure would cause a lateral movement at the waling beam level of 6 mm and a resultant deflection at the capping beam of about 15 mm. Bearing this effect in mind, a back analysis of the wall without prestress in the raking props predicted a wall deflection of up to about 60 mm. In this analysis no other shortcomings in the temporary works were modelled.

The sub-contractor says there were gaps between the waling beam following clearing out works prior to the sprayed concrete being applied. It is considered that further shortcomings in the temporary works, including factors such as the lack of continuous support between the piles and the waling beam, give a reasonable explanation for the recorded 100 mm deflection of the wall.

There is no evidence to suggest that the design, materials and workmanship in the permanent works contributed to the undue movements of the retaining wall.

11.4 Conclusions

The undue movements of the retaining wall and consequent damage to the canal resulting in increased water ingress into the basement were the result of shortcomings in design and construction of the temporary works to support the retaining wall. Responsibility for the construction sequence including the temporary works rested with the sub-contractor.

The temporary works design engineer ought to have advised the sub-contractor and developer that the application of the shotcrete would lead to a build up of water pressure behind the wall, which would generate further undue movements of the retaining wall. It appears that the temporary works designer failed to consider the effect of the increased water pressure on the deflection of the wall. Consequently, following the sprayed concrete being applied to the retaining wall, significant further damage was caused to the canal.

Overall, the temporary works design engineer's work fell below the standard to be expected of a reasonably competent temporary works design engineer.

The movements of the retaining wall and consequent damage to the canal at this site were completely avoidable. The project should have been set up so that the management of the risk associated with the movements of the retaining wall with the potential to damage the canal fully involved all the stakeholders so that when the construction sequence, including the temporary works, was devised and adopted adequate consideration was given to the movements of the retaining wall. Further, all the stakeholders should have embraced the strategy within the project for the timely implementation of suitable contingency measures if and when the monitoring of the movements of the wall during construction revealed they were required.

Whilst the project proceeded without using the lowest possible risk solution for the form of retaining wall, it ought to have been possible to design and construct the basement using a contiguous bored pile retaining wall whilst minimising the movement and damage to the canal. However, on this project the party responsible for devising the sequence of construction failed to follow the advice given to them in respect of having contingency measures such as additional props to hand on site: in fact when the deflection of the retaining wall exceeded the specified limit no action was taken. Further, no adequate consideration was given to the deflections of the wall during construction, which was the key criterion to minimise damage. It also seems that only the load capacity of the temporary works was considered.

It seems that the sub-contractor essentially self-certified their works as-constructed, so many of the shortcomings in the temporary works were only identified after undue deflections of the retaining wall and significant water ingress had occurred. If the structural engineer and/or the temporary works design engineer had supervised or monitored the works, it is likely the shortcomings could have been corrected and addressed at an early stage before the canal was significantly damaged.

This case study indicates that careful consideration of all aspects of works, engineering principles of equilibrium and compatibility with appropriate material properties, and the use of a relatively simplified back analysis method, can provide an adequate means of assessing the causes of a complex engineering problem.

References

Borin DL (2002) WALLAP anchored and cantilevered retaining wall analysis program. Geosolve, London

British Standards Institution (BSI) (1985) BS 8110, Part 2: structural use of concrete. Code of practice for special circumstances, BSI, London

British Standards Institution (BSI) (2001) BS 5950 Part 1: 2000, structural use of steelwork in building. Code of practice for design, rolled and welded sections. BSI, London

Burland JB, Standing JR, Jardine FM (2001) Building response to tunnelling: case studies from construction of the jubilee line extension. CIRIA and Thomas Telford, London

Chapter 12
Failure Mechanism of Anchored Retaining Wall Due to the Breakage of Anchor Head

Kazuya Itoh, Naotaka Kikkawa, Yasuo Toyosawa, Naoaki Suemasa and Toshiyuki Katada

Abstract In this research, a case history of temporary earth support collapse is first illustrated briefly and the mechanisms of accident occurrences are introduced, with the results showing that the shallow penetration of piles mainly caused the sequences of collapse. In order to understand these failure characteristics and mechanisms, centrifuge model tests using an in-flight excavator were carried out. The failure mechanism of the retaining wall in this labour accident was first demonstrated using centrifuge model tests by Toyosawa et al. Failure mechanism ofanchored retaining wall, 667–672 (1996). In this paper, we added some viewpoints regarding the mechanism of the retaining wall and it was thus clarified that the active and passive earth pressures in the retaining wall increased during excavation and then the anchor head exceeded the capacity with respect to tensile stress. As a result, the retaining wall and ground behind the wall collapsed suddenly.

Keywords Anchored retaining wall · Failure mechanism · Centrifuge modelling · Labour accident

12.1 Introduction

Accidents due to collapse frequently occur during ground excavation. The need to decrease these accidents is currently a major concern not only in Japan, but also in several other countries as well.

K. Itoh (✉)
Department of Urban and Civil Engineering, Tokyo City University
(formerly at NationalInstitute of Occupational Safety and Health), Tokyo, Japan
e-mail: k-ito@s.jniosh.go.jp

N. Kikkawa · Y. Toyosawa
Construction Safety Research Group, National Institute of Occupational
Safety and Health Japan (JNIOSH), Tokyo, Japan

N. Suemasa · T. Katada
Department of Urban and Civil Engineering, Tokyo City University, Tokyo, Japan

V.V.S. Rao and G.L. Sivakumar Babu (eds.), *Forensic Geotechnical Engineering*,
Developments in Geotechnical Engineering, DOI 10.1007/978-81-322-2377-1_12

First, a case history has been described in which a temporary earth support collapsed and the ground behind the support failed. A total soil volume of 200 m^3 with a maximum thickness of 4 m slid into an excavation area and as a result five workers were killed and two others were injured.

Second, in the laboratory, the model ground in front of the anchored retaining wall was excavated using an in-flight excavator in a centrifugal field. Based on the results of two centrifuge model tests, the deformation characteristics, the earth pressure and failure mechanisms are discussed in this paper.

12.2 Case History of the Temporary Earth Support Collapse

12.2.1 Outline of the Construction Work

This accident occurred at the building construction site which had an excavation area width of 17.6 m and length of 30.3 m as shown in Fig. 12.1. Figure 12.2 shows the results of a boring log geological survey.

The main piles were used as a temporary earth support system with the lateral wooden sheeting method. The length and the interval of the main piles were 15 and 1 m, respectively. The diameter of the ground anchors was 117 mm, the length was 20.5 m (effective length 5.0 m), the angle was 45° and the horizontal interval of the anchors was 2.5 m.

12.2.2 Situation Relating to the Accident Occurrence

In the morning that the accident occurred, at around 10:00 a.m., workers noticed that some H-steel piles on the east side had not penetrated below the excavation

Fig. 12.1 The ground plan and location for the boring investigation

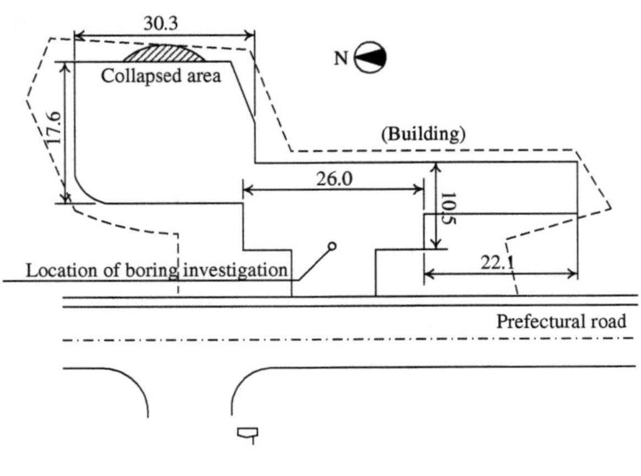

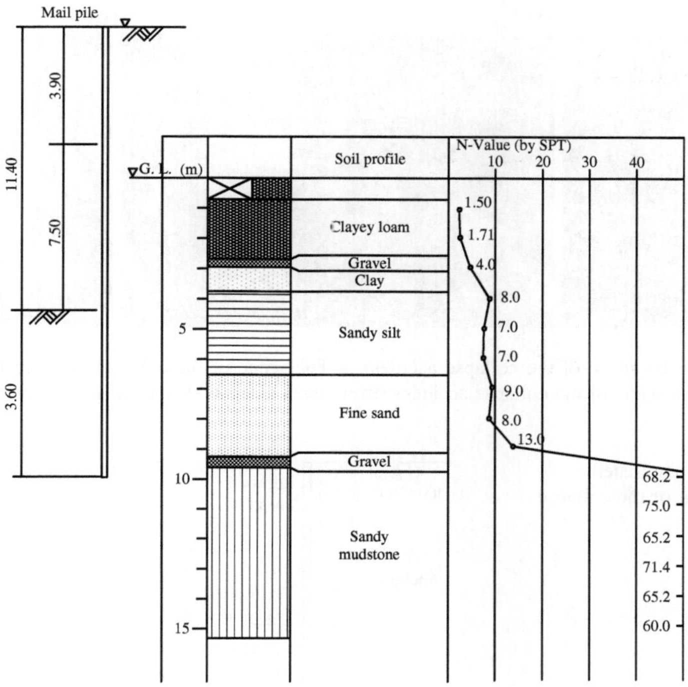

Fig. 12.2 Boring log of the construction site

bottom. The engineer in charge felt there was a danger of a collapse, and he tried to reinforce the main piles with additional remedial piles and blind concrete.

At around 1:15 p.m., with a loud noise, the central 15 m temporary earth support structure on the east side collapsed suddenly with the ground behind the support (Fig. 12.3). The ground slid into the excavation area with a soil volume of approximately 200 m^3 and a maximum thickness of approximately 4 m.

Due to this collapse, eleven H-steel piles toppled. Four workers at the bottom of the excavation site were buried alive and three workers were struck by H-steel piles. As a result, five workers were killed and two others were injured.

12.2.3 Estimation of the Mechanism of the Collapse

The sequence of collapse is shown schematically in Fig. 12.4. The mechanism of collapse is summarised as follows;

(1) Remedial work was being performed on the excavation bottom despite insufficient penetration of the main piles. It is assumed that, due to insufficient penetration, the earth pressure caused the displacement of the main piles.

Fig. 12.3 The scene of the collapse accident. **a** The scene of the collapse accident (from south side). **b** The scene of the collapse accident (from north side). **c** Damaged H-steel for anchor head

Fig. 12.4 Estimated mechanism of the collapse

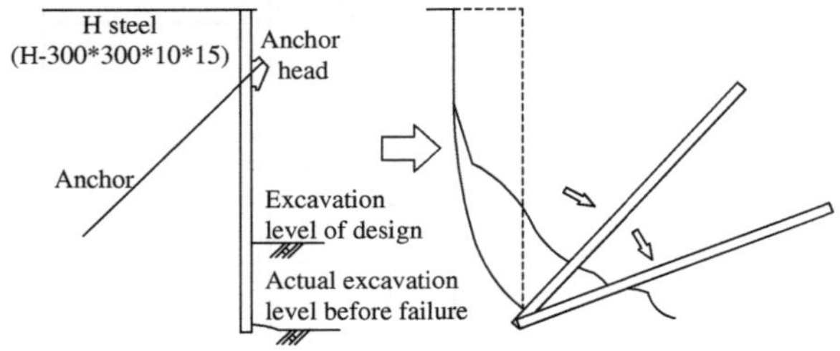

(2) The main piles were supported by the ground anchors installed at an angle of 45° from the horizontal plane. Downward force provided from the anchor's tensile force made the connection between the wale which supported the anchor heads and the anchor heads ineffective.

Finally, the unstable earth support structure led to a huge-volume soil collapse.

(3) The cause of insufficient penetration of the main piles was shortage of excavation at the upper site; due to an engineering error which disregarded the plane of design as shown in Fig. 12.5.

12.2.4 Monitoring

Instrumentation is usually required to monitor the performance of a sheet pile structure during construction. Measurements of movements and pressures furnish valuable information for use in verifying design assumptions. Most importantly, the data may forewarn of a potentially dangerous situation that could affect the stability of the structure. In this construction site, however, only visual inspection was used, such as checking for unusual surface signs.

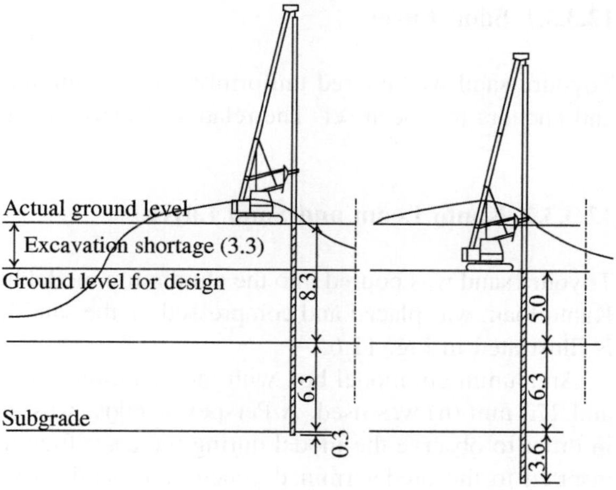

Fig. 12.5 The cause of a shortage of embedded depth of sheet pile walls (unit: *m*)

12.3 Centrifuge Model Tests

12.3.1 JNIOSH NIIS Mark II Centrifuge

Centrifuge model tests were conducted to examine the failure mechanism of this accident, such as the retaining wall supported to ground anchors. All the tests described here were conducted using the JNIOSH NIIS Mark II Centrifuge (Horii et al. 2006).

12.3.2 In-flight Excavator

The in-flight excavator developed by Toyosawa et al. (1998) was used in this study. In this test, for the excavation, this excavator can rake the soil horizontally using a cutting blade. The movement of the in-flight excavator was controlled manually from the centrifuge operation room which is in the upstairs of the centrifuge. The model ground was excavated in steps up to the occurrence of collapse, where the excavated height of each step was about 5 or 6 mm.

12.3.3 Model Preparation

From the boring log of the construction site shown in Fig. 12.2, the excavation site consists mainly of Kanto loam (a kind of volcanic cohesive soil in Japan), sand and silt layers. In this paper, (1) a unit layer of sand and (2) a strata of Kanto loam and sand layer were tested.

12.3.3.1 Sand Layer

Toyoura sand was poured uniformly into an aluminium model box in which wall and anchors had been set. The relative density of sand was about 90 %.

12.3.3.2 Kanto Loam and Sand Layer

Toyoura sand was poured into the aluminium model box to a designated level. Then Kanto loam was placed and compressed on the sand layer. The profile of the model is illustrated in Fig. 12.6.

An aluminium model box with internal dimensions of 100 mm (w), 450 mm (l) and 272 mm (h) was used. A Perspex window was installed in one side of the box in order to observe the model during the centrifuge test. The model sheet pile was inserted to the predetermined penetration depth. The model anchors, which were 1.0 mm diameter wire ropes through Teflon tubes, were installed at 45° and the ends of wires were connected to the load cells. Figure 12.7 shows the model of sheet pile (2 mm thick aluminium) with eight earth pressure sensors and six strain gages.

Fig. 12.6 Cross section of profile of two layers

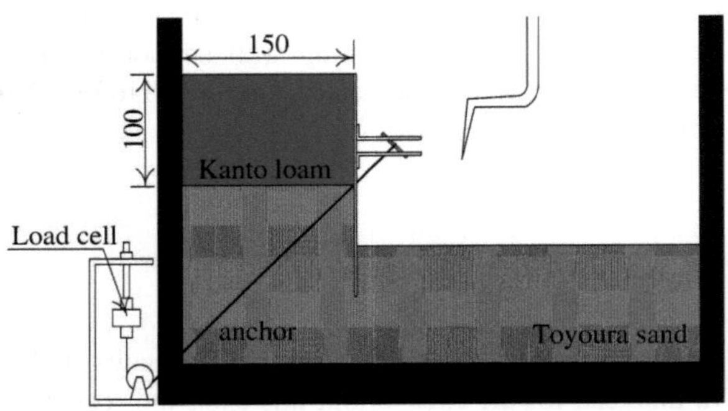

Fig. 12.7 Model of sheet pile wall

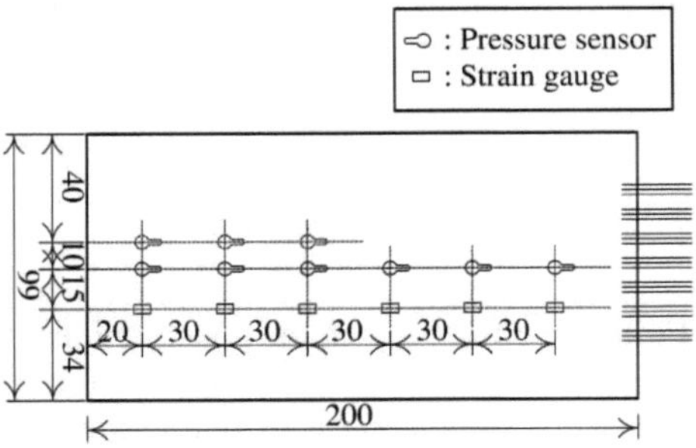

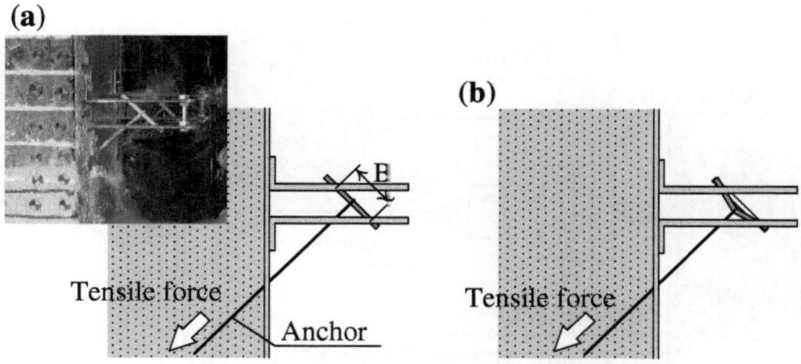

Fig. 12.8 Physical modelling of broken anchor head. **a** Initial condition. **b** Anchor head itself broken

12.3.4 Modelling of Broken Anchor Head

Figure 12.8 gives the experimental system for the broken anchor head used in this study. The mechanism of the broken anchor head is as follows. It is a problem of a simply supported beam of length *B*, and cross section, ϕ, carrying a tensile force. A series of pullout tests is carried out to clarify the brittleness of the anchor head model as shown in Fig. 12.9a. Figure 12.9b shows typical examples of how the beam of length *B*, would change the load-displacement curves for the case of wood (ramin) and cross section, ϕ, of 3 mm. Comparing the data obtained from the beam of length *B*, it can be clearly seen that there is a decreasing maximum load as *B* increases. In this study, the model anchor head used was wood (ramin), *B* of 20 mm, ϕ of 3 mm from tensile force 90 N at failure by Toyosawa (1998).

12.3.5 Test Procedure

Figure 12.10 shows the model setup in the aluminium model box. The sensor's pressure value and bending moment on the model sheet pile were initialized prior to the model setup. Preloading for each anchor was adjusted to 9.8 N before testing. The centrifugal acceleration was then increased to 50 g for each test. Once every sensor's value was stable, the excavation started. To simulate the real accident, the ground in front of the model sheet pile was excavated by the in-flight excavator to a depth of 9.5 m near the end of the sheet pile.

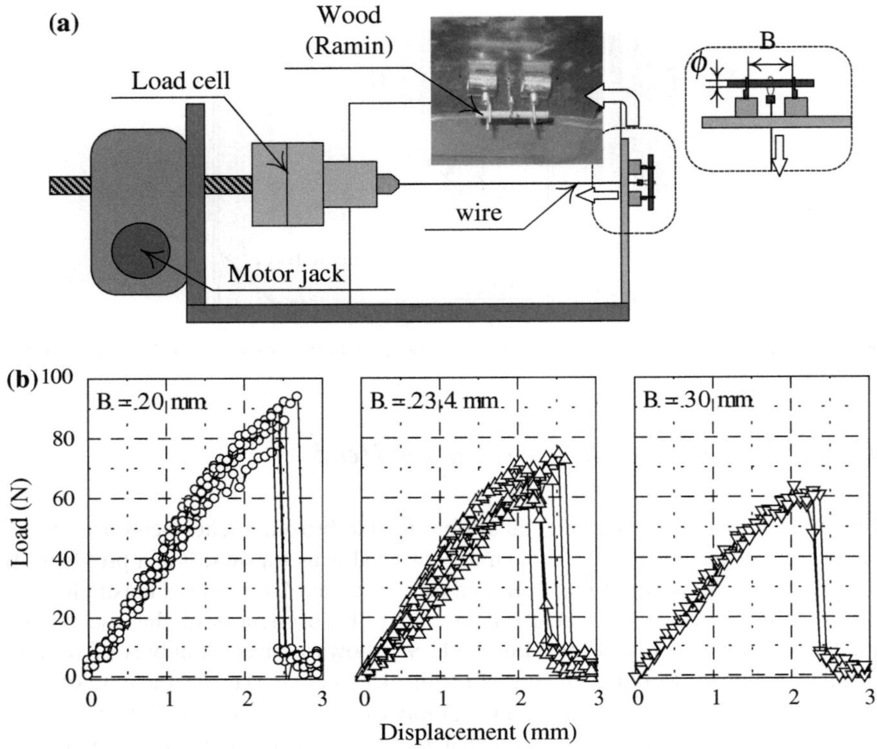

Fig. 12.9 Pullout test. **a** Pullout test apparatus. **b** Load—displacement curves in pullout test

Fig. 12.10 Model setup

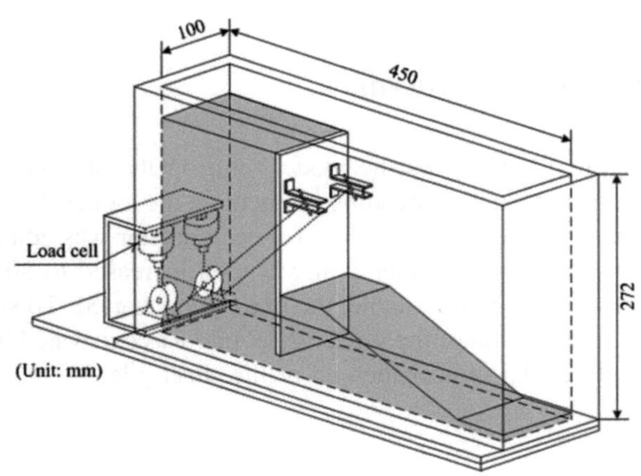

12.4 Test Results and Discussions

12.4.1 Sequence of Failure

Figure 12.11 shows the profile of each model after failure. Figure 12.11a shows the model with sand layer after failure. During the excavation process, the penetration depth of the wall decreased step by step as the excavation proceeded. Failure occurred when the penetration depth of the wall reached about 35 mm and continuous failure with a small displacement emerged with each excavation step. Figure 12.12 shows the ground displacement field behind the model sheet pile

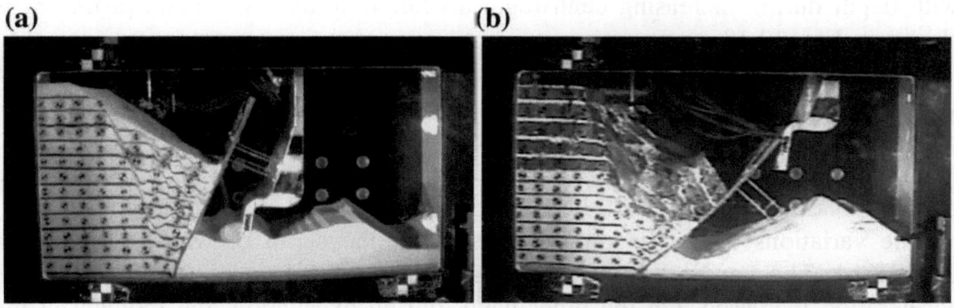

Fig. 12.11 Profile of models after failure. **a** Sand layer. **b** Kanto loam and sand layer

Fig. 12.12 PIV analysis in the case of the sand layer

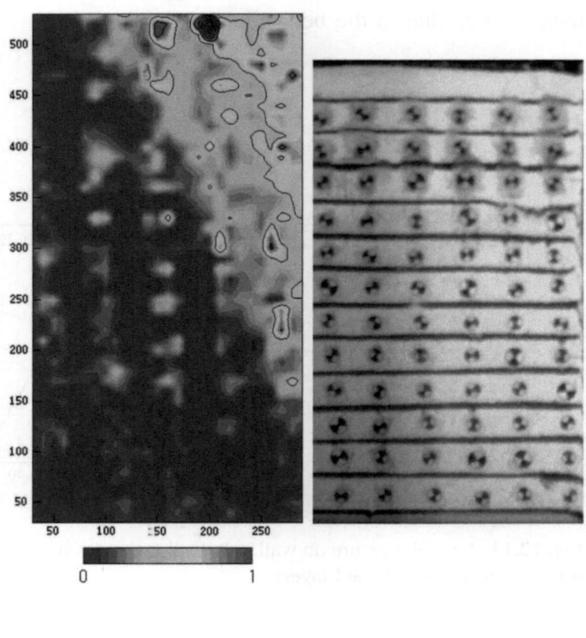

at the final excavation step using PIV (Particle Image Velocimetry, Kikkawa et al. 2006). The failure line angle was about 65°.

The model consisting of Kanto loam and sand layers is shown after collapse in Fig. 12.11b. The failure happened suddenly when the penetration depth reached about 60 mm. The failure line's angle in the sand was about 65°.

12.4.2 Earth Pressure and Bending Moment on Sheet Pile Walls

The measured earth pressure (total lateral stress) on the model sheet pile together with depth during increasing centrifugal acceleration and excavation processes is shown in Fig. 12.13.

The active and passive earth pressures increased during excavation rather than with increasing centrifugal acceleration. For the sand layer model (Fig. 12.13a), both pressures rapidly increased at an excavation depth of 35 mm. For the Kanto loam and sand layer (Fig. 12.13b), the active pressure exceeded 60 kN/m^2 and the passive pressure exceeded 120 kN/m^2 at an excavation depth of 60 mm.

The variations of the bending moment with the depth of wall are shown in Fig. 12.14. The bending moment increased almost uniformly as the centrifugal acceleration increased. On the other hand, during excavation, the moment at the middle of the pile rapidly increased while the moment of the pile at the excavation surface decreased as the excavation depth increased. Therefore, the pile was bent significantly as the upper part of the pile stuck out while the pile at the excavation surface did not move much. This suggested that the anchor head suffered severe tensile force due to the bent pile.

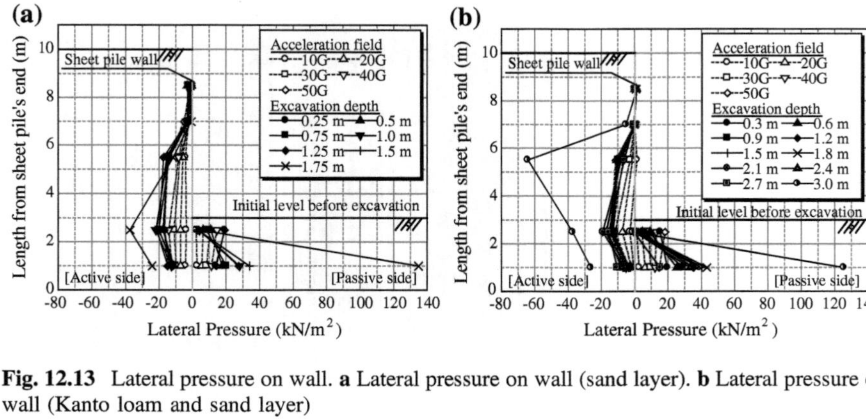

Fig. 12.13 Lateral pressure on wall. **a** Lateral pressure on wall (sand layer). **b** Lateral pressure on wall (Kanto loam and sand layer)

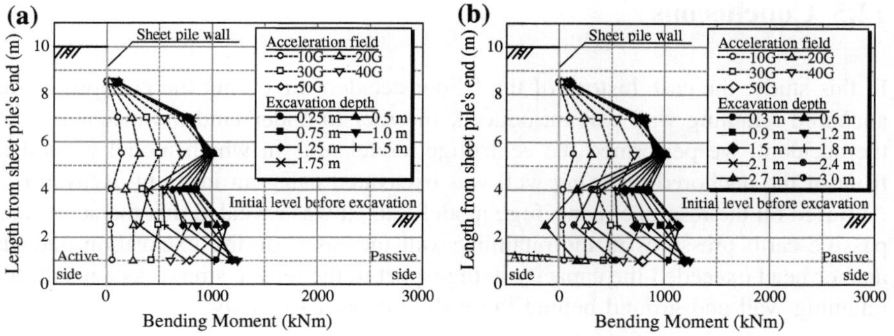

Fig. 12.14 Bending moment at wall. **a** Bending moment at wall (sand layer). **b** Bending moment at wall (Kanto loam and sand layer)

12.4.3 Tensile Force Acting on Anchor

Figure 12.15 shows the measured tensile force acting on the anchor with increasing centrifugal acceleration and excavation steps. The values in this figure were the average of the two anchors which were set in one model.

The tensile force increased with increasing acceleration in each test. During the excavation process, tensile force increased gradually prior to the failure in both cases. It should be noted that those tensile forces were almost the same as the capacity of the anchor head as shown in Fig. 12.9, or had already exceeded the capacity.

Therefore, it is believed that the active and passive earth pressures increased during excavation and then caused the anchor head to break. As a result the retaining wall and the ground behind the wall collapsed suddenly.

Fig. 12.15 Tensile force with increasing centrifugal acceleration field and excavation steps

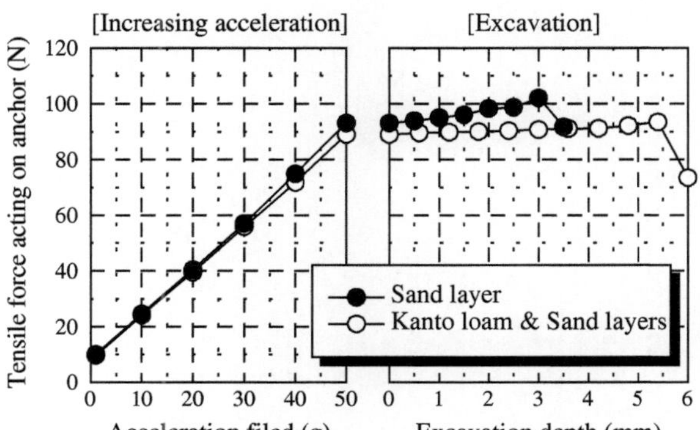

12.5 Conclusions

In this study, the case history of the labour accident involving the collapse of the anchored retaining wall was introduced. In order to understand the mechanism of the accident; we performed the centrifuge model tests in which model ground in front of the anchored retaining wall was excavated using an in-flight excavator.

Based on the results of centrifuge model tests, it was revealed that the active and passive earth pressures in the retaining wall increased during excavation and the anchor head exceeded the capacity with respect to the tensile stress. As a result, the retaining wall and ground behind the wall collapsed suddenly.

References

Horii N, Itoh K, Toyosawa Y, Tamate S (2006) Development of the NIIS Mark-II geotechnical centrifuge. In: Ng, Zhang and Wang (eds) The 6th international conference on physical modelling in geotechnics. Taylor & Francis, London. pp 141–146

Kikkawa N, Nakata Y, Hyodo M, Murata H, Nishio S (2006) Three-dimensional measurement of local strain using digital stereo photogrammetry in the triaxial test. In Hyodo, Murata and Nakata (eds.). Taylor & Francis, London. pp 61–67

Toyosawa Y, Horii N, Tamate S, Suemasa N (1996) Failure characteristics of sheet pile wall in centrifuge tests. In: Proceedings of geotechnical aspects of underground construction in soft ground, Balkema, pp 225–230

Toyosawa Y, Horii N, Tamate S, Suemasa N, Katada T (1998) Failure mechanism of anchored retaining wall. In: Proceedings of the International Conference on centrifuge 98 Tokyo, Rotterdam, Balkema, pp 667–672

Chapter 13
Backwards Problem in Geotechnical Earthquake Engineering

S. Iai

Abstract The author's initial thought on the backwards problem in geotechnical earthquake engineering is presented through an example of damage to caisson quay wall during earthquakes. Both a simplified and detailed dynamic analyses are presented. It is essential to confirm, at the outset, that the backwards problem is well defined. There should be sufficient geotechnical data and earthquake data to match the analysis procedure used for solving the backwards problem. Ill-defined backwards problem, either due to lack of required geotechnical or earthquake motion data, should be corrected before the trial solution of the backwards problem.

Keywords Backwards problems · Failure mode · Simplified analysis · Dynamic analyses

13.1 Introduction

Backwards problem in geotechnical earthquake engineering is to identify the process to the residual states (often in terms of excessive deformation rather than complete collapse) of geotechnical structures as the most likely scenario that is explained through geotechnical and earthquake data. Backwards problem usually consists of three steps. The first step is to define the problem in terms of failure mode and extent of deformation. If the problem involves the function of the geotechnical structure, then the degree of damage to the function (indirect damage) should also be defined in terms of performance objectives. The second step is to assume possible processes of scenarios from the initial state (before the earthquake) to the residual state (after the earthquake). The final step is to select the most likely scenario(s)-based available data from geotechnical investigations and earthquake motion recording.

S. Iai (✉)
Disaster Prevention Research Institute, Kyoto University, Kyoto, Japan
e-mail: iai@geotech.dpri.kyoto-u.ac.jp

V.V.S. Rao and G.L. Sivakumar Babu (eds.), *Forensic Geotechnical Engineering*,
Developments in Geotechnical Engineering, DOI 10.1007/978-81-322-2377-1_13

This paper presents the author's initial thought on the backwards problem in geotechnical earthquake engineering using the example of a caisson-type quay wall, a relatively simple soil structure system.

13.2 Simplified Dynamics Analysis

A gravity quay wall is made of a caisson or other rigid wall put on a seabed, and maintains its stability by a friction at the bottom of the wall. Typical failure modes during earthquakes involve a seaward displacement, settlement and a tilt. For a quay wall constructed on a firm foundation, an increase in the earth pressure from the backfill plus the effect of an inertia force on the body of the wall result in the seaward movement of the wall as shown in Fig. 13.1a. If a width to height ratio of the wall is small, a tilt may also be involved. The past case histories of gravity quay walls often belong to this category (e.g., Iai et al. 1998). When the subsoil below the gravity wall is loose and excess pore water pressure is increased in the subsoil; however, the movement of the wall is presumed to be associated with a significant deformation in the foundation soil, resulting in a large seaward movement involving a tilt and settlement as shown in Fig. 13.1b.

The former mode of failure, shown in Fig. 13.1a, has been conventionally analyzed by evaluating the seismic active earth pressures using Mononobe–Okabe equation (Mononobe 1924; Okabe 1924). This equation is derived by modifying Coulomb's classical earth pressure theory to account for inertia forces. In the uniform field of horizontal and (downward) vertical accelerations, $k_h g$ and $k_v g$, the

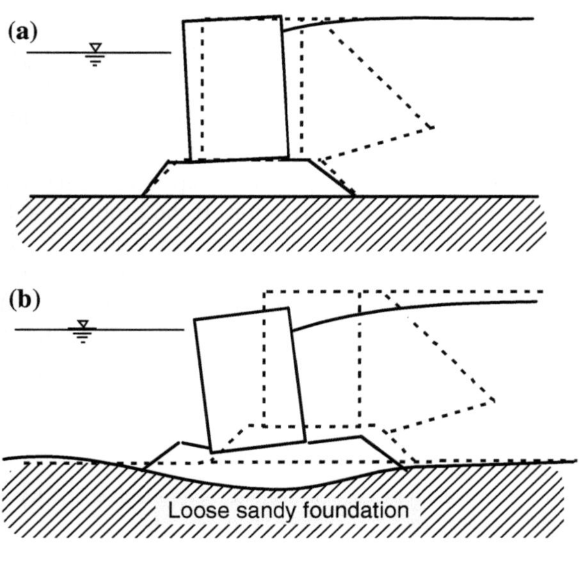

Fig. 13.1 Deformation/ failure modes of gravity quay wall. **a** On firm foundation. **b** On loose sandy foundation

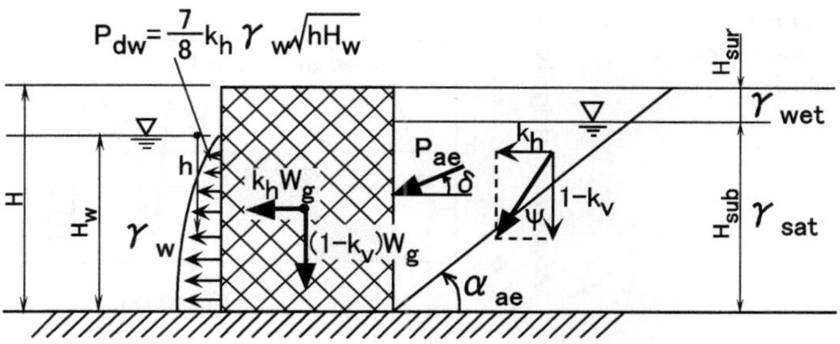

Fig. 13.2 Active earth pressure and other actions on a gravity caisson during earthquakes

body force vector, originally pointing downward due to gravity, is rotated by the seismic inertia angle, ψ, defined by (see Fig. 13.2).

$$\psi = \tan^{-1}\left[\frac{k_h}{1-k_v}\right] \tag{13.1}$$

The Mononobe–Okabe equation is, thus, obtained by rotating the geometry of Coulomb's classical solution through the seismic inertia angle, ψ, and scaling the magnitude of the body force to fit the resultant of the gravity and the inertia forces.

Simple, straightforward methods have been developed for evaluating the permanent displacements of a sliding block (Newmark 1965). In this analysis, first, the stability of the wall and the backfill are evaluated using the lateral earth pressure theory based on Mononobe–Okabe's equations. The threshold acceleration is determined by the value resulting in a factor of safety of unity for sliding of the wall-backfill system. For example, the threshold acceleration, at, for a vertical retaining wall is given by the expression (Richards and Elms 1979).

$$a_t = \left(\mu_b - \frac{P_{ae}\cos\delta - \mu_b \sin\delta}{W_g}\right)g, \tag{13.2}$$

where μ_b is the coefficient of interface friction between the wall and the foundation rubble or soil, P_{ae} is the active earth thrust computed using the Mononobe–Okabe method, δ is the wall-backfill interface friction angle, W_g is the weight of the wall per unit width, and g is the acceleration of gravity. It should be noted that a_t must be known in order to calculate P_{ae}, therefore an iterative procedure is necessary.

Once the threshold acceleration has been determined, then a set of acceleration time histories is selected for sliding block analysis. Displacements calculated through sliding block analysis are sensitive to the characteristics of acceleration tie history used in the analysis. The acceleration time histories should be representative of the seismic condition in both duration and frequency content. When the ground motion acceleration exceeds the threshold acceleration, a_t, the wall-backfill system begins to move by translation along the base of the wall and the failure plane through the backfill. By double integrating the area of the acceleration time history

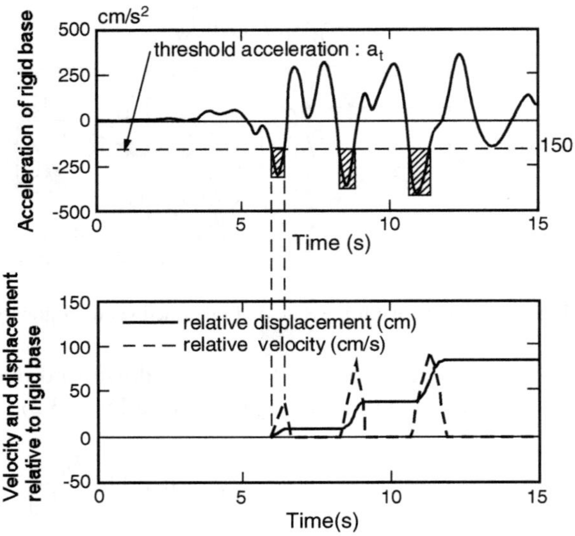

Fig. 13.3 Example of computing displacement in the sliding block analysis

that exceeds a_t, and continuing the time integration until the sliding stops, the displacement of the wall relative to the firm base below the failure plane can be determined as shown in Fig. 13.3. This computation can be easily performed using common spreadsheet routines or a simple computer code.

13.3 Detailed Dynamic Analysis

The latter mode of failure, shown in Fig. 13.1b, had not received wide attention until the Great Hanshin earthquake of 1995. Many of the caisson walls in Kobe port were constructed on a loose saturated backfill foundation of decomposed granite, which was used for replacing the soft clayey deposit in Kobe Port to attain the required bearing capacity of foundation. Shaken with a strong earthquake motion having the peak accelerations of 0.54 and 0.45 g in the horizontal and vertical directions, these caisson walls displaced toward the sea about 5 m maximum, 3 m on average, settled 1–2 m, and tilted about 4° toward the sea. Figure 13.4 shows a typical example of the cross section and the deformation after the earthquake (Inagaki et al. 1996). Although the sliding mechanism could explain the large horizontal displacement of the caisson walls, this mechanism did not explain the large settlement and tilt of the caissons. Reduction in the stiffness of foundation soils due to an excess pore water pressure increase, then, was speculated as a main cause of the damage to the caisson walls at Kobe Port.

This speculation was confirmed by a series of effective stress analyses based on the strain space multiple mechanism model (Iai et al. 1998). The model parameters were evaluated based on the in situ velocity logging, the blow counts of Standard

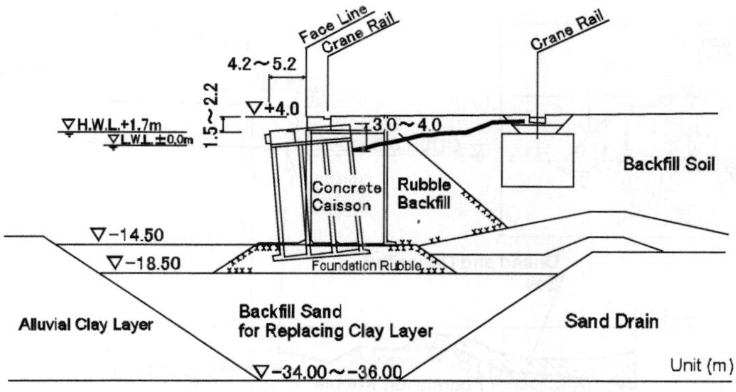

Fig. 13.4 Cross section of gravity quay wall at Kobe port and deformation/failure during 1995 Great Hanshin earthquake, Japan

Penetration Tests (SPT N-values) and the results of the cyclic triaxial tests. The specimens used for the cyclic triaxial tests were undisturbed samples 60 cm long with a diameter of 30 cm obtained by in situ freezing technique. The input earthquake motions were those at the Port Island site successfully recorded by the Kobe City Government. The analysis domain used for the finite element analysis covered a cross-sectional area of about 220 m by 40 m in the horizontal and vertical directions.

The effective stress analysis resulted in the residual deformation shown in Fig. 13.5. As shown in this figure, the mode of deformation of the caisson wall was to tilt into and push out the foundation soil beneath the caisson. This was consistent with the observed deformation mode of the rubble foundation shown in Fig. 13.6, which was investigated by divers. Note that a significant deformation was induced in the foundation soil beneath the caisson wall. The order of displacements of the wall was also comparable to that observed and shown in Fig. 13.4.

In order to evaluate the overall effect of geotechnical conditions on the displacements of a gravity wall, the following three analyses were performed as a

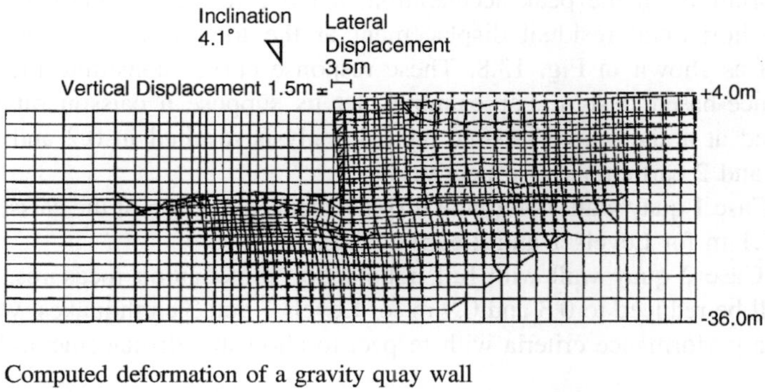

Fig. 13.5 Computed deformation of a gravity quay wall

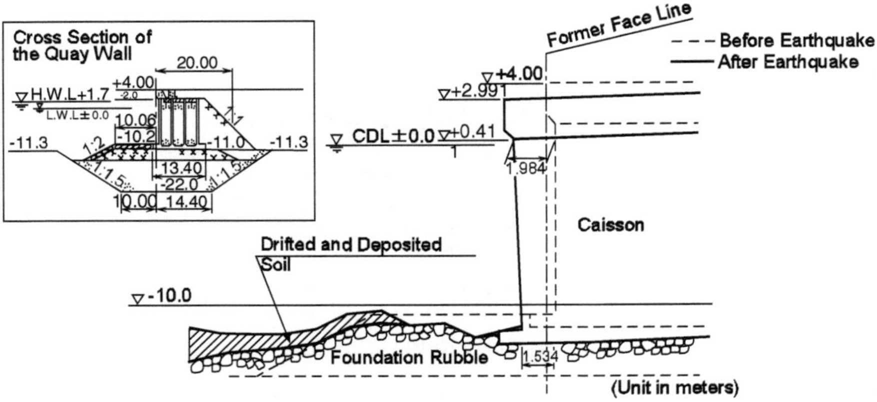

Fig. 13.6 Deformation of rubble foundation of a quay wall investigated by divers

parameter study. The parameter study included a virtual soil model, to be called nonliquefiable soil, which has the same properties as those used in the aforementioned analysis but without the effect of dilatancy. To distinguish the cases in the parameter study, the case which dealt with the actual quay wall during the earthquake described earlier is designated as Case-1.

Cases-2 through 4 are defined depending on the extent of the nonliquefiable soil relative to the caisson wall as shown in Fig. 13.7.

The major results of the analysis are summarized in Table 13.1 including those of Case-1. These results indicate that the deformation of the gravity wall may be reduced up to about one half of that at the earthquake if the excess pore water pressure increase were prevented in the subsoil as Case-2. In particular, the effect of the pore water pressure increases in the foundation soil beneath the caisson wall is about twice as that of the backfill as understood from the comparison of Cases-3 and 4. These results were partly confirmed by the seismic performance of the quay walls at Port Island (phase II), where a caisson wall had been put on the foundation improved with a sand compaction pile technique at the time of the earthquake (Iai et al. 1998).

For variations in the peak acceleration of the earthquake motion input at the base, the horizontal residual displacement at the top of the caisson wall was computed as shown in Fig. 13.8. These response curves constitute the basis for performance-based design. For example, let us suppose a caisson quay wall is constructed at a site where seismic hazard analysis resulted in 0.2 and 0.3 g for Levels 1 and 2 earthquakes, respectively.

For a Case 1 quay wall, horizontal displacements at the top of the caisson will be 1.3 and 2.1 m for Levels 1 and 2 earthquakes.

For a Case 2 quay wall with full liquefaction remediation measures, displacements will be reduced to 0.3 and 0.9 m for Levels 1 and 2 earthquakes, which may satisfy the performance criteria with respect to allowable displacement. The effect

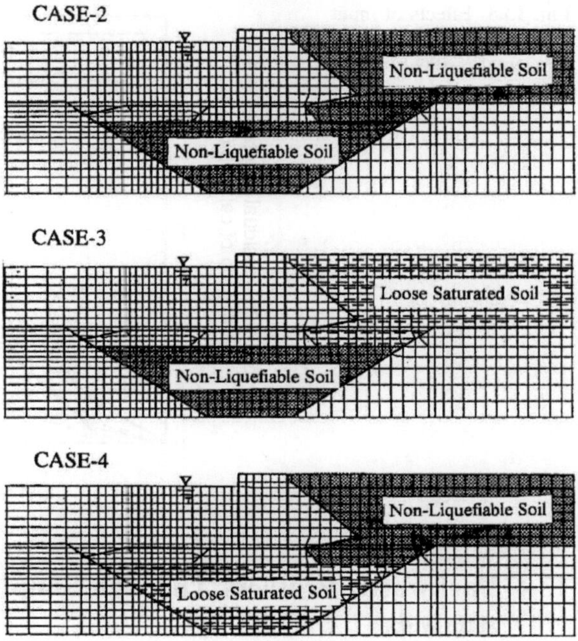

Fig. 13.7 Conditions assumed for parametric study, Cases 2 through 4

Table 13.1 Major results of parametric study for gravity quay wall

Case	Residual displacements of caisson		
	Horizontal (m)	Vertical (m)	Tilt (°)
Case 1	3 5	1.5	4.1
Case 2	1 6	0.6	2.4
Case 3	2.1	0.7	3.1
Case 4	2.5	1.1	2.2

of the vertical component of input acceleration time histories was studied by varying the peak acceleration of the vertical component of acceleration ranging from zero to 0.40 g (Ichii et al. 1997). The peak acceleration of the horizontal component was unchanged from the original value of 0.54 g. The phase difference between the vertical and the horizontal components were also varied. This parametric study suggested that the effect of the vertical component is to increase the residual displacement less than about 10 %.

As discussed in this section, geotechnical conditions significantly affect the performance of a gravity quay wall. They govern the mode and extent of the deformation/failure. In particular, the failure mode associated with a loose foundation soil should be more thoroughly evaluated in the design practice. Appropriate characterization of these conditions and a suitable seismic analysis will lead to a reasonable seismic design of a gravity quay wall.

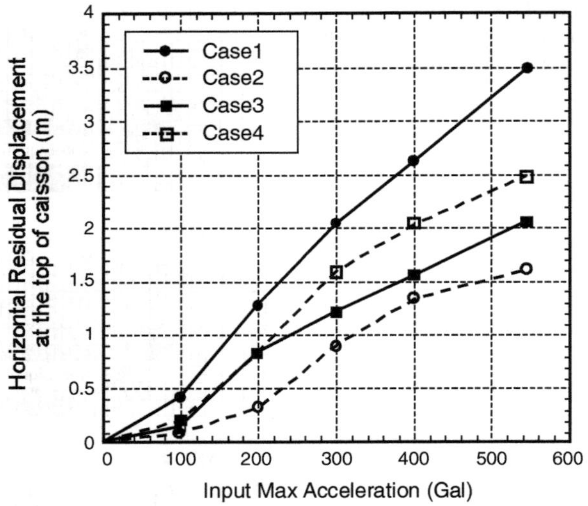

Fig. 13.8 Effects of input acceleration levels on horizontal residual displacement

13.4 Backwards Problem Through Simplified Dynamic Analysis

If a caisson quay wall is constructed on firm foundation and if the failure mode of the caisson quay wall is confirmed in translational sliding mode, then the backwards problem becomes relatively simple. The crucial steps in this simple backwards problem can be reduced to evaluate the threshold acceleration for sliding defined by Eq. (13.2) and to evaluate the acceleration time history during the earthquake. As shown in Eqs. (13.1) and (13.2), geotechnical parameters that should be evaluated are internal friction angle of backfill, the coefficient of interface friction between the wall and the foundation rubble or soil, and the wall-backfill interface friction angle.

In practice, the acceleration time history during the earthquake is often missing. In this unfortunate case, a large uncertainty will be left in the backwards problem even in this simplest backwards problem.

13.5 Backwards Problem Through Detailed Dynamic Analysis

If a caisson quay wall is constructed on loose sandy foundation, then the backwards problem needs detailed dynamic analysis by taking account geotechnical conditions of backfill and foundation soil. The crucial steps in this backwards problem follow the procedure in solving the typical boundary value problem: (1) initial condition, (2) boundary condition, (3) input earthquake motions, (4) appropriate constitutive modeling of soil, structure, and interface element, and (5) appropriate numerical solution procedure through the use of finite element technique.

Initial condition for a soil structure system in geotechnical earthquake engineering is often computed through a static analysis using the same mesh and constitutive equation for dynamic analysis but by applying gravity acceleration. Boundary conditions should take into account the transmission of the seismic wave in and out from the analysis domain for finite element analysis. The constitutive equation is utmost importance and the applicability of this equation for various loading conditions should be preconfirmed.

If all the above procedures are satisfied, then the essential problem in backwards problem becomes the procedure to evaluate the geotechnical conditions. In the example presented earlier for the caisson wall in Kobe Port, the model parameters were evaluated based on the in situ velocity logging, the blow counts of Standard Penetration Tests (SPT N-values), and the results of the cyclic triaxial tests. The specimens used for the cyclic triaxial tests were undisturbed samples 60 cm long with a diameter of 30 cm obtained by in situ freezing technique. These sets of procedures actually performed are considered the most ideal case. As a result, the primary cause of the damage to the caisson quay wall was identified as shown earlier. In particular, the summary of the parameter study shown in Table 13.1 indicate that inertia effect is about 50 % (=1.6 m/3.5 m) (Case-2), and the effect of the liquefaction in foundation is about 30 % (=(2.5–1.6 m)/3.5 m) (Case-4) and the rest is due to the effect of the liquefaction at backfill (Case-3). The liquefaction of backfill and foundation soil (Case-1 and 3) also takes account of the primary cause of tilting.

In practice, only crude geotechnical investigation may be possible. In this practical case, there will be large uncertainty in geotechnical conditions. Availability of the earthquake ground motion is another matter of question in practice as discussed earlier. It might be possible to arrive at the correct answer by putting the two wrong answers together, one with wrong geotechnical condition, the other with wrong earthquake motion data. In this case, the backwards problem is ill-defined. The backwards problem has to be correctly defined at first, then the efforts to solve this problem will have a chance to give us a correct answer and contribute to enhancing our engineering knowledge in geotechnical earthquake engineering.

13.6 Conclusions

The author's initial thought on the backwards problem in geotechnical earthquake engineering is presented through an example of caisson-type quay walls during earthquakes. Both the simplified and detailed dynamic analyses are presented.

Backwards problem in geotechnical earthquake engineering is to identify the process to the residual states (often in terms of excessive deformation rather than complete collapse) of geotechnical structures as the most likely scenario that is explained through geotechnical and earthquake data. Backwards problem usually consists of three steps. The first step is to define the problem in terms of failure mode and extent of deformation. If the problem involves the function of the geotechnical structure, then the degree of damage to the function (indirect damage) should also be defined in terms of performance objectives. The second step is to assume possible processes of scenarios from the initial state (before the earthquake) to the residual state (after the earthquake). The final step is to select the most likely scenario(s)-based available data from geotechnical investigations and earthquake motion recording.

It is essential to confirm, at the outset, that the backwards problem is well defined. Ill-defined backwards problem, either due to lack of required geotechnical or earthquake motion data, should be corrected before the efforts are put into the solution of the backwards problem.

References

Iai S, Ichii K, Liu H, Morita T (1998) Effective stress analyses of port structures. In: Soils and foundations, special issue on geotechnical aspects of the January 17 1995 Hyogoken-Nambu earthquake, no 2, pp 97–114

Ichii K, Iai S, Morita T (1997) Effective stress analyses on the performance of caisson type quay walls during 1995 Hyogoken-Nambu earthquake. Report of Port and Harbour Research Institute 36(2):41–86 (in Japanese)

Inagaki H, Iai S, Sugano T, Yamazaki H, Inatomi T (1996) Performance of caisson type quay walls at Kobe port. In: Soils and foundations, special issue on geotechnical aspects of the January 17 1995 Hyogoken-Nambu earthquake, pp 119–136

Mononobe N (1924) Considerations on vertical earthquake motion and relevant vibration problems. J JSCE 10(5):1063–1094 (in Japanese)

Newmark NM (1965) Effects of earthquake on dams and embankments, (5th Rankine lecture). Geotechnique 15(2):139–160

Okabe N (1924) General theory on earth pressure and seismic stability of retaining wall and dam. JJSCE 10(6):1277–1323

Richards RJ, Elms D (1979) Seismic behavior of gravity retaining walls. J Geotech Eng ASCE 105 (GT4):449–464

Chapter 14
Failure Analysis of Underground Construction—Lessons Learned from Taiwan

B. Hsiung and T. Sakai

Abstract Underground construction is widely adopted in urban area but sometimes the construction might cause severe failure. In this paper, successful interpretation of instrumental data is addressed and several examples are demonstrated since geotechnical monitoring is thought to be an effective tool to avoid failure. Further, the failure may also be caused by misinterpretation of responsibilities of parties relating underground construction, so responsibilities distribution in site investigation, design, construction, and geotechnical monitoring are discussed. It is recommended that adequate data of site investigation shall be provided by the client and shall not leave all responsibilities to the contractor. Well-experienced experts shall be invited to evaluate the quantity and quality of data of site investigation before the project goes to tender.

Keywords Site investigation · Instrumentation · Monitoring · Data analysis

14.1 Introduction

Underground construction is widely adopted for the need of underground space due to fast development of urban area in Taiwan but unfortunately disastrous failure sometimes might also be induced. In this paper, successful interpretation of instrument data for underground construction is addressed and several examples are demonstrated based on underground metro in both Taipei and Kaohsiung, Taiwan, since geotechnical monitoring is commonly thought to be an effective tool to avoid failure. Further, the failure may be caused by misinterpretation of responsibilities of

---

B. Hsiung (✉)
Department of Civil Engineering, National Kaohsiung University of Applied Sciences, Kaohsiung, Taiwan
e-mail: benson@cc.kuas.edu.tw

T. Sakai
Maeda Corporation, Taipei, Taiwan

V.V.S. Rao and G.L. Sivakumar Babu (eds.), *Forensic Geotechnical Engineering*, Developments in Geotechnical Engineering, DOI 10.1007/978-81-322-2377-1_14

parties relating underground construction so responsibilities sharing among the clients, consultants, and contractors in site investigation, design, construction, and geotechnical monitoring are discussed.

14.2 Underground Works in Taipei and Kaohsiung

As underground metros in two major cities in Taiwan, Kaohsiung and Taipei were adopted as the background for this paper and a general view of geology and network in both cities is described first.

Taipei is located in northern Taiwan and it is the center of the political and economic activities of the country. The city is mainly located in Taipei Basin which formed by river deposits from three major rivers, Tahan Creek, Hsintien Creek, and Keelung River. Three geological zones surround the Taipei Basin. The Tantu volcano group is to the north, Linkuo Tableland is to the west, and the Tertiary sedimentary rocks to the southeast. The total area of the 243 km^2 Taipei basin has an altitude less than 20 m above the mean sea level. The average elevation of the Taipei Basin is about 5 m above mean sea level. As mentioned above, the basin was filled with river deposits. The base of the Taipei basin mainly consists of sedimentary rock with minor amounts of volcanic rocks of the Tatun volcanic group in the north region of the basin. In cores obtained from a 260 m deep exploratory hole near the city center of Taipei, Tertiary rock was found at 213 m below ground level (Wang Lee and Lin 1987). Hence, the depth of rock bed in the center of the Taipei Basin is estimated to be around 250 m.

The sedimentary material above the rock may be divided into Hsinchuang, Chingmei, and Sungshan Formations. Among them, the Hsinchuang Formation is only found in the western part of the Taipei Basin and includes the alluvial deposits from Tahan Creek. The Chingmei Formation is the alluvial deposit from Hsintein Creek, and consists of 50–140 m thickness of gravel. The Sungshan Formation overlies the Chingmei Formation and it was formed of sands and clay deposited by Tahan Creek, Hsintein Creek, and Keelung River.

Due to a need of the city, the mass rapid transit system in Taipei (TRTS) was initiated at the end of 80s and the network is still expanding. Most of construction works for TRTS were delivered underground. For underground stations and crossover, the cut-and-cover method using reinforcement concrete diaphragm wall as retaining structures with internal props was adopted and shield machines were selected for bored tunnels.

Kaohsiung is located in southern Taiwan and it is the second largest city on the island. Figure 14.1 represents the geology of Kaohsiung City. The city is situated at the mouth of three rivers, Dien-Pao River in the north, Love River in the middle, and Chien-Jen River in the south, and as a result the ground conditions in Kaohsiung city are mainly sandy and silty with clay, as depicted in Fig. 14.1.

The network of Kaohsiung mass rapid transit system (KRTS) was planned at late 90s and fully commenced to construct in 2002. KRTS now has two lines and most

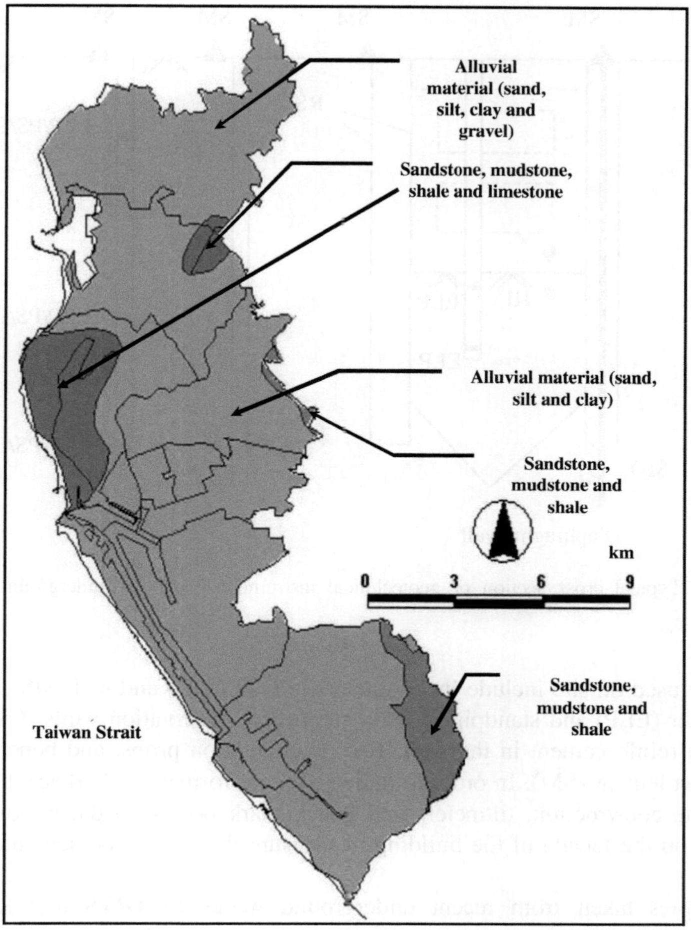

Fig. 14.1 Geology of Kaohsiung

of them were constructed underground. Similar to TRTS, the cut-and-cover method was adopted for stations and crossover and shield machines were chosen for bored tunnels.

14.3 Interpretation of Instrument Data

Instruments aim to provide useful information to prevent geotechnical failure in advance. Figure 14.2 presents a typical cross section of geotechnical instruments used for underground works in Taiwan. As presented in Fig. 14.2, instruments

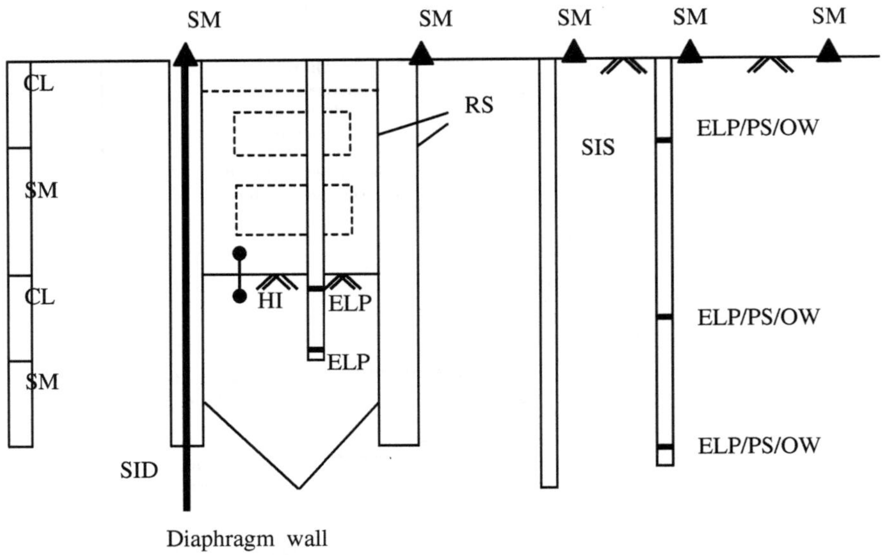

Fig. 14.2 Typical cross section of geotechnical instruments used for underground works in Taiwan

generally used on site include inclinometers in wall (SID) and soil (SIS), electrical piezometer (ELP) and standpipe piezometers (PS), observation wells (OW), strain gauge on reinforcement in the wall (RS), load cells on props, and benchmark for surface settlement (SM). In order to indicate the performance of adjacent buildings during the construction, tiltmeters and bench mark points on the buildings were installed on the façade of the building to measure the tilting and settlement of the building.

Examples taken from recent underground works in TRTS and KRTS are demonstrated in this paper to express how instrument data are interpreted. In addition, some special measurements taken on site are also reported.

Wall deflections were measured at an excavation located in eastern Taipei. The maximum excavation depth of the site is 22.85 m and it was retained by a 43 m deep and 1.2 m thick diaphragm wall. Eight-level internal bracing were adopted to provide additional horizontal supports. It is also noted that depth of bedrock on site was various, from 28 to 40 m below surface level. It was found that a large wall displacement (up to 115–135 mm) was measured and Fig. 14.3 presents a comparatively poor quality of wall constructed on the site. It was constructed using hydraulic grab-type machine and it was reported by the contractor that appearance of rock affects the quality of the wall. Observations here did reflect the quality of wall and this might not be pointed out during the design stage.

Settlements of some buildings near an excavation of TRTS also located in eastern Taipei were observed. It was found that all activities related to excavations were completed but these buildings continued to settle and the reason is unknown.

Fig. 14.3 Quality of diaphragm wall

Change of piezometric levels was further explored and it was found that piezometric levels in clay still gradually raise up but not reached the level before commence of the excavation. It was thus recommended that additional effective ground stress due to change of piezometric level should be the reason for additional buildings settlement and these buildings may not remain stable unless piezometric levels fully recovers.

Figure 14.4 presents the settlement of the buildings next to excavation of O6 Station in KRTS. Similar to the example taken from Taipei, activities of excavation were completed but the building continued to settle. As indicated in Fig. 14.5, it was observed that pore pressure of ground continued to decrease, though the excavation was completed long time ago and this was thus thought to be the main reason for generation of additional settlements.

A large-scale collapse due to construction of cross passage between two running tunnels in Kaohsiung Metro occurred in 2004. An underground pass exactly above the metro tunnel was also damaged and has to be rebuilt. It was confirmed that an unforeseen ground characteristic of silty sand in Kaohsiung in relation to soften of soil strength due to disturbance from the excavation is the main reason for the accident.

Some emergency rescue mitigation measures were taken immediately after the accident but reconstruction of the whole project was started approximately 6 months after the accident.

As shown in Fig. 14.6, the central section of the tunnel was constructed using cut-and-cover method and 1.5 m thick, 60 m deep reinforcement diaphragm wall were installed first as retaining structure. The excavation was then conducted by 11 stages. The maximum excavation depth is 30.2 m. After excavation reached final excavation level, precast reinforcement concrete segments were erected on site to build the tunnel as well as cross passage between two running tunnels. Backfill was delivered using the controlled low-strength material (CLSM) from final excavation level to bottom of base slab of underground pass to fill space outside tunnels and reconstruction of underground pass was carried out segment by segment.

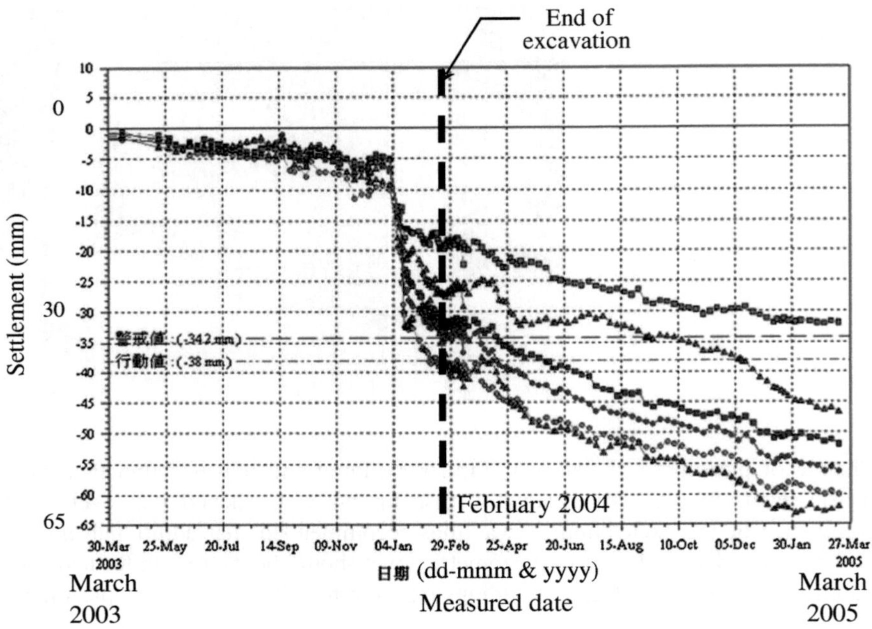

Fig. 14.4 Settlements of the buildings

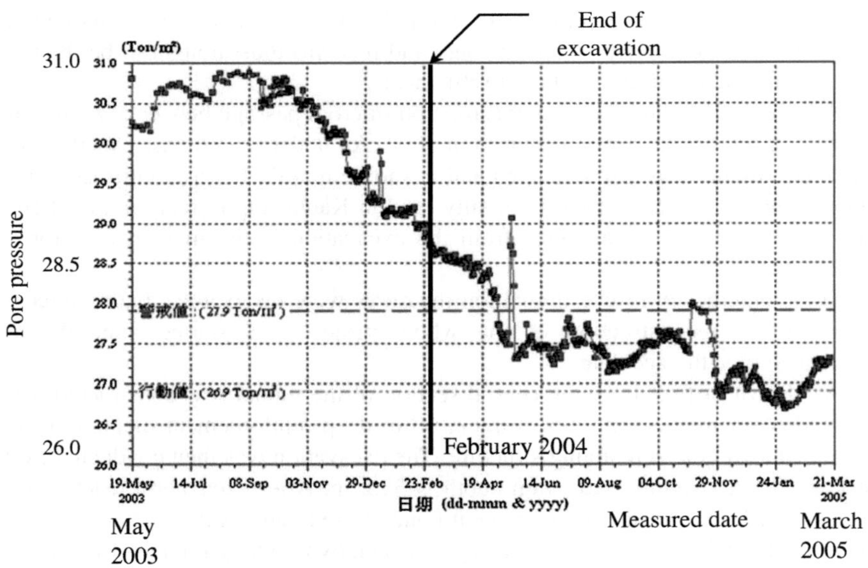

Fig. 14.5 Change of pore pressure

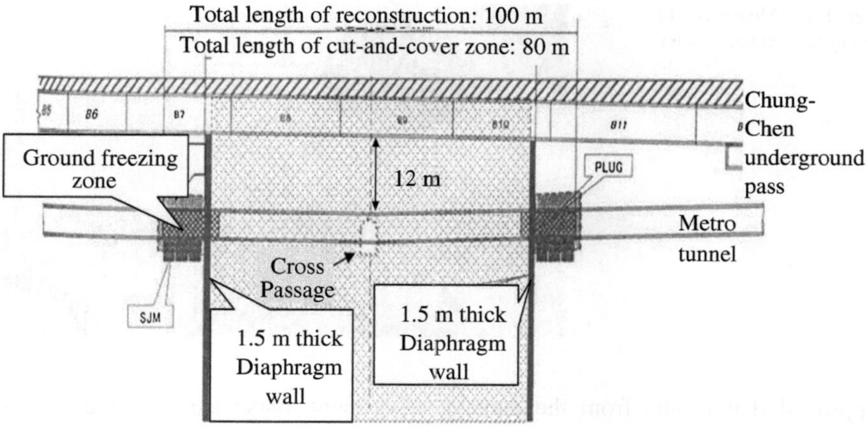

Fig. 14.6 Cross section of excavation in longitudinal direction

Also depicted in Fig. 14.6, 36 tunnel segments were also affected by accident and have to be replaced but all of them were located outside the cut-and-cover zone, ground freezing method was adopted to freeze the ground in order to stop flow of groundwater and then these segments were replaced.

Thermometers were installed on site in order to confirm the ground freezing was fully completed. Most of thermometers were installed up to 12.0–20.8 m below surface level but some of them were installed even deeper, to top of tunnel or into inside the tunnel though a punching hole on tunnel segment. There were 40 thermometers in total and interval between two thermometers varies between 1.8 and 4.0 m. Such measurement did provide useful indication during stages of ground freezing and replacing tunnel segment.

At last, the other example also in Kaohsiung Metro was raise here to demonstrate the effectiveness of instrument. A 37-year-old building stood 1–4 m from a 20.1 m deep excavation and the excavation was retained by 0.8 m thick and 42.0 m deep diaphragm walls. The horizontal props using H-type steel were adopted to stiffen the excavation. Except monitoring of ground performances caused by the deep excavation, structure performance was measured and taken as base of damage assessment. Boscardin and Cording (1989) recommended that the damage to the structure is closely connected with horizontal strain and angular distortion and the device-like horizontal tape extensometer, as shown in Fig. 14.7 was used to measure the distance between two nails on façade of the building and expansion or shortening in comparison with previous measurement for the distance between two nails could thus be determined. Horizontal tensile or compressive strain could be interpreted also.

Similarly, the angular distortion was calculated based on measurements of settlements of buildings from various points of the foundation. Hsiung (2009) has

Fig. 14.7 Measurement using tape extensometer

approved that results from the damage assessment associated with the method suggested by Boscardin and Cording (1989) were satisfied with observations from the site.

14.4 Discussions

It is no doubt that success of geotechnical monitoring could prevent failure but failure might also be made because of misinterpretation of responsibility-sharing associated with observations from underground works in Taipei and Kaohsiung. Therefore, responsibility-sharing between the client, designer and contractor in geotechnical investigation, design and monitoring is discussed associated with lessons learned from recent construction of Taipei and Kaohsiung Metro.

First of all, the responsibility of site investigation is discussed, especially in Design-Build and Turnkey (DBT) model construction project. DBT model is a fast-track model for constructing underground work and it has become more and more popular all over the world. In DBT model, the employer (the client) is mainly responsible for land acquisition, access of the site, termination and extension of the contract and payment and leaves most of responsibilities of working performance to the contractor. As the design is also delivered by the contractor itself, the employer is not involved with the design. It is seen commonly that in many DBT model underground projects that the employer conducted very limited site investigations and testing so a clearly picture of ground characteristic could not been detected before commence of construction. However, in Clause 4.11 of the Conditions of Contract for Design-Build and Turnkey prepared by the Federation Internationale des Ingenieurs-Conseils (FIDIC) "Unforeseeable Sub-Surface Conditions," it states that "If sub-surface conditions are encountered by the Contractor which in his opinion were not foreseeable by an experienced contractor, the Contractor shall give notice to the Employer's Representative so that the Employer's Representative can inspect such conditions....if such conditions were not foreseeable by an experienced contractor, proceed in accordance with Sub-Clause 3.5 to agree or determine:

(a) any extension of time to which the Contract is entitled under Sub-Clause 8.3, and
(b) the additional Cost due to such conditions, which shall be added to the Contract Price.

In fact, ground characteristics might be suddenly changed in a short period due to some external impacts and accident might occur before the contractor could give any notice to the employer or employer's representative. The after-event investigation could also indicate the accident was caused by an unforeseen ground condition and an well-known and experienced contractor are not suppose to be blamed in both cases, though the employer definitely push the responsibilities to the contractor. Moh and Hwang (2007) reviewed some major accidents of underground metro in Asia Pacific and it was found that accidents in DBT model projects might lead 6–12 months delay of the project and repair cost could be up to USD 70 million in one project. Many of them were induced by unforeseen ground risk. Knights (2005) presented 15 projects over the period 1994–2004, which all faced major ground-related problems with financial losses, in total, of more than 500 million US dollars. Clayton (2001) addressed that evidence from the past shows that construction cost overruns are significantly reduced as expenditure on site investigation is increased. An example taken from Kaohsiung Metro was chosen for further commentary. Figure 14.8 shows all boreholes of O6 Station of Kaohsiung Metro and it is seen there are eight holes in the range of 200 m. Only six of them

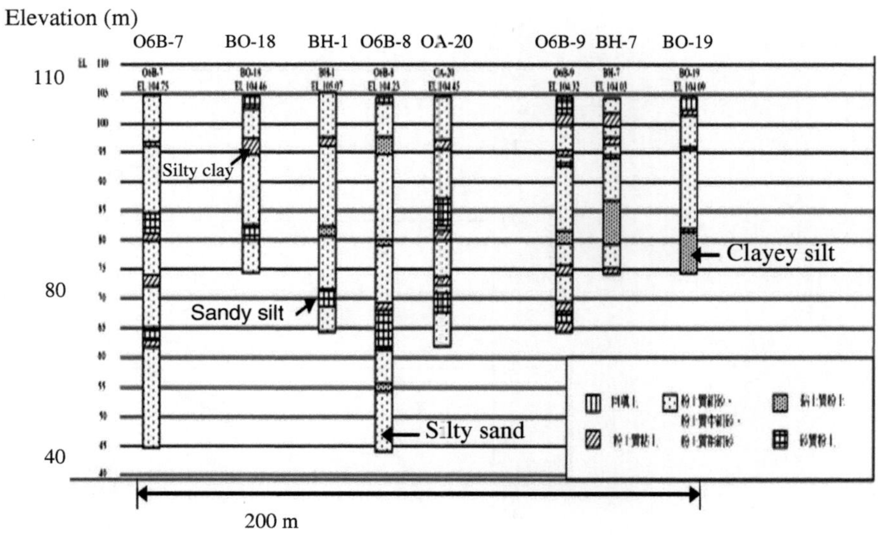

Fig. 14.8 Borehole information of O6 Station

were conducted before and during tender stages (in the name of "BO", "BH", and "OA") and a full picture of ground conditions for excavations at O6 Station could not be defined by information given here. Due to the limit of time in preparation stage, a large-scale site investigation cannot be conducted by the contractor so such data have to be well-prepared by the employer. Inadequate data of site investigation may induce the accident and both of the employer and the contractor will suffer from the loss of accident so a "partnership" shall be formed between the employer and the contractor in order to prevent such matter, especially for an underground project constructed in the place never has a similar project before. Considering reasons stated above, adequate data of site investigation shall be provided by the client and shall not leave all responsibilities to the contractor. Well-experienced experts shall be invited to evaluate the quantity and quality of data of site investigation before the project goes to tender.

Second, the responsibility of design is discussed. It is thought reasonably that the responsibility of design shall belong to the designer, no matter recruited by the employer directly or by the contractor. Due to fast development of computer hardware and software, computer-aided design is popularly adopted for geotechnical design of deep excavations and tunnel. Associated with exactly same soil parameters, Fig. 14.9 presents outcome from two-dimensional and three-dimensional analytical results of lateral wall deflections with observation of a

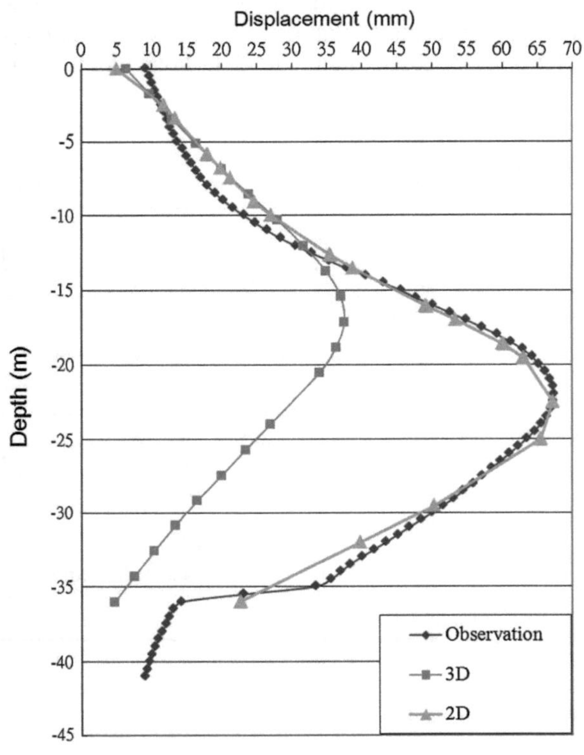

Fig. 14.9 Analytical wall deflections

19.7 m deep excavation. Difference here might be contributed by different definition of interface used in the software and boundaries and dimensions selected for analyses. Considering the observations above, it is concluded that analytical results might be different due to various assumptions of the analytical mesh and model, though the same soil parameters were given. In a large-scale underground construction project, reasonable soil and structure parameters shall be provided by the employer before commence of design as this could possibly increase the reliability and minimize the risk in design, though different assumptions stated above may still induce the difference.

Further, as shown in Fig. 14.3, the quality of wall could not be predicted before commence of the design but could possibly be detected after wall installation. The design of the excavation should thus be revised after wall installation because of the quality of wall. Moreover, conducting independent check could increase the reliability of the design and it cost less for design change before commence of construction of an underground project. It is thus recommended that independent checks shall be delivered and revision of design should be concerned from time to time in order to reduce the risk. The employer might not be capable to carry out tasks stated above and it is suggested that special consultants with expertise can be recruited for it.

It is anticipated that most of the responsibility of construction shall belong to the contractor. Recent problems for underground construction in Taiwan include: (1) it is hard to precisely evaluate cost and time of construction before commence of the project as the tender preparation period is very short; (2) low standard of prequalification adopted and lowest bid in a very competitive market lead to very limited budget of a project. These are key factors for defects, overrun, and delay of construction of underground projects in Taiwan.

At last, the responsibility of monitoring is discussed. In engineering practice in Taiwan, the employer does have a monitoring plan but the contractor is the final decision maker for type, location, depth, and quantity of instruments but has to submit the monitoring plan to the designer and the employer to ask for approval. During the construction, a subcontractor for monitoring is responsible for all site measurements and has to pass all data to the contractor for the first review and then to the designer and the employer for reviews also under certain circumstance. Design change and protection measures of adjacent structures proposed by the contractor are adopted if movements and stress of ground or structure are beyond the warning levels given by the employer. Refer to examples from Taipei and Kaohsiung stated in this paper, the contractor was suppose to provide the reason and solutions but it seems the party was not capable to do so. It is thus suggested that geotechnical experts shall be invited to provide recommendations based on monitoring data from time to time.

As stated above, geotechnical failure might occur suddenly so automatic, real-time monitoring system can possibly pass message to all related parties shortly in order to extend time of response. Hsiung et al. (2011) described an intensive monitoring system using several automatic monitoring instruments for bored tunnels constructed beneath operating taxiway inside the international airport but the reliability of the instrument has to be concerned.

14.5 Conclusions

Some conclusions can be drawn based on findings from this paper.

First, successful interpretation of instrument data is addressed and several examples are demonstrated since geotechnical monitoring is thought to be an effective tool to avoid failure. Reliable instrument data can reflect the reality on site.

Second, adequate data of site investigation shall be provided by the client and shall not leave all responsibility to the contractor due to any reason.

Third, in a large-scale underground construction project, reasonable soil and structure parameters shall be provided by the employer before commence of design as this could possibly increase the reliability and minimize the risk. Moreover, conducting independent check could increase the reliability of the design.

At last, it is suggested that geotechnical experts shall be invited to provide recommendations based on monitoring data from time to time. Geotechnical failure might occur suddenly so automatic, real-time monitoring system can possibly pass message to all related parties in order to extend time of response but the reliability of the instrument has to be concerned.

Acknowledgments The authors want to thank Shanshin Design Corporation for providing data and information of ground freezing in Kaohsiung Metro. The authors also want to thank the efforts of Mr. Fong-Chen Hsieh and Mr. Ching-Hau Chang, Master students of National Kaohsiung University of Applied Sciences for paper preparation.

References

Boscardin MD, Cording EG (1989) Building response to excavation-induced settlement. J Geotech Eng, ASCE 115(1), 1–21

Clayton CRI (2001) Managing geotechnical risk. Published by Thomas Telford Publishing, London

Hsiung BCB (2009) Field performance of an excavation using sleeve grouting. Proc Inst Civ Eng, Ground Improv 162(GI4):175–183

Hsiung BCB, Yamamoto S, Shoda S, Chang WC (2011) A case record of tunnels bored in gravel based on the Taoyuan international airport link project. J GeoEng, Taiwan Geotechnical Society (under review)

Knights M (2005) Risk management of tunneling works. Presentation to advisory board, 23 Nov, Hogeschool Zeeland, Vlissingen

Moh ZC, Hwang RNH (2007) Lessons learned from recent MRT construction failures in Asia Pacific. (Opening Keynote). In: Proceedings of the 16th SEAGC. Subang Jaya, Malaysia

Wang Lee CM, Lin TP (1987) The geology and land subsidence of the Taipei basin. Memoir Geol Soc China 9:447–464

Chapter 15
Failure Analysis of a Highway Dip Slope Slide

Wei F. Lee, H.J. Liao and C.H. Wang

Abstract A catastrophic dip slope failure incident took place at 14:33 of April 25, 2010, at Da-Pu section of National Freeway No. 3 in northern Taiwan. It resulted in a nearly 200,000 m^3 debris mass on the freeway and caused five deaths. In an effort to identify root causes of such a disastrous failure, a forensic investigation committee was formed to conduct a comprehensive failure analysis. Progress of completed forensic investigation is reported in this paper. Purpose of this paper is hoped to introduce the applications of forensic investigation technologies in similar forensic investigations, as well as appropriate means of adapting numerical modeling in a forensic investigation.

Keywords Failure · Ground anchors · Numerical models · Field and laboratory tests

15.1 Case Information

No. 3 Freeway is one of the main vehicle arteries running north–south through Taiwan. It cut through middle of the failed dip slope and resulted in a maximum 35 m deep excavation to accommodate eight lanes of traffic. The disturbed dip slope lies in the east–west direction. Tip of the slope sits in the west and stretches toward

W.F. Lee (✉)
National Taiwan University of Science and Technology, Taipei 10607, Taiwan
e-mail: weilee@mail.ntust.edu.tw

H.J. Liao
Professor, Department of Construction Engineering, National Taiwan
University of Science and Technology, Taipei 10607, Taiwan
e-mail: hjliao@mail.ntust.edu.tw

C.H. Wang
Department of Civil Engineering, Chuo University, Tokyo, Japan
e-mail: coleus@gmail.com

V.V.S. Rao and G.L. Sivakumar Babu (eds.), *Forensic Geotechnical Engineering*,
Developments in Geotechnical Engineering, DOI 10.1007/978-81-322-2377-1_15

the west in a “cheese cake” shape wedge. As a result of excavation, south bound (west) slope was a typical dip slope cut which failed during the incident. The excavated dip slope was designed with three tiers with 1:1 to 1:1.5 slopes. All three tiers were reinforced by ground anchors which were designed with 60 ton anchor load and a horizontal spacing of 2.6 m. Total 572 anchors were constructed in order to stabilize the excavation. These ground anchors were installed with an inclined 20° angle and an identical fix-end length of 10 m. Free length of these ground anchors ranged from 8 to 24 m depending on their locations and elevations (Lee et al. 2013).

15.1.1 Event on April 25, 2010

On 14:33 on April 25, 2010, a severe dip slope slide occurred at 3k + 100 of National Freeway No. 3 (Fig. 15.1). The slide has a length of approximately 185 m from the top to roadside, and a width of about 155 m at the bottom along the freeway. Total area of the slide exceeds 14,000 m^2. Thickness of the sliding mass was in a range of 15–20 m. Debris generated was in an amount of near 200,000 m^3. Such massive debris ran across Freeway No. 3 within seconds and crushed four vehicles. Also shown in the figures, dip angle of the failure plane was in a range of 12°–15° which was in line with the dip angle of sandstone and shale–sandstone alternation layers. There were no rainfall or earthquake events recorded prior to the failure for weeks. During the slide, major amount of ground anchors were ripped off by massive forces generated by debris. However, serious corrosion was observed on damaged ground anchor components indicating that deterioration of anchors was quite aggressive inside the studied slope.

Fig. 15.1 Freeway No. 3 Landslide (after Liao and Lee 2011)

Immediately after the incident, a forensic investigation committee was formed, and extensive investigation works including field explorations, laboratory testing, and numerical analyses were conducted in detail to possibly access genetic causes and mechanisms of the landslide. At the moment of landslide, it was reported by eye witness that there were unconfirmed number of cars buried under the landslide debris. So the National Freeway Bureau was under tremendous pressure to dig victims out as quickly as possible and also let the freeway opened to traffic again. It took 3 days to recover the victims and 7 days to remove all the landslide debris. Unfortunately, most of the evidences needed to find out the causes of this landslide were also destroyed in the process. The investigation work then relied very much not only on the geologic and engineering information gathered but also on the phenomena observed at the landslide site and the photos taken during construction period.

15.2 Geological Condition

Figure 15.2 shows the geological cross section of this dip slope formation (1:50000 scale) based on the boring log data and field observation after the event. The major geological formations include Ta-Liao and Si-Te formations in northern Taiwan which are mainly massive sandstone standing on shale, sandstone, and alternative shale and sandstone layer (Hung 2002). Dip angles of the bedding layers underlying the slope are ranging from 12° to 15°. As indicated in the figure, sliding mass was composed by massive sandstone block and thin shale–sandstone alternation which

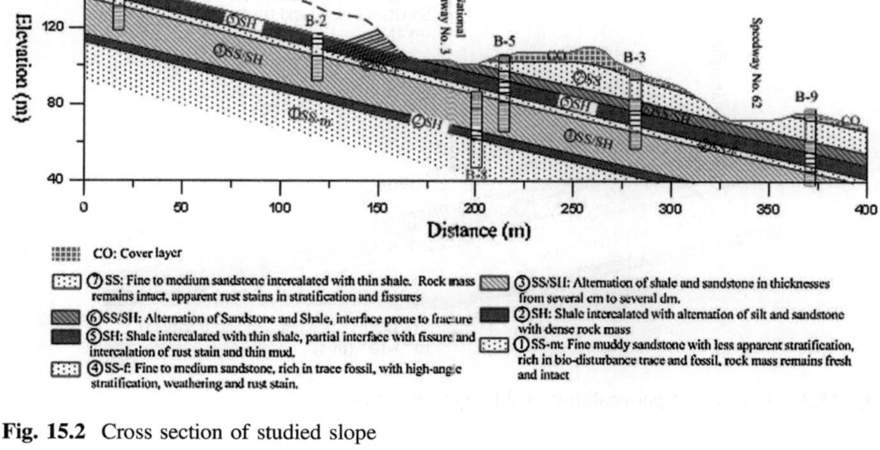

Fig. 15.2 Cross section of studied slope

is prone to fracture. The sliding surface was located at the interface between shale (Layer No. 5) and thin shale–sandstone alternation (Layer No. 6). Several conjugated fissures and joint cracks were found parallel to the southern edge of the slope. The brownish stain was observed in the surface of crack indicating that groundwater had been flowing through these openings for a long time. However, very limited information on the weathering condition and development of joints and cracks before the failure was available. For the upslope of Freeway No. 3, the sliding surface was exposed by slope cutting. The strike of sliding surface has an angle of about 20° from the strike of cut slope face on the southern portion of this dip slope. It makes the dip slope sliding kinematically feasible. However, the strike turns parallel to the Freeway on the northern portion and intercepted at the road level for about 50 m. In other word, the sliding surface daylighted on the slope face on the south but plunged under the road level at the northeastern corner of the slope. In addition, the northern edge of the "cheese cake" shaped slope was a free face which provided no constraint to sliding. So, this slope is still kinematically feasible for a dip slope slide when the ground anchors lost most of the tieback capacity due to corrosion, a dip slope sliding was actually happened and provided the kinematically feasibility of the dip slope (Liao et al. 2013).

The permeability of the different geological layers was determined with the in situ Lugeon test. As shown in Fig. 15.3, the permeability of shale and interlayered shale and sandstone was less than 10^{-8} m/s; while the permeability of sandstone varies from 10^{-4} to 10^{-8} m/s depending on the extent of fracture inside. For sandstone of Layer No. 4 underlying the shale of Layer No. 5, its permeability is in the range of 10^{-4}–10^{-5} m/s (Lee et al. 2013). It is the major groundwater-bearing layer in the slope. Finding of this water-bearing layer was reconfirmed by the results of electrical resistivity imaging profile (RIP) shown in Fig. 15.4. Several zones of low electrical resistance (i.e., high water content) were found under the

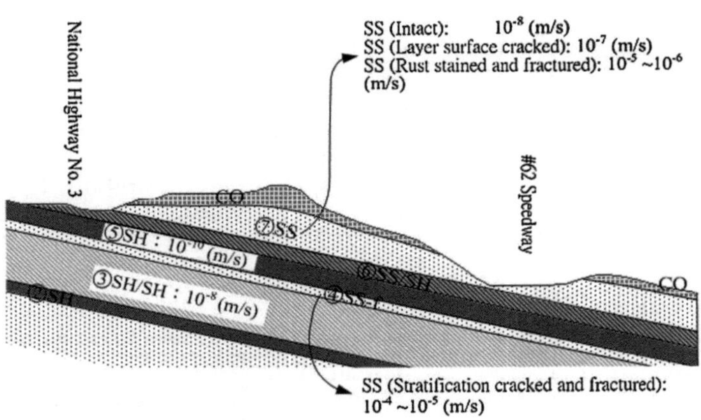

Fig. 15.3 Summary of permeability of different formations

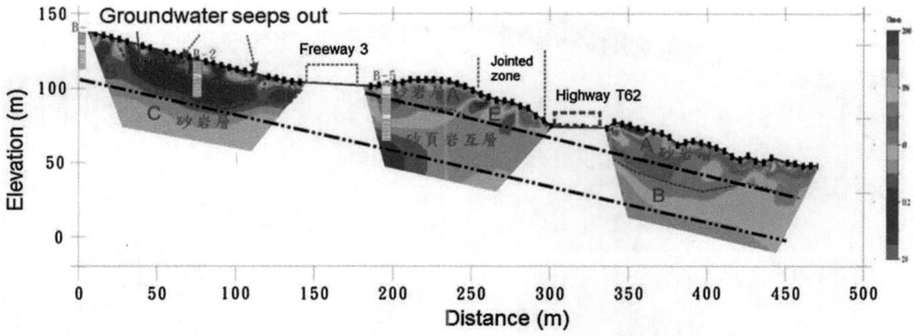

Fig. 15.4 Resistivity imaging profile of dip slope formation after landslide (after Liao and Lee 2011)

exposed sliding surface of Freeway No. 3. Some of the low electrical resistance zones even extended to the face of sliding surface. It shows that the sandstone and alternative shale–sandstone layers are the major water supply layers and groundwater could flow up and down through the cracks and unsealed anchor holes inside the slope. It can be summarized from the geologic and groundwater conditions of the slope that the long-term stability of Freeway No. 3 slope was close to marginal.

15.3 Issues of Ground Anchors

During the process of slope sliding, a large amount of ground anchors were ripped off by the massive forces generated from the sliding mass. Serious corrosion observed in the ground anchor components indicated the abundance of groundwater in the slope (Fig. 15.5). There were no anchor found detached from the fixed end.

By measuring the length of remained steel strands left on the sliding surface, three types of steel strands breakage can be categorized as shown in Fig. 15.6. The "Red" category stands for the anchors of which strand breakage was closely under the anchor head. The "Yellow" category stands for the strand breakage in between anchor head and sliding surface. The "Green" category stands for the strand breakage near the sliding surface. The "Blue" category stands for the anchors remained on the slope face. Since the "Blue" anchors located on the not moving portion of the slope, they have no direct link with this landslide. If neglecting the anchors still remained on the slope face (Blue category), approximately 40 % of the broken ground anchors were in Red category. As indicated by the distribution of Red marked anchors, the groundwater level could rise to 5–7 m above the outcrop line on the slope face. When the slope mass slid down, the steel strands of anchors in Red category broke at a location very close to the anchor head. It implies that steel strands were corroded under the anchor head for the Red category. Above the

Fig. 15.5 Failure plane of the studied slope (Lee et al. 2013)

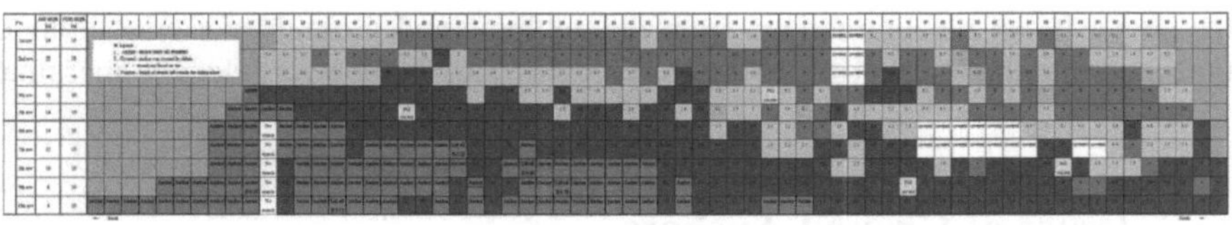

Fig. 15.6 Summary of inspection result of anchors (Lee et al. 2013)

Red zone was Yellow zone where steel strands were broken in the free end, but were some distance away from the anchor head. Since no anchor was found pulled out from its fixed end, it can be summarized that all the failed anchors were resulted by breaking steel strands at different locations of free end rather than by being pulled out from the fixed end.

Investigation of ground anchor construction records indicates that, during anchor construction, the standard operation procedure to let the entire anchor hole filled with cement grout first and then inserted the tendon assembly to the hole. However, it was found from the anchor holes left on the sliding surface (Fig. 15.7), the annular space between anchor hole and plastic sheath in the free length was not fully sealed with cement grout. Nor was the space inside the plastic sheath. Obviously, some cement grout leaked out through the cracks and joints inside the slope or simply a malpractice during anchor construction. Since all the anchors were inclined downward, the annular space easily became the storage space for the perched groundwater inside the slope. Figure 15.8 illustrates that a schematic drawing of voids under anchor head could be resulted by poor anchor construction

Fig. 15.7 Unsealed free anchor end found from anchor remained on sliding surface

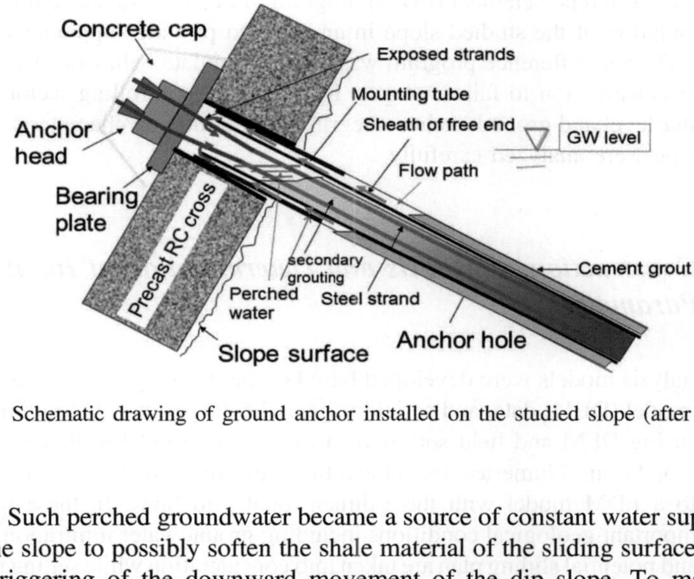

Fig. 15.8 Schematic drawing of ground anchor installed on the studied slope (after Liao et al. 2013)

practice. Such perched groundwater became a source of constant water supply from inside the slope to possibly soften the shale material of the sliding surface and then enable triggering of the downward movement of the dip slope. To make thing worse, no serious corrosion protection measure was taken to protect the steel strands under the anchor head. So the existence of perched water in the anchor hole caused the steel strands to be constantly exposed to or submerged by groundwater. As a result, the steel components of anchor corroded very quickly. Figure 15.9 shows a picture taken from the back of a fallen precast RC cross. Obviously, the steel strands under the anchor head suffered a very serious corrosion problem.

Fig. 15.9 Seriously corroded steel strands found from the back of a fallen anchor head

15.4 Numerical Analysis

The 3D finite difference element (FDM) program FLAC3D was adapted to conduct failure simulation of the studied slope in an effort to provide supplement verification. The 3D finite difference program was able to simulate behavior of the studied slope from construction to failure stages. Influences of controlling factors such as groundwater level and ground anchor integrity on deformation characteristics of the studied slope were analyzed carefully.

15.4.1 Construction of Models and Determination of Input Parameters

The 3D analysis models were developed based on field survey data, 5 × 5 m digital elevation model (DEM) data, and aerial images. The 3D geomorphology model was first built using DEM and field survey data as the base model to define modeling boundary conditions. Numerical model was then developed by transforming the base model into a FDM model with three-dimensional grid lines. In these numerical models, important geological conditions including groundwater infiltration, outcrop location, and potential sliding plan are taken into consideration while setting up model details. Photos preserved at construction and failure stages are valuable information for building the models. Figure 15.10 indicates that failure plane was truly developed along the sandstone–shale alternation layer. Such a potential sliding layer could also be observed during construction as shown in Fig. 15.11 (Lee et al. 2013).

Input parameters of these numerical models were obtained from both field and laboratory tests. Table 15.1 summarizes the rock strength properties obtained by performing direct shear tests on cored specimens (Lee et al. 2013). Results show

Fig. 15.10 Aerial photo of dip slope failure (Lee et al. 2013)

Fig. 15.11 Exposed potential sliding plane during construction (Lee et al. 2013)

that peak friction angle and residual friction angle of intact rock material on site are, respectively, between 22.5°–28.5° and 14.1°–19.8°. Peak cohesion of tested rock materials is in a range from less than 1 kg/cm^2 of sandstone and shale alternation to 3 kg/cm^2 of shale. As shown in the table, no cohesion was found on onsite rock material at residual state. Modulus of rock materials used in 3D numerical analyses was determined based on stress–strain curves of both direct shear tests and

Table 15.1 Strength properties of rock material tested

Test No.	Bore No.	Depth (m)	Peak strength		Residual strength		Lithology	Shear face
			Cp (kg/cm^2)	ϕ_p °	Cr (kg/cm^2)	ϕ_r °		
RDS(D)-4	B-6	16.60–17.00	0.28	22.5	0.0	19.8	Alternations of sandstone and shale	Lamination
RDS(D)-5	B-7	18.00–19.00	3.2	28.5	0.0	22.7	Shale	Intact rock core
RDS(W)-3	B-3	16.00–17.00	0.9	27.7	0.0	23.2	Alternations of sandstone and shale	Lamination
RDS(W)-4	B-5	16.00–17.00	1.1	26.2	0.0	14.1	Shale	Intact rock core

unconfined uniaxial loading tests on cored rock specimens. Input permeability values of the 3D FDM model were adapted from in situ permeability test results as shown in Fig. 15.3. Because of such a big permeability difference between sandstone layer and sandstone–shale alternation, critical groundwater tables were defined to stay above the potential failure planes to account for possible effective stress reduction and material strength reduction caused by seepage inflow and surface runoff infiltration. Anchor loads were modeled as external point loads for such 3D analysis. 3D FDM analyses conducted in this study are aimed to simulate possible sliding process and to verify the sliding mechanisms. However, soil elements within the 3D numerical model are still assigned with Mohr–Coulomb failure criteria. Very simple built-in constitutive correlation was adapted to calculate stress–strain distributions at each modeling stages. Moreover, rock formation underneath potential sliding plane was simplified as homogenous layer to accelerate iteration because failure plane of this slide was clearly exposed. Fracturing of rock mass was not simulated individually in the analyses because limited information was available before the failure. Effects of discontinuities and weathering of rock mass were taken into consideration by applying overall strength reduction to the built-in material elements.

15.4.2 Failure Analysis Using 3D Finite Difference Method

Figure 15.12 shows the base model with element grids for 3D FDM analysis. As shown in the figure, geomorphologic appearance before freeway construction and actual boundary conditions were simulated. Figure 15.13 illustrates locations of the applied anchor loads and the controlled point for deformation monitoring. Detailed stress–strain information of the investigated slope at construction stage or under different material degradation conditions could be observed thanks to such FDM analyses. Simulation result of construction stage could be found in the completed forensic report (Liao and Lee 2011). To examine the effect of groundwater level to

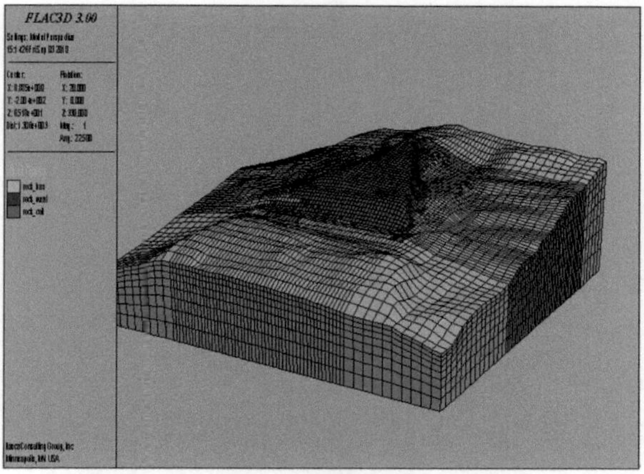

Fig. 15.12 Base 3D FDM model with element grids

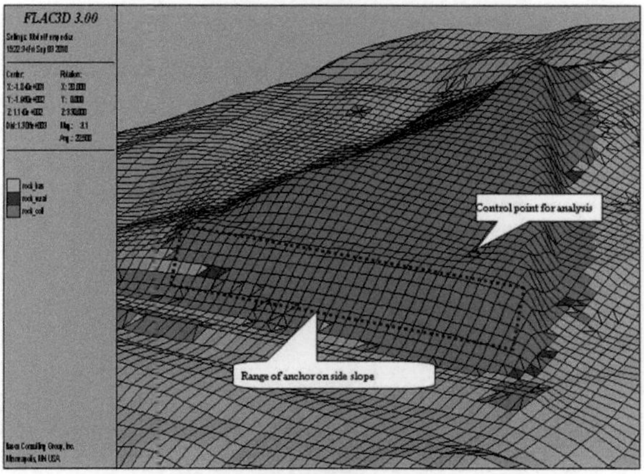

Fig. 15.13 Sketch of locations of ground anchors and control point for monitoring deformation

the slope stability, similar reduction on strength properties of rock stratum was adapted to those elements submerged under groundwater table. Figure 15.14 shows the analysis results in displacement contour obtained by raising the groundwater level from elevation 128–137 m. As shown in the figure, southern side of the slope posses larger displacements. This finding coincides with the field observation indicating that sliding started from the southern side of the slope toward north. Figures 15.15 and 15.16 present displacement and shear strain vectors along pre-defined slide plane. The displacement and shear strain contours verify the activation of failure under high groundwater level condition.

Analysis was further conducted by setting the cohesion of potential sliding plane to 0 and systematically reducing friction angles of that from 15° to 12°. In addition to strength properties of rock, anchor loads were also decreased systematically from original design level to only 10 % anchor load remaining. Figures 15.17 and 15.18

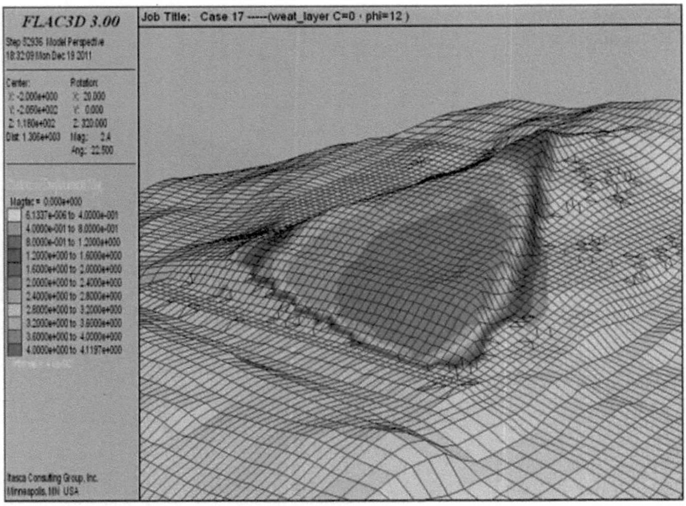

Fig. 15.14 Deformation contour of high groundwater level case

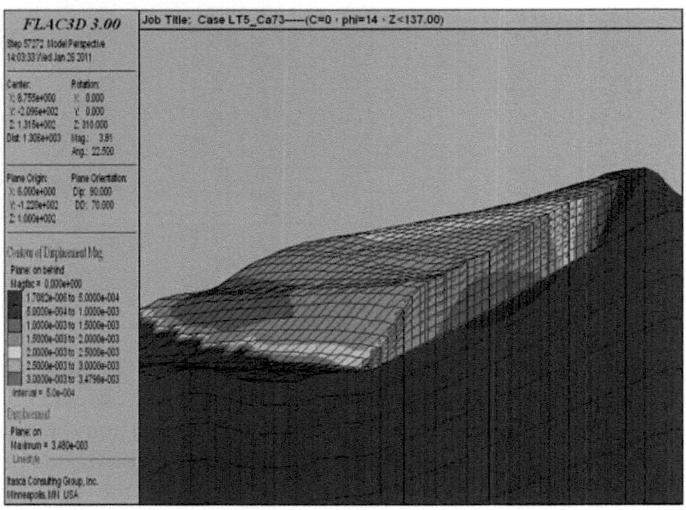

Fig. 15.15 Displacement result of high groundwater level at predefined cross section

show the analysis results along predefined critical sections of the case with highway groundwater table and 50 % anchor load reduction. As shown in the figures, lateral deformation could be clearly observed as both deformation and shear strain vectors inclined toward the toe of the slope. In order to summarize results of 3D analyses, displacements of the designated control point as indicated in Fig. 15.13 with

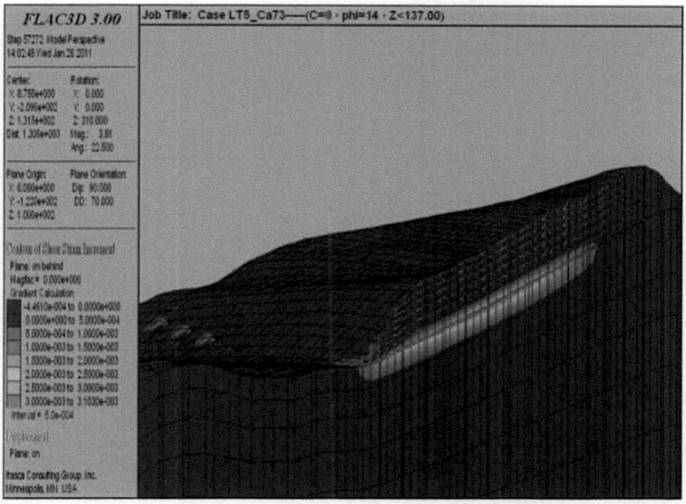

Fig. 15.16 Shear strain result of high groundwater level at predefined plane

different rock friction angles and anchor load reductions are summarized in Fig. 15.19 to further illustrate possible failure hypotheses. The analysis results indicate that the studied slope intends to initiate large displacement when the rock friction angle is lower than 14° and anchor load loss reaches more than 50 %. Moreover, the slope would exert progressive failure when the friction angle is lower than 12° even with minimum amount of anchor load loss. It is concluded that material deterioration and anchor load reduction would probably be the major factors contributing to the investigated slope failure.

15.5 Forensic Discussions

The studied dip slope (cheese cake shape wedge) was actually cut through by both Freeway No. 3 and Highway T-62 in the late 1990s. The former used extensive tied back system to hold back the cut slope; the latter adopted very limited slope stabilization measures. Ironically, the cut slope of Highway T-62 remained stable. During construction, two different types of bridge construction method were used to connect the cut slope with the elevated road on the both ends for the Freeway No. 3 and Highway T-62. The former used the incremental bridge construction method; while the latter used the traditional bridge construction method. The traditional method built a temporary structure from the ground up to support the formwork and the in-place fabrication of bridge structure. In comparison, there was no need for the ground up supportive structure for the incremental bridge construction method. So it can speed up the construction process. But to carry out the incremental bridge

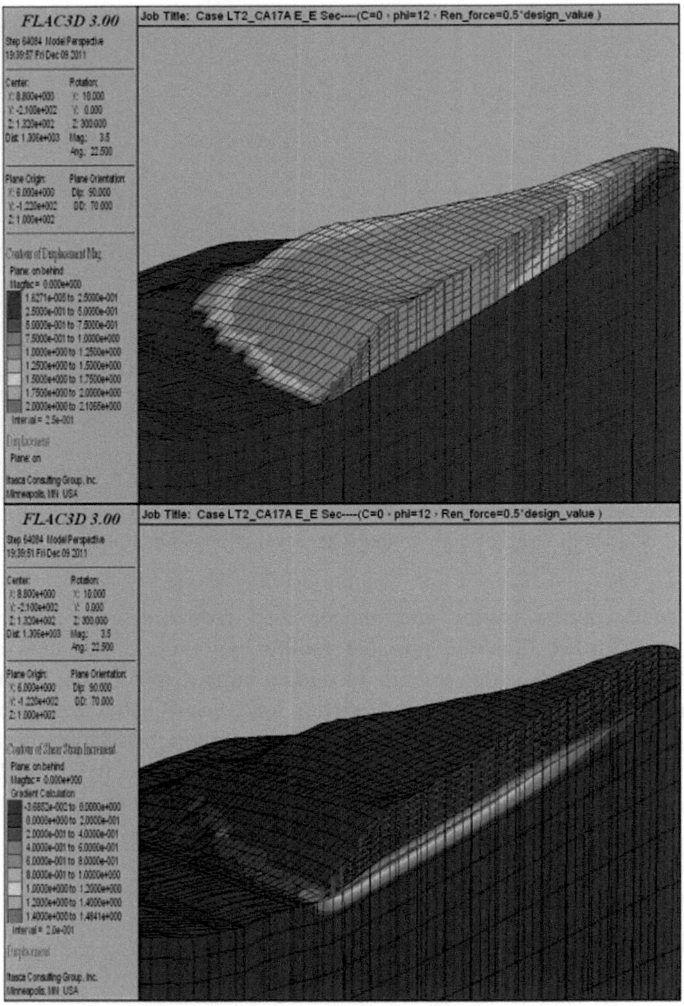

Fig. 15.17 Analysis results of 50 % anchor load reduction

construction, it needed a launching pit for the fabrication of precast bridge segment. The pit was located near the southeast corner of the slope (Figs. 15.20 and 15.21) and had the dimension of 5.5 m in depth and about 50 m in length (including the driveway). Since the pit excavation was very close to the toe of this cut slope, a 5.5 m deep pit excavation removed the sliding stop portion of this dip slope and caused a complete daylighting of the sliding surface on the face slope. After removing this sliding stop block, it initiated the downward movement of the dip slope. As shown in Fig. 15.22, shortly after the excavation of launch pit, the

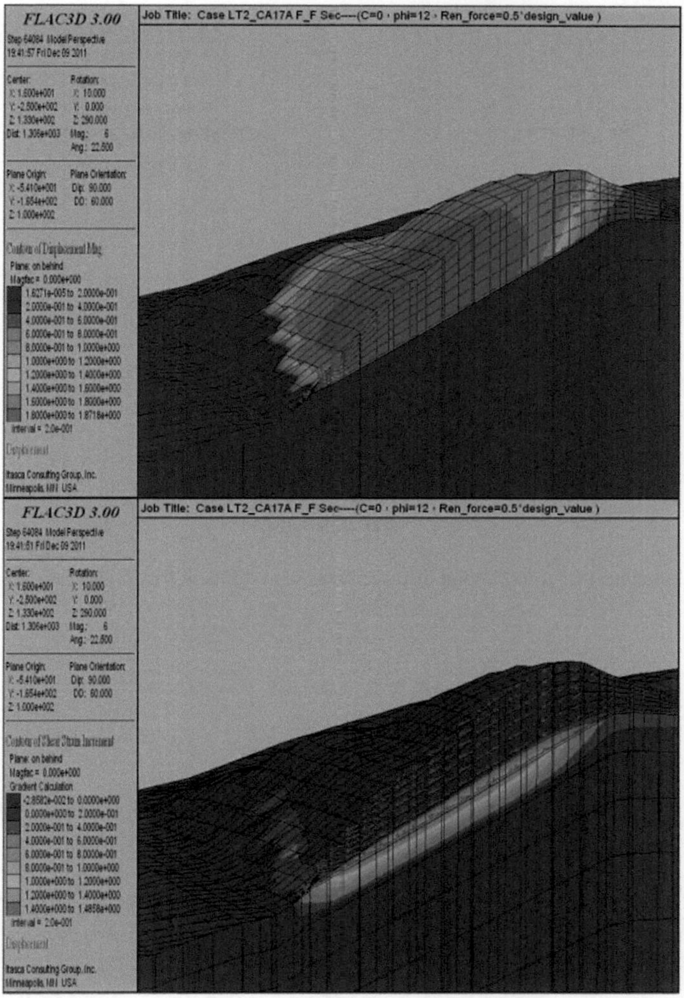

Fig. 15.18 Analysis results of 50 % anchor load reduction

readings of anchor load cells (LC 2–4, location shown in Fig. 15.20) mounted on the slope jumped up from around 70 ton to 76–87 ton, respectively. Such a jump of load cell reading was a clear sign of slope movement. In other words, the downward movement of this dip slope had been triggered during the construction stage. To stop the downward sliding of the slope, 18 tieback anchors in 2 levels (10 and 8 each) were installed on the upslope of the launch pit (Fig. 15.21) at that time. Although the anchor load increase (LC 2–4) was slowed down after the installation of additional anchors, the load cell readings still increased slowly to 96 ton (LC-4) and 84 ton (LC-2 & 3) at the end of construction (Fig. 15.22). Therefore, it can be

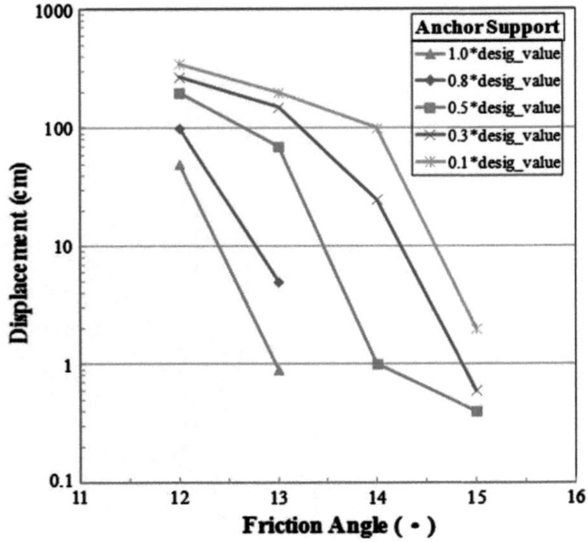

Fig. 15.19 Summary of control point displacements with different friction angles and anchor load reductions

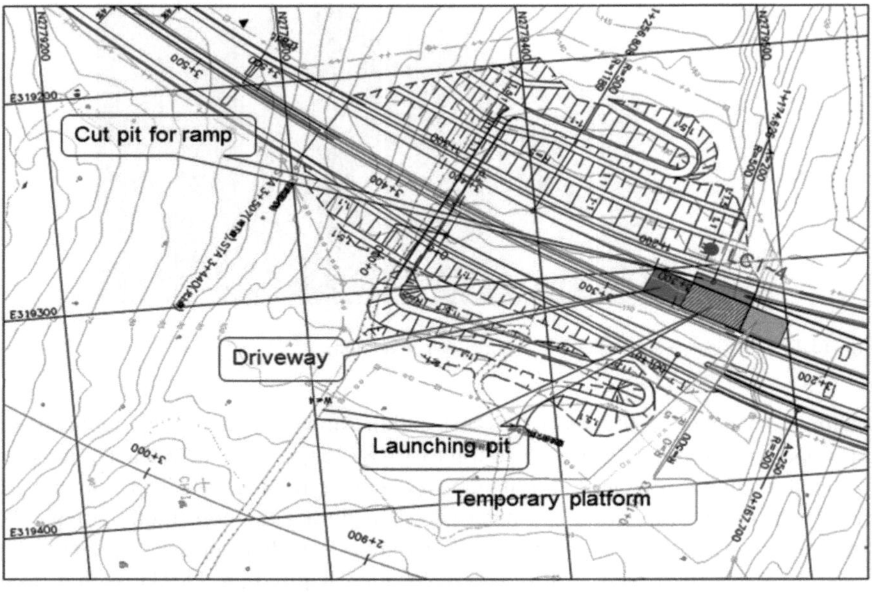

Fig. 15.20 Slope toe cutting by launching pit

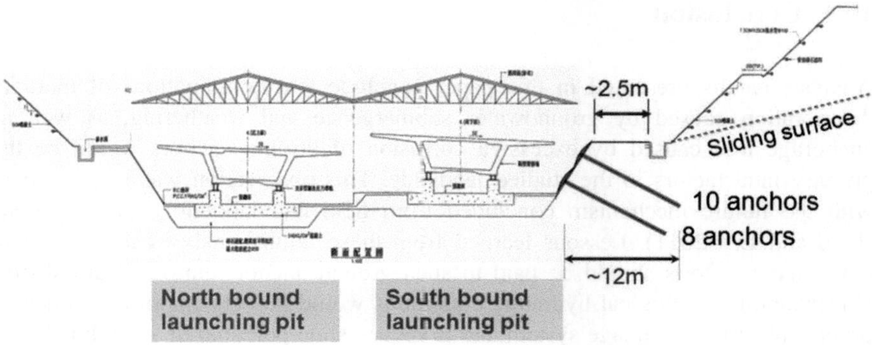

Fig. 15.21 Relative condition between launching pit and the cut slope

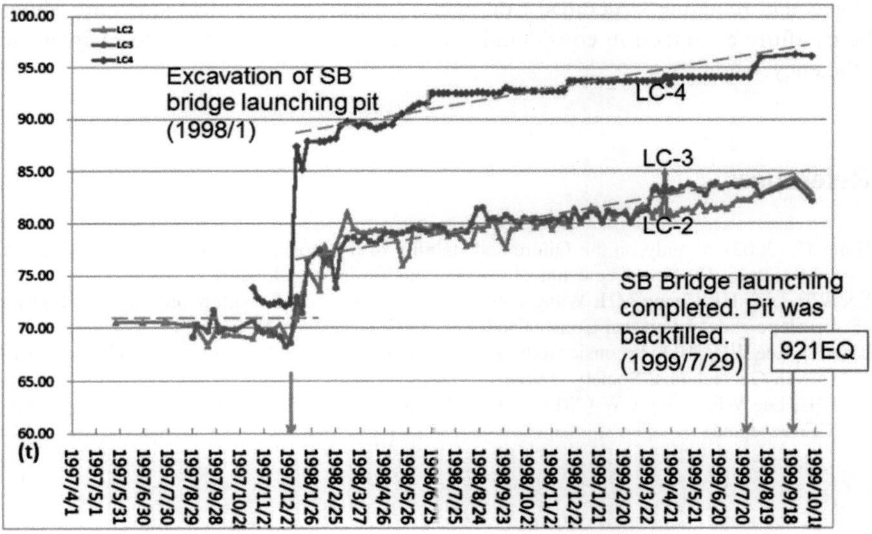

Fig. 15.22 Load cell readings during the period of slope cutting and the time of launching pit excavation

said that the movement of slope had been initiated since then. So after the triggering of initial sliding, it would create some open up of voids on the sliding surface. It in turn allowed groundwater to seep to the sliding surface through these voids easily. With the presence of groundwater, it accelerated the deterioration process of shale material on the sliding surface and also the corrosion of steel strands of tieback anchors. In comparison, no launching pit was excavated during the bridge construction of Highway T-62. Without launching pit excavation, the sliding surface which caused the landslide of Freeway No. 3 remained under the cutting level of Highway.

15.6 Conclusion

Analysis results presented in this paper conclude that combinations of material deterioration caused by groundwater submergence and weathering, as well as anchorage loss caused by excessive corrosion of ground anchors would be the primary fatal factors to the studied landslide. This observation was in agreement with the failure mechanism concluded from field and laboratory investigations (Liao and Lee 2011). Lessons learned from these failure analyses are that more maintenance efforts should be paid to such ground anchor reinforced dip slopes. Understanding geological hydraulic conditions would be very important to design proper subsurface drainage system and to reduce slide potential of such dip slopes. Moreover, long-term corrosion prevention measures of ground anchors should be considered when designing anchored supported dip slope cuts. Most importantly, health inspection and monitoring programs should be imposed to similar manmade slopes and retaining structures. Life cycles of such geotechnical structures should be carefully examined in considerations of material deterioration and maintenance efficiency.

References

Hung JJ, (2002) A study on the failure and stability of dip slopes. Sino-Geotechnical Magazine, vol 94, pp 5–18, Sino-Geotechnical Foundation, Taipei (in Chinese)

Lee WF, Liao HJ, Chang MH, Wang CW, Chi SY, Lin CC (2013) Failure analysis of a highway dip slope slide. J Perform Constr Facil ASCE (To be appeared)

Liao HJ, Lee WF (2011) Forensic study on the dip slope failure of freeway no. 3 at chainage 3.1 k. Taiwan Geotechnical Society, Taipei

Liao HJ, Lee WF, Wang CW (2013) A tale of twin cut slopes in Taiwan. Forensic Engineering, ICE (accepted)

Chapter 16
Characterisation of Failure at a Large Landslide in SE Queensland by Geological Mapping, Laboratory Testing, Instrumentation and Monitoring

D.C. Starr, J. Woodsford and D.F. Marks

Abstract Geotechnical studies undertaken for Cuts 3 and 4 of the South West Transit Corridor (SWTC) at Redbank Plains revealed complex geology, comprising trachyte flows extruded on top of sediments and basalt flows. The road cuttings form part of the Centenary Highway Extension, Springfield to Yamanto Package. The paper describes the investigation of a large-scale landslide encompassing Cuts 3 and 4, the nature and extent of which was revealed after the original design of the earthworks were substantially complete. In order to characterise the landslip mass, the geometry of the landslip mass and rate of movement was investigated using a large number of drill holes with inclinometer installations, which were monitored together with surface movement measurements. Instrumentation also included piezometers and observation wells. Laboratory testing was undertaken on material recovered from the shear surface, and compared with shear strength parameters derived from back analyses. Results of stability analyses are briefly described, and the monitoring data is presented and discussed. A Web-based data management system for providing an early warning system for potential future movements is described.

Keywords Cut slopes · Landslide · Instrumentation · Monitoring

D.C. Starr (✉)
Golder Associates, 147 Coronation Drive, Milton, Queensland 4064, Australia
e-mail: david_starr@bigpond.com

J. Woodsford · D.F. Marks
Department of Transport and Main Roads, Brisbane, Australia

V.V.S. Rao and G.L. Sivakumar Babu (eds.), *Forensic Geotechnical Engineering*,
Developments in Geotechnical Engineering, DOI 10.1007/978-81-322-2377-1_16

16.1 Introduction

Cuts 3 and 4 are located between Chainage 9450 and Chainage 10,250 of Package B-Springfield to Bundamba Creek Section of the SWTC. The original design of the earthworks at Cuts 3 and 4 comprised benched cuts of 63 and 46 m height, respectively, with batters of 1:1.5 to 1:2, to be determined on conditions actually encountered (see Photo 16.1: Oblique aerial view of Cut 3 and Cut 4 during construction, July 2007).

In July 2007 while bulk earthworks were nearing completion, instability occurred within a section of the benched profile of the northern face of Cut 4.

At that time, Cuts 3 and 4 were in the process of being excavated to original design, with the proposed carriageway level at RL82–RL83 m. The excavations were mostly completed for both the Cuts except for along the lower southern bench, batters and carriageway of Cut 3.

To assess the possible extent and nature of the instability, additional investigation boreholes were undertaken in August and September 2007 and a report recommending remedial works on the basal southern batters of Cut 3 and Cut 4 was presented in November 2007. The works, which involved installation of nails and deep drain holes along the basal batters and surface shotcreting, were subsequently commenced in late 2007.

In February 2008, following significant rainfall periods, and while the remedial works were still underway, slumping and failure occurred along a section of the basal batter slope of Cut 3 south between Chainage 9620 and Chainage 9650.

In association with the slumping, some relative displacement was also evident along a geological contact which tracked south-west up the slope (Refer Photo 16.3). Within a few days, shear cracking became evident in the asphalt of the road pavement at the base of the cut (see Photo 16.2), as well as an extension of the shear

Photo 16.1 Oblique aerial view of Cut 3 and 4. July 2007

Photo 16.2 Shear movement in road pavement, Cut 3. February 2008

Photo 16.3 Movement along contact zone, Bench 2 to Bench 3, Cut 3 south. February 2008

movement in the Cut 3 north roadside batter. At this stage, it was recommended that movement monitoring of surface monuments in proximity to the contact and the slumped area be commenced, together with installation of inclinometers to detect the suspected deep-seated slip surface.

The surface survey confirmed that surface movements were occurring in Cut 3, and initial results from the first inclinometers installed indicated that deep-seated movements were also occurring. A more detailed investigation programme was then proposed, which developed into a wide-ranging and extensive programme of fieldwork, testing and analysis. In particular, data have been gathered to relate rates of movement with rainfall events and groundwater levels.

16.2 Investigations Undertaken

The landslip features have been studied by a combination of techniques, as summarised below.

- Air photo studies by Golder and external consultants. Use has been made of stereo air photography sourced from Sunmap, and a series of oblique air photos taken during construction, provided by the Department of Transport and Main Roads.
- Drilling/geological mapping/test pitting. A comprehensive programme of additional drilling and test pitting has been undertaken during the period 2007–2009. Approximately, 4.8 km of core was recovered. The geological map includes data from 56 test pits excavated outside of the slope remedial work area, to the east, west and north of Cuts 3 and 4.
- Surface level monitoring. A survey team from the DTMR has undertaken monitoring of the level and location of a large number of surface monuments, commencing in April 2008, and read at approximately 2 weekly intervals. The reading interval was increased once every 2 months after the first year. Attempts were also been made to monitor surface movements by means of radar and laser surveys.
- Inclinometers. A total of 46 inclinometers have been installed and monitored by the DTMR. A number of installations have been duplicated where excessive movements have rendered the original installations unusable (see Fig. 16.1—location of instrumentation, and Fig. 16.2).

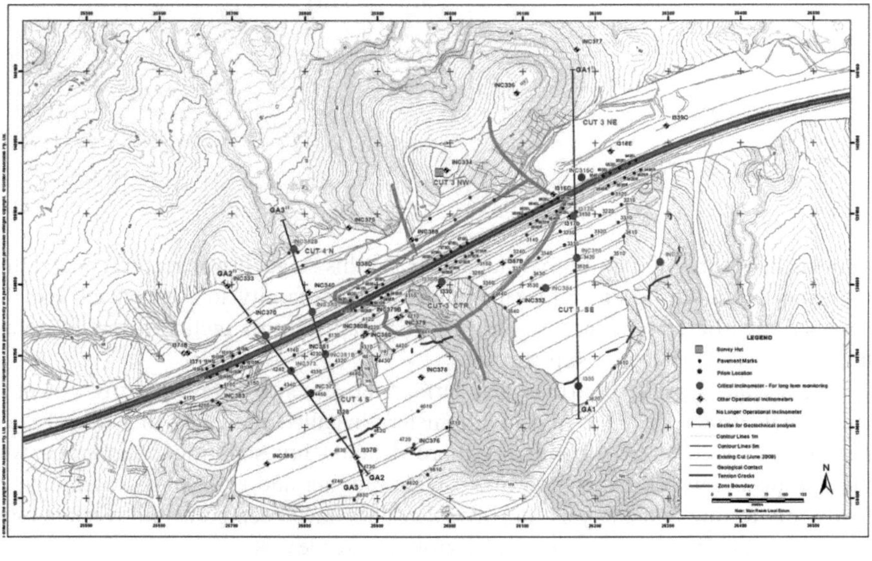

Fig. 16.1 Cuts 3 and 4, geotechnical investigation and location of instrumentation as at opening of road in 2009

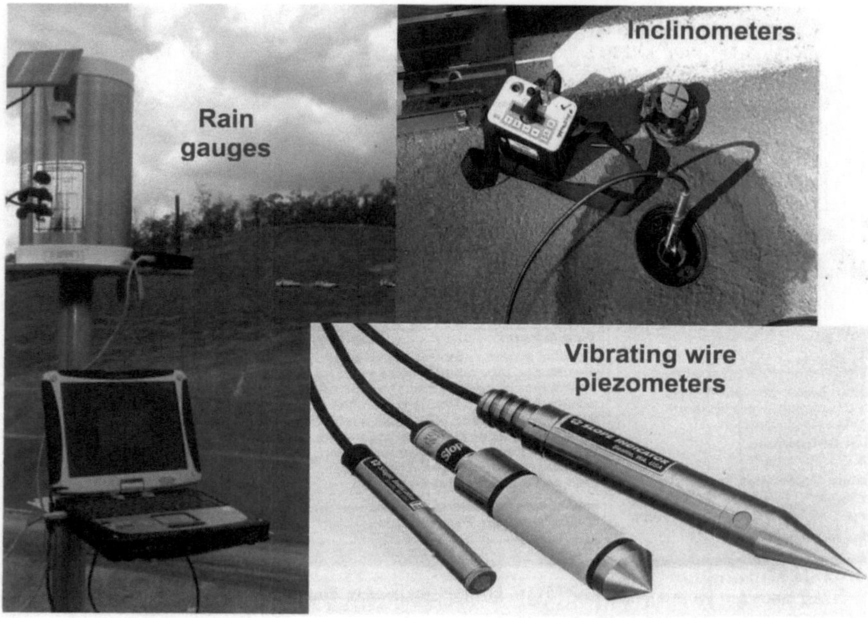

Fig. 16.2 Geotechnical instrumentation

- Groundwater monitoring from Vibrating Wire Piezometers. Piezometers installed in boreholes by the DMR have been read using data loggers (see Fig. 16.2).
- Geophysical surveys. As part of the additional geotechnical investigations, the stratigraphy and structures in the vicinity of Cuts 3 and 4 have been investigated by a Ground Magnetics Survey conducted by Terra Search Pty Ltd. The results have been incoroprated within the Geological Plan of the landslide area (Fig. 16.3).
- Laboratory testing. A series of ring shear tests were undertaken on remoulded material recovered from the shear surface sampled from core drilling.

16.3 Geology

Published geological maps and memoirs prepared by the Department of Mines indicate that the rocks exposed in Cuts 3 and 4 are part of the Silkstone Formation, of the Tertiary Booval Group, which is overlain by trachyte of the Tertiary Hypabyssal Intrusions. The effects of the trachyte intrusions are a significant factor in an understanding of the extremely complex geology of Cuts 3 and 4.

Interpretation of the geology was carried out with the aid of a three-dimensional model of the geology, developed using the geological modelling and mining software VULCAN. A typical cross section produced from the VULCAN model and screen capture from the model itself are shown as Figs. 16.4 and 16.5.

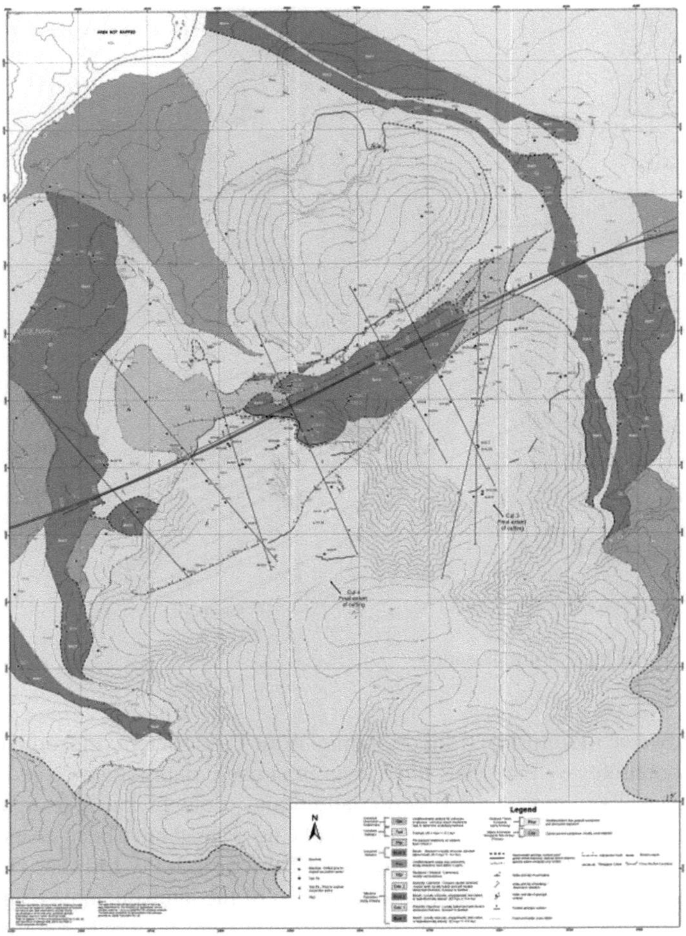

Fig. 16.3 Geological plan of the landslide area

A detailed description of the geology is beyond the scope of this paper, but it is of interest to note the following features.

- The area is on the south-east margin of the Redbank Plains basin. After uplift and erosion of the Ripley Road Sandstones this area became a depositional basin over 53 Ma. Initially, reworked deposits derived from erosion of the Ripley Road Sandstones were deposited as fine grained labile sandstones. Over this layer, laminated mudstones and siltstones were deposited.
- The presence of overlying basalt and dolomitic claystone sequences suggest a fairly uniform depositional environment, consisting of lacustrine lakes with periods of aridity and volcanic activity. The two basalts have been age dated at

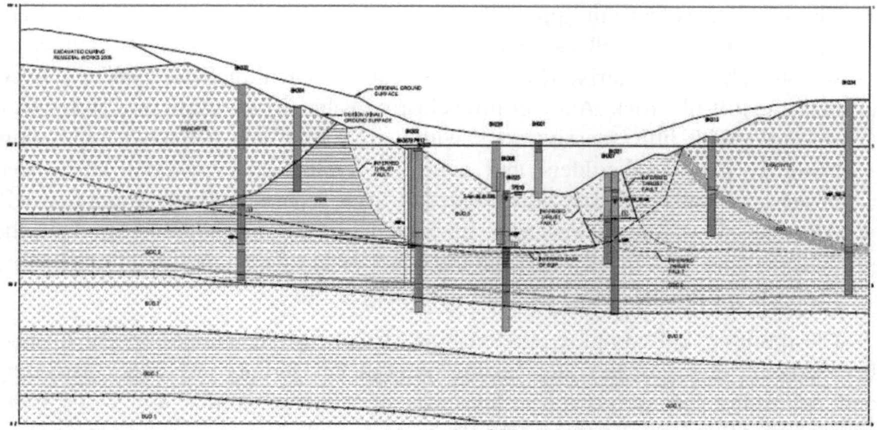

Fig. 16.4 Typical cross section through Cut

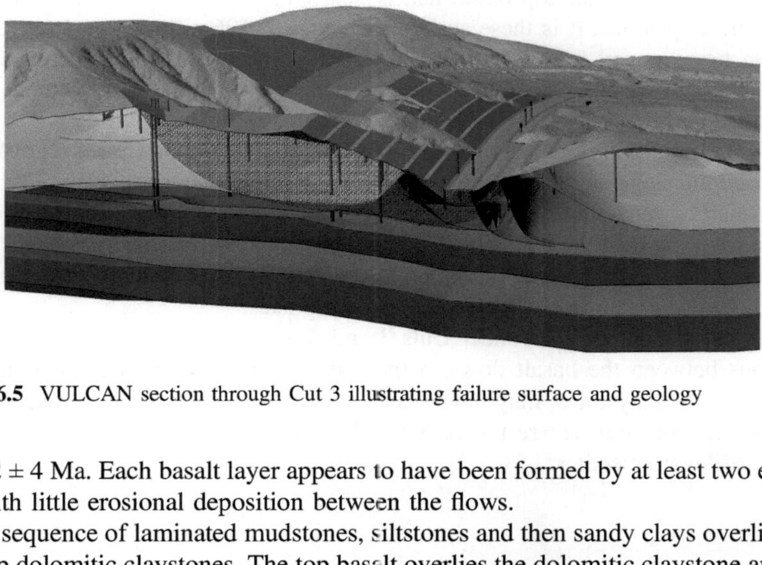

Fig. 16.5 VULCAN section through Cut 3 illustrating failure surface and geology

52 ± 4 Ma. Each basalt layer appears to have been formed by at least two events with little erosional deposition between the flows.

- A sequence of laminated mudstones, siltstones and then sandy clays overlies the top dolomitic claystones. The top basalt overlies the dolomitic claystone and the sediments. This basalt has been age dated at 26.3 ± 0.3 Ma and is likely to indicate a reactivation of volcanicity with flows over an eroded undulating topography. There is also evidence that this basalt may have been deposited within water, as evidence by poorly developed pillow lava structures identified within the flow.
- Soon after extrusion of this top basalt (within about 0.3 million years according to the age dating) the trachyte volcanism occurred. There is evidence that erosion of the top basalt followed by deposition of thin sediments occurred prior to the initial trachyte placements. This could indicate rapid uplift and erosion.

- The trachyte volcanism appears to have occurred in at least two events. The trachyte to the north of Cut 3 is highly fractured with large clasts of trachyte within a shattered matrix. This could indicate flow within water causing a hyaloclastite tuff-like rock. Another flow clearly visible on the south of Cut 4 is well shattered with little massive rock within it. Above this a trachyte with flow banding, massive boulders and areas of hyaloclastite material have been observed. Finally, a clearly flow-banded trachyte is observed in the south of Cut 3.
- An area of weathered vesicular trachyte in Cut 3 south (designated as the "Cappuccino rock"), may indicate the effect of hot gases and steam originating from the baked sediments beneath the trachyte flow.

The multiple and complex relationships of the trachyte masses appear to have been a result of intrusion and extrusion through the Tertiary sequence, displacing and deforming the surrounding claystones and altered basalts above the more competent dolomitic claystones. These viscous and thick flows appear to have bulldozed the softer and more ductile sediments.

It is postulated that these complex trachyte flows have created initial sheared surfaces well below the top basalt and typically at the top of the more competent dolomitic claystones. It is these surfaces which, after 26 million years (or less) of erosion have moved following the removal of some of the surrounding Tertiary sediments and basalts, creating conditions which have initiated movement along weakened surfaces below the base of the trachyte.

Construction activities following the initial excavation of Cuts 3 and 4 have increased the slope angle and further reduced the safety factors, and caused the rate of downslope movement to increase. Stripping of the vegetation and topsoil and exposure of the trachytes directly to precipitation, may have also increased surface water ingress rates, causing rapid build-up in water levels in the trachyte and reducing effective stresses on the inferred slip planes.

The movement surfaces under Cuts 3 and 4 appear to be within the claystone horizons between the basalt flows in the Silkstone Formation. The mineralogical content of the claystones may have contributed to the low residual strength properties of this material. The results of X-ray diffraction analyses indicate the presence of mixed-layer smectites.

16.4 Instrumentation and Monitoring

Monitoring of slope movements is being carried out using surface targets (the system has been upgraded to include prisms as targets), and monitoring of displacements in inclinometers. Monitoring of displacements in some locations commenced in late March 2008, and additional monitoring locations were added later in 2008 and in the period up to 30 June 2009. Figure 16.6 shows the relationship between groundwater levels and rates of movement.

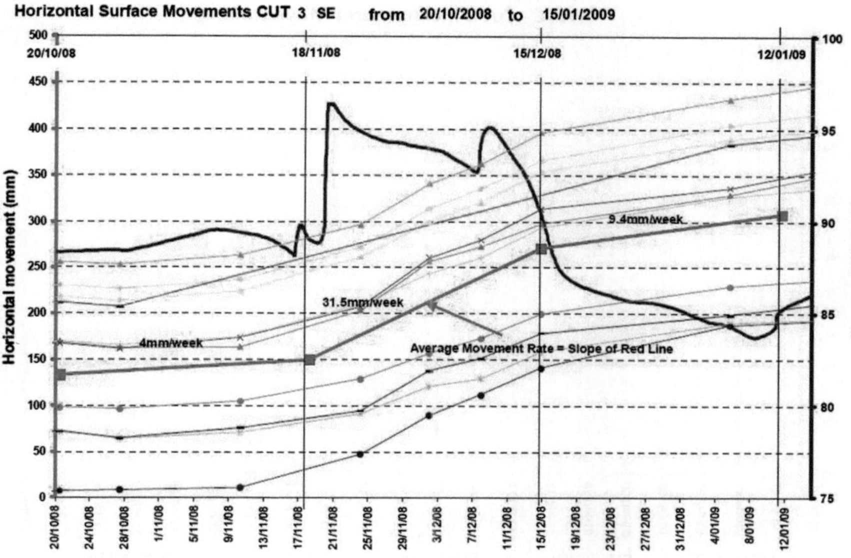

Fig. 16.6 Relationship between groundwater level and rate of movement

Figures 16.7 and 16.8 plot the direction of movement and rates of movement derived from inclinometer monitoring. The inclinometer readings indicate the depth at which shear displacement is occurring in each of the inclinometers. This data was very useful for modelling the geometry of slip surfaces for numerical stability analyses.

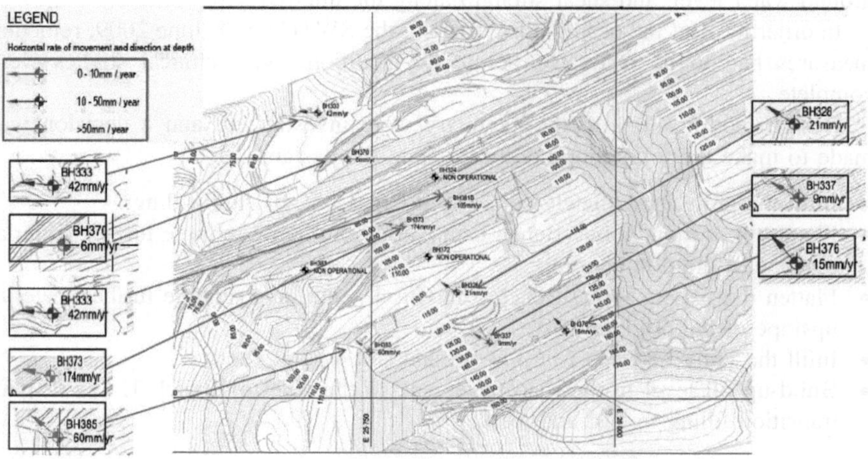

Fig. 16.7 Inclinometer movements (November 2010–March 2011)

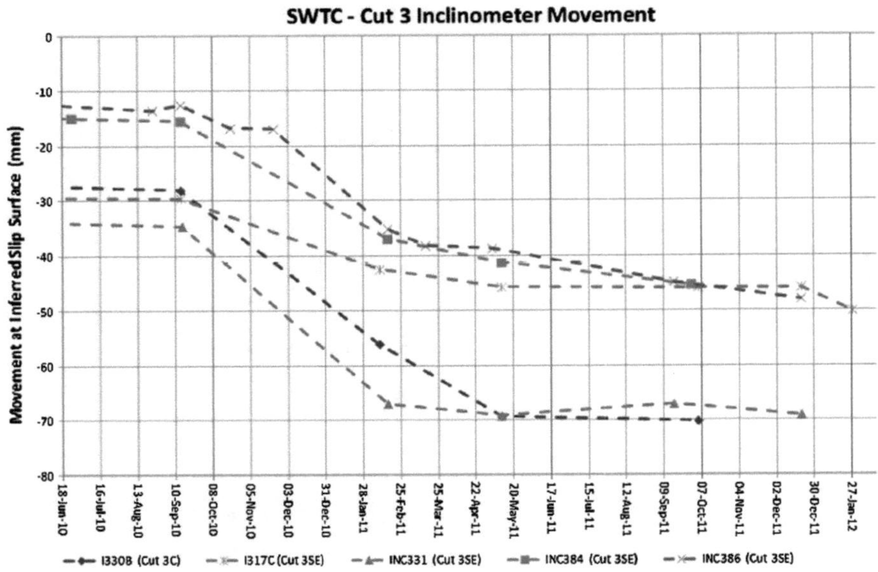

Fig. 16.8 Magnitude of inclinometer movements—June 2010–January 2012

16.5 Remedial Measures

Stability analyses undertaken from mid 2008 demonstrated that the stability of the high batters forming Cut 3 south and Cut 4 south are related to the geometry of the surface and the inferred slip plane (particularly shape—planar vs. circular), the ground water level, and shear strength along the inferred slip plane.

In order to meet the deadline for opening the SWTC on 28 June 2009, remedial measures had to be instigated before the additional geotechnical studies were complete.

Design modifications to the geometry were investigated, and a decision was made to make the following modifications to the design:

- Infill the gully immediately adjacent to Cut 3 NE, by bulk filling;
- Flatten the batters above Bench 6 in Cut 3 south, and remove further material upslope of the original crest;
- Flatten the batters above Bench 5 in Cut 4 south, and remove further material upslope of the original crest;
- Infill the gully between Cut 3 south and Cut 4 south; and
- Build-up fill level in the road pavement, to the level of Bench 1, Cut 3, with transition filling to east and west.

Photo 16.4 Cut 3 south and Cut 4 south, after crest removal March to June 2009

In addition, permanent borehole pumping facilities have been installed from batters at selected locations in Cut 3 south and Cut 4 south. An oblique air photo showing remedial works commenced in March 2009 is presented as Photo 16.4.

16.6 Numerical Analyses

Stability analysis for Cuts 3 and 4 have been carried out using two-dimensional cross-sectional models. For both Cuts, the models have been set up to represent the conditions along cross sections that are oriented in the general direction of maximum observed movement. The model geometry adopted for Cuts 3 and 4 is illustrated in Fig. 16.9.

Inclinometer monitoring indicates that for both Cuts 3 and 4, the failure surface passes through a slickensided zone that is present beneath the trachyte and near the surface of the dolomitic claystone that underlies the cut. Stability modelling was constrained by explicitly modelling the measured geometry of the failure surface. Preliminary analyses were carried out as back analyses to assess the angle of friction for the material at the failure plane.

For Cut 3, monitoring results were used to develop a relationship between the rate of movement and the average groundwater level in the trachyte mass that was moving, for the original design slope geometry. The relationship is illustrated in Fig. 16.6. The available data suggest that the rate of movement reduces to zero (i.e. Factor of Safety equal to 1.0) for an average groundwater level of RL 85 m. Using this groundwater level and the geometry illustrated in Fig. 16.9, back analysis indicates a friction angle of 8° for the material at the failure plane. This compares to a value of 8.8° measured in a ring shear test carried out on a sample of claystone.

Forward analyses were then carried out to assess Factors of Safety as a function of groundwater level, for the revised geometry following the remedial works

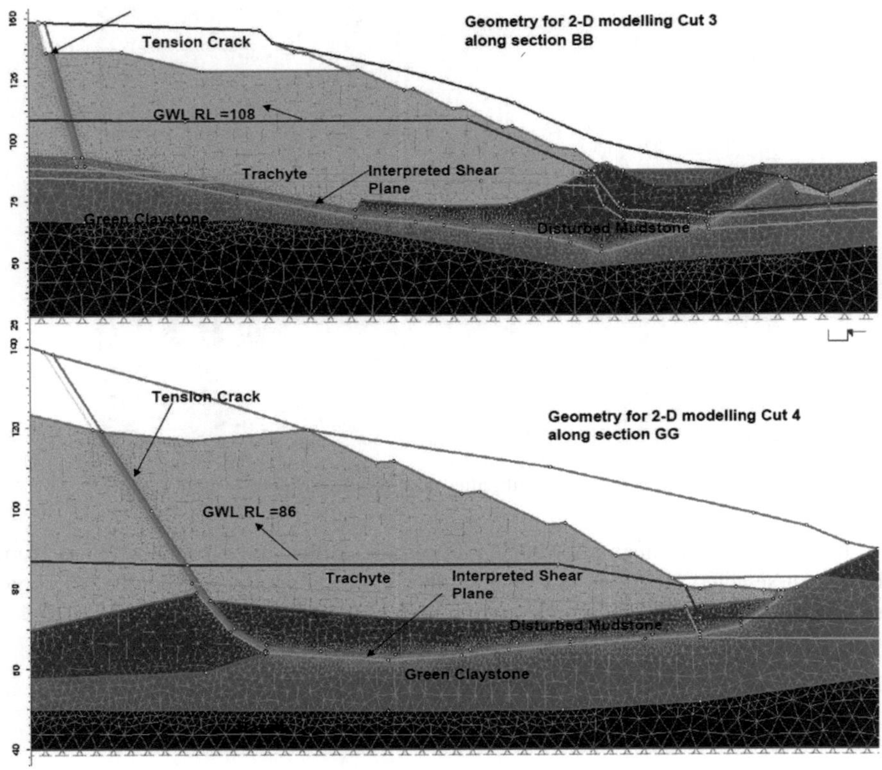

Fig. 16.9 Geometry used in stability analysis

discussed above. The results of these forward analyses indicate that for the revised geometry, the average groundwater level in Cut 3 would need to increase to RL 108 m for the Factor of Safety to reduce to 1.0, whereas it would need to increase to approximately RL 95 m in Cut 4.

Management protocols have been established to allow measures such as reduced speeds or road closures in the event that groundwater levels increase and predicted Factors of Safety therefore decrease to unacceptably low levels. The management system is based on real-time monitoring of rainfall and groundwater levels, the results of which are available through a Web-based data retrieval system.

16.7 Web-Based Monitoring System

A Web-based data retrieval system has been set up, based on a modem to IP transmittal of real-time monitoring information. The system captures and stores the following information:

- Inclinometer movements
- Surface movements, based on an automatic total station laser acquisition system
- Rainfall from automatic rain gauges
- Water levels from VWP piezometers
- Pump well performance

The database platform is based on an SQLL server, with an interface with ESRI ArcGIS Server.

Acknowledgments The authors wish to acknowledge all of the parties involved in the design and construction of the Centenary Highway Extension, Package B. The design was undertaken by Sinclair Knight Merz, with Connell Wagner acting as Superintendent during construction. Project Management was undertaken by Transport and Main Roads, SEQ Projects. DTMR Geotechnical Section provided advice and encouragement during the investigations, and invaluable expertise in the installation of inclinometers and piezometers. Leighton Contractors undertook construction of Package B and the remedial works for Cuts 3 and 4. Golder, was the geotechnical consultant for the project, and an Independent Review was undertaken by Professor Steve Hencher, University of Leeds, UK.

Chapter 17
Post-liquefaction Data Collection and Analyses for Earthquakes in New Zealand

Md. Mizanur Rahman and T.G. Sitharam

Abstract Post-liquefaction data collection is a long, time consuming and laborious process. After Darfield earthquake on 4 September 2010 in New Zealand, national and international post-liquefaction reconnaissance survey teams were formed within weeks. However, huge information was available within days in social electronic media such as blogs, Facebook, etc. These information were collected and used in initial reconnaissance planning and early reports. These new methods of data collection are discussed in this paper. Then, site investigation processes such as lateral spreading, SWS (Swedish Sounding), CPT (Cone Penetration Test), etc., are discussed briefly. The post-liquefied CPT data reveal that soils are still liquefiable even after initial liquefaction during Darfield earthquake. This explains why some sites were re-liquefied during aftershocks. This paper also discusses about post-liquefaction data collection and analysis for identifying re-liquefiable sites.

Keywords Post-liquefaction reconnaissance · Initial planning · Data collection · CPT data

17.1 Introduction

A Mw 7.1 inland earthquake struck South Island of New Zealand on 4th September at 4.35 am. The epicenter of the earthquake was approximately 10 km below (focal depth) and 10 km southeast of the town of Darfield which is about 40 km west of

Md. Mizanur Rahman (✉) · T.G. Sitharam
Department of Civil Engineering, Indian Institute of Science, Bangalore, India
e-mail: Mizanur.Rahman@unisa.edu.au

Md. Mizanur Rahman
School of Natural and Built Environments, University of South Australia, Adelaide 5095, Australia

V.V.S. Rao and G.L. Sivakumar Babu (eds.), *Forensic Geotechnical Engineering*, Developments in Geotechnical Engineering, DOI 10.1007/978-81-322-2377-1_17

the city of Christchurch. Then, two other major earthquakes of Mw 6.3 and Mw 6.0 struck Christchurch on 22 February, 2011 and 13 June, 2011, respectively. These three earthquakes caused widespread liquefaction and lateral spreading in Christchurch and the town of Kaiapoi. Several international groups such as GEER (Geo-engineering Extreme Event Reconnaissance) USA, JGS (Japanese Geotechnical Society) members, etc., arrived in Christchurch within weeks and participated in reconnaissance survey. However, many national groups were formed within days from Auckland University, University of Canterbury and many other institutions to conduct reconnaissance survey and to collect post-liquefaction data. The evidence of liquefaction and lateral spreading in Christchurch are most widely documented and contributed in a large number of scientific articles (Cubrinovski et al. 2010, 2012; Kaiser et al. 2012). While some international and national groups were well organised and equipped, they started week after the earthquake. However, many individuals were collecting scientific evidence right after the earthquake and posting them in electronic social media and blogs. This information was important and significantly assisted in planning reconnaissance survey and in writing early reports and articles. This was a new dimension in forensic data collection after earthquakes in New Zealand. The first author was also involved in post-liquefaction data collection soon after the 4th September earthquake to the end of December 2010, and was one of the few who shared Darfiled earthquake photos specific to soil liquefaction in social media such as Facebook (Rahman 2010) few hours after the first earthquake which attracted great attention to public and geotechnical engineers. This article discusses about the role of different electronic media in sharing scientific information and their contribution towards early scientific reports and articles. This paper also discusses about post-liquefaction site investigation and explains why soils in Christchurch re-liquefied during moderate aftershocks.

17.2 Christchurch Geology

The city of Christchurch and the nearby town of Kaiapoi (17 km from CBD) are situated near to the east coast of Canterbury plains, which is the largest flat land in New Zealand, formed by glacial fed rivers (Cubrinovski et al. 2010). Christchurch and Kaiapoi are 450 and 5 km^2 (approx.) in size with a population of 350,000 and 5,000 (approx.), respectively. The city was mainly swamp land and regularly flooded by Waimakarri River prior to the establishment of the city in 1850. The Waimakirri River had two major branches: South and North. The position of branches were significantly changed over time and contributed to upper geological formation. The position of the south branch of Waimakirri River as in 1965 is superimposed in present day Google map by Cubrinovski et al. (2010) as shown in Fig. 17.1. The build-up area on old river channel in Kaiapoi has suffered significant

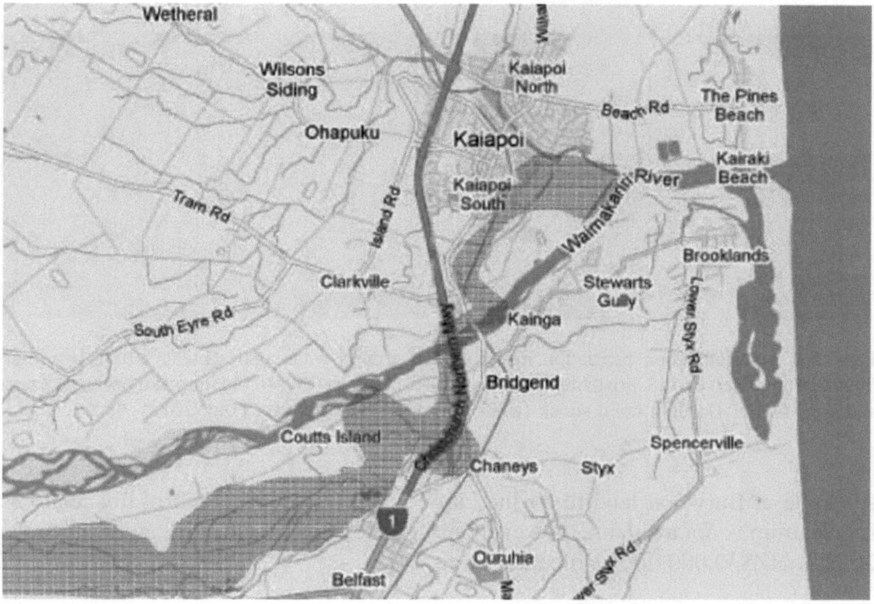

Fig. 17.1 River channel position in 1865 (in *red*) is presented in present day Google map (Cubrinovski et al. 2010)

liquefaction and lateral spreading during the earthquakes. A very similar ground damages were found near to the Avon River which passes through the city of Christchurch.

17.3 Damages Due to Liquefaction

Christchurch was relatively well prepared for soil liquefaction. The liquefaction prediction map of Christchurch before earthquake was available to resident and public use by ECan as shown in Fig. 17.2a. The observed liquefied areas after earthquake are shown in Fig. 17.2b for comparison. Although the city was relatively well prepared, approximately 15,000 houses and residential properties were severely damaged due to liquefaction and almost half of them beyond economic repair (Kaiser et al. 2012). A large amount of silt (sand with fines) and water ejected to surface due to liquefaction during earthquakes which took around 2 months for clean-up after each event. An estimation of over 500,000 tonnes of silt has been

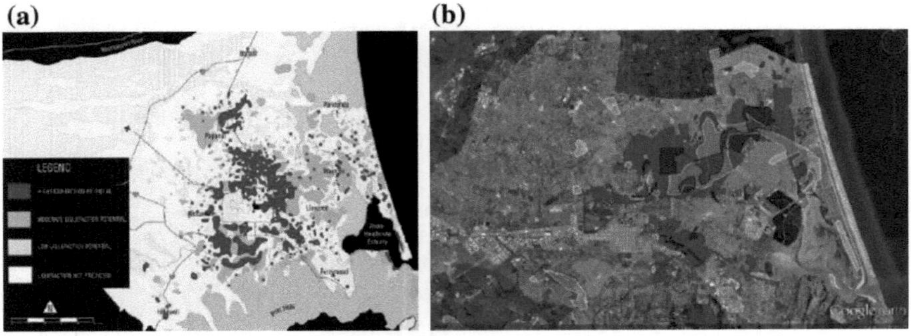

Fig. 17.2 **a** Liquefaction prediction map for high ground water table (ECan 2010), **b** primary liquefaction observed in Christchurch due to the 22 February Mw 6.3 earthquake (*colored areas*) and the Mw 7.1 Darfield main shock (*white contours*) (Cubrinovski et al. 2012)

stockpile at Burwood landfill by the city council. A detail record of financial cost for clean-up is not available, however an estimated cost of clean-up alone at March, 2012 is NZ$30,000,000 (Villemure et al. 2012).

17.4 Data Collection from Electronic Media

17.4.1 Greendale Fault and Blogs

It is likely that Greendale fault was formed during crustal extension more than 50–60 million years ago (Cubrinovski et al. 2010), however it was previously

Fig. 17.3 **a** Greendale fault (GNS); **b** one of many Greendale fault photos that published in Professor Petley's (2010) blog in 18th of September, 2010

unidentified prior to Darfield earthquake in 2010. The Greendale fault is shown in Fig. 17.3a. The fault is about 28 km long and the surface rapture reveal up to a maximum of 4.6 m displacement. Some important information and photo graphic evidence appear as early as in 18th of September, 2010 in electronic blog by Petley (2010) for public use. The photos from the blog revealed that the displacement dominated by dextral (right lateral) movement (Fig. 17.2b). The information in this blog was also significantly contributed in early reports such as Cubrinovski et al. (2010).

17.4.2 Liquefaction Evidence and Social Media

A wide spread silt ejection as found in eastern part of the city, however, some sites were cleaned up from early morning of 4th September before geotechnical experts arrived into the sites. Thus, it was difficult to identify liquefaction in less affected build-up area. However, a large number of photos were shared in social media such as Facebook from 4th September (Rahman 2010; Taylor 2010) and they not only caught great attention to public and engineers, but also helped in identifying liquefaction affected build-up areas. An example of these photos is shown in Fig. 17.4. A large number of these public photos also manually tagged in geographic map to identify failure pattern; a classic example of Greendale fault is shown in Fig. 17.5. A modified version of this information was later published as a journal article (Quigley et al. 2012). Smart phone with GPS camera (e.g. iPhone) was also used to take photo and automatically placed in Google map and later, they contributed significantly to identify liquefied sites.

Fig. 17.4 A photo of one of the liquefaction sites that shared on Facebook o 4th of September (Rahman 2010)

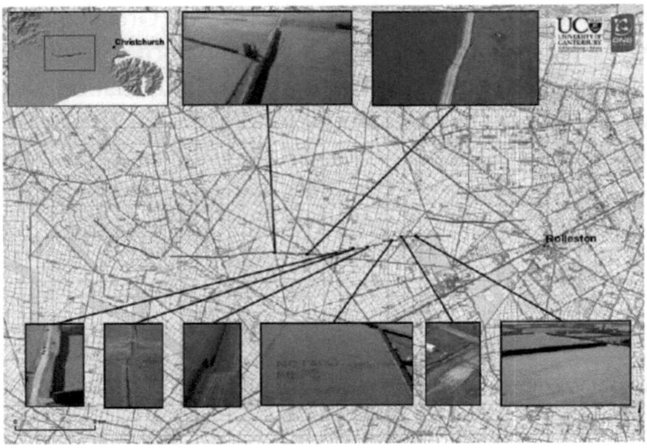

Fig. 17.5 Photo of Greendale fault and their location in geographic map (Quigley 2010)

(a) (b)

Fig. 17.6 Large liquefied site at Hoon Hay, Christchurch, New Zealand; **a** viewed as in Google Earth on 28 of September, 2010 by first author (Google 2010), **b** as viewed on Google map on 15 of December, 2012

17.4.3 Liquefaction Evidence and Google Map (Earth)

The severity of liquefaction was also captured by Google Eye (2010). The images of liquefied sites of Christchurch and Kaiapoi were available in Google Earth week after the earthquake. It gave an opportunity to see ariel view of large liquefied site as shown in Figs. 17.6 and 17.7. The Google Earth, along with ariel photography, was later used in identifying large liquefied site. The light colour of ejected sand

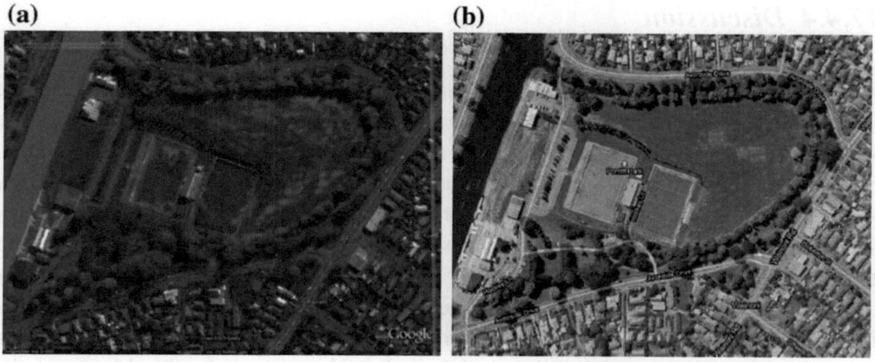

Fig. 17.7 Lateral spread site at Charles st., Kaiapoi, Christchurch, New Zealand; **a** viewed as in Google Earth (Pender et al. 2012), **b** as viewed on Google map on 8 of July, 2011

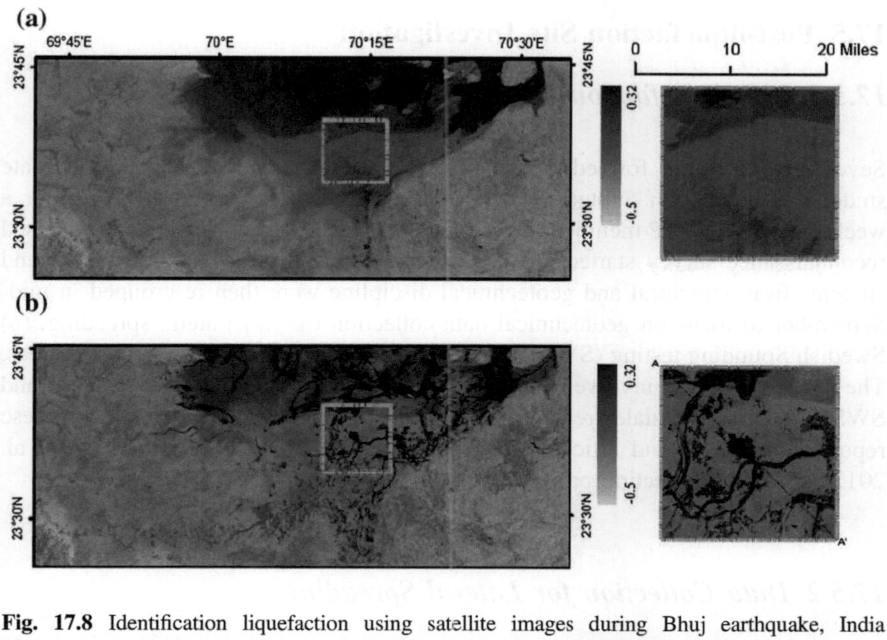

Fig. 17.8 Identification liquefaction using satellite images during Bhuj earthquake, India (Oommen et al. 2011). **a** Pre-event tasseled cap wetness image (05-Nov-2000). **b** Post-event tasseled cap wetness image (09-Feb-2001)

and water contrast very well with green land scape of New Zealand. A detail about this technique was published in Pender et al. (2012). Based on the similar principle, large liquefied sites due to Bhuj earthquake in India were identified using satellite images by Oommen et al. (2011) as shown in Fig. 17.8.

17.4.4 Discussion

As shown in the above discussion that voluntary and freely available public information in electronic media such as blogs, Facebook, Google, etc., could be very useful in identifying and characterising physical characteristic of a natural disaster such as earthquake induced liquefaction. It would be very useful for early stage reconnaissance survey planning and setting strategies. This could also play significant role in developing countries where access to the site and advanced engineering equipment for post-disaster investigation is not easily available. Early data or photos tagged with geo-referenced location map would give very essential information. A little guide from authorities to public could help in developing a great wealth of information from social media. However, one should screen the information and be cautious in using them in engineering practice.

17.5 Post-liquefaction Site Investigation

17.5.1 Post-liquefaction Data Collection

Several groups were formed from University Canterbury staff and postgraduate students after the 4th September earthquake for reconnaissance survey. After a week later GEER, JSP members, etc., joined national groups. Thus, more organised reconnaissance survey started one week later. University of Canterbury staff and students from structural and geotechnical discipline were then re-grouped in mid-September to focus on geotechnical data collection of—(a) Lateral spreading, (b) Swedish Sounding testing (SWS) and (c) Settlement of buildings in liquefied areas. The first author was involved in collecting lateral spreading data at Kaiapoi and SWS testing in Avondale area up to the end of October, 2010. The details of these reported in reports and articles (Cubrinovski et al. 2010, 2012; Robinson et al. 2012). The data collection process is explained below.

17.5.2 Data Collection for Lateral Spreading

An example of lateral spreading is shown in Fig. 17.9. The site was close to the shed as shown in Fig. 17.10b. Electronic distance metre was used for recording the lateral spreading opening and settlement. GPS, as shown in Fig. 17.10a, was used during this study to keep trace of transects as shown in blue line in Fig. 17.10b. First an intersect was set in the field to interest most of the lateral spreading.

Fig. 17.9 Lateral spreading at South Kaiapoi, Christchurch, New Zealand

(a) **(b)**

Fig. 17.10 Lateral spreading data collection; **a** GPS used in data collection, **b** lateral spreading data super imposed in Google map, published in Cubrinovski et al. (2010)

Then, while tracing along the transect lateral spreading was measured. The data were then plotted in Google map to create the Fig. 17.10b. Numerous analysis of these data have been done and compared with early literature (Cubrinovski et al. 2012; Robinson et al. 2012). The Swedish Sounding (SWS) data on post-liquefied site confirmed that Christchurch soil contains sand with a varying amount of fines.

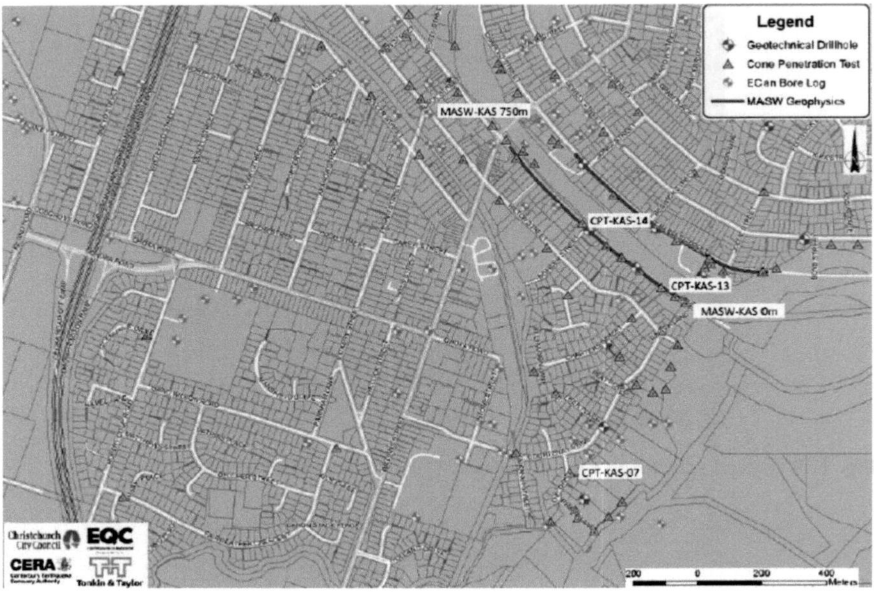

Fig. 17.11 Location of ground investigation by EQC (Earthquake commission) following the 2010–2011 earthquakes

SWS data has a very good correlation with SPT (standard penetration test) and CPT (cone penetration test) test and thus can be used for liquefaction screen as outlined in Youd et al. (2001). Analysis of SWS data for varying fines content showed that very low liquefaction resistance up to 5 m of top soils and this explained why liquefied site re-liquefied during aftershocks (Taylor and Cubrinovski 2011).

17.6 Post-liquefied CPT Test Data

A large number of post-liquefaction site investigations were performed at Kaiapoi by CERA (Canterbury Earthquake Recovery Authority). Some of the borehole and CPT test sites, as shown in Fig. 17.11, are available through their website. The post-liquefied CPT data of site CPT-KAS-07 was analysed as proposed by Youd et al. (2001). For this analysis $a_{max}/g = 0.30$ was taken from Kaiapoi North School site (43°22′12″S–172°36′24″E), water table = 1.5 m and Mw = 6.5. The data showed that the top soil layer up to 5 m depth is highly liquefiable, as shown in Fig. 17.12, even though the sites suffered significant liquefaction during previous earthquake.

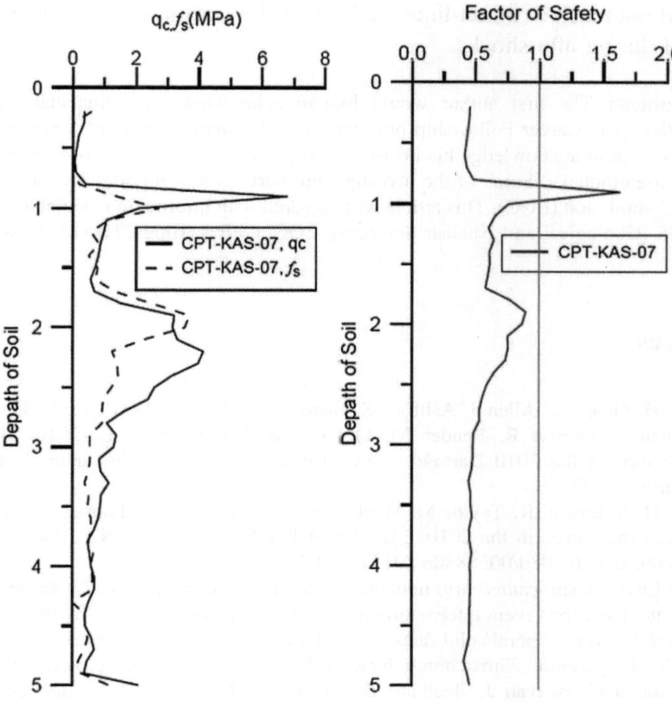

Fig. 17.12 Liquefaction screening from CPT data on site CPT-KAS-07 as suggested Youd et al. (2001)

17.7 Recommendation and Conclusions

Three major earthquakes struck Canterbury plains during September 2010 to June 2011. A wide spread soil liquefaction in the forms of sand boil and lateral spreading were observed in Christchurch and Kaiapoi. National and international reconnaissance survey teams were formed within weeks to better understand liquefaction behaviour. However, some scientific information were available in electronic media such blogs, Facebook, Google, etc., from the first day. These public information in social media were successfully used in initial planning for reconnaissance survey, early reports and even scientific articles. From this experience, it can be recommended that a publicly available repository cloud develop a large amount of information after a natural disaster. However, one should be very careful in collecting and using these public information.

It is generally believed that once a site liquefied as a form of sand boil, soil reached a denser state and may not be liquefied in future earthquake event. However, site investigation (SWS, CPT) exhibited that soil in Christchurch and

Kaiapoi did not reach to a non-liquefiable state. This explains why some sites were re-liquefied during aftershocks.

Acknowledgments The first author would like to acknowledge the financial support from Australia–India Early Career Fellowship program to visit Indian Institute of Science, Bengalore. He would also like to acknowledge his Post-doctoral position at University of Canterbury during 4th September earthquake. Some of the investigation works were supported by the New Zealand Earthquake Commission (EQC). This article was prepared with information available in website of GeoNet, GNS (Geological and Nuclear Sciences), EQC, CERA, Blogs, Facebook and Google.

References

Cubrinovski M, Green R, Allen J, Ashford S, Bowman E, Bradley BA, Cox B, Hutchinson T, Kavazanjian E, Orense R, Pender M, Quigley M, Wotherspoon L (2010) Geotechnical reconnaissance of the 2010 Darfield (New Zealand) earthquake. University of Canterbury, Christchurch, p 173

Cubrinovski M, Robinson K, Taylor M, Hughes M, Orense R (2012) Lateral spreading and its impacts in urban areas in the 2010–2011 Christchurch earthquakes. NZ J Geol Geophys 55 (3):255–269. doi:10.1080/00288306.2012.699895

ECan (2010) Environment canterbury: liquefaction potential hazard map for Christchurch for high ground water table (pre-event information provided to resident and public by ECan. http://ecan.govt.nz/publications/General/solid-facts-Christchurch-liquefaction.pdf. Accessed 15 Dec 2012

Google (2010) Google earth: Christchurch, New Zealand. http://www.google.com/earth/index.html

Kaiser A, Holden C, Beavan J, Beetham D, Benites R, Celentano A, Collett D, Cousins J, Cubrinovski M, Dellow G, Denys P, Fielding E, Fry B, Gerstenberger M, Langridge R, Massey C, Motagh M, Pondard N, McVerry G, Ristau J, Stirling M, Thomas J, Uma SR, Zhao J (2012) The Mw 6.2 Christchurch earthquake of February 2011: preliminary report. NZ J Geol Geophys 55(1):67–90. doi:10.1080/00288306.2011.641182

Oommen T, Baise LG, Gens R, Prakash A, Gupta RP (2011) Collaborative research with Tuft University and Universty of Alaska Fairbanks: application of satellite data for post-liquefaction reconnaissance. USGS award number: G10AP00025 and G00010AP00026

Pender M, Wotherspoon L, Cubrinovski M, Bowman E, Orense R (2012) Evidence of earthquake-induced liquefaction obtained from GeoEye-1 images. Geotech Lett 2:49–53. doi:10.1680/geolett.11.00042

Petley D (2010) Daveslandslidesblog. http://daveslandslideblog.blogspot.com/2010/09/images-of-darfield-canterbury.html. Accessed 10 Sept 2010

Quigley M (2010) Recent canterbury earthquakes: greendale fault: September 4 2010 earthquake. http://www.geol.canterbury.ac.nz/earthquake/map_photos.pdf. Accessed 15 Sept 2012

Quigley M, Duffy B, Woodhead J, Horton T, Haig D, Moody L (2012) U-Pb dating of a terminal Pliocene coral from the Indonesian Seaway. Mar Geol 311–314:57–62. doi:10.1016/j.margeo.2012.01.004

Rahman MM (2010) Christchurch earthquake, September 4, 2010. http://www.facebook.com/album.php?aid=81111&id=1513056304&l=d70559a8bd. Accessed 4 Sept 2010

Robinson K, Bradley BA, Cubrinovski M (2012) Analysis of liquefaction-induced lateral spreading data from the 2010 Darfield and 2011 Christchurch earthquakes. In: Proceedings of the New Zealand society for earthquake engineering: 2012 annual technical conference (NZSEE), Christchurch, 13–15 April 2012

Taylor M (2010) Christchurch earthquake 4 September 2010. http://www.facebook.com/album.php?aid=271722&id=600410089&l=4c13294a1a. Accessed 7 Sept 2010

Taylor ML, Cubrinovski M (2011) Preliminary assessment of liquefaction in urban areas following the 2010 Darfield earthquake. In: Proceedings of the ninth pacific conference on earthquake engineering building an earthquake-resilient society, NZSEE, Auckland, New Zealand, p 056, 14–16 April 2011

Villemure M, Wilson TM, Bristow D, Gallagher M, Giovinazzi S, Brown C (2012) Liquefaction ejected clean-up in Christchurch during the 2010–2011 earthquake sequence. In: Proceedings of the 12 NZSEE conference, New Zealand, Paper number 131

Youd TL, Idriss IM, Andrus RD, Arango I, Castro G, Christian JT, Dobry R, Finn WDL, Harder LF, Hynes ME, Ishihara K, Koester JP, Liao SSC, Marcuson WF, Martin GR, Mitchell JK, Moriwaki Y, Power MS, Robertson PK, Seed RB, Stokoe KH (2001) Liquefaction resistance of soils: summary report from the 1996 NCEER and 1998 NCEER/NSF workshops on evaluation of liquefaction resistance of soils. J Geotech Geoenviron Eng 127(10):817–833

Chapter 18
Forensic Geotechnics—Some Case Studies from Singapore

C.F. Leung

Abstract Being an island state, there is abundance of thick marine clay present in various parts of Singapore. This soft marine clay often poses great challenges in geotechnical projects resulting in severe geotechnical problems in a good number of instances. This paper presents three case studies from Singapore on the applications of forensic engineering to geotechnical problems. The investigations on the causes of geotechnical failures highlight that the process of forensic geotechnics is different from the conventional design and analysis in geotechnical practice. The first case study involves a relatively straightforward case of excavation failure in which the consultant who carried out the temporary work design was accused of incompetency. Subsequent analysis revealed that the temporary work contractor was actually at fault as the excavation method statement had not been followed. The second and third cases are more complex as the causes of failure need some efforts to identify.

Keywords Soft soils · Excavation failure · Bund failure · Highway failure

18.1 Introduction

According to the Concise Oxford Dictionary, one definition of forensic engineering is 'dealing with engineering aspects of legal problems'. As such, forensic geotechnical engineering would involve engineering aspects of geotechnical problems or failures in court, arbitration or mediation proceedings. Knowledgeable geotechnical engineers are often involved as expert witnesses either for the plaintiff, defendant or the presiding judge to provide inputs on the causes of geotechnical problems in legal cases. Unlike conventional design and analysis of geotechnical projects, forensic analysis involves very different methodologies due to a variety of

C.F. Leung (✉)
Centre for Soft Ground Engineering, Department of Civil and Environmental Engineering, National University of Singapore, Singapore 117576, Singapore
e-mail: cvelcf@nus.edu.sg

V.V.S. Rao and G.L. Sivakumar Babu (eds.), *Forensic Geotechnical Engineering*, Developments in Geotechnical Engineering, DOI 10.1007/978-81-322-2377-1_18

circumstances such as lacking vital information, uncertainties in ground conditions, as well as roles and responsibilities of personnel in a project etc.

As Singapore consists of one main island surrounded by many other small islands, thick deposits of soft marine clay are commonly encountered. This clay is of recent deposits in geological time and termed geologically as the Kallang Formation (Pitts 1984). The marine clay can be up to 50 m thick with undrained shear strength as low as 10 kPa onshore or 5 kPa along the coast of offshore islands and underneath the seabed. The presence of thick, soft marine clay poses significant challenges to geotechnical engineers as upon excavation, the soil movements can be excessive even with very strong retaining supports. This paper presents three case studies from Singapore on the investigation of causes of geotechnical failures involving soft marine clay. The first case is fairly straightforward while the second and third cases are complex. The methodologies in these cases are presented in this paper.

18.2 First Case on Excavation Failure

During the cause of temporary excavation for the construction of basement in marine clay in a project in Singapore, significant soil movements (see Fig. 18.1) were observed at one side of the site when the excavation reached a depth of 2.7 m. A number of inclinometers installed at the site recorded over 200 mm lateral soil movement while settlement markers revealed soil settlement of over 100 mm at some places. Inspection of the excavation site revealed a diagonal corner strut had buckled, see Fig. 18.2. Remedial action taken at the site included additional lateral supports to stabilise the soil movements. Excavation subsequently continued till the intended basement level. Repair of adjacent properties in the vicinity needed to be carried out which involved substantial cost.

After the project has been completed, the temporary work contractor went bankrupt. To recover the large repair cost for the adjacent properties, the developer of the project went to court to file claims from the consultant accusing its faulty design of the temporary work. Figure 18.3 shows the proposed method statement of the excavation scheme. It stated that excavation was first carried out until a berm remained to support the retaining wall, see Fig. 18.3a. An inclined steel strut as shown in Fig. 18.3b should then be installed before the remaining excavation could be carried out. The consultant knew that they were not at fault as the temporary work contractor did not follow the proposed method statement. For convenience, the contractor simply removed the soil layer by layer horizontally during excavation.

The consultant engaged a geotechnical expert witness to defend the case. The duty of an expert witness is to provide an unbiased independent interpretation on the case, regardless which party engages him/her. The expert then conducted a finite element (FEM) analysis to identify the cause of the excavation failure. The outcomes of the FEM analyses are illustrated in Fig. 18.4. If the method statement is

Fig. 18.1 Significant ground movement observed at an excavation site

Fig. 18.2 Problem of diagonal corner steel strut

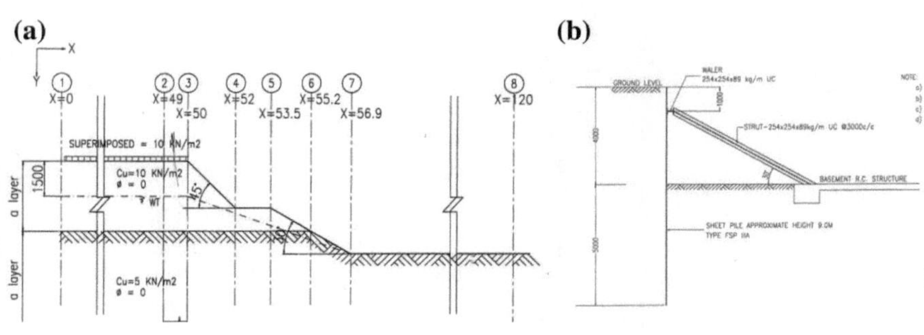

Fig. 18.3 Design soil excavation scheme. **a** Soil berm. **b** Inclined steel strut

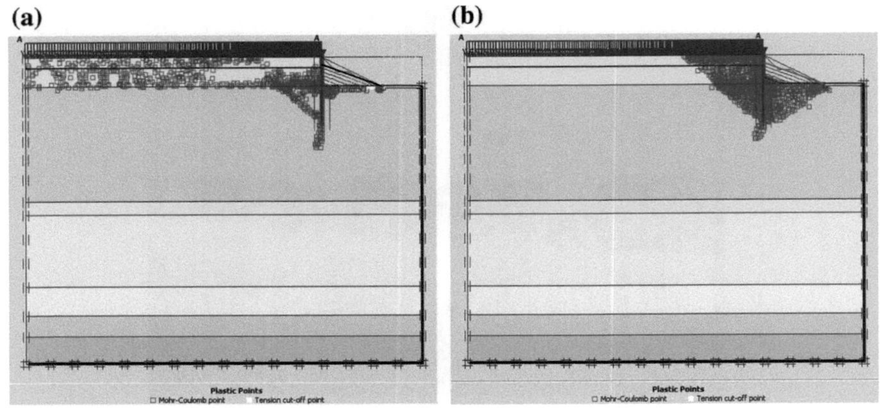

Fig. 18.4 Extent of soil yielding: **a** design scheme and **b** contractor scheme

followed for the excavation, Fig. 18.4a clearly shows that the yielding of soil around the retaining wall is insignificant. The corresponding soil and wall movements as well as induced bending moment on the wall are well within the desired limits. On the other hand, if indeed the soil had been removed horizontally layer by layer, significant soil yielding is noted indicating significant wall movement and bending moments, see Fig. 18.4b.

To further illustrate the cause of failure, the results of horizontal-layered excavation are presented in Fig. 18.5. It is evident from the FEM analysis that the factor of safety of the retaining wall decreased considerably with increasing excavation depth. The wall eventually reached its failure (factor of safety being unity) at an excavation depth of 2.7 m, the same as that observed in the field. With such convincing evidence, the case was settled out of court speedily.

Fig. 18.5 Variation of factor of safety against unsupported excavation depth

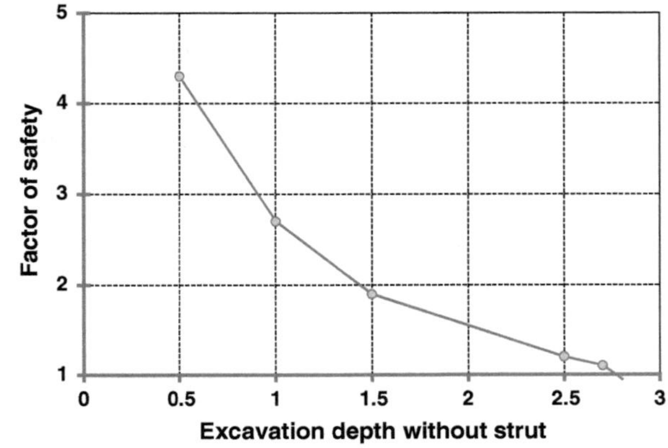

18.3 Second Case on Reclamation Bund Failure

To cater for its rapid economic and population growth, Singapore has increased its land area by over 20 % through land reclamation since its independence in 1965. In fact, land reclamation activities already started in the nineteenth century under the British rule with the port and the river estuary areas being reclaimed. As an example, a seafront road which was named 'Beach Road' is now at least 1 km from the present shore line. Prior to the placement of sand fill, a perimeter bund (Fig. 18.6) which may stretch tens of kilometres in length need to be constructed. The foundation of the bund is either reinforced by installing sand compaction piles (SCP) in the in situ soft soil stratum or by replacement of the in situ soft soil with sand.

It has been reported that the stability of the seaward bund slope is critical during fill placement (Leung and Shen 2004). The gradient of the slope should not be too gentle as it would greatly increase the cost of bund construction especially for reclamation in deep waters. The placement of the sand fill should not be too slow as this would significantly lengthen the construction period resulting in higher cost of construction while too rapid a placement of fill or too steep a bund slope may cause slope instability. Failures of reclamation perimeter bunds had been reported in Singapore and millions of dollars had been spent to fix these failures. A balance must be struck between the stability of the bund slope and cost/time of bund construction.

During the course of construction of the perimeter bund for land reclamation in the western part of Singapore, the bund suddenly failed at one part of the reclaimed land site on one fine day. Millions of dollars had been spent to stabilise the bund and the contractor made insurance claim to partially recover the cost of repair. A geotechnical expert is engaged by the insurance adjuster to identify the causes of the bund failure. The expert first established the soil profile and properties of the site. By examining the site investigation reports prior to and post failure, it is established that the marine clay has a unit weight of 16 kN/m^3 and undrained shear strength of 5 kPa at the top with strength increasing by 1.32 kPa/m with depth. The

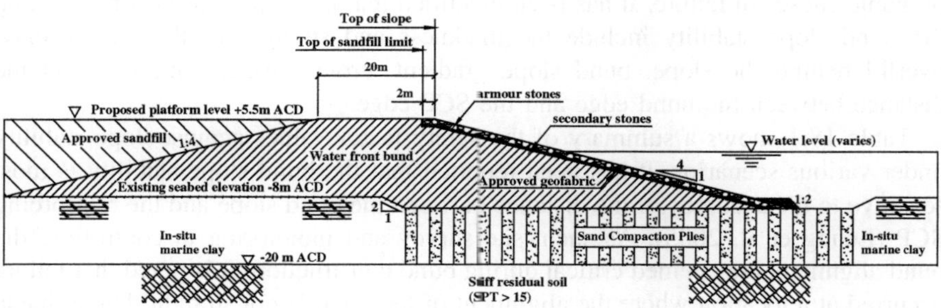

Fig. 18.6 Configuration of perimeter bund for land reclamation (after Leung and Shen 2004)

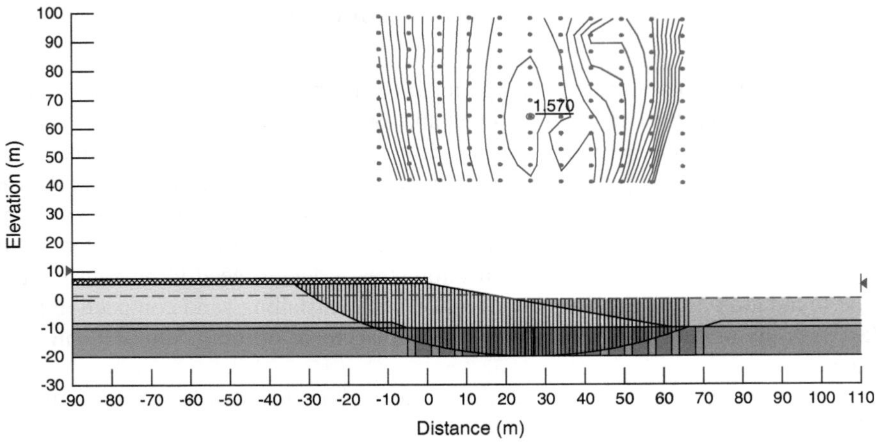

Fig. 18.7 Verification of original design scheme

strength increase is calculated by the product of the effective soil unit weight of 6 kN/m^3 and the coefficient of strength increase upon effective overburden pressure of 0.22. Using proportional area approach (Priebe 1978), the composite soil, consisting of SCP with marine clay in between, is established to have an average unit weight of 17.2 kN/m^3, apparent cohesion of 3.5 kPa and friction angle of 20°. The sand fill has a bulk unit weight of 18 kN/m^3 above water table, 20 kN/m^3 below water table and a friction angle of 30°. The water level in the slope is taken to be 1.5 m above the seawater level.

Using the above independently evaluated soil parameters, the consultant's original design is checked. Figure 18.7 shows that the factor of safety against slope failure is determined to be 1.57 which is not far off from the value of 1.55 given in the consultant's design calculations. This verifies that the design has achieved the desire safety factor of 1.5 for the bund design. With such a relatively high safety factor, the bund slope should not fail even if certain parameters and situations become less favourable. Parametric studies are hence conducted to evaluate the probable causes of failure. It has been identified that the important factors affecting the bund slope stability include the thickness and strength of the marine clay, overfill behind the slope, bund slope gradient, ground water variation, and the distance between the bund edge and the SCP edge.

Table 18.1 shows a summary of the safety factor values against slope stability under various scenarios. It has been reported that the bund slope stability is most sensitive to the accuracy of the distance between the bund slope and the reinforcing SCP (Leung et al. 2005). As such, site survey and monitoring in controlling the bund alignment are deemed critical during bund construction. It is noted that failure occurred at a location where the alignment of the bund is not straight. Under harsh marine conditions, the accuracy of the bund alignment can be very difficult to

Table 18.1 Variations of safety factor against bund failure

Parameter	Particular	Factor of safety	Degree of severity
Original	Original design	1.570	Similar to consultant's value of 1.55
Bund slope alignment	Distance between bund edge and SCP foundation edge shifted by 5–30 m	1.391–0.946	High
Bund gradient	1:3.5 instead of 1:4	1.387	Medium
Soft clay strength	1 kPa/m depth increase instead of 1.32 kPa/m	1.346	Medium
Overfilling	Overfill by 2.5 m	1.335	Medium
Water table	3 m instead of 1.5 m above sea level	1.434	Low
Soft clay thickness	3 m thicker	1.487	Low

control. In addition, overfilling behind slope, variations in soft soil strength and thickness and bund slope gradient may not be the same as planned or intended. Table 18.1 reveals that each of the above factors has various degrees of severity on the stability of the bund slope. It is evident that the distance between the base of the bund slope and the edge of the SCP foundation is most influential affecting the bund stability. This is followed by the soft clay strength, overfilling behind the bund and the bund gradient. On the other hand, the change in water table within the bund and increasing soft clay thickness appear to be less critical on the slope stability.

It can be established from Table 18.1 that a single factor alone could not trigger the bund failure as the adopted safety factor against slope failure of 1.5 has relatively high margin against variations in the different parameters and configurations. Similar to other major geotechnical factors, the cause of this bund failure is due to a combination of multiple factors. As an example, if certain parameters in Table 18.1 occur simultaneously at the same location, namely the underlying soft marine clay is weaker than the typical value, the bund alignment deviates landward for say 13 m relative to the SCP, and the bund has been overfilled too fast to +7.5 mCD without proper control, the bund slope would fail with the factor of safety dropping below unity, as shown in Fig. 18.8. The geotechnical expert thus established that the simultaneous occurrence of several adverse conditions had triggered the failure of the bund slope at the location with the leading cause being the bund alignment.

To examine the mechanism of the slip and the extent of soil yielding, finite element analyses was carried out under a variety of configurations. Figure 18.9 shows the extent of soil yielding for the case of thickness of marine clay varying from 10 to 15 m. By evaluating the extent of the soil yielding, the finite element analyses reveal that the misalignment of the relative distance between the edges of the bund and the compaction pile foundation system would affect the bund stability the most, as compared to other factors such as soft clay strength and thickness,

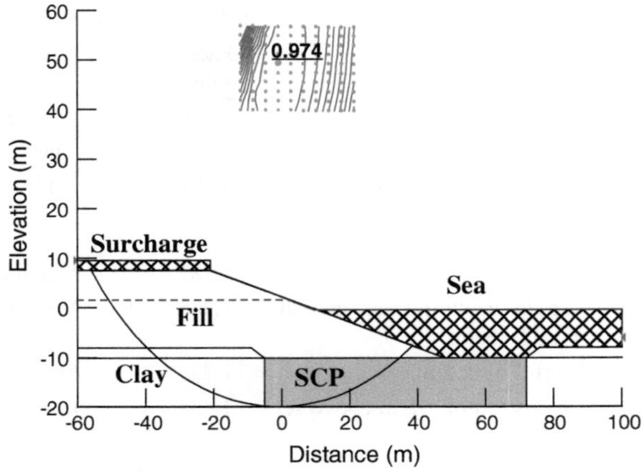

Fig. 18.8 Bund stability under a combination of adverse changes

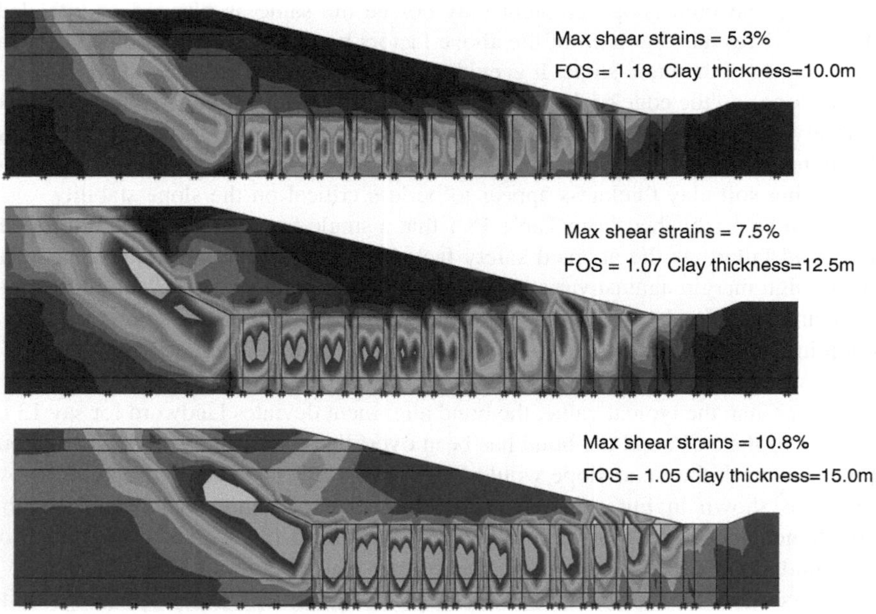

Fig. 18.9 Extent of soil yielding for soft clay thickness ranging from 10 to 15 m

overfilling and bund angle. Although the failure of the bund is triggered by multiple adverse factors, it can be established that the misalignment between the edges of the bund and sand compaction piles is the major factor. Such misalignments might occur at the location which is of curved shape in plan. However, the misalignment alone could not trigger the bund failure. The other major contributing factors include the soft clay strength, bund gradient and overfilling behind the bund slope.

18.4 Third Case on Nicoll Highway Failure

To improve the transport system for its residents, Singapore has been building underground railways since the mid-80s. The construction of the subway station is often carried out by cut and cover method. Such excavation in thick soft marine clay often causes significant lateral soil movements. The original Nicoll Highway Station location has a marine clay stratum as thick as 50 m deep and land reclamation was carried out in 1950s. The station was constructed by excavating down to 36 m with retaining walls using diaphragm with 10 levels of lateral steel struts as well as the soft clay beneath the station base being improved by jet grouting prior to excavation. On 20 April 2004 and upon reaching an excavation depth of 33 m with 9 level struts and top jet grouting layer beneath excavation level removed, a massive retaining wall failure occurred, see Fig. 18.10.

The Singapore government commissioned an enquiry and the report of Commission of Enquiry (COI 2005) was published about a year later. A total of 17 expert witnesses in different areas were engaged by various parties to establish the causes of the massive excavation failure. The instrument observations prior to failure and the findings of the COI are elaborated elsewhere (Leung and Tan 2007). The position of the COI is that the failure did not happen suddenly but progressively. It was concluded that the main causes of the collapse included the underdesign of strut/waler connection detail and the underdesign of the diaphragm wall due to incorrect use of the computer software in the geotechnical finite element analysis. The problems with instrumentation and monitoring are part of the contributing factors for the

Fig. 18.10 Eyewitness's photograph on Nicoll Highway Failure

failure. These design errors resulted in the failure of the waler connection at the 9th level steel strut at about 33 m below the ground. As the load is transferred to the higher struts, the struts failed one by one upward progressively (from 8th level strut to 7th level strut and so on) resulting in the total collapse of the retaining wall system.

The development of the diaphragm wall deflection with time prior to collapse is shown in Fig. 18.11. The south side of the wall was noted to deflect very much more than those on north side after 23 February 2004. As such, the assumption of symmetrical wall sections in the FEM analysis is no longer valid, implying that the soils on the south side are much weaker than those on the north side. As excavation progressed, the south side of the wall continued to displace much more than the north at each stage of the excavation. By 5 March 2004, the first back-analysis trigger level of 176.9 mm was exceeded. By 25 March 2004, the first back-analysis design level of 252.7 mm was exceeded. A second back-analysis was done to justify increasing the trigger level to 251.3 mm and design level to 359 mm. These levels were exceeded on 25 March and 17 April 2004, respectively. The wall eventually collapsed at 3.30 pm on 20 April 2004 upon the removal of the upper JGP and before installation of the final 10th level struts.

The failure of the waler-strut connection at the 9th level strut was examined in detail (Tan et al. 2005). It was established in February 2004 that the plate stiffener (Fig. 18.12) in the waler-strut connection was replaced by the C-channel stiffener (Fig. 18.13). After the failure, laboratory tests were conducted by structural expert to evaluate the performance of the two types of stiffeners. Figure 18.14 shows that

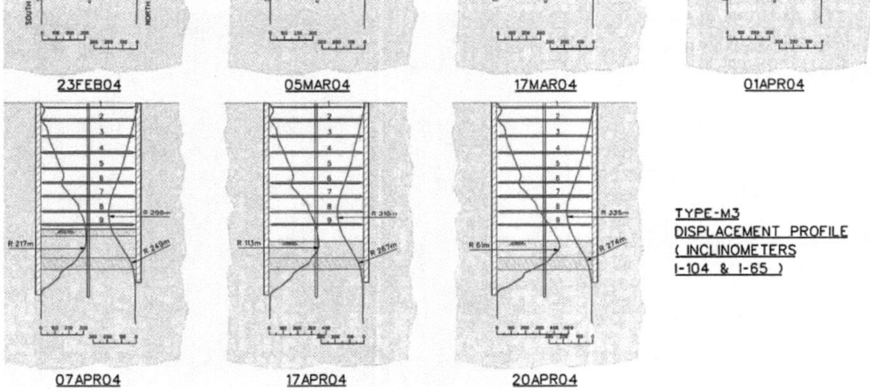

Fig. 18.11 History of wall deflection prior to collapse (after COI 2005)

Fig. 18.12 Plate stiffener (COI 2005)

Fig. 18.13 C-channel stiffener (COI 2005)

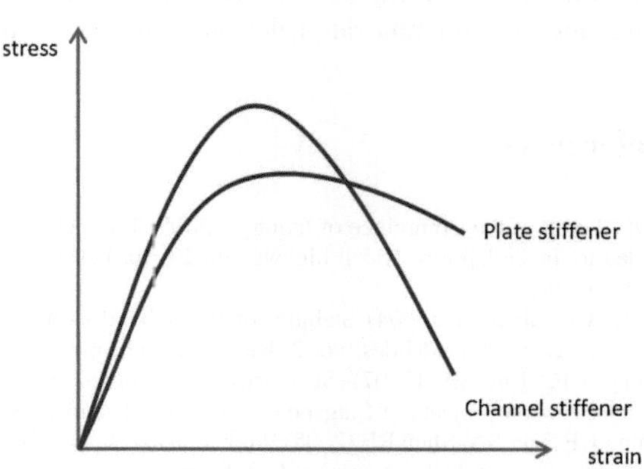

Fig. 18.14 Performance of plate versus C-channel stiffener

the C-channel stiffener would undergo a major reduction in resistance under high deformation as it is generally brittle in nature while the load-displacement response of the plate stiffener is ductile in nature. The replacement of plate stiffener by the channel stiffener is deemed to be a major factor of the waler failure leading to the progressive failure of the entire retaining system.

Beside design errors identified above, other technical and administrative faults were identified by the COI. There were evidently human errors with inadequate care since early days of the construction. Personnel at the site did not take the warning signs seriously. A multiplicity of events led to the position where design, construction, instrumentation, management, and organisational system failed to perform defensively. Even as excessive soil movements were revealed by the instruments at site, there were no proper and appropriate design reviews, as well as inadequate contingency and remedial means. There were also abuse of the back analyses and failure to institute a regular, close and effective monitoring scheme for construction.

18.5 Concluding Remarks

Using three case studies on geotechnical failures in soft marine clay from Singapore, this paper demonstrates the techniques of forensic geotechnical engineering. Although the approach to determine the causes of failure is similar, each and every case has to be dealt with differently depending on a variety of factors. In the Nicoll Highway collapse case study, all the experts met but failed to agree on many factors that would trigger the massive retaining system collapse. In the expert meetings, some common factors on the causes of failure were agreed upon and they are highlighted in this report. This paper also demonstrates that unlike the design and analysis of traditional geotechnical projects, forensic geotechnical engineering is a form of reverse engineering. Different experts looking at the same event may come out with different rationales and opinions on the case.

References

COI (Report of the Committee of Inquiry) (2005) The incident at the MRT circle line worksite that led to the collapse of Nicoll highway on 20 April 2004, vols. I and II. Ministry of Manpower, Singapore

Leung CF, Shen RF (2004) Stability of water bund slope. In: Proceedings of 9th international symposium on landslides, vol 2. Rio de Janeiro, pp 1031–1036

Leung CF, Tan SA (2007) Successes and failures of instrumentation programs in major construction projects in Singapore. ASCE Geotechnical Special Publication 175

Leung CF, Tan SA, Shen RF (2005) Predictions versus performance of land reclamation bund. In: Proceedings of 16th international conference on soil mechanics and geotechnical engineering, vol 5. Osaka, pp 3540–3541

Pitts JP (1984) A survey of engineering geology in Singapore. Geotech Eng 15:1–20

Priebe H (1978) Abschatzung des scherwiderstandes eines durch stopverdichtung verbesserten Baugrundes. Die Bautechnik 8(55):281–284

Tan SA, Liew JY, Helmut S, Vermeer P (2005) MOM experts on lessons learnt from the incident at the MRT circle line C824 that led to the collapse of the Nicoll highway. MOM experts report

Chapter 19
Technical Shortcomings Causing Geotechnical Failures: Report of Task Force 10, TC 302

R.A. Jessep, L.G. de Mello and V.V.S. Rao

Abstract A selection of around 100 geotechnical failures, from Asia, South America and Europe, have been analysed to establish the primary shortcoming causing the failures. In broad terms, some half of the failures were caused by shortcomings in design, about a fifth were caused by inadequate site investigation, about a fifth were due to unforeseen phenomena and occurrences and about a seventh were caused by shortcomings in construction. In order to limit the future occurrence of such failures, the following is recommended:

- Carrying out a desk study on every project, which includes identifying an outline geological and geomechanical framework.
- Ensuring adequate funding is available for site investigation.
- Improving the education and training of engineers.
- Ensuring designs are completed by competent engineers who appreciate the limits of their own knowledge and experience.
- Ensuring that all designs are checked by a suitably qualified professional and preferably subject to an independent peer review.
- Ensuring appropriately qualified professionals manage engineering risk and monitor quality during construction.
- Ensuring construction methods are selected by appropriately qualified people who appreciate the limits of their own knowledge and experience.
- Carrying out independent peer reviews of chosen construction methods.
- Publishing and learning from case histories of past failures.
- Increasing investment in investigation, testing, instrumentation, monitoring and research.

R.A. Jessep (✉)
Robert Jessep & Partners Ltd, Horsham, UK
e-mail: rob@robertjessep.co.uk

L.G. de Mello
Vecttor Projetos S/C LTDA, Sao Paulo, Brazil

V.V.S. Rao
Nagadi Consultants (P) Ltd, Delhi, India

V.V.S. Rao and G.L. Sivakumar Babu (eds.), *Forensic Geotechnical Engineering*, Developments in Geotechnical Engineering, DOI 10.1007/978-81-322-2377-1_19

Keywords Technical shortcomings · Design · Investigation · Construction

19.1 Introduction

The brief for the task force was to assess the technical shortcomings that have caused a number of selected geotechnical failures, to analyse why these occurred and to highlight the considerations in the different stages of projects to avoid the key shortcomings identified and hence, in so far as possible, future failures.

For the purpose of this task force, a geotechnical failure is considered to be any undesirable occurrence in ground works and structures relating to both serviceability and ultimate limit states: hence it includes exceedance of the strength of materials, collapse, undue cracking and excessive deformation of the ground or structure including compression, settlement and tilting.

The precise reasons for a particular failure may be extremely complex and involve a large variety of interconnecting factors. However, it is generally possible to define a primary shortcoming that led to a particular failure. The shortcomings that caused failures have been broadly categorised as follows:

- Inadequate site investigation including desk study,
- Inadequate design including the ground model and selected parameters,
- Inadequate construction,
- Shortcomings in engineering knowledge, ability and understanding, i.e. unforeseeable occurrences and phenomena.

A sample of around 100 geotechnical failures from three regions of ISSMGE (Asia, South America and Europe) was assessed to establish the primary shortcomings that caused those failures. The case histories from Asia were all from India (after Rao and Sargunan 2010). The data on the shortcomings causing geotechnical failures in the three regions has been analysed to identify the areas in which there is a global and regional vulnerability and whether or not there are similarities or differences between the regions.

In this report, we also discuss how these shortcomings can be addressed to reduce the number of failures arising from geotechnical engineering with outline advice for clients, engineers and other relevant parties.

19.2 Analysis—Overview of Results

A summary of the analysis of the shortcomings in the case studies are included in the Appendix (Tables 19.1, 19.2 and 19.3) for Asia, South America and Europe. The broad categorisation of the shortcomings from the selection of analysed failures is shown in Fig. 19.1. In broad terms, the data shows in all regions considered that

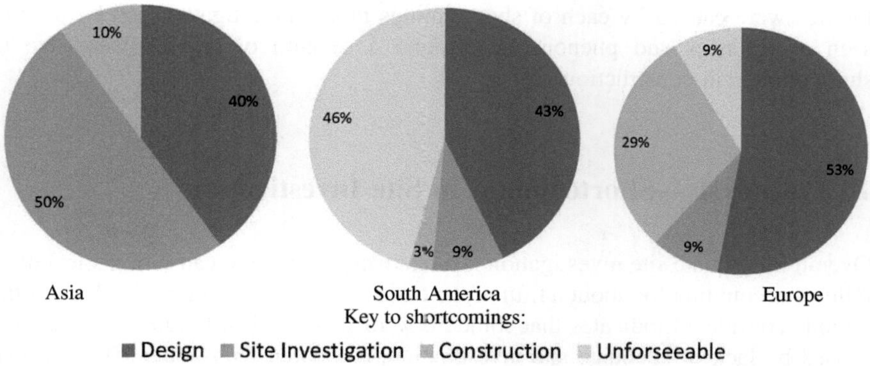

Fig. 19.1 Broad categorisation of shortcomings—regions

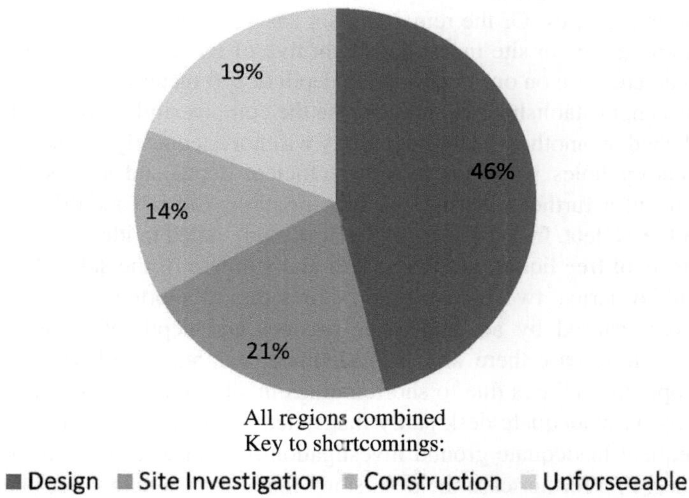

Fig. 19.2 Broad categorisation of shortcomings—all cases

shortcomings in design were a significant cause of geotechnical failures whereas shortcomings in site investigation were a significant cause of such failures in Asia but only minor causes in South America and Europe. Shortcomings in construction were a minor cause of geotechnical failures in Asia and South America but accounted for a relatively significant portion of the failures in Europe. In Asia, no failures were attributable to shortcomings in knowledge, understanding and ability with unforeseeable occurrences and phenomena whereas in South America this was a significant cause of geotechnical failure and in Europe it was a minor cause.

The overall picture from the combined analysis of the categorisation of the shortcomings causing geotechnical failures is presented in Fig. 19.2. A little under a half of all failures were caused by shortcomings in design and around a fifth of failures were caused by each of shortcomings in site investigation and by unforeseen occurrences and phenomena. Around a seventh of failures were due to shortcomings in construction.

19.3 Analysis—Shortcomings in Site Investigation

Overall inadequate site investigation is a relatively significant cause of geotechnical failures accounting for about a fifth of the cases analysed. Analysing the data for the regions considered indicates that some 50 % of geotechnical failures in Asia were caused by lack of adequate site investigation, compared with under 10 % in both South America and Europe.

In Asia in nine cases of geotechnical failure (or 60 % of failures due to shortcomings in site investigation), no site investigations were carried out at all and this resulted in the failures. Of the remaining six cases of geotechnical failures in Asia due to shortcomings in site investigation, in five of these the ground was not adequately characterised: on one occasion the depth of soft material that would be prone to scour was not established, in another case the compressibility of the subsoils was not established, in another the hydrogeology was not adequately assessed, in another the exploratory holes were not sunk to sufficient depth and so a soft layer was missed, and in a further case the site investigations did not identify voids in the ground or their effect. In a single case, the desk study failed to identify the risk posed by the effects of free liquid on the strength and stiffness of the subsoil.

In South America, two of the three failures due to shortcomings in site investigation were caused by an inadequate number and depth of exploratory holes whilst in the third case there was an inadequate desk study and walkover survey.

In Europe, the failures due to shortcomings in site investigation were caused in one case by an inadequate desk study that failed to identify past failures of slopes and subsequent inadequate ground investigation that failed to identify pre-existing shear planes. In a further case the desk study failed to adequately consider the very deep source of clayey fill material and the subsequent ground investigation did not assess the swelling potential of this fill. In a third case no boreholes were sunk into the founding rock to assess the strength and compressibility of this stratum, which was used to support a building.

19.3.1 Discussion

It is likely that most if not all of the failures related to shortcomings in site investigation could have been avoided if adequate desk studies and exploratory site investigations were carried out.

The evidence from the case histories shows that in many cases in Asia little or no site investigation was carried out. This could be due to a lack of understanding of the engineers involved in the particular projects, possibly arising from the lack of involvement of a specialist, or due to reluctance amongst the clients to commit any or any sufficient funding to site investigation.

In Europe, there are a large number of standards, codes of practice and guidance for site investigation (including for clients) and it seems consequently the occurrence of shortcomings in site investigations causing failures is proportionally relatively small; in South America and Europe, none of the reported failures were due to there being a complete absence of site investigation.

Some of the failures caused by shortcomings in site investigation occurred due to inadequate desk studies. Thorough desk studies are essential at the conceptual stage of projects and should consider the geotechnical aspects of the project. The desk study should search for published information including on geology, relevant site investigations, soil properties and groundwater as well as site history and occurrences and phenomena that are relevant to the site/subsoils. It should also involve a walkover survey to assess relevant features of the site and seek to identify hazards. Once gathered, this information can then be reviewed to identify an outline geological and geomechanical framework and risks at a site, so the engineers can consider the further information that is required for the design. The nature and scope of the desk study can be adjusted to the size and complexity of the project and ground conditions so geotechnical risks can be managed in an economic but effective way. If a desk study is not performed it is a substantial lost opportunity.

The next stage of site investigation should be to undertake ground investigation, sampling and testing using appropriate techniques. A relatively significant proportion of the failures due to shortcomings in site investigation are caused by inadequate categorisation of the ground which can be due to a lack of the number and/or depth of exploratory holes.

When designing site investigations, engineers should properly consider the scope of the site investigation and the variability of the ground encountered: it is possible that several phases of investigations will be required if the ground is significantly variable or particular features are observed and then they need to be investigated further. Engineers should ensure that exploratory holes are of sufficient depth to go beyond the influence of foundations and also consider other factors such as voids or deeper soft layers.

In some instances inappropriate methods of ground investigation and testing are adopted or key parameters required for the design are not assessed by direct measurement in soil tests. Engineers must therefore adequately consider the properties of the ground that are important for a particular project such as strength, stiffness, permeability, drainage paths, shrinkage, swelling, softening, shear planes and zones, erosion, piping, liquefaction, heterogeneity, voids and other risks. The methods of exploration, sampling and testing adopted must be appropriate for the ground conditions present at a particular site and to categorise the relevant subsoils and obtain the required parameters. Full details of these considerations are beyond the scope of this report but there is extensive published guidance on these matters.

Overall the cost of site investigation is small in relation to the overall project cost and it is essentially correct to say that "you pay for site investigation whether you

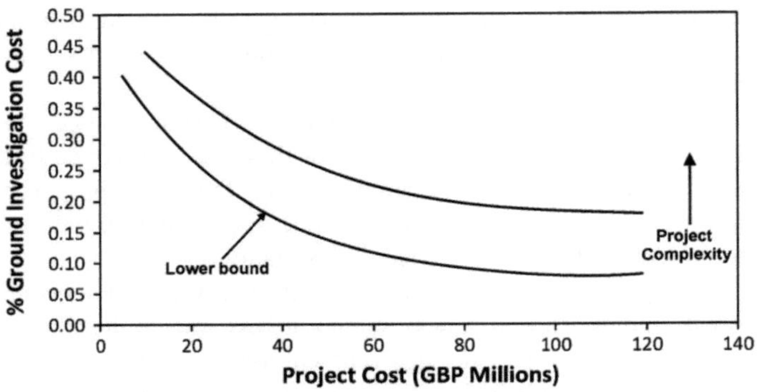

Fig. 19.3 UK cost of site investigation

have one or not" (Littlejohn 1991): in practice the outturn cost of a geotechnical failure is probably far greater than the cost of an adequate site investigation. A study in the USA concluded that spending up to 3 % of a project's cost on site investigation could be worthwhile in managing risks and reducing overall project cost, and hence by inspection avoiding failures (Threadgold 2002). Similarly, it has been shown that the percentage of increase in the total construction cost during projects in the UK can be limited by increased expenditure on site investigation (Clayton 2001): all reported projects that had an increased project cost by in excess of 50 % had spent less than 2.5 % of the outturn project cost on site investigation.

It is imperative to educate clients of the need to properly fund site investigation. In the UK the Association of Geotechnical Specialists has published indicative information on the relative cost of ground investigation (exploratory holes, testing and factual reporting only) compared against project cost (Cook 2005): this is presented in Fig. 19.3. This could be used as a rough guide of the relative proportions of project expenditure to be spent on ground investigation. In certain ground conditions and for certain types of projects that have significantly increased complexity, the cost of adequate site investigation will be substantially increased.

19.4 Analysis—Shortcomings in Design

In all three of the regions that were analysed, shortcomings in the design were a significant cause of geotechnical failure, accounting for approximately half of all the failures analysed.

The particular nature of the shortcomings causing the failures is relatively diverse. There are, however, some common causes. One common issue is the inappropriate selection of soil and/or groundwater parameters for design which caused three failures in Asia, six in South America and four in Europe. This frequently involved a lack of appreciation by the designer of the behaviour of the soil and groundwater and changes in these over time, which could lead to a reduction in strength or increase in loading.

In South America, four cases of geotechnical failures were the result of inappropriate or incorrect design solutions, in particular in terms of support methods, drainage design and protection measures and the specified construction method. In Europe, ten of the failures analysed were caused by shortcomings in the design solution, which did not adequately consider either, strength, deformation and compatibility, or adequately account for the behaviour of the ground such as the full depth of compressible strata and the mechanisms of behaviour affecting certain soils.

The inappropriate selection of a design or calculation method was the root cause of three geotechnical failures in Asia and two in Europe. In the Asian cases, the methods chosen for bearing capacity calculation for strip footing and raft foundations were incorrect and one method of slope stability analysis was incorrect. In Europe, in one case, the calculation method adopted for a working platform capacity was incorrect and in the other the method used to calculate settlements was incorrect.

In two cases in Asia and one in South America where failure was due to shortcomings in the design, the ground model used in the design was too simplistic and did not adequately account for the variability found in the subsoils.

In both Asia and South America, in two cases, the shortcomings in design leading to the geotechnical failure were related to the incorrect selection of load cases considered in the design.

In all three regions, a single case of failure that was analysed was caused by an absence of slope stability analysis when a slope was being designed.

Of those cases analysed, a single case of geotechnical failure in both Asia and Europe was caused by a lack of settlement calculations having been carried out to assess the settlements and hence was a lack of consideration of serviceability limit state.

One case of failure caused by shortcoming in design in South America was caused by improper specification which did not limit the grouting pressure.

19.4.1 Discussion

Overall shortcomings in design are a significant cause of geotechnical failures that is common to all the regions assessed.

It is essential that when selecting a design or calculation method, parameters and ground model, the engineer responsible for the design appreciates the factors that are important in the design and has sufficient understanding to identify the correct load cases, possible failure mechanisms, and the behaviour of the soil, rock and groundwater, considering strength, deformation and compatibility within the whole system, which, where appropriate, could be through a soil structure interaction analysis. Other issues such as vibrations and changes in groundwater also need to be considered.

The interpretation and selection of soil and rock parameters must consider factors including appropriate stress range, anisotropy, loading direction, sample disturbance, appropriate constitutive model, and the effect of the inevitable simplifications that are necessary when interpreting data. Similarly the development of a ground model for design must give adequate consideration to factors including the variability and heterogeneity of the ground despite the need to simplify the strata to undertake modelling of the ground.

In short, a full design risk assessment is required and is an essential part of the design process but this relies on the skills of the engineers responsible for the design.

The adequacy of a design relies to a large extent on the skills and experience of the designer. The root cause of the failures caused by shortcomings in design seems to be engineers (or others) working beyond or outside their skills and experience. To prevent, or at least limit, future failures caused by shortcomings in design, there are three ways this should be addressed.

The first is to train engineers (or others) to appreciate their own limitations in respect of knowledge, experience and competence so they do not attempt design solutions beyond their capability or use inappropriate methods but instead seek outside assistance. The second is continuous education of engineers to improve their knowledge and understanding of the wealth of experience and ability geotechnical engineers have built up over time. The third necessary step is arranging for the design to be checked by a suitably experienced professional and, preferably, then a review of the design carried out by an independent peer reviewer: the use of independent peer reviews has been successful in certain projects in the UK over the past 15 years. The first and second steps seek to prevent design errors at source whereas the third seeks to identify and address such errors before projects proceed to construction.

There is an enormous wealth of published information across the globe on design-related information: this is beyond the scope of this report but suffice to say, it is relatively easy nowadays to carry out a search to obtain information on design methods, case histories and other relevant information that could be used by design engineers.

19.5 Analysis—Shortcomings in Construction

Overall about a seventh of the geotechnical failures analysed were the result of shortcomings in construction. Shortcomings in construction only accounted for a minor proportion of the cases of geotechnical failure analysed in Asia (three or around 10 %) and South America (one or around 3 %) whereas in Europe they accounted for a relatively significant proportion (ten or around 29 %).

Inadequate construction practice accounted for two failures in Asia which were caused by the fill material not being placed or compacted properly and the base of piles not being cleaned properly. The single case in South America related to a failure caused by the use of excessive blasting energy in a drill and blast tunnel.

In Europe, seven cases of those analysed were caused by inadequate construction practice including two with an inappropriate method and application of backfilling, a single case where the specified sequence of backfilling was not followed, another where the base of the excavation was not cleaned properly, a further where the piles that were installed were too short, another where the excavation was too deep and berms were too small and finally, a single case where the thickness of the working platform was not adequately maintained during construction.

Selection of an inappropriate construction method accounted for a single case of failure in Asia where stone columns were not constructed properly using a vibroflot. In Europe, three cases were caused by the use of inappropriate construction methods, one relating to the installation of sheet piles by percussive and vibratory techniques in loose water-bearing sand and two further cases where the augured piling methods adopted were not appropriate, leading to flighting and piping failures, respectively.

19.5.1 Discussion

In all regions analysed, geotechnical failures occurred due to shortcomings in construction. The relative proportions of the failures due to shortcomings in construction are, however, seemingly quite variable. Nevertheless, most if not all of the failures caused by shortcomings in construction are likely to be avoidable.

The occurrence of failures due to inadequate construction practice essentially relate to lack of quality control during construction. It is essential to undertake testing and monitoring to validate the design and construction and useful in many cases to carry out back analysis to confirm design parameters and principles. It is also necessary for those responsible for supervising or monitoring works to an overview of dimensional accuracy and material strengths, densities and/or stiffness.

Appropriate professionals should be responsible for managing quality and engineering risks during construction: they must have sufficient understanding of the design intent, the specification and the risks. For this reason, generally this should be carried out by engineers who not only should be capable of managing risks and monitor quality during construction, but can also look for problems arising that were not expected or anticipated in the design and could lead to failure. Self-certification by contractors is seldom appropriate.

Further, there is often an opportunity during construction to address shortcomings in site investigation and design or even due to problems arising from shortcomings in knowledge, understanding and ability, which manifest or become apparent during construction: often there will be minor or modest failures during construction, or other indicators such as excessive movement, which provide an opportunity to learn and hence mitigate losses and possibly prevent failures or further more significant failures. If monitoring data, which must be assessed in a short space of time after capture, shows up unexpected behaviour, procedures should be in place to review the situation, react to the issues and effect changes which may prevent failures. If suitably qualified and experience professionals are used to manage risks and monitor quality, there will be an opportunity to react to issues arising on site and possibly to avert some failures.

It is essential that suitable construction methods are adopted, which preferably have a proven track record. If new methods are to be used, or existing methods used in new conditions, then trials should be conducted to establish the efficacy and suitability of the intended technique. The method of construction needs to be chosen by suitably qualified and experienced persons, who have due cognisance of the limitations of their own knowledge and experience. In many cases, independent peer review of the construction methods would be a useful additional safeguard to prevent failures due to the inappropriate selection of construction methods.

19.6 Analysis—Shortcomings in Knowledge, Understanding and Ability of Society

In South America, 16 (about 46 %) of the failures analysed were attributed to shortcomings in mankind's knowledge, understanding and ability with unforeseeable phenomena and occurrences. These related to six (about 17 %) cases with unforeseeable hydrogeological conditions, three of the cases which involved substantial localised flows of groundwater and three of which involved high pressures in narrow weak layers or fractures. A further nine cases of failure (about 26 %) were caused by unknown behaviour of rock (and some soil) which had weak layers, on one occasion with sinuous foliation, elastic-brittle failure, retrogressive behaviour, decomposition or high residual stresses which led to bursting. The final case involved the unforeseen effect of logs impacting on a dam and leading to failure.

In Europe, three cases of failure (about 9 %) from those analysed were attributed to shortcomings in mankind's existing knowledge, understanding and ability. These related to a previously unknown phenomenon with particular ground conditions and bacteria adversely affecting concrete, locally unknown occurrence of high sulphates and a failure in some relatively aged works due the collapse of an unknown dissolution feature. Conversely, none of the failures analysed in Asia were attributed to such shortcomings in knowledge, understanding and ability.

19.6.1 Discussion

It is likely that shortcomings in our collective knowledge, understanding and ability will cause geotechnical failures throughout the world. Whilst a large proportion of the failures in South America were attributable to shortcomings in knowledge, understanding and ability, particularly with unexpected ground conditions and soil/rock behaviour, particular ground conditions on the projects that suffered these failures were very challenging. This relatively high occurrence in South American cases may either be due to these particularly challenging ground conditions, which are not necessarily a function of the region but may simply be a function of the particular projects chosen, or they may be due to particular types of projects which have been reported on, many of which relate to tunnelling.

We can reduce the occurrence of failures due to shortcomings in our collective knowledge, understanding and ability by learning from past failures: case histories of failures present a very significant opportunity to learn from such occurrences. An increased expenditure on site investigation, testing, instrumentation, monitoring and research will assist the identification and the development of our understanding of phenomena and behaviour and identification of risks posed in the ground. As an example, the risk of dissolution features was historically known on one analysed site in Europe but the particular feature of importance was not identified: nowadays it may have been possible to identify the feature with ground penetrating radar whilst an increased density of boreholes may also have identified the feature.

However, regardless of the amount of expenditure, the risks of unknowns in the ground remains significant and failures due to unforeseen ground conditions, phenomena and occurrences are likely to continue.

19.7 Recommendations

In summary, to prevent recurrence, or at least limit the occurrence, of geotechnical failures, the following procedures and processes in construction projects should be implemented:

Site Investigation:

- Carry out a desk study at the concept stage of every project, which includes identifying an outline geological and geomechanical framework.
- Ensure adequate funding is available for site investigation.
- Improve the education and training of engineers in relation to carrying out desk studies and designing and specifying adequate site investigations.

Design:

- Improve the education and training of engineers in relation to design methods, ground models, parameter selection and other relevant design considerations.
- Ensure designs are completed by competent engineers with appropriate skills for the particular design and educate engineers (and others) to appreciate their own limits.
- Ensure that all designs are subject to checking by a suitably qualified professional.
- Carry out independent peer reviews of designs.

Construction:

- Ensure appropriately qualified professionals manage engineering risk and monitor quality during construction.
- Ensure appropriately qualified people select construction methods and educate those people to appreciate the limits of their own knowledge and experience.
- Carry out independent peer reviews of chosen construction methods.

Shortcomings in knowledge, understanding and ability will always be likely to cause failures but it is important that we learn from case histories of past failures and conduct wider investigation, testing, instrumentation, monitoring and research, and together with increased experience, the occurrence of such failures should diminish over time. It should follow that the more investment there is in research, investigation and design, the fewer the number of failures that will occur.

19.8 Conclusions

A selection of around 100 geotechnical failures, from Asia, South America and Europe, have been analysed to establish the primary shortcoming causing the failures.

Overall, nearly half of the failures analysed were caused by shortcomings in design, about a fifth were caused by inadequate site investigation, around a further fifth were due to unforeseen phenomena and occurrences and about a seventh were caused by shortcomings in construction. Save for failures caused by shortcomings in design, which were significant in all three regions considered, there was a significant variation in the relative proportions of the types of shortcomings causing failures in the three regions analysed. However, the recommendations to limit the number of future failures are valid worldwide. These can be summarised as follows.

In respect of adequate site investigation, these include carrying out a desk study at the concept stage of every project, ensuring adequate funding is available for site investigation and improving the education and training of engineers in relation to carrying out desk studies and designing and specifying adequate site investigations.

In respect of adequate designs, these include improving the education and training of engineers in relation to design methods, ground models, parameter selection and other relevant design considerations, ensuring designs are completed by competent engineers with appropriate skills for the particular design who are educated to appreciate the limits of their own knowledge and experience, and ensuring that all designs are subject to checking by a suitably qualified professional and preferably also subject to an independent peer review.

In respect of adequate construction, ensure that appropriately qualified professionals manage engineering risk and monitor quality during construction, appropriately qualified people select construction methods and they are educated to appreciate the limits of their own knowledge and experience and carrying out independent peer reviews of chosen construction methods.

Finally, in respect of shortcomings in our collective knowledge, understanding and ability, we should seek to learn from case histories of past failures and increase investment in investigation, testing, instrumentation, monitoring and research to continue to develop our skills.

Appendix

See Tables 19.1, 19.2 and 19.3.

Table 19.1 Task force 10: Technical shortcomings causing geotechnical failures (Region—Europe)

No.	Project/structure	Foundations	Subsoil profile	Distress/failure	Primary shortcoming	Other shortcomings	Notes	References
1.1	Secant piled wall formed of displacement and cased continuous flight auger piles	Working platform of 500 mm of crushed rubble	1 m of soft to firm clay fill over 1.5 m of clayey peat and 2 m of firm clay over chalk to depth. water from 1 m	Bearing capacity failure of the working platform and subgrade resulting in a piling rig overturning	Inadequate design	Inadequate construction	Inappropriate design method resulting in substantial shortfall in platform thickness contributed to by a shortfall in platform thickness as-laid	Report by R A Jessep for a private client
1.2	Railway tunnel	–	Tertiary clay and claystone, with variable weathering and slickensides	Tunnel collapse	Inadequate design		There was insufficient bearing capacity of the tunnel foundation due to non-closure of the invert	See 1 below
1.3	Construction of very large diameter bored piles in dry bores	N/A	6 m of medium dense well-graded gravel over heavily overconsolidated stiff to very stiff-fissured clay with water from 3 m	Collapse of the ground surrounding a pile bore during construction	Inadequate construction		A borehole was backfilled with arisings creating a water path from the gravel into the clay, leading to softening and consequent failure of the ground into tho opon boro during later piling construction	Report by R A Jessep for a private client
1.4	New 15 m deep basement with contiguous bored piles retaining wall adjacent to a canal	Contiguous piles embedded into bedrock	3.5 m of silty sand over 3 m of very weak sandstone, 6.5 m of very weak mudstone and sandstone to depth. perched water above mudstone	Undue excessive lateral movement of the retaining wall led to rupturing of the canal lining and consequent substantial leakage into the excavation	Inadequate construction		The temporary berms were smaller and the excavations deeper than considered in the design whilst the installation of temporary propping was incomplete	See 2 below
1.5	Railway cutting widening with soil nailed slopes	N/A	Overconsolidated silty clay with groundwater within 2 m of ground level	Large excessive deformation of the front face with wedges of soil sliding downwards	Inadequate site investigation		Site investigation did not identify steep pre-existing shear planes with low shear strength	Report by R A Jessep for a private client

(continued)

Table 19.1 (continued)

No.	Project/structure	Foundations	Subsoil profile	Distress/failure	Primary shortcoming	Other shortcomings	Notes	References
1.6	Precast concrete tunnel lining	N/A	Well-graded gravel over stiff-fissured clay. Perched water in places in gravel	Cracking and deformation of a concrete tunnel lining	Shortcoming in knowledge, understanding and ability		Unknown phenomenon with bacteria in causing concrete deterioration in groundwater pocket in clay depression	Report by L G de Mello et al. for a private client
1.7	Construction of a deep basement with sheet piled retaining wall	N/A	Loose becoming medium dense (below 20 m) fine rounded sand. Groundwater from 4 m depth	Lateral movement, settlement and cracking of adjacent masonry building on strip footings during sheet pile installation	Inadequate construction		Sheet piles driven with vibratory and percussive method with high continuous vibration causing settlement of loose sand	Report by R A Jessep for a private client
1.8	Steel-framed building with brick spandrel masonry and glazing above	Pad foundations and ground beams to frame with ground bearing slab	Up to 3 m of fill derived from incinerator waste over firm to stiff clay	Expansion of fill leading to movement of the slab by up to around 50 mm and cracking of the slabs and walls	Inadequate design		The fill material was approved despite not being subject to appropriate testing to assess swelling potential required by the specification/standards	Report by R A Jessep for a private client
1.9	Highway embankments	Natural ground	Several metres of limestabilised stiff clay fill over stiff clay, with rising groundwater levels	Up to 80 mm heave of the road pavement causing cracking	Shortcoming in knowledge, understanding and ability		Unknown feature present in ground: the site won fill locally contained high sulphates that reacted with the lime when the groundwater rose in the fill	See 3 below
1.10	A buried pipeline constructed in a trench up to 8 m deep and attached to a buried tank	Founded in bedding and surround within a trench in the soil	Stiff to very stiff overconsolidated clay. No groundwater	Leaks from joints in the pipes after backfilling due to offsets of about 20 mm between adjacent pipe lengths	Inadequate design		No consideration was given to the longitudinal deformation of the pipeline consequent on the settlement of the ground after backfilling the deep trenches	Report by R A Jessep for a private client

(continued)

Table 19.1 (continued)

No.	Project/structure	Foundations	Subsoil profile	Distress/failure	Primary shortcoming	Other shortcomings	Notes	References
1.11	New external concrete slab hardstanding	Ground bearing slab on replaced and compacted ground	1 m of fill over very soft organic clay to 10 m depth over very stiff clay. Groundwater at 1 m	Excessive settlement of surfacing	Inadequate design		Only the upper few metres of the clay was improved and remaining highly compressible soils settled significantly	Report by R A Jessep for a private client
1.12	Construction of a 4-storey steel-framed office building	Ground bearing floor slab	Up to 3 m of fill comprising lumps of mudstone in a matrix of firm clay over mudstone. Groundwater at 1 m	Very significant heave of ground bearing slab	Inadequate site investigation		The fill came from a 100 m deep excavation: the lumps of mudstone are therefore prone to long-term swelling due to water absorption, which was not considered in the desk study and site investigations	See 4 below
1.13	Steel portal-framed building	Pad foundations to frame and ground bearing slab founded on 0.3 m granular sub-base and 4 m long stone columns	Made ground comprising 4 m of demolition rubble over 4 m of organic silt derived from settlement lagoons over chalk to depth. Groundwater at 2 m	Undue settlement causing differential settlement of the building and ground slab by over 200 mm	Inadequate design		The demolition rubble was improved using the stone columns but these did not penetrate the very highly compressible silt, which consolidated and crept significantly under the loads from the building and ground	Report by R A Jessep for a private client
1.14	Two-storey masonry office building	Concrete strip footings	20 m mudstone fill over mudstone. Site underdrained by old adits	Very large settlements and differential settlements of building with consequent distortion and cracking	Inadequate design		Collapse settlement due to water entering the fill was not adequately considered when foundation types and bearing capacity were recommended	Report by R A Jessep for a private client
1.15	Steel-framed building with ground bearing slab and blockwork walls	Frame founded on piles and slab founded on ground improved with stone columns	7 m of very soft to soft clay over mudstone to depth. Groundwater at 1.5 m	Differential settlement of the building causing cracking of the slab and walls	Inadequate design		The stone columns were 5 m long thereby not reaching the mudstone: as a result the loads on the slab caused settlement of the clay relative to the pile foundations	Report by R A Jessep for a private client

(continued)

Table 19.1 (continued)

No.	Project/structure	Foundations	Subsoil profile	Distress/failure	Primary shortcoming	Other shortcomings	Notes	References
1.16	Reinforced concrete reservoir about 50 years old founded at 4 m depth	Ground bearing reinforced concrete slab	Typically 12 m (up to 30 m) of sand, with a consistent upper surface, over chalk with an irregular surface due to dissolution features	Sinkhole formed after water released into a soakaway caused sand to flow into a void, causing settlement and cracking in the reservoir	Shortcoming in knowledge, understanding and ability		A competent ground investigation with boreholes had been carried out but the relatively narrow dissolution feature leading to a void had not been picked up	Report by R A Jessep for a private client
1.17	Cut and cover tunnel of precast concrete arches	Rotary-bored piles	Stiff-fissured overconsolidated clay	Collapse of the arch	Inadequate construction		The sequence of backfilling of the arch was not followed	See 5 below
1.18	Precast concrete piles installed to support a concrete slab	380 mm diameter driven cast in situ concrete piles on a 3.3 m square grid	About 1.5 m of fill over typically 9 m of very soft clay over medium dense sand interbedded with soft clay with bedrock at 20 m	Substantial settlement of the piles and slab	Inadequate construction	Inadequate design	Many of the piles failed to reach the sand when they were driven. In places settlements were contributed to by a lack of consideration in the design of the effect of soft clay layers within the sand	Report by R A Jessep for a private client
1.19	A bridge with a pier in a waterway	15 m diameter circular concrete pad foundation on rock, formed in a cofferdam	4 m of water over 5 m of very soft organic clay and silt over very weak sandstone	Undue and excessive settlement of the foundation	Inadequate construction		Layers of soft silt were left in the cofferdam before placement of the concrete. These layers settled significant when loaded	Report by R A Jessep for a private client
1.20	Three-storey office building with piled foundations	Small diameter driven piles founded in the stiff clay. Floor slab spanning to pile caps but cast on soil	Made ground comprising waste from an incinerator to 2 m depth over soft clay and stiff clay	Cracking of the walls and ground slabs due to expansion of fill	Inadequate design		The fill was approved despite not being subject to appropriate testing, to ascertain it had swelling potential, as required by specification and standards	Report by R A Jessep for a private client

(continued)

Table 19.1 (continued)

No.	Project/structure	Foundations	Subsoil profile	Distress/failure	Primary shortcoming	Other shortcomings	Notes	References
1.21	Construction of a new building with a retained masonry facade and a new deep basement	Perimeter secant bored pile wall and internal piled foundations to support columns	Loose to medium dense fine rounded silty sand to 12 m depth over stiff clay. Groundwater at 4 m depth	Significant settlement and consequent cracking of the retained facade due to volume loss	Inadequate construction		Piping failure occurred when the male secant piles were constructed with the casing only slightly ahead of the auger with no drilling fluid used	Report by R A Jessep for a private client
1.22	Cutting widening to allow new train lines to be laid	N/A	Mudstone weathered to a residual soil at the surface and adjacent to layers of limestone with running water	Delayed cutting failure with deep sliding failures	Inadequate design		No consideration was given in the design for softening of the mudstone after excavation and unloading or the weak layers of clay soils present in the slopes that were revealed in the ground investigation	Report by R A Jessep for a private client
1.23	Temporary slopes formed to construct a new basement in open cut	N/A	Highly compressible soft to firm clay with layers of loose sand and very soft Silt to depth. Groundwater at 2 m	Very large lateral displacement and sliding failures of the temporary slopes causing damage to nearby structures	Inadequate design		No adequate consideration was given to deformation of the slopes or bands of soil with little strength, which would fail in sliding failure	Report by R A Jessep for a private client
1.24	Deepening of a cutting, containing a small river, to 8 m depth, using sheet piles at the toe of the slope	N/A	Heavily overconsolidated firm to very stiff-fissured clay with groundwater typically at 1 m	Slope failure with rotational slip on weak shear planes formed in the clay and failure of the sheet piles	Inadequate design		The design of the deepened cutting did not consider the effect of softening of the clay over time following excavation, resulting in the delayed failure of the slope	Report by R A Jessep for a private client
1.25	Construction of continuous flight auger piles next to a building on pad footings	N/A	Very loose to medium dense fine rounded silty sand to 20 m depth with groundwater at 2 m	Flighting of sand into the auger caused substantial volume loss settlement, movement and cracking of the neighbouring building	Inadequate construction		The piling construction technique was not suitable in the ground conditions found at this site	Report by R A Jessep for a private client

(continued)

Table 19.1 (continued)

No.	Project/structure	Foundations	Subsoil profile	Distress/failure	Primary shortcoming	Other shortcomings	Notes	References
1.26	New building with a basement adjacent to an existing masonry building, that was underpinned	The existing building was founded on strip footings at 4 m depth	3 m of medium dense sand over soft to firm clay to 6 m depth over sandstone. Groundwater at 2 m	Settlement, distortion and cracking of the masonry building during underpinning and basement construction	Inadequate design		The underpinning design did not adequately consider the bearing pressures during the works devised and did not provide lateral support to the existing building	Report by R A Jessep for a private client
1.27	Masonry retaining wall and hardstanding built on fill	0.3 m deep footings and 150 mm of granular fill	3 m of firm to stiff clay fill from mudstone underlain by mudstone	Significant subsidence and cracking of wall and hardstanding due to collapse settlement	Inadequate construction		The fill was not adequately compacted and contained large lumps. No drainage holes were provided in the wall	Report by R A Jessep for a private client
1.28	Reinforced concrete tanks and associated pipework	Driven precast concrete piles to the tanks and pipework in trench fill	Up to 4 m of fill over 3 m of very soft organic clay fill, over 8 m of very soft clay over stiff clay groundwater at 2 m	Significant settlement of the ground, due to filling, damaging to the pipework	Inadequate design	Inadequate site investigation	The estimated settlement was grossly optimistic with creep not considered. Samples of the soils were not subject to testing to assess the compression characteristics	Report by R A Jessep for a private client
1.29	Large diameter continuous flight auger piles	working platform of geotextile and 0.6 m of well graded crushed concrete	Well-graded medium dense gravel with lenses of medium dense sand. Groundwater at 1.5 m	Piling rig overturned as it extracted its full auger following a bearing failure of the piling platform	Inadequate construction		The platform was not properly maintained and reduced to half the design thickness, and the ground around piles was disturbed locally reducing the subgrade strength	Report by R A Jessep for a private client
1.30	Modified vertical alignment of a railway line requiring widening of a cutting	N/A	Mudstone completely weathered to a residual soil at the surface	Major cutting failures after the mudstone softened to very soft clay	Inadequate design	Inadequate site investigation	There were no calculations to justify the slope angle used and insufficient site investigation to characterise the ground conditions	Report by R A Jessep for a private client

(continued)

Table 19.1 (continued)

No.	Project/structure	Foundations	Subsoil profile	Distress/failure	Primary shortcoming	Other shortcomings	Notes	References
1.31	Anchored-bored pile wall for an industrial building	N/A	Sand and gravels and tertiary clay	Excessive lateral movements of the anchored bored pile wall, with distress of adjacent structures	Inadequate design	Inadequate site investigation	The water pressures considered in the wall design were inappropriate. The site investigation did not identify strata with artesian water	Report by L G de Mello for private client
1.32	New concrete framed building with 12 m deep basement and no drainage	Bored piles	Very soft clay to 5 m over medium dense gravel to 8 m depth and stiff to very stiff clay groundwater at 3 m	Basement slab punching shear failure with consequent flooding of the basement	Inadequate design		The groundwater pressure was not properly considered in the design	Report by R A Jessep for a private client
1.33	25 m deep 12 m diameter cofferdam constructed as a caisson and then by underpinning	N/A	Firm to stiff clay containing boulders with thin bands of sand. Groundwater at 1 m	During underpinning there was a substantial washout from a sand band, causing uneven and significant settlement of the cofferdam	Inadequate design		The ground was not dewatered prior to construction so as the excavation approached a sand layer there was a bottom failure and piping failure	Report by R A Jessep for a private client
1.34	Construction of a 4-storey steel-framed office building	Driven hollow steel tube pile foundations of 244 mm diameter founded at about 5 m	Up to 3 m of firm to stiff clay fill over mudstone, completely to highly weathered to about 10 m depth	Excessive settlement and differential settlement of the building causing cracking in the finishes and glazing	Inadequate site investigation	Inadequate construction	No boreholes were sunk to assess weathering of the bedrock. Piles were driven and subject to dynamic testing only with no allowances made for dynamic effects or long-term relaxation	See 6 below

Table 19.2 Task force 10: Technical shortcomings causing geotechnical failures (Region—South America)

No.	Project/ structure	Foundations	Subsoil profile	Distress/failure	Primary shortcoming	Other shortcomings	Notes	References
2.1	Metro underground station	N/A	Biotite—Gneissic rock with dipping subvertical foliation	Tunnel collapse, daylighting, and reaching the access shaft	Shortcoming in knowledge, understanding and ability		Unforeseeable behaviour of subvertical bands of weathered biotite with elastic-brittle properties resulting in sudden collapse	See 1–9 below
2.2	Tunnel constructed with TBM	N/A	Heterogeneous piedmont deposits, with boulders and pockets of water	TBM failures that daylighted in a park	Inadequate site investigation		Limited site investigation had not found the boulders or pockets of water, so they were not considered in the design	Report by L G de Mello for private client
2.3	Metro station constructed with NATM	N/A	Granit-gneissic residual soils, saprolites, and sound rock	NATM Tunnel face failure	Inadequate design		The side drift was of un-favourable cross section, without invert and crown support or protection	Report by L G de Mello for private client
2.4	Pipeline	N/A	Alluvial soils	Pipeline collapse	Inadequate design		Too simplistic geological characterization and pipe design ignored tolerances and variations of the soil stiffness	Report by L G de Mello for private client
2.5	TBM excavated tunnel	N/A	Metamorphic rocks (schists) and igneous intrusions (gneiss, granitic or granodioritic rocks)	Rock mass failure associated with a large inflow of water	Shortcoming in knowledge, understanding and ability		Unforeseeable hydrogeological conditions with a large inflow of water not associated with any of the faults or other known geological features	Report by L G de Mello for private client
2.6	Highway embankment	Natural ground	Embankment over residual soil/faulted and fractured basalt rock	Slope failure	Shortcoming in knowledge, understanding and ability		Unforeseeable geological conditions with extremely high artesian water pressure forces acting in the impervious contact between the earthfill and the fractured rock mass	Report by L G de Mello for private client
2.7	Highway embankment	Natural ground	Up to 17 m of compacted fill over 5 m of soft silty clay, over residual soil	Excessive settlement of the embankment	Inadequate site investigation		Investigation failed to identify a drain which was blocked by the fill causing saturation and embankment failure	Report by L G de Mello for private client

(continued)

Table 19.2 (continued)

No.	Project/ structure	Foundations	Subsoil profile	Distress/failure	Primary shortcoming	Other shortcomings	Notes	References
2.8	Viaduct	Caissons at base of a slope	Slope of colluvium over residual soils and sound rock	Slope failure, reaching and affecting the viaduct foundation	Inadequate design		The slope failed due to a high water table following intense rainstorm and weak soils over the rock which were not allowed for in the design	Report by L G de Mello for private client
2.9	Viaduct	Caissons at base of a slope	Natural slope of soft residual soils and gneissic weathered rock	Slope failure, reaching and affecting the viaduct foundation	Inadequate design		The slope failed due to a high water table following intense rainstorm and weak soils over the rock which were not allowed for in the design	Report by L G de Mello for private client
2.10	Embankment	Natural Ground	Embankment fill over 20 m of soft organic clay	Embankment slope failure	Inadequate design		The design used an unrealistic high stiffness for the soil underneath the fill	Report by L G de Mello for private client
2.11	Steel mill	N/A	Residual soil of clayey/silty fine sand	Superficial erosions	Inadequate design		Intense rainstorms caused erosion of unprotected slope with inadequate drainage	Report by L G de Mello for private client
2.12	Power station	N/A	Colluvial and residual soils, over weathered rock	Slopes failures	Inadequate design		Inadequate use of the "observational method"	Report by L G de Mello for private client
2.13	Power station	N/A	Granite gneiss intercalated with weathered biotite schists and amphibolites bands	Rock slope failure	Shortcoming in knowledge, understanding and ability		The behaviour of the critical stratum was unforeseen although the stratum was identified during investigations	See 10–14 below
2.14	Highway slope	N/A	Weathered biotite-quartz, schists, gneisss, basic dike and quartzitic lenses	Slope Failure	Shortcoming in knowledge, understanding and ability		Unforeseeable conditions with a retrogressive failure due intense and concentrated rainstorms	See 10, 11, and 15–17 below

(continued)

Table 19.2 (continued)

No.	Project/ structure	Foundations	Subsoil profile	Distress/failure	Primary shortcoming	Other shortcomings	Notes	References
2.15	Viaduct	Deep caissons foundations	Colluvium and residual soils, over weathered biotite gneiss	Slow deep displacements, associated with water-level fluctuations	Shortcoming in knowledge, understanding and ability		Unforeseeable hydrological conditions locally with water infiltration causing movements	See 10, 11, 18 and 19 below
2.16	Dam	Natural ground	Colluvium and residual soil, with presence of blocks, over sound rock	Overtopping and upstream cofferdam failure	Inadequate site investigation		Insufficient site investigation with greater thickness of colluvium/residual soil present than expected	See 10 below
2.17	Dam	Natural ground	Heterogeneous rock mass, with anastomosis discontinuities	Rock slope failure	Inadequate design		Allowance for changes in the rock mass characteristics due to blasting were not made in the design	Report by L G de Mello for private client
2.18	Tunnel	N/A	Volcanic rocks; inter-bedding riolites and dacites, afanitic andesites	TBM tunnel loss due to rock mass collapse	Shortcoming in knowledge, understanding and ability		Unforeseeable high residual stresses in the rock mass caused bursting failures	Reports by L G de Mello for private client
2.19	Tunnel	N/A	Volcanic rocks, riolites, dacites, andesites	TBM damage due to rock mass collapse	Shortcoming in knowledge, understanding and ability		Unforeseeable high residual stresses in the rock mass caused bursting failures	Reports by L G de Mello for private client
2.20	Highway embankment	Natural ground	Earthfill embankment, over colluvium above basalt saprolite and sandstone	Embankment fill/slope failure	Inadequate design		High waterpressure forces acting in the impervious earthfill embankment were not considered in the design	Report by L G de Mello for private client and References 10
2.21	Tunnel	N/A	Granite/Gneissic residual soil, saprolite and rock	Tunnel crown failure	Shortcoming in knowledge, understanding and ability		Unforeseeable local weak veins combined with high water pressures at a depression of the sound rock interface, resulted in failure despite forepollings and drains being installed on the roof	Report by L G de Mello for private client

(continued)

Table 19.2 (continued)

No.	Project/ structure	Foundations	Subsoil profile	Distress/failure	Primary shortcoming	Other shortcomings	Notes	References
2.22	Piled embankment	Piles	8–12 m layer of peat and soft to very soft clay	Excessive settlements, failure of foundations and the embankment fill	Inadequate design		The design did not adequately consider the horizontal loads in the peripheral piles	Reports by L G de Mello for private client
2.23	Dam	Natural ground	Subvertical/ subhorizontal fractured basalt and colluvium	Penstock slope failure	Inadequate design		The design did not consider adequately the presence of rock discontinuities identified in the site investigation	Report by L G de Mello for private client
2.24	Anchored diaphragm wall	N/A	Overconsolidated clay	Damage of adjacent structures during grouting of anchors	Inadequate design		Design did not limit the grouting pressure resulting in horizontal hydrofracturing	Report by L G de Mello for private client
2.25	Dam	Natural ground	Residual and saprolite soils of sandstone, over friable sandstone	Spillway failure caused by piping process	Shortcoming in knowledge, understanding and ability		Unpredictable heterogeneous behaviour of the local sandstone rock	Report by L G de Mello and others for private client
2.26	Dam	Natural ground	Colluvium above sandy saprolite, over gneiss rock	Natural slope failure	Shortcoming in knowledge, understanding and ability		Unforeseeable behaviour of a band of the saprolite	Report by L G de Mello and others for private client
2.27	Dam	Natural ground	Basalt volcanic rocks	Dam floodgates collapse and diversion tunnel failure	Shortcoming in knowledge, understanding and ability		Unpredictable impact of tree logs on floodgate concrete structure caused failure	Report by L G de Mello and others for private client
2.28	Pond	Natural ground	Silty clay	Dam failure	Inadequate design		Design did not specify filter criteria between gabions and in situ silty clay leading to internal piping	Report by L G de Mello for private client
2.29	Highway	Natural ground	Thick residual soil, over gneiss rock	Rock slope failure	Shortcoming in knowledge, understanding and ability		Unforeseeable event due to thin soil layer at 15 m depth and excessive rainfall causing pore-pressure rises	Report by L G de Mello for private client

(continued)

Table 19.2 (continued)

No.	Project/ structure	Foundations	Subsoil profile	Distress/failure	Primary shortcoming	Other shortcomings	Notes	References
2.30	Highway	Natural ground	Saprolitic soil, with the presence of relict structures and discontinuities	Slope failure, after several years of operation	Shortcoming in knowledge, understanding and ability		Unforeseeable event due to excessive rainfall causing pore-pressure rises in soil with weak relic shear surfaces	Report by L G de Mello for private client
2.31	Highway	Natural ground	Phyllite saprolite	Slope failure, after several years of operation	Shortcoming in knowledge, understanding and ability		Unforeseeable event due to very complex geology on phyllite saprolite, presenting sinuous foliation	Report by L G de Mello for private client
2.32	Slope	Natural ground	Granite residual soil	Slope failure	Shortcoming in knowledge, understanding and ability		Unforeseeable event due to unexpected geological behaviour	Report by L G de Mello for private client
2.33	Dam	Natural ground	Quaternary sands, overlying sound gneissic rock	Dam failure during construction due to high compaction pore pressures in clay fill	Inadequate design		No stability analysis was carried out and the specification for compacting the alluvial clay soil was inadequate	See 20 below
2.34	Drill and blast tunnel	Natural ground	Volcanic rocks, riolites, andesites	Tunnel crown failure	Inadequate construction		The blasting energy applied, exceeded the design blasting charge	Report by L G de Mello for private client
2.35	Anchored diaphragm wall	N/A	Embankment fill, over residual and saprolite soil	Anchored diaphragm wall failure, causing the rupture of part of an adjacent building's foundation	Inadequate design	Inadequate construction of anchors	The design underestimated the loads on the wall and had a low factor of safety	Report L G de Mello and others for private client

Table 19.3 Task force 10: Technical shortcomings causing geotechnical failures (Region—Asia)

No.	Project/structure	Foundations	Subsoil profile	Distress/failure	Primary shortcoming	Other shortcomings	Notes	References
3.1	Two-storey building	Spread foundations	Fill comprising sandy to clayey Silt in an abandoned quarry pit	Cracking and tilting of the building and separation from the ground	Inadequate site investigation		No site investigation was carried out to establish the ground conditions	See 1 below
3.2	Industrial storage sheds	Double under reamed piles	Grey to yellowish brown plastic Clay	Extensive cracking after a storm	Inadequate design		The design of the piles did not consider the forces arising during a storm	See 2 below
3.3	Bridge	Well foundations	Sedimentary deposits below bed level	Cracking in the staining after the first flood	Inadequate site investigation		The site investigations failed to establish the likely scour depth or the high flood level	See 3 below
3.4	Submersible bridge	Well foundations	2 m of soil over hard Rock	The foundation could not be constructed to the design depth	Inadequate site investigation		No site investigation was carried out to establish the ground conditions	See 4 below
3.5	Pre-stressed concrete bridge	Well foundations through a cofferdam	Very fine soft marine silty clay	Well could not be sunk due to the continuous sinking of cofferdam island	Inadequate site investigation		The site investigation did not adequately identify the soil properties including compressibility	See 5 below
3.6	Hillside slope	Natural ground	Poorly cemented sandstone, with layers of clayey silts	Slope failure	Inadequate design		No slope stability analysis was carried out to assess the stability of the slope	See 6 below
3.7	Cutting slope	Natural ground	Highly jointed and banded rock	Slope failure	Inadequate design		The design of the slope did not consider the limited strength on the joints found in the rock	See 7 below
3.8	Five-storey building	Shallow foundations on improved ground	Soft clay to 30 m depth	The improvement by pre-loading using sand drains was ineffective	Inadequate site investigation		The site investigation did not adequately categorise the ground conditions, drainage paths and soil permeability, so the preloading was ineffective	See 8 below
3.9	Deep buried reinforced concrete tank	Effective raft	Laminated bands of sandstone with clay and shells	Landslide and significant settlement of the tank	Inadequate design		The soil parameters and design methods used in the design were inappropriate	See 9 below

(continued)

Table 19.3 (continued)

No.	Project/structure	Foundations	Subsoil profile	Distress/failure	Primary shortcoming	Other shortcomings	Notes	References
3.10	Suspension bridge	Block for abutment	Massive sandstone	Wide cracks formed in the rock mass	Inadequate site investigation	Inadequate design	No site investigation was carried out and the design did not consider the effect of the stresses applied to the rock	See 10 below
3.11	Natural hillside slope	Natural ground	Glacial deposits	Landslides	Inadequate design		The preliminary slope stability assessment was deficient	See 11 below
3.12	Residential buildings	Spread foundations	Silty sand underlain by silt to clay of low compressibility	Extensive cracking occurred in the buildings	Inadequate site investigation		No site investigation was carried out to establish the ground conditions	See 12 below
3.13	Vertical extension of community centre building	Strip footings	Poorly graded fine to medium sand	Excessive settlements	Inadequate design		The method of analysis used to assess the capacity was not appropriate	See 13 below
3.14	Residential buildings	Strip footings	Medium plasticity silty clay	Excessive settlements	Inadequate site investigation		No site investigation was carried out to establish the ground conditions	See 14 below
3.15	Aeration tanks	Raft foundation	Boiler ash mixed with murram (laterite)	Extensive cracking of the tanks	Inadequate design		The selected soil parameters and the design methods were inappropriate	See 15 below
3.16	Water treatment plant construction of reinforced concrete walls and slabs	Raft foundation	Gravelly sand underlain by hard rock with a large pocket of soft silty clay	Differential settlements and tilting of the structures	Inadequate site investigation		The site investigation failed to identify the soft silty Clay as it was terminated at the surface of the rock	See 16 below
3.17	Overhead water tank	Ring beam shallow footing	Very soft blackish grey silty clay	Tilting	Inadequate site investigation	Inadequate design	No site investigation was carried out to establish the ground conditions and the design of the ring beam was faulty	See 17 below
3.18	Coal conveyor system	Shallow foundations	Foundations laid on non-engineered fill	Large settlements	Inadequate construction		The fill was not placed and compacted in layers	See 18 below
3.19	Industrial building	Driven cast in situ piles	Marine clay	Undue settlements	Inadequate design		The design did not consider the settlement of the pile group	See 19 below

(continued)

Table 19.3 (continued)

No.	Project/structure	Foundations	Subsoil profile	Distress/failure	Primary shortcoming	Other shortcomings	Notes	References
3.20	Earthen dam	Soil improved using stone columns	Residual soil of micaceous Silt underlain by fractured and weathered rock	Improved ground not of sufficient stiffness	Inadequate construction		The stone columns did not achieve the required performance and needed to be improved by vibrofloatation	See 20 below
3.21	Factory building	Shallow footings	Compacted fill	Excessive settlement during construction	Inadequate design		The loads during construction were greater than considered in the design	See 21 below
3.22	Four-storey residential building	Shallow footings	Soft cohesive fill overlying rice mill waste	Excessive settlements and sinking	Inadequate site investigation		No site investigation was carried out to establish the ground conditions	See 22 below
3.23	Seven-storey hotel building	Strip footings	Unknown	Excessive settlements resulting in splitting of the building	Inadequate site investigation		No site investigation was carried out to establish the ground conditions	See 23 below
3.24	Eight-storey apartment building	Strip footings	Unknown	Excessive settlement of the building due to bearing capacity failure	Inadequate design		The allowable bearing pressure of the foundations had not been properly assessed	See 24 below
3.25	Deep cutting slopes for a canal	Natural ground	Expansive clay with bands of sand	Slope failures	Inadequate design		The stability analysis did not adequately consider the strata	See 25 below
3.26	Residential buildings	Under reamed piles	Silty Clay	Excessive settlement and low capacity found during pile load tests	Inadequate construction		The base of the pile bored below the under ream was not cleaned	See 26 below
3.27	Cutting for a railway line	Natural ground	Laterite with hard crust over softer soil	Slope failure	Inadequate design		The soil parameters chosen were inappropriate	See 27 below
3.28	Overhead water tank	Raft foundation	Alluvial soil	The structure suffered excessive tilting	Inadequate site investigation		No site investigation was carried out to establish the ground conditions	See 28 below

(continued)

Table 19.3 (continued)

No.	Project/structure	Foundations	Subsoil profile	Distress/failure	Primary shortcoming	Other shortcomings	Notes	References
3.29	Steel tank for storing molasses	Base of tank acting as a raft on surface of the ground	Cohesionless soil	Ground was too wet/ soft for the tank to be placed	Inadequate site investigation		Molasses stored in open pits nearby caused the soil to become waterlogged and soften	See 29 below
3.30	Single-storey buildings	Strip footings	Laterite with ant hills causing voids and poor drainage	Cracking occurred in the building	Inadequate site investigation		The site investigation did not identify the voids and problems caused by them	See 30 below

References

Clayton CRI (2001) Managing geotechnical risk: time for change. Geotech Eng 149(1):3–11. Institution of Civil Engineers, Thomas Telford, London

Cook J (2005) Gaining client cooperation for high quality site investigations and risk assessment. In: Proceedings of piling & foundations conference, ground engineering, London

Littlejohn GS (1991) Inadequate site investigation. Institution of Civil Engineers, Thomas Telford, London

Rao VVS, Sargunan A (2010) Technical vulnerabilities in geotechnical failures—indian experience. In: Proceedings of international symposium on forensic approach to analysis of geohazard problems, ISSMGE, Mumbai, 2010

Threadgold L (2002) Site investigations within the development process. Geotechnics Ltd, Coventry, UK

Chapter 20
Effects of Vibration by Demolition on Nearby Machine Shop Floor—Wave Measurement for Dynamic Property of Ground

Y. Iwasaki and K. Nakagawa

Abstract This paper describes a case study of damage to machine shop floor caused by shock vibrations during demolishing work of adjacent warehouse. Based on field measurements of ground motion by dropping weight on the ground at the site, dynamic properties of the ground were characterized. Fall of the warehouse elevator's counterweight could have caused damage to the steel-meshed, 30 cm-thick cement mortar floor. It was estimated that the shock waves that propagated from the dropped counterweight caused larger tension strains than allowable and resulted in cracks in the mortar slab.

Keywords Vibrations · Field measurements · Damages · Cracks

20.1 Introduction

Dismantling an old house is usually strictly controlled in order to protect living environment in the area nearby.

Legally, it is required to keep the vibration and sound from construction work within a limited level. In 2009, the author was asked by Osaka District Court to consider the effects of precise machines that were installed in a workshop due to dismantling work at a nearby structure (Fig. 20.1).

Y. Iwasaki (✉)
Geo-Research Institute, 4-3-2, Itachi-Bori, Nishi-Ku Osaka 550-0012, Japan
e-mail: yosh-iw@geor.or.jp

K. Nakagawa
Osaka City University, 3-3-138, Sugimoto, Sumiyoshi Osaka 558-8585, Japan
e-mail: naka@jck-net.org

V.V.S. Rao and G.L. Sivakumar Babu (eds.), *Forensic Geotechnical Engineering*, Developments in Geotechnical Engineering, DOI 10.1007/978-81-322-2377-1_20

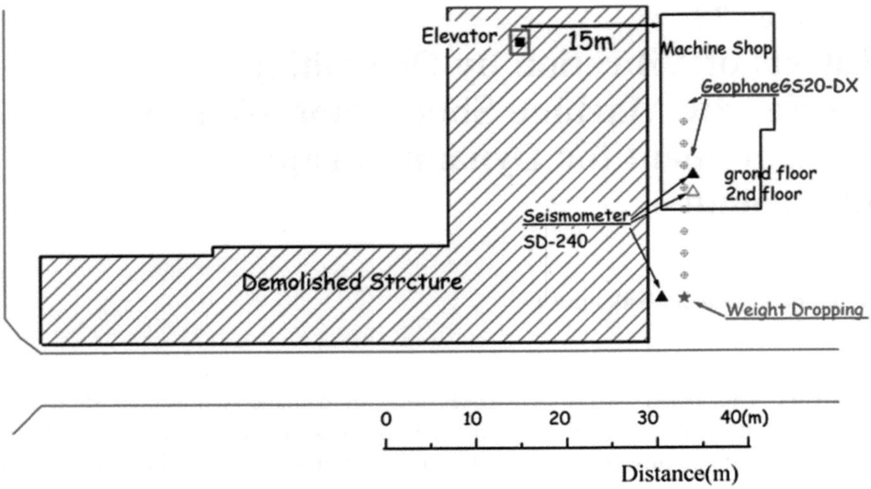

Fig. 20.1 Site map

20.2 Site and Probe

The site was located in the southeastern part of Osaka where the geotechnical condition is thick with soft clayey soils of some 16 m above dense gravel layer. A local machine shop equipped with precise machine tools was shocked by a sudden vibration during demolishing work in an adjacent warehouse. The warehouse was a two-storeyed steel frame structure with lightweight concrete panels. The roof was supported by a lattice beam with H-shaped steel elements. The shock was identified as due to snapping of the wire attached to the elevator's counterweight resulting in the drop of counterweight (Fig. 20.2).

The precise machine tools that were controlled by digital programs started malfunctioning after the accident. Other strong shocks several times from the demolishing work were from falling of the H-shaped steel beams. After completion of the demolishing work, many cracks were found in the machine shop floor as shown in Fig. 20.3.

The floor was of cement mortar 20 cm thick with steel mesh to prevent tension crack above a foundation of gravel layer as shown in Fig. 20.4.

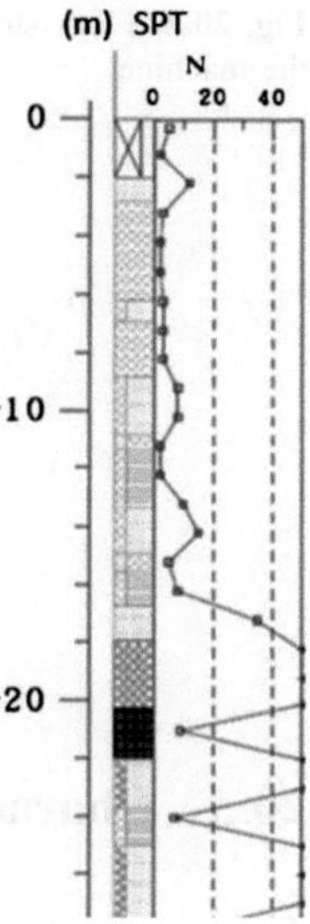

Fig. 20.2 Boring log

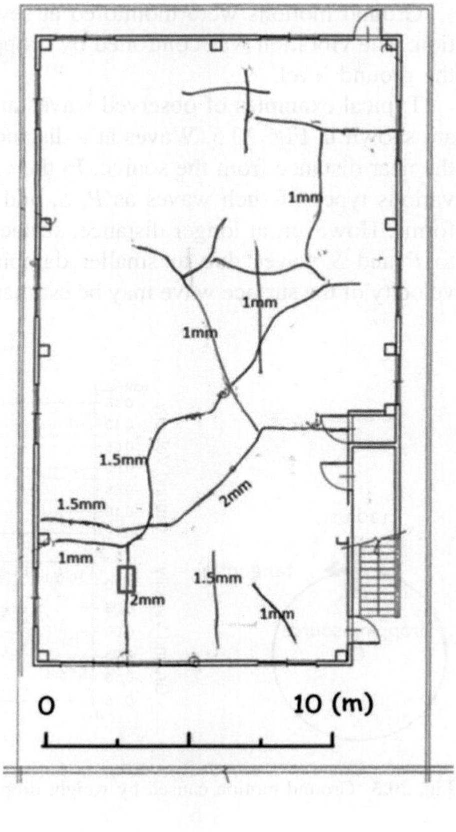

Fig. 20.3 Cracks in the floor of machine shop

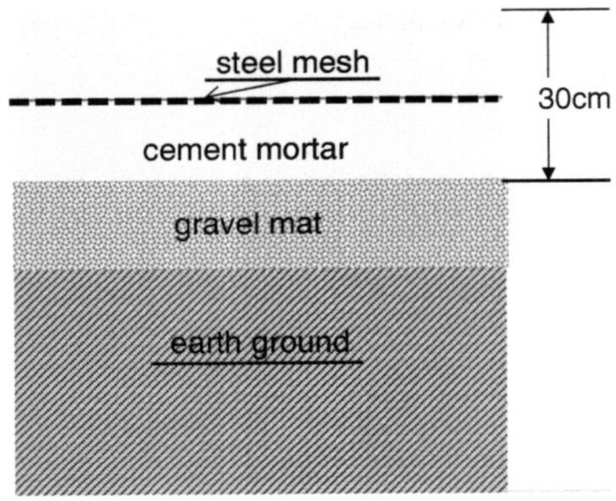

Fig. 20.4 Floor structure of the machine

20.3 Characterization of Dynamic Properties of Ground

To estimate the effects of vibrational effects caused during demolishing of warehouse, weight drop experiment was performed at the site.

Ground motions were monitored at several distances from the source of vibration. The vibration was controlled by dropping weights from different heights from the ground level.

Typical examples of observed waves at two different distances of 2 and 12.5 m are shown in Fig. 20.5. Waves at a distance of 12.5 m become much simpler than the near distance from the source. In the vicinity of the source, the combination of various types of such waves as *P*, *S*, and surface waves make complicated wave forms. However, at longer distance, surface waves become predominant compared to *P* and *S* waves due to smaller damping nature of surface waves. The wave velocity of the surface wave may be estimated as the distance between two receivers

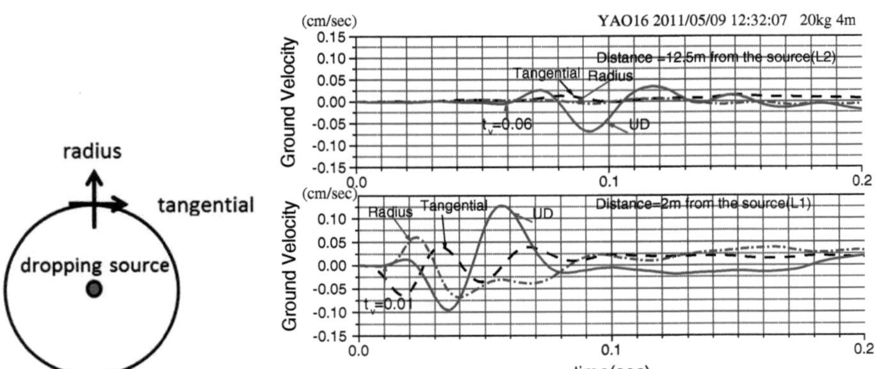

Fig. 20.5 Ground motion caused by weight dropping

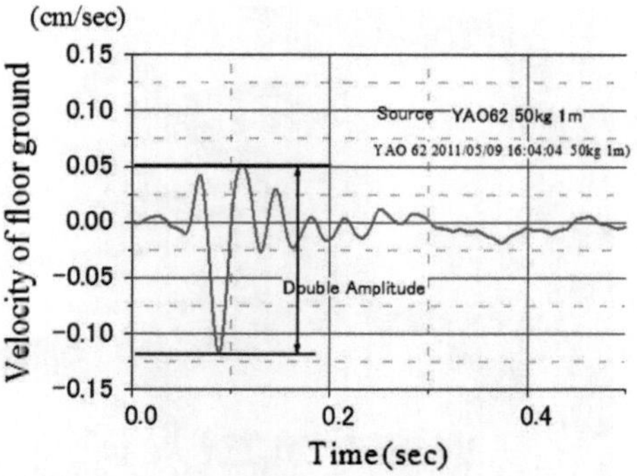

Fig. 20.6 Double amplitude

divided by the difference in the arrival time of initial rise of the waves as follows (Fig. 20.6 and Photo 20.1):

$$V = (12.5 - 2.0)\,\mathrm{m}/(0.06 - 0.01)\,\mathrm{s} = 190\,\mathrm{m/s}$$

If the weight is dropped at the center of the array under homogeneous conditions of the ground, horizontal motion will be predominant in the radius component. The recorded wave contains some comparable levels of the tangential component. This may be due to the incomplete symmetry of the dropping mass as shown in Photo 20.2 and some inevitable small setting errors of seismometer as well as inhomogeneity of the present ground.

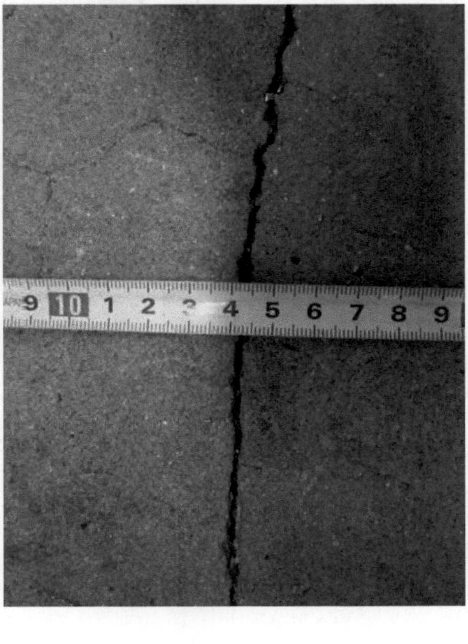

Photo 20.1 Crack in the concrete floor

Photo 20.2 Experiment of dropping weight

Photo 20.3 Dropping weight

The apparent width of the pulse wave of the vertical component is found as 0.035–0.04 s at the distance of 2 and 12.5 m from the source (Photo 20.3).

The maximum double amplitude values of ground velocity at 12.5 m from the source are listed in Table 20.1 for various combinations of different falling heights for masses of 20 and 50 kg.

Table 20.1 Experimental results

Test no.	Weight mass	Falling height	Double amplitude	Falling velocity at the ground	Test no.	Weight mass	Falling height	Double amplitude	Falling velocity at the ground
	20 kg	(m)	(cm/s)	(cm/s)	18	50 kg	0.5	0.42	313
3		0.5	0.22	313	19		0.5	0.47	313
4		0.5	0.175	313	21		1	0.58	443
5		0.5	0.255	313	22		1	0.61	443
6		0.5	0.26	313	23		1	0.66	443
2		1	0.29	443	24		1	0.86	443
7		1	0.26	443	25		2	0.86	626
9		2	0.37	626	27		2	0.76	626
10		2	0.365	626	28		2	0.8	626
11		2	0.41	626	29		3	0.75	767
13		3	0.51	767	30		3	0.7	767
14		3	0.46	767	31		3	0.92	767
15		4	0.48	885	32		4	0.98	885
16		4	0.48	885	33		4	0.92	885

Fig. 20.7 Double amplitude of vertical component of ground velocity at 12.5 m

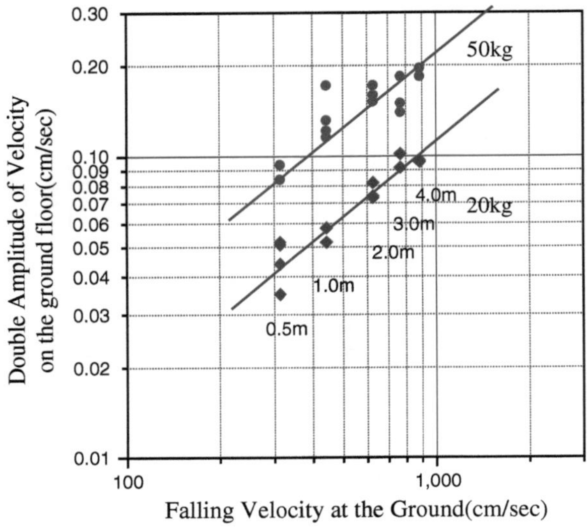

On the other hand, falling velocity at the ground for each case are calculated by and also listed in the table. The velocity is calculated by the following equation.

$$V = \sqrt{vgh} \tag{20.1}$$

As shown in Fig. 20.7, the log of the double amplitude shows linear relationship with the log of the falling velocity of the weight at the ground surface.

$$v = aV^{0.834} \tag{20.2}$$

As shown in Fig. 20.7, the ratio of the double amplitude for two different masses may be assumed constant as

$$v = \mathrm{b}\mathrm{M}^{0.756} \tag{20.3}$$

In a more general expression, double amplitude may be expressed as a function of dropping weight mass and falling velocity of the mass at the ground.

$$v = cM\,0.756V\,0.834 \tag{20.4}$$

The constant c is obtained as

$$c = 3.5 \times 10^{-5}$$

Equation (20.4) is obtained for surface velocities at 12.5 m from the source of dropping weight.

The wave amplitude generally attenuates with distance from the source of dropping point. Velocity sensors were set at different distances from the source as shown in Fig. 20.8. It should be noted that the floor of cement mortar locates

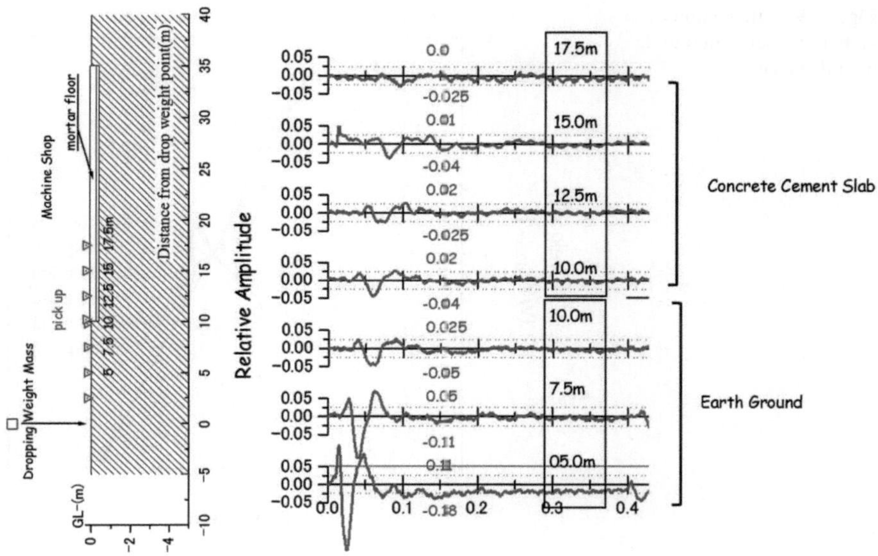

Fig. 20.8 Attenuation of velocity amplitude of wave propagation from dropping mass

beyond the distance of 10 m. The amplitude decays were observed with distance as shown in Fig. 20.8.

The peak-to-peak velocity amplitude of the vertical component at various distances from the source were divided by the value at the reference distance of 12.5 m from the source and plotted in Fig. 20.9.

It was found that there is little difference between these vibrations on the ground and on the floor of the mortar cement at a boundary of 10 m from the source.

The pulse width of velocity wave of 20 kg mass weight was compared to 50 kg mass from the same falling height in Fig. 20.10. The pulse width was about $T = 0.045$ s which corresponds to an apparent frequency of 22 Hz with little difference between the two weight masses and falling heights. The peak amplitude is found to be 0.75 times double the amplitude peak-to-peak value.

20.4 Falling Effect of Counterweight to Nearby Structure

20.4.1 British Standard

The effect of the impact of warehouse elevator's counterweight falling is estimated in this section.

The mass of the counterweight of the warehouse's elevator in the second floor was 1030 kg. Falling height was 300 cm. The distance between the falling point of

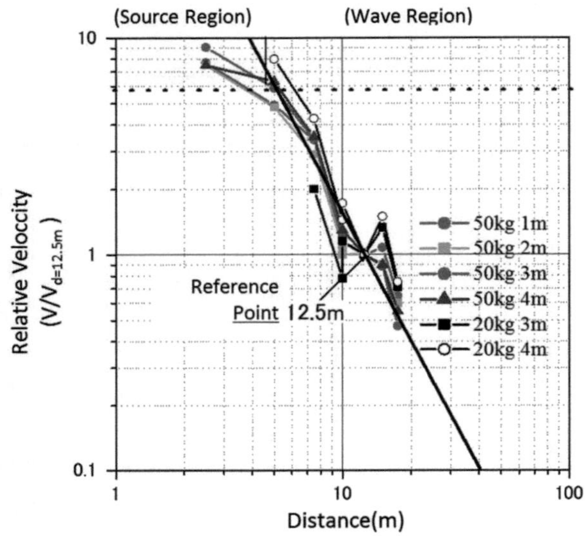

Fig. 20.9 Attenuation of vertical velocity amplitude with distance

the counterweight and the edge of the machine shop floor was 15 m. Based on the equations obtained in the former section, the falling velocity was calculated as 767 cm/s. The peak-to-peak velocity amplitude estimated at 12.5 m was 1.7 cm/s. and from 1.5 to 2.6 cm/s, respectively, at 15 m from the dropping point. These

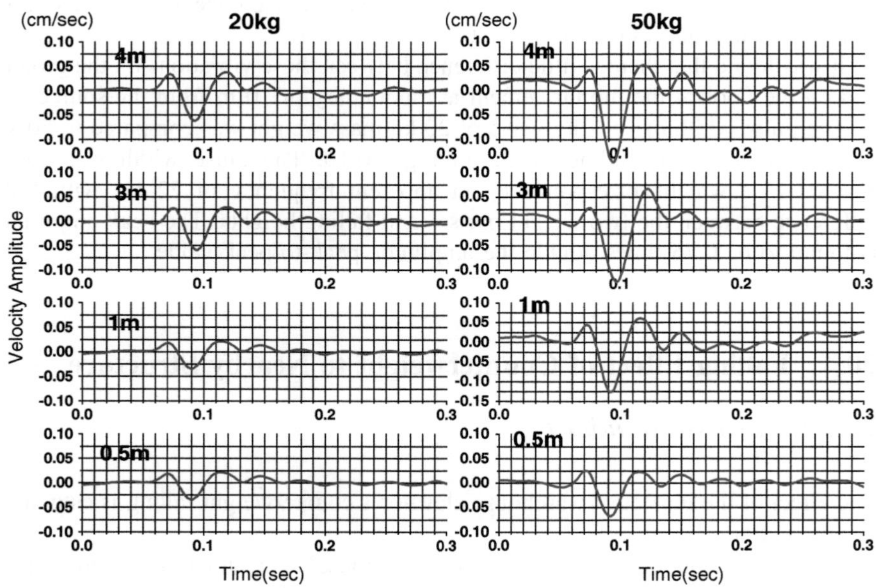

Fig. 20.10 Change of pulse width by different weight mass

Table 20.2 Estimated double amplitude by the falling of countermass

Counterweight mass (kg)	1030
Falling height (cm)	300
Falling velocity at the ground surface (cm/s)	767
Distance (m)	15
Peak to peak velocity amplitude (cm/s) at 12.5 m	1.7
Peak to peak velocity amplitude (cm/s) at 15 m	1.5−2.6
Peak velocity amplitude (mm/s)	20

Table 20.3 Transient vibration guide values for cosmetic damage (BS 7835-2 1993)

Line see Fig. 20.11	Type of building	Peak component particle velocity in frequency range of predominant pulse	
		4–15 Hz	15 Hz and above
1	Reinforced or framed structures	50 mm/s at 4 Hz and above	
	Industrial and heavy commercial buildings		
2	Unreinforced or light framed structures	15 mm/s at 4 Hz increasing to 20 mini's at 15 Hz	20 mm/s at 15 Hz increasing to 50 mm/s at 40 Hz and above
	Residential or light commercial type buildings		

Note 1 Values referred to are at the base of the building (see Sect. 6.3)
Note 2 For line 2, at frequencies below 4 Hz, a maximum displacement of 0.6 mm (zero to peak) should not be exceeded

values are listed in Table 20.2. Table 20.3 shows the guide values of transient peak component particle velocity for cosmetic damage of buildings based on the British Standard 7385-2 (1993). The guide level is divided into two groups for reinforced and unreinforced structures. These levels are plotted in Fig. 20.11.

The frequency of transient wave by counterweight of 1030 kg might be lower than the experimental dropping mass of 20 and 50 kg. However, the same frequency is assumed as the experimental value of 22 Hz.

The peak threshold value of the particle velocity is obtained as about 30 mm/s based on Fig. 20.11. The expected peak velocity by the experiment is 20 mm/s, which is smaller than the threshold value but within the same level.

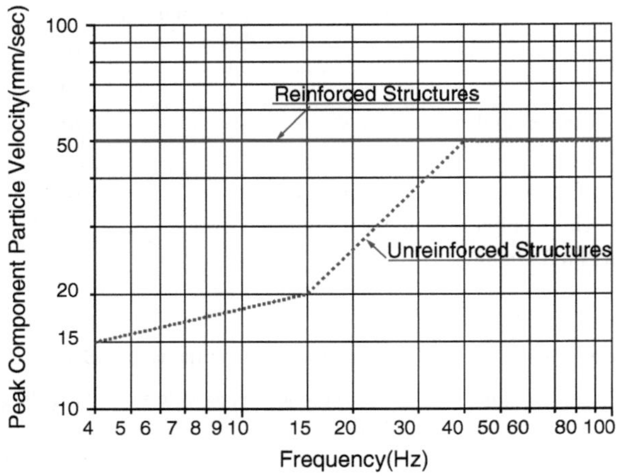

Fig. 20.11 Guide levels of vibration to cause damage of cosmetic part of the structure

20.5 Bending Crack

Since the pulse is traveling in the mortar floor, the floor surface deforms as the wave travels with a velocity of about 190 m/s.

The cement mortar floor deforms as the displacement of traveling waves. Figure 20.12 shows the velocity with time, displacement with distance, as well as inverse of the curvature of the floor plate.

The mortar plate is expected to deform in bending mode, where the tension and compression strains are induced at the surface of the plate.

The amount of strain ε is related to radius R of the curvature of bending deformation as shown in Fig. 20.13.

$$\varepsilon = t/R$$

where t: one half of thickness of the floor.

The tension strain at failure of unreinforced mortar was obtained by Ueda et al. as $\varepsilon = 0.0001$ (Ueda et al.1993).

The thickness of the mortar floor is 40 cm and the critical radius R_c corresponding to tension failure is 0.15 m/0.0001 = 1500 m. Curvature of a mortar plate less than $R_c = 2000$ m is very likely to induce tension crack on the mortar surface.

Figure 20.12 shows the minimum curvature that takes place at around 16 m distance point with the curvature of $R = 500$ m, which is smaller than R_c and tension strain increases to $\varepsilon = 0.003$ and accordingly the tension failure was very much likely to take place.

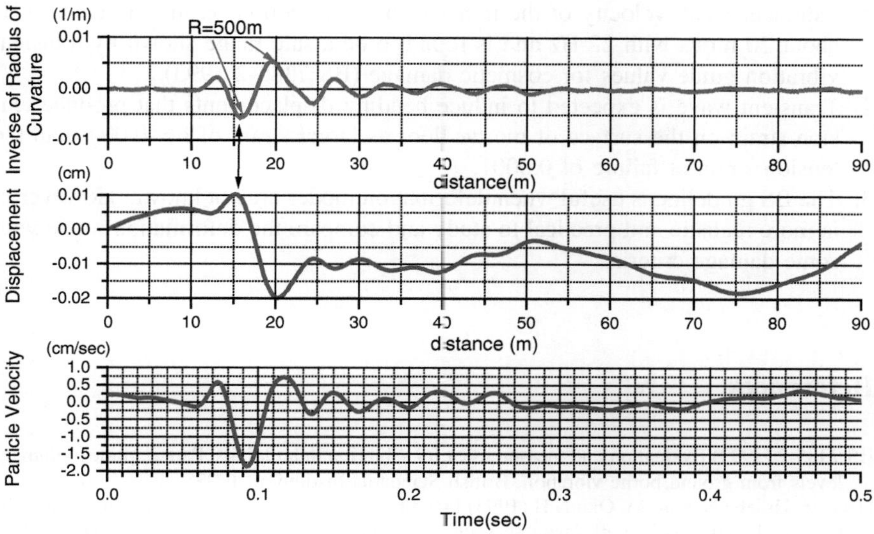

Fig. 20.12 Deformation of cement mortar floor with Inverse of radius of curvature

Fig. 20.13 Strain and curvature

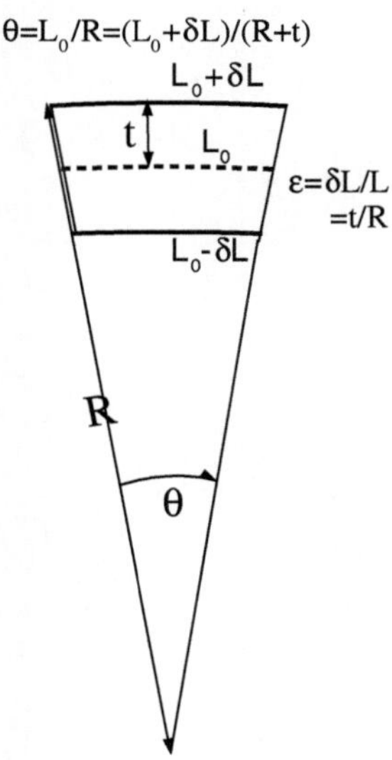

20.6 Conclusions

Effects of the falling counterweight of an elevator during a warehouse demolition to the adjacent mortar floor in a machine workshop was studied by dynamic characterization of field experiments of falling mass.

1. Estimated peak velocity of the transient particle motion on the mortar floor is about 20 mm/s with 22 Hz and is found to be a safe range shown by transient vibration guide values for cosmetic damage (BS 7385-2 1993).
2. Transient wave is expected to induce bending displacements that produce tension strain on the surface of mortar floor as larger strain of $\varepsilon = 0.004$ than the tension strain at failure of 0.0001.
3. The BS guideline is useful when deformation modes are not known. However, it is more realistic and practical to study and compare the deformation mode with some damage criteria.

References

BS 7385-2 (1993) Evaluation and measurement for vibration in buildings. Part 2: guide to damage levels from ground borne vibration, British Standards Institution, 1993

Ueda M, Hasebe N, Sato M, Okuda H (1993) Fracture mechanism of plain concrete under uniaxial tension. In: Proceedings of Japanese society of civil engineering, vol 466, pp 69–78, May 1993, JSCE (in Japanese)

Chapter 21
A Case of Piping in Sand Under a Dam and Its Back Analysis

P. Sembenelli

Abstract The paper reports a groundwater-flow-related failure occured to a homogeneous sandfill dam, 20 m high, and describes in detail the analysis of the leak occured during the first impounding under a head of 3.6 m only. After an examination of various hypothesis, piping as the failure mechanism is proposed and checked with numerical analysis derived from Schmertmann's theory as well as with a field test. Back analysis is extended to explain what happened after an uninterrupted pipe established between reservoir and the toe of the dam.

Keywords Piping · Failure · Seepage · Foundation · Sand · Dam · Safety

List of Symbols

B	Horizontal segment of seepage path (ft)
B_{ur}	Bulk unloading—reloading modulus
C_i	Flow concentration coefficient (overall)
C_{jt}	Flow concentration coefficient for a pipe
C_{IP}	Flow concentration coefficient of the flow net in the horizontal plan
C_{IS}	Flow concentration coefficient of the flow net in the vertical plan
C_{2n}	Flow concentration coefficient at outlet, normal to the axis of the pipe
C_{ap}	Flow concentration coefficient at outlet, parallel to the axis of the pipe
€3	Flow concentration coefficient for 3D flow condition
C.ir	Flow concentration coefficient for a 3D pipe
Cr″I	Required cohesion
C,s	Shape factor for the pipe's cross section
GI,	Uniformity coefficient
D	Thickness pipeable
d_{oo}	Diameter 10 % passing
d_{fto}	Diameter 60 % passing
d_{ave}	Characteristic diameter controlling the deposition process

P. Sembenelli (✉)
Member ASCE, President and Managing Director SC—Sembenelli Consulting, 3/5 via Santa Valeria, 20123 Milan, Italy
e-mail: info@scsembenelli.com

V.V.S. Rao and G.L. Sivakumar Babu (eds.), *Forensic Geotechnical Engineering*, Developments in Geotechnical Engineering, DOI 10.1007/978-81-322-2377-1_21

D_{ur}	Edometric (constrained) unloading–reloading modulus
e	Void ratio
F_3	Proportionality factor for a semicircular pipe (ft/min)
FSH	Factor of safety against hydrofracturing
FSP	Factor of safety against piping
FSS	Factor of safety against shear
F_v	Water volume factor representing the dilution of the water/soil suspension
G_s	Specific gravity of grains
h	Head (ft)
H_I	Actual maximum head existing along any given pipe segment (ft)
H_{niax}	Maximum allowable head (ft)
I	Seepage gradient
i_c	Critical seepage gradient
i_v	Vertical seepage gradient
k	Coefficient of permeability (ft/min)
k_h	Horizontal coefficient of permeability (ft/min)
k_j	Coefficient of permeability of generic segment (ft/min)
k_{rp}	Coefficient of permeability proper of the reference segment in a direction parallel to the pipe axis (ft/min)
k_v	Vertical coefficient of permeability (ft/min)
k_{an}	Coefficient of permeability normal to the slope a (ft/min)
k_{ap}	Coefficient of permeability along the slope a (ft/min)
L	Pipe segment length
L'	Transformed pipe segment length taking as reference the coefficient of permeability k_{rp} of the reference segment
L^2	Transformed pipe penetration length
L YD	Thickness factor
I'/L'	Pipe development
LJ	Length of generic segment
n	Manning's roughness coefficient
N_F	Number of flow channels
NI I	Number of head drops
p	Dynamic water pressure
q	Flow
R	Hydraulic radius (ft)
R_c	Lane's weighted creep ratio
RH	Scour velocity reduction factor
S	Degree of saturation
S	Slope of piezometric surface
T	Overall time for piping
t	Vertical segments of seepage path
v	Velocity of the flow in the pipe (ft/s)
Vti3	Lower bound transport velocity (ft/min)

vs	Velocity required to scour a uniform particulate material (ft/s)
vt	Velocity required to drag a uniform particulate material (ft/s)
w	Water content
z	Pipe diameter (mm)
DH	Minimum head difference needed to maintain the lower bound flow velocity (ft)
(DH/L′) cr	Critical local gradient
AT	Time for the pipe to move across each segment L of the path (min)
a	Dip from horizontal
f	Friction angle
Y_d	Dry unit weight
y_t	Total unit mass
Y	Unit mass
s_h	Horizontal stress
a_v	Vertical stress

21.1 Introduction

Field investigation techniques, laboratory tests, and analytical tools available to dam designers, to handle strength-related problems, are far more developed, sharp and conclusive than those available to deal with seepage phenomena.

Aspects related to strength, stresses, and deformations received maximum attention in past years. A waste number of exercises can be performed in this respect that satisfies both authors and reviewers of a given design, to the point that supporting a design with elaborated numerical proofs some times approaches self complacency. Seepage and seepage-related problems, on the contrary, normally receive lesser attention and a far coarser processing even in a thorough design work. Checks and analyses, required by Codes, reflect this state of things.

A design tuned on the arsenal of available numerical analyses, rather than focused on providing a sound solution to the issues vital to the dam life, may cause the actual hierarchy of problems posed by the physical conditions of a site, to fall off-view. Severe flaws may remain unchecked or underestimated, particularly on the side of fluid flow through particulate media. Jet, by far the largest number of accidents to embankment dams can be related to the flow of water through foundation soils and/or rocks.

This paper deals with a groundwater-flow-related failure of a dam. The failure was produced more than 10 years ago. Analyses, testing, repairs, and a controlled filling of the reservoir up to reaching the original design pool level, required several years to complete. Reporting the entire story was not considered until after the dam was restored to proven normal service conditions.

The dam (see Fig. 21.l) is a homogeneous sandfill structure, 20 m high. Its crest length exceeds 5 km. A scour preventing pervious bituminous facing covers the 4 and 2.5H/1V upstream slope. A 25 m wide horizontal drain is provided under half of the 2.4H/1V downstream slope. The drain is processed gravel enclosed in a nonwoven, heat-bonded, polypropylene fiber geotextile. A gravel transition and a uniform coarse rock toe (made of laterite chunks) were provided for the downstream of the drain. Concrete pipes, 200 mm in diameter, run along the edge of the drain, they were intended to collect seepage and to convey it to gauging chambers set at regular intervals along the dam's toe.

The embankment is generally founded over a natural clay blanket. However between Ch. 1+200 and 2+000 approximately, this clay layer was covered by about 1 m of alluvial and eolic deposits described, in the design documents, as "sandy silt."

During construction, discussions between the designer, the engineer and the contractor, led to:

(i) placing a 1.5 m thick artificial clay blanket over the sandy silt
(ii) extending the upstream blanket, made of low permeability siltysand, to a final length of 100 m.

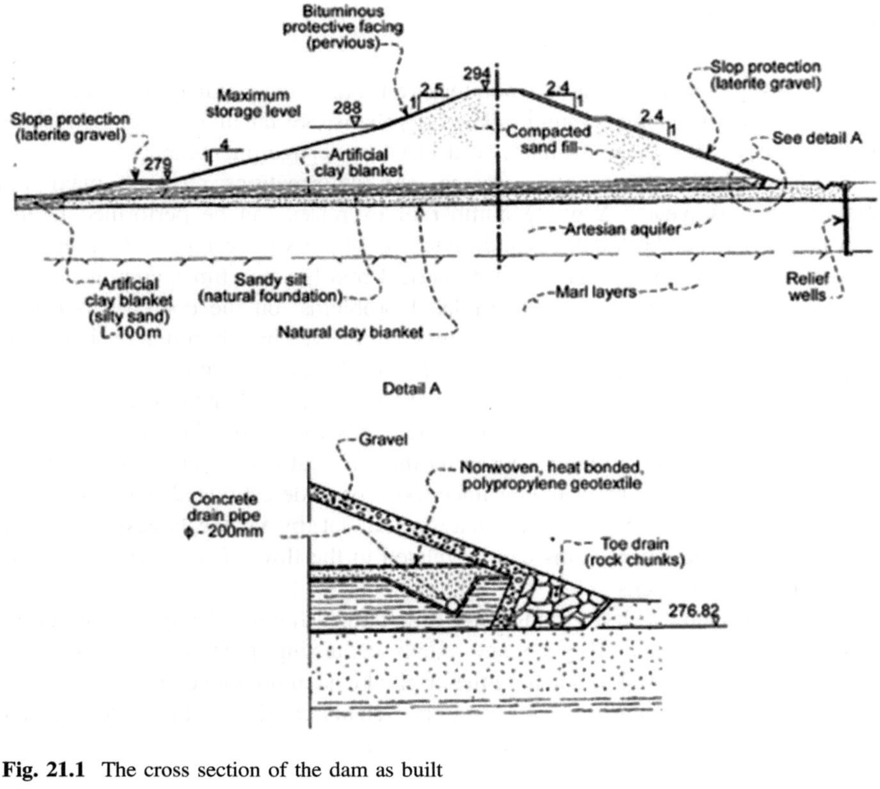

Fig. 21.1 The cross section of the dam as built

Between Ch. 1+880 and 1+960, the lowest part of the dam, and the artificial clay blanket, were built in different periods. A construction joint, most probably running across the entire embankment width, was formed at Ch. 1+900. Persons in charge of construction describe this and other construction joints as gently sloping so that equipment could move across. Throughout the work, each lift of the artificial clay blanket was compacted with more than 8 passes of a CAT 825 C vibratory tamping roller. The foot end area was 18,300 mm^2 and the compaction pressure applied was 880 kPa. According to quality controls on record, the joint at Ch. 1+900 was built in a satisfactory manner. The characteristics of the clay used for the blanket are such that intense cracking would have developed if the clay remained exposed too long. Climate conditions prevailing during placement might have favored some desiccation. Signs of lamination were later observed in exploratory shafts, possibly the effect of repeated passes of loaded scrapers.

21.2 The First Impounding and the Leak

Reservoir impounding started on June 11, when the water level was 277.63, high enough to inundate the upstream blanket. By June 20, the water topped the u/s berm and, thereafter, remained around el. 279.5, until July 25. A fast rise in the reservoir level is produced in 15 days from July 25 to August 10, when water elevation reached 281.18. On August 15, the reservoir water level was still 281.19. A water head of 3.6 m (representing 35 % of the design head) insisted over the upstream blanket when a leak becomes evident at the downstream toe. A very small head had thus been experienced for a time, not shorter than 5 days and no longer than 15 days.

The leak was later located near Ch. 1−900. One of the instrumented sections foreseen by the design had been located at Ch. 1+940. It was equipped with three vibrating wire piezometers which never produced reliable readings.

On the morning of August 15, the drain collector pipes exiting at Ch. 2+000 showed an increased discharge. The water color was described as whitish and the combined discharge of the two V notches was gauged in 3.4 1/s. At 19 h seepage daylighted along the toe drain near Ch. 1+940 (el. 276.7). At 22 h the drain pipe discharge was gauged as 21 1/s and the water color was described as darker than it appeared in the morning.

On August 16, at 8 h, the drain pipes discharged 22 1/s while water daylighted at three different points along the toe drain between Ch. 1+890 and 1+910. The combined discharge from toe drain and drain pipes was gauged, at a Parshall flume downstream, as 57 1/s. Water was initially muddy and carried small clay lumps. Later the water became dark-gray and carried sand, too. Between 10 and 12 a.m. of August 16, the quantity of water rushing from the toe drain diminished appreciably to suddenly increase to a much larger flow, at 13 h about. Figure 21.2 shows the

Fig. 21.2 The exit point of the leak as it appeared in the afternoon of August 16, 1984. Water surfaces under a substantial head and carry large quantities of suspended solids. The picture gives an idea of the grading of the latrine chunks used in the toe drain

water loss as it could be seen on August 16: the water sprung under a substantial head. Manual soundings indicated that the water was passing between the artificial clay blanket and the transition material next to the toe drain. In the evening of August 16, the total water loss was gauged as 300 1/s.

In the morning on August 17, the water loss was 400 1/s and increasing. In the afternoon, the estimated flow underpassing the dam, exceeded 1300 1/s and an eddy appeared 55 m u/s from the dam axis. Later, during the night, the d/s slope caved-in over a roughly circular area 10 m in diameter, centered 32 m d/s from the dam axis near Ch. 1+900. On August 18, the seepage inlet could be plugged with sand bags and well-graded sand and gravel. Depressions appeared both on the bituminous facing of the upstream slope and on the downstream slope, just above the toe drain.

Early in September, the reservoir water level was raised again to el. 281.2 and it remained above el. 281 for the entire month of October. In November, the reservoir was emptied for diagnosis and repair.

21.3 Investigating the Leak

Investigations started soon after the leak and were continued up to mid-February of the following year.

A total of 45 boreholes were drilled in the area around Ch. 1+900 along nine lines parallel to the dam axis. All holes were deep enough to enter the natural foundation at least 1 m, with a maximum of 2 m. All holes were wash-bored with nearly continuous SPT testing and spoon sampling. Field and laboratory tests carried out consisted of 359 in situ water contents, 43 Atterberg's limits, 12 grading curves, and 42 determinations of the fraction passing #200 sieve (0.074 mm). Except for the water content determinations, the majority of the tests were performed on foundation soils existing above the natural clay blanket.

Borings allowed to detect, along 5 of the 9 lines, an approximately elliptical cavity, 2–3 m wide and 0.5–1.5 m high. At both ends, the cavity ran within the clay of the artificial blanket. In the center portion near the dam axis, the cavity was lower, reaching into the sandy silt of the natural foundation. The cavity seemed to run almost straight along Ch. 1+900 with only one visible bend 10 m from its d/s end. Part of the cavity was standing and was filled with muddy water. Clay lumps were found over its bottom at two sections.

On January, the entire portion of the dam affected by the leak was gradually removed from crest to foundation. While demolition proceeded, nine backhoe trenches were opened. The d/s end of the cavity was uncovered and excavated with hand work while the u/s toe was removed by machine. The excavation of the cavity allowed taking disturbed and undisturbed samples for laboratory testing. Geometrical and pictorial records of the cavity cross sections were possible only at those sections where the cavity had been filled by the sand and gravel used to plug the leak. Figure 21.3 shows the cavity as it appeared 50 m u/s from the dam axis. Nearby, a finger-like extension, only 0.3 m high, extended sideways in excess of 3 m from the cavity's wall. By February the largest part of the cavity roof had collapsed and the cavity could therefore be surveyed in plan, only. Figure 21.4

Fig. 21.3 The cavity across section 50 m u/s from the dam axis. The wine red material is the sand +gravel used to plug the leak. The cavity *upper half* is within the artificial clay blanket. The *lower half* of the cavity has been cut through the natural foundation

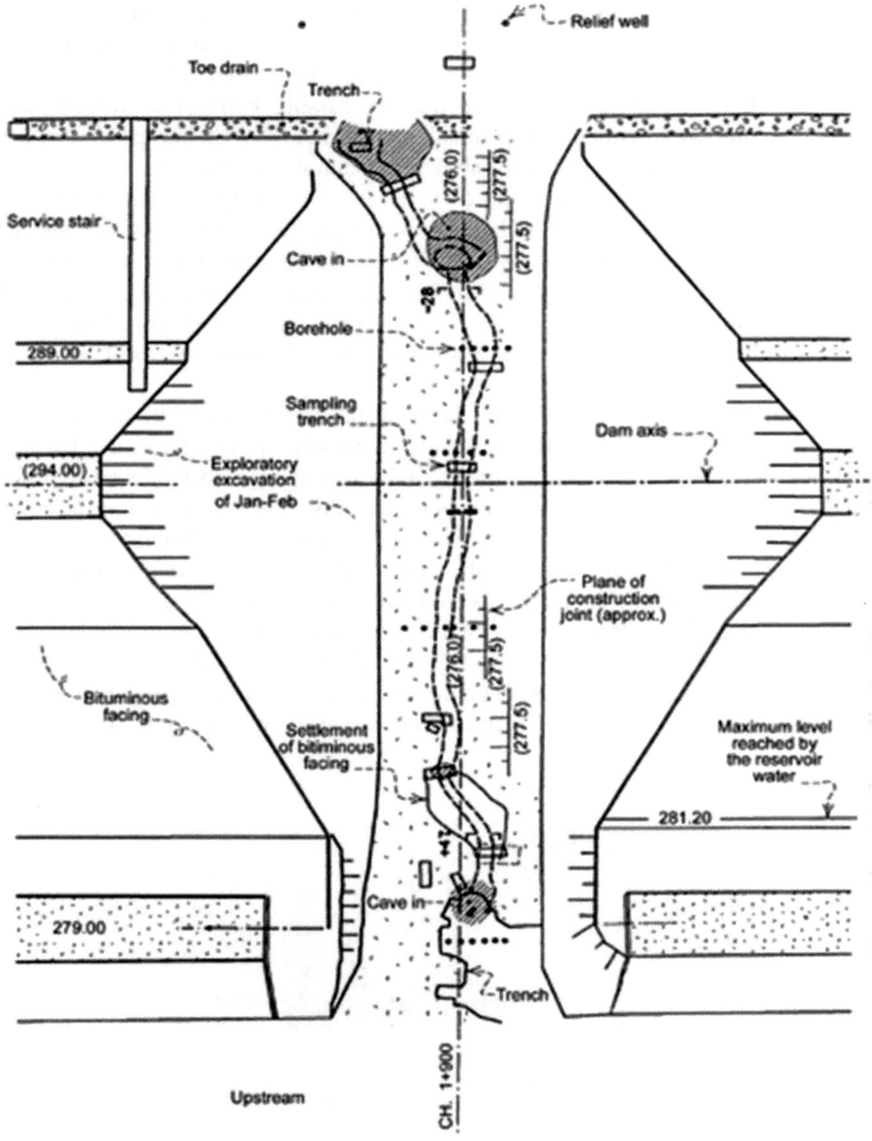

Fig. 21.4 Horizontal cross section at el. 277 of the cavity produced by the leak. BH locate some of the exploratory boreholes, T locate exploratory trenches

shows a horizontal section of the cavity at el. 277. The cavity had a nearly constant width of about 2 m and a winding path with maximum offsets of 4 m from an ideal line at Ch. 1+900.

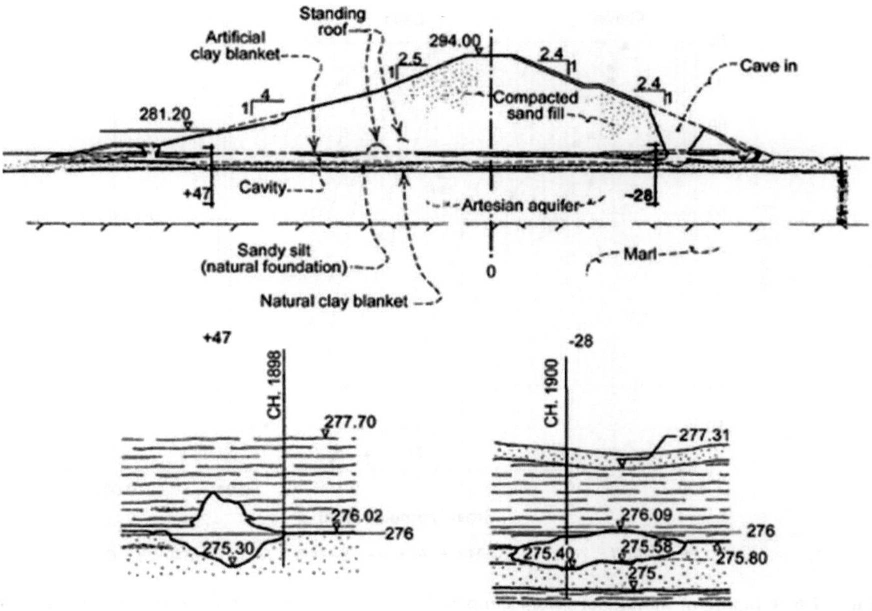

Fig. 21.5 Cross section of the dam at Ch. 1900 approximately showing the cavity resulting from the leak. BH locate some of the exploratory boreholes

Figure 21.5 shows the vertical section of the cavity and dam at Ch. 1+900 approximately. The cavity bottom reaches some 0.5–0.8 m below the original foundation grade, into the natural sandy silt formation. Apparently, 10 m of the cavity near its d/s end and 15 m of the cavity near the u/s end run within the artificial clay blanket.

Observations made at 13 positions along the cavity showed that the "sandy silt" layer immediately under the dam was not a uniform deposit but rather a superposition, in layers, streaks, and strings, of three different materials: (i) sandy silt with up to 60 % finer than silt size (0.075 mm), (ii) fine uniform sand containing almost no silt but up to 20 % of clay fines (−0.002 mm), (iii) coarse sand with less than 5 % silt (−0.075 mm) interspersed with quartz gravel elements in the 1/4″–1″ size. Figure 21.6 shows the gradings of the three materials actually forming the so-called "sandy silt." The sand grains were mostly quartz in loose or medium dense conditions.

The natural clay blanket, considered a noncontinuous feature, was invariably found where the exploration was carried deep enough (nine positions). Such clay blanket had, to all appearances, a thickness ranging from 0.2 to 0.6 m and formed, at least at Ch. 1+900, a continuous horizon below el. 274.5.

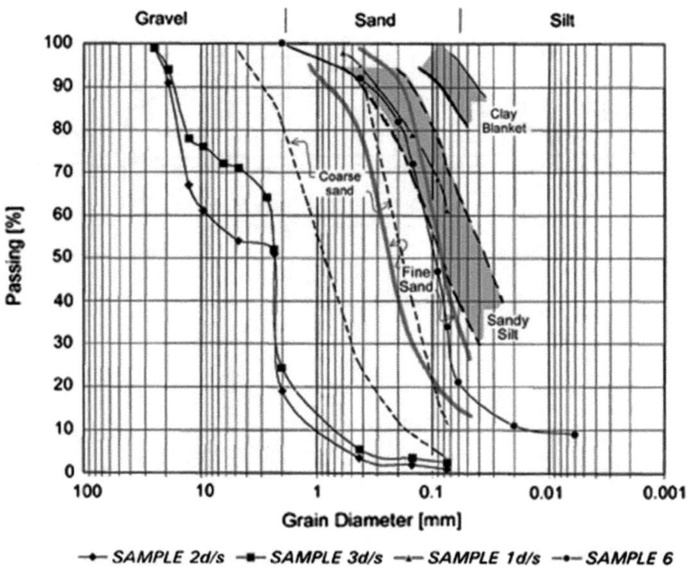

Fig. 21.6 Grading of individual layers sampled in the "sandy silty" foundation above the natural clay blanket, at Ch. 1900

21.4 The Leak Back Analyzed

Once the cavity defined in detail, the point was to explain and to prove how the leak could produce across an embankment 115 m wide, under a head of 3.6 m only and in less than 60 days after impounding started. The particular location of the cavity (contained within the clay blanket at both ends of the pipe) and its peculiar shape (horizontal fingers extending more than 3 m sideways, at places) had to fit with the explanation. Different mechanisms were initially considered: improper construction (faulty construction joint), unforeseen behavior of the materials (collapse or dispersive soils), or excessive seepage across the embankment with suffusion and transport of soil particles (piping).

21.4.1 First Hypothesis: Faulty Joint

A construction joint was formed at Ch. 1+900 in the artificial clay blanket. Most likely the joint run on a sloping plane rising roughly 4H/1V toward Ch. 1+800.

Some 14 quality controls, performed on the man-made clay blanket, between Ch. 1+880 and Ch. 1+950, indicated a minimum of 92 % of fines (−0.075 mm),

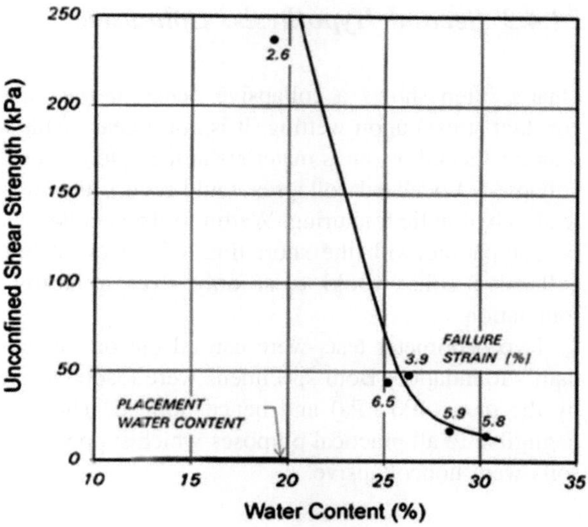

Fig. 21.7 Reduction of the unconfined shear strength of the compacted clay from the artificial clay blanket placed under the dam, with increasing water content. Note that placement water contents were between 12 and 20 %

in situ water contents between $w = 14.4$ and 19.1 % (ave. 16.9, st. dev. 1.25 %) and in situ dry unit weights between $ya = 15.2$ and 17.2 kN/m^3 (ave. 1.64 st. dev. 0.06 t/m^3). All figures fall well within the specified and accepted placement limits for this material. Drying up, and the consequent scarifying-wetting-rolling operations, recorded on December 15, 28 and 29, indicated a close control by the Supervision on placement operations.

The undrained cohesion versus water content relationship for the clay of the artificial blanket, given in Fig. 21.7, allows calculating that a vertical crack cannot stand open under the load of the center portion of the dam. The limit condition for an open crack to exist was assumed to be an = 0. The related minimum cohesion required was computed as creq = (crv − CTH)/2 = $y * h/2$. A total unit mass $y_t = 18.5$ kN/m^3 and the maximum embankment height at Ch. 1+800, $h = 18$ m, result in a required cohesion creq = 166 kPa. Such undrained cohesion level may exist only if the clay water content stays below $w = 22$ %.

Some softening would certainly take place in a "dry" joint through a clay having an initial placement content up to $w = 19$ % and supposedly filled with water for 60 days. The undrained cohesion most probably was less than required to support an open crack. Moreover, the pipe path does not match with the straight line geometry of the construction joint. As a matter of fact, the pipe departs from the joint plane by more than 5 m between 15 and 40 m u/s from the dam axis and turns away from the joint plane by more than 10 m in the last 30 m (25 % of the entire pipe length). There is no visible relation between the pipe cross section and the construction joint plane.

21.4.2 Second Hypothesis: Collapsive Soils

Quartz often shows a collapsive behavior (i.e., a reduction in volume under a constant stress) upon wetting. It is not unreasonable to suppose that predominantly quartz silts and/or sands materials in the natural foundation, once wetted, may have collapsed. Localized collapses could reduce the stress level in the foundation and lead to hydraulic fracturing. Wetting-related collapse phenomena are fast enough to be compatible with the short time of the event but it was hard to imagine that collapsive soils would exist only over a narrow strip of such an extended foundation.

Two oedometer tests were carried out on undisturbed samples taken from the dam's foundation. Both specimens were loaded to the overburden pressure applied by the dam (400 kPa) and hence flooded. The collapse following flooding was negligible to all practical purposes which supported the assumption that foundation soils were noncollapsive.

21.4.3 Third Hypothesis: Dispersive Soils

Clays with a particularly high Sodium content are easily dispersed in the permeating water and can be removed from the soil skeleton. Dispersion and removal of the clay fraction will increase the voids of the soil and may progress to the point where continuous pipes, connecting the reservoir to the d/s toe, are created.

The results of two Pinhole tests carried out with the procedure suggested by Sherard et al. (1976), classified the clay used in the artificial clay blanket, as dispersive. Also the clay fractions present in the silty sand were dispersive and proved to be very erodible.

However, the time elapsed between the starting of the filling and the leak is against the hypothesis of dispersion. Dispersion phenomena are generally slow and require a fully established water flow to develop. The degree of saturation measured on several samples taken from the foundation sandy silt ranged from $S = 0.52$ to 0.93 thus suggesting that a generalized permeation was not yet existing under the dam. This would tend to exclude appreciable dispersion processes.

21.4.4 Accepted Hypothesis: Piping

In consideration of the dam cross section and of the profile and characteristics previously described for its foundation, shown in Figs. 21.5 and 21.6, piping or retrogressive erosion, promoted by excessive exit gradients, was considered as a possible failure mechanism. Several features that would favor a piping process existed like:

- the presence of a fine-grained soil over the entire foundation
- the cohesive, impervious layer placed over it (a pipe in cohesionless soil requires a roof with some cohesion to stay open)
- the absence of a positive cut-off reaching the natural clay blanket a coarse toe drain made of fine-free chunks of laterite, directly in contact with the "sandy silt" and/or the fine sand of the foundation
- the presence of streaks of pervious, coarse sand under and/or within the "sandy silt"
- the existence of a lower cohesive impervious horizon (the natural clay blanket) which isolated hydraulically the top meter or so of the foundation and helped confining seepage flow.

The start and progress of the piping was visualized as follows: a "no flow" piezometric level gradually builds up in the coarse and very coarse sands within the sandy silt. The exit gradient might have been close to the full head divided by the depth of the silt (i = 3/0.5), well in excess of threshold levels, and might have initiated piping at the bottom of the toe drain. The silt is not retained by the blocks of the toe drain and the pipe grows downward. The pipe is a minute opening (see Fig. 21.12) and its discharge quite small so that it finds its way along the toe drain and cannot be detected. A flow condition now establishes through the "sandy silt." With some lag, the equipotential lines are essentially the same in the pervious sand and in the overlying sandy silt. With a sand clock process, the pipe moves up through the sandy silt until it gets under the clay roof. Figure 21.8 describes the main stages of the process. If the grading (permeability) of the "sandy silt" remains unchanged under the remaining part of the embankment, the pipe will propagate toward u/s at an increasing rate.

21.5 Piping Verified

A quantitative check of the piping mechanism was done using Schniertmann's theory (1980). It may be appropriate to recall that the associated method for numerical analysis is based on an ideal, parallel flow taking place under constant gradient (uniform permeability).

To get the input data needed for the analysis, a number of operations are necessary. For each piping path to be investigated, the first operation is the identification of uniform pipe segments and hence computing the relevant transformed path lengths L' according to the position reached the pipe's end:

$$L = aL_1 = k_{rp}/k_i \tag{21.1}$$

so that they might be assigned a coefficient of permeability identical to the coefficient of permeability k_{rp}, proper to the soil of the reference segment (k are those

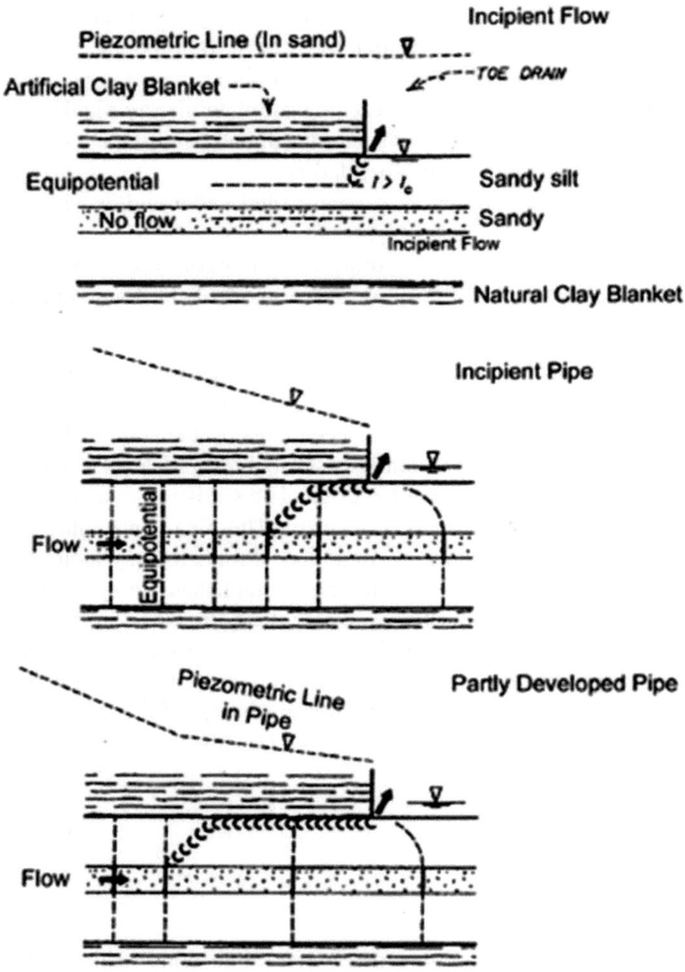

Fig. 21.8 Idealized early stages of a retrogressive erosion (piping) at the toe drain of the dam and its extension toward upstream within the "sandy silt" layer. The presence of a level of coarser and pervious sand and the absence of filtering action are two conditions for the process

applicable to the direction parallel to the pipe axis). A similar operation must be done on the lengths 1′ penetrated by the pipe at successive times.

The next operation is correcting the actual flow net to compensate for nonparallel (converging/diverging) flow conditions. A set of two concentration factors C_{lp} and C_{ls} must be derived from flow nets in the horizontal plan and in the vertical one, the overall correction being:

$$C_i = C_{1p} * C_{1s} \tag{21.2}$$

The convergence of the flow approaching the end of the pipe, for different pipe penetrations (position of the pipe's end along the selected path), is taken into account with 2 more Concentration Factors C_{2n}, correcting for the convergency of flow in the vertical plane, and C_{2p}, correcting for the flow component parallel to the axis of the pipe. A further Concentration Factor C_3 is introduced to take into account 3D flow conditions.

Schmertmann has derived the values of C_2 from a number of flow nets. The value of C_a is taken as $C_3 = 2$ for an asymmetric pipe's end in a homogeneous soil and as $C_3 = 4.6$ when an impervious roof is present and the pipe cross section becomes semicircular. The relevant numerical values of the coefficients are shown in Tables 21.1 and 21.2.

The simple geometry of the problem permits working with a simple model for the piping process consisting of only 2 uniform sections along a piping path leading horizontally from toe to toe (segment 0–1) and upward into the reservoir (segment 1–2).

The conditions at the inner end of a pipe are considered to be the same as those existing on a submerged and level sand, subject to both moving water and upward seepage. Vertical gradients ($i_v > 0$) into a level, cohesionless soil, reduce significantly the water's horizontal velocity required to initiate scouring. According to Schmertmann, the "Scour Velocity Reduction Factor" R'' can be expressed as:

$$R_\alpha = \frac{\text{scour velocity when } i_v > 0}{\text{scour velocity when } i_v = 0} = \sqrt{1 - \left(\frac{i_v}{i_c}\right)} \tag{21.3}$$

The critical seepage gradient i_c, defined in an ideal vertical upward seepage field, is the gradient necessary to set a surficial soil particle in a "quick" state (i.e., ready to move even with a nominal water velocity). Because the velocity of water at the head of a growing pipe is very small, a scour velocity close to zero $R_a = 0$ has been assumed by Schmertmann as the condition required to sustain a growing pipe. The expression of R_a, along either a level or a sloping pipe is:

$$R_a = \sqrt{\left[1 - \frac{(C_1 * C_{2p} * \sin\alpha + C_1 * C_{2n} * C_3 * \cos\alpha) * (\Delta H/L)}{i_c}\right]} \sqrt{[*]} \left(\cos a + \frac{\sin\alpha}{\tan\phi}\right) \tag{21.4}$$

All terms in Eq. (21.4) are related to the geometry of the problem and to the position of a particular segment along the pipe path. Given a certain path geometry, the only term that can be modified to make $R^\wedge = 0$ is the gradient (AH/L') and a trial procedure is used to find the critical local gradient (AH/L')cr making $R_a = 0$ along each segment of the selected pipe path.

Table 21.1 Geomatry

Segment (point)	Dip (fro m hor.)	Angle of repose	Crit. grad.	Thick ness pipe able	Segment length	Hor. perm. coeff.	Vert perm. coeff.	Reference perm. coeff.	Trans formed path length	Trans formed pipe penetration	Thickness factor	Pipe devel.
(1)	α(*) (2)	ϕ(*) (3)	i_c (4)	*D* (ft) (5)	*L* (ft) (6)	k_h (Cm/s) (7)	k_v (Cm/s) (8)	k_{rp} (cm/s) (9)	*L'* (ft) (10)	*l'** (ft) (11)	*L'*/D (12)	*l'*/*L'** (13)
0	0	30	1.10	4.9	–	–	–	–	397	1	81	0.01[a]
0–1	0	30	0.95	4.9	395	10^{-5}	5×10^{-6}	1×10^{-5}	397	395	81	0.50
1–2	0	35	1.00	00	4.9	5×10^{-5}	2×10^{-5}	2×10^{-5}	795	–	162	0.9

[a]Assumed values

$10^{-5} = k_{rp}$ coefficient of permeability proper of the reference segment in a direction parallel to the pipe axis

Table 21.2 Concentration factors

Segment (point)	Concentration horizontal	Concentration vertical	Concentration overall	Concentration normal	Concentration parallel	Concentration 3D
(1)	C_{1p} (2)	C_{1s} (3)	C_1 (4)	C_{2n} (5)	C_{2p} (6)	C_3 (7)
0	1	2.5	2.5	1.5	3.0	Roof 4.6
0–1	1	1.5	1.5	6.5	6.5	Roof 4.6
1–2	1	1.0	1.0	14.0	~13.0	Roof 4.6

The factor of safety against piping along any given pipe segment FSP is computed as the ratio of the maximum allowable (safe) head H_{max} derived from (AH/L')cr to the actual maximum head H_I between the segment's ends.

$$FSP = \frac{H_{max}}{H_I} \tag{21.5}$$

The values of H_{max} and FSP, at different points along the pipe, are summarized in Table 21.3. The fact must be stressed that the value of FSP is different for each segment of the pipe and for each pipe path considered. More often than it is actually realized, pipes may initiate at some downstream point and progress substantially but fail to extend beyond the point where conditions leading to FSP > 1 exist. An overall FSP for the whole foundation has no actual meaning.

The above values of the FSP prove that piping could actually take place under the head existing in August and that the pipe could advance from point 0 to 1 and extend further to segment 1–2 thus connecting the reservoir to the downstream toe.

As a check, Lane's weighted creep ratio R_c: was computed.

$$R_c = \frac{1/3 + \sum t}{h} \tag{21.6}$$

Substituting the geometrical data of the problem B = 395 ft, Et = 4.9 ft, h = 16.9 ft (head existing on August 15) R_c = 8 can be obtained which is slightly below the value of R_c = 8.5 recommended as a safe limit for silts. This provides an overall empirical confirmation of the above result.

The linear growth of a pipe is the result of soil scour at the pipe's end followed by transport of the removed soil particles by channeled water flow in the pipe. As shown by the plot of Fig. 21.9, for particles having a d$o size in the range of

Table 21.3 R_a and FSP

Segment (point)	Trial grad.	Scour. vel. reduction factor	Critical grad.	Max. all. head.	Avail, head[a]	Safety factor against piping
(1)	($\Delta H/L'$) (2)	R_α (3)	$(\Delta H/L')_\sigma$ (4)	$H_{TM\ Ic}$ (ft) (5)	H_I (ft) (6)	FSP (7)
0	0.026	0.429	0.032	12.7	13.11	0.96
	0.028	0.349				
	0.030	0.243				
	0.032	<0				
0–1	0.019	0.321	0.022	8.73	13.11	0.66
	0.020	0.236				
	0.021	0.092				
	0.022	<0				

[a]Available head complied as (281.15 – 276.82) = 5.33 m * 3.028 = 13.11 ft

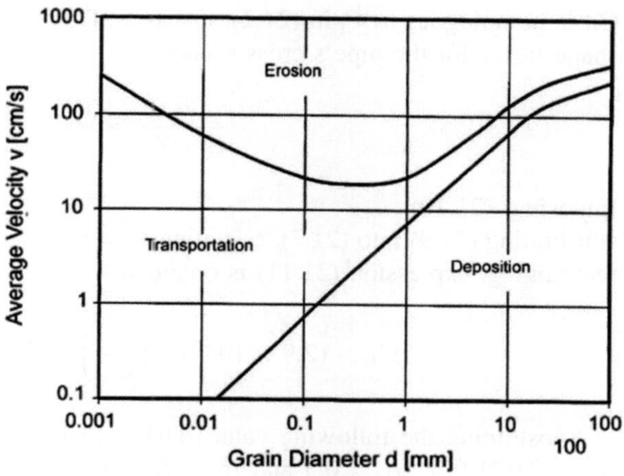

Fig. 21.9 Average flow velocities required to start sedimentation or erosion for various size of grains (from Krumbein and Sloss 1963)

0.1 mm, flow velocities sufficient to initiate erosion (scour) are about 50 times larger than those required to ensure transport. The lowest entropy principle suggests that once scour initiated, the flow velocity in the pipe will be the lowest compatible with transport and the size of the pipe the smallest compatible with the equilibrium (non-erosion) of its wall. Because the supply of water is practically independent from the actual size of the pipe, larger than optimum pipes will cause the flow to slow down (which would produce deposition of the largest soil particles thus reducing the pipe size) while smaller than optimum pipes will cause the flow to accelerate (which would produce some erosion on the pipe's wall thus enlarging the pipe).

The maximum size of the pipe that can still discharge the eroded soil can be derived equating the velocity of the flow, computed from the flow net, to the lower bound velocity in the pipe required for the flow to transport the soil particles present at the head of the pipe, according to plot in Fig. 21.9. Expressions (21.7), (21.8) and (21.9) are used: (i) for the total flow from an unisotropic soil into a 3D pipe, (ii) for the lower bound transport velocity as function of particle size and (iii) for the permeability coefficient as function of particle diameter (Hazen's formula):

$$\frac{q}{C_s * z^2} = \left(\frac{N_1}{N_u}\right)^* \frac{\Delta H * z * C_{3f} * \sqrt{k_{ap} * k_{an}}}{C_s * z} \tag{21.7}$$

$$V_{d3} = 0.16 * \left(\frac{d_{ave}}{0.001}\right) * F_3 \tag{21.8}$$

$$k = 2 * d_{10}^2 \tag{21.9}$$

A new set of Concentration Coefficients Cif and Car must be introduced and derived for the conditions of an established flow toward a partly penetrating pipe.

Cif is homologous to C_i in (21.2), C$f is a 3D flow concentration factor. C_s is the shape factor for the pipe's cross section.

$$\frac{q}{C_s * Z^2} > V_{d3} \tag{21.10}$$

Imposing: (21.10)
substituting (21.9) into (21.7), assuming $C_s = p/2$, $C_{3f} = 7.1$ $F_3 = 0.5$ $d_{ave} = d_{60}$ and rearranging, expression (21.11) is obtained:

$$\Delta H > (2.9 \times 10^{-3}) * \left(\frac{z}{d_{10}}\right) * \frac{C_u}{(N_F/N_H) * C_{11}} \tag{21.11}$$

Substituting the following values DH = 13.11 ft, $d_{10} = 0.04$ mm, $C_u = 2$, $N_F/N_H = 0.029$ (for $1/L = 0.5$ and for $L/D = 80$), $C_{1f} = 2.5$, a value $z = 6.5$ mm is derived as the maximum size of the pipe. Such a size may appear quite small, yet it is 10–100 times larger than the grains moving inside it.

As a check, the velocity of the flow in the pipe was computed using Manning's formula:

$$v = \frac{1.486}{n} * \sqrt[1.5]{R} * \sqrt{S} \tag{21.12}$$

solving with the following values $n = 0.02$ (which compares to $n = 0.017$ for rough concrete), $R = n * r/(2 * p + 4) = 0.0065$ (ft), $S = \text{DH}/L = 13.11/395 = 0.0332$, a velocity $v = 0.15$ m/s is obtained, which is close to the velocity required to scour a uniform particulate material with $d_{10} = 0.04$ mm ($v_s = 0.30$ m/s) and about 30 times the velocity required to drag the same material ($Vt = 0.004$ m/s).

The time for the pipe to move across each segment L of the path is obtained by dividing the volume of the pipe (multiplied by a water volume factor F_v representing the dilution of the water/soil suspension), by the total flow into that pipe segment:

$$\Delta t = \frac{\frac{\pi}{2} * z * F_v}{(N_F/N_H) * \Delta H * C_{3F} * \sqrt{k_{ap} * k_{an}}} \tag{21.13}$$

The overall time for piping being T = SDt, the sum of the individual times required for the pipe head to move across each pipe segment.

Substituting the following values: $z = 6.5$ mm = 0.0213 ft, $L = 3.95$ ft, $F_v = 3$, $k_{ap} = 7.5 \times 10^{-5}$ cm/s = 1.476×10^{-4} ft/min, $k_{an} = 5 \times 10^{-5}$ cm/s = 0.984×10^{-4} ft/min, DH = 13.11 ft, $C_{1f} = 2.5$, $C_{3f} = 7.1$, $N_F/N_H = 0.029$, a piping time $T = 33$ days is obtained, close enough to the 20 days during which the reservoir level was above el. 279.5. It must be noted, however, that time figures obtained using expression (21.13) vary in direct proportion to water volume factor F_v and with the root of parallel and normal permeability coefficients. Schmertmann suggests $F_v = 5$ while $F_v = 3$ was used by the Author. Relatively minor and reasonable modifications in the above values would reduce the piping time to about 10 days or lengthen it up to nearly 5 months.

21.6 The Piping Tested

Piping as described and quantified above appeared as a realistic mechanism. To gain further evidence an attempt was made of reproducing the piping process in a roughly 1–2 scale. An exploratory pit was excavated on the upstream blanket, 53 m from a temporary cofferdam retaining the reservoir water at el. 278.03. The pit bottom was lowered to el. 276.3 (head = 1.7 m), within the "sandy silt" natural deposit (blue–gray at this particular spot). An Auger hole nearby allowed locating the coarse sand within the silt, at el. 275.8 say 0.5 m below the pit bottom grade.

About 2 min after excavation completed, water surfaced on the bottom of the pit. The flow soon concentrated at four small pipes with visible transport of very fine sand and suspended clay particles imparting to the water a milky appearance. The sequence of Figs. 21.10, 21.11 and 21.12 shows one pipe as it evolved in the lapse of 10 min. Visual observation suggested that the size of each pipe was not larger than 10 mm.

Computing the weighted Lane's creep ratio R_c for this model case with (21.6) and the following problem values: $B = 53$ m, St = 0.7 m, resulted in $R_c = 10.8$, about 20 % higher than the one computed for the dam on August 15. No attempt was made to apply Schmertmann's method to this model considering the limited information available about the soil condition along the flow path.

Fig. 21.10 The exploratory pit where a model piping was observed. The gray material is the sandy silt present over most of the foundation under the dam. The picture was taken immediately after excavation completed. The pit was excavated with backhoe and was about 90 cm wide

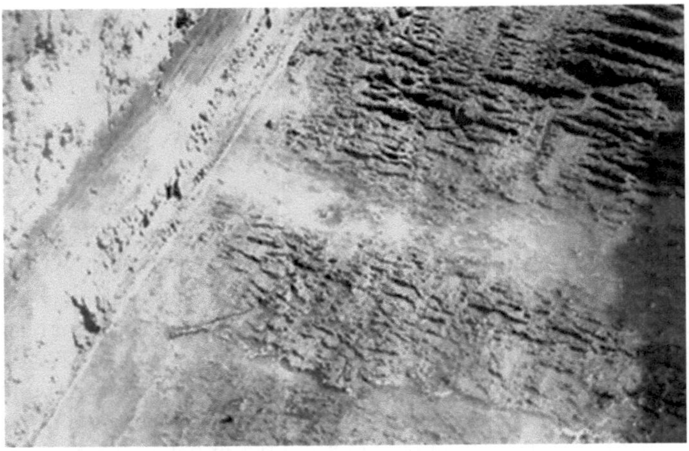

Fig. 21.11 The early aspect of pipe forming through the sandy silt under excessive gradient. Picture taken 4 min after picture of Fig. 21.10

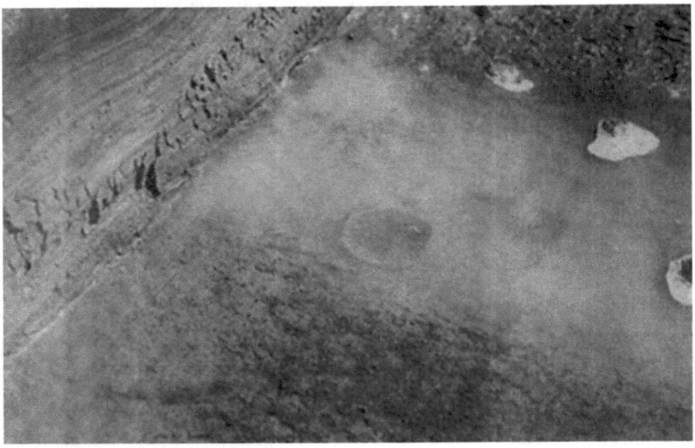

Fig. 21.12 A fully developed pipe in sandy silt. The dimensions of the pipe are below 1 cm. Picture taken 10 min after picture of Fig. 21.10

21.7 From Pipe to Cavity

Reconstructing what happened after an uninterrupted pipe established between the reservoir and the toe of the dam, under the artificial clay blanket, if of lesser practical importance is not straightforward. It was however considered necessary extending the back analysis to explain how the cavity could have propagated within

Table 21.4 Factors of safety against hydrofracturing under the artificial clay blanket

Station[a]	Vert stress	Hor. stress	Water pressure	FSH
(1)	σ_v (kg/cm^2) (2)	σ_h (kg/cm^2) (3)	p (kg/cm^2) (4)	(5)
+75	6.56	3.89	5.20	0.74
+70	6.56	3.89	4.98	0.78
+65	7.30	4.30	4.76	0.90
+60	8.04	4.76	4.55	1.04
+55	8.04	4.76	4.33	1.09
+50	9.49	5.63	4.11	1.36

[a]From the dam axis toward upstream

the clay blanket. Table 21.4 presents the computed levels of horizontal and vertical stress and the dynamic water pressure levels in the pipe, at different distances from the upstream toe of the dam. Comparing the (minor) horizontal stress <7 h with the corresponding dynamic water pressure p, the factor of safety against hydro fracturing FSH can be computed. It can be seen that hydraulic fracturing is a possibility at several locations.

$$\mathrm{FSH} = s_h/p \tag{21.14}$$

The clay of the artificial blanket was placed over the in situ sandy silt which, in its normal state, was partly saturated, and therefore capable of applying some suction to the clay pore water. If the water content of the clay is reduced, the clay undergoes a volume change and the horizontal stresses get reduced by approximately:

$$\Delta\sigma_n = \left(\frac{G_s * \Delta W}{1+e}\right)^{0.33} * B_{\mathrm{ur}} \tag{21.15}$$

Two odometer tests were run on the clay of the blanket. For a vertical stress level in the order of 50 MPa the unloading-reloading constrained D_{ur} and bulk B_{ur} moduli were measured to be in the order of D_{ur} = 3.5 MPa and B_{ur} = 1.5 MPa. Using $G_s = 2.65$ and $e = 0.55$, it was possible to compute that a reduction of about 0.5 % of the clay water content, induced by suction, would reduce the horizontal stress by Ds_h = 300 kPa. Conversely, a capillary suction of about 10 kPa (a normal level for a silt with d_{10} = 0.04 mm), could reduce the horizontal stresses by some Ds_h = 200 kPa, enough to bring the safety factor against hydraulic fracturing FSH below the threshold of unity at all points located further than 50 m from the dam's centerline.

Unconfined compression tests made on undisturbed samples taken at different elevations across the artificial clay blanket, near the cavity, showed a marked softening near its bottom. The results of such tests are given in Fig. 21.13.

Based on the above considerations, it is not unreasonable assuming that the clay blanket could be hydraulically fractured under the reservoir head of August 15 near

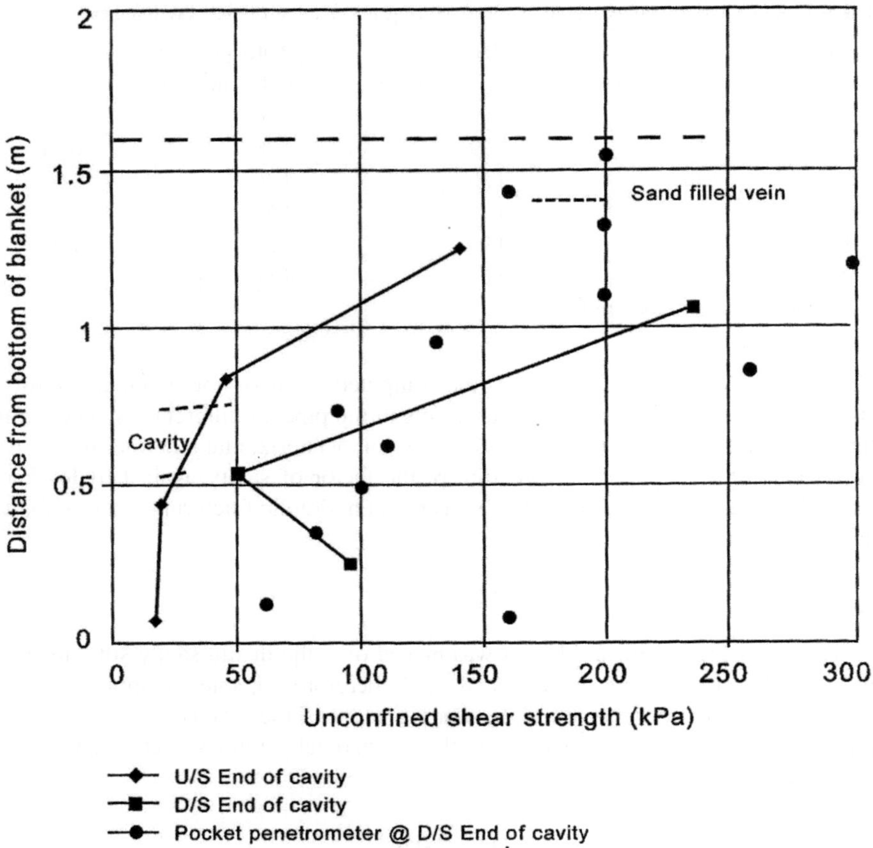

Fig. 21.13 Profiles of undrained shear strength of the clay from the artificial clay blanket obtained from unconfined compression tests carried out on soil specimens sampled close to the cavity wall near its upstream and downstream ends

both ends of the pipe were the overburden was low, thus allowing the pipe to find its way into the clay. Foundation piping followed by hydraulic fracturing at the edges of the artificial clay blanket, would take the water of the pipe in contact with the drain. At Ch. 1+900 the drain collector pipe original elevation was 276.61, about 0.20 m lower than ground level (276.79). This explains why the rising water level found an early outlet through the pipe. The geotextile, given its low cross-transmissivity (a heat-bondedgeotextile is seldom used for drainage), probably limited the amount of water that could enter the drain and the water found an overflow outlet at Ch. 1+940 where ground level was 276.70. As flow increased, water found new and more direct outlets. The presence of cave-ins, close to the toe on both slopes, appears a further evidence of the proposed hydrofracturing and collapse mechanism.

The horizontal fingers departing from the main cavity, seen near the downstream and upstream toes, were also explained as resulting from hydrofracturing. They were credibly cracked open by the overpressures produced in the conduit when some portion of the roof caved-in or when the d/s exit became temporarily clogged by the low permeability geotextile of the collapsing drain. The sudden reduction of flow noticed on the morning of August 16 is, most probably, the visible result of something of this sort.

21.8 Concluding Remarks

Failures induced by underground water flow differ from strength-related failures because of the mass of material involved and the processes taking place before and after failure:

- a strength-dictated failure is not significant unless it affects a substantial volume of soil. Stress redistribution and full mobilization of the available strength take place before a declared failure. Such effects often prevent a full-fledged collapse and the unstable soil volume may self heal. After failure, the mass attains a new equilibrium configuration so that it may be reclaimed and made usable again.
- a seepage-dictated failure may be critical even if a relatively small volume of soil is actually affected. Once a pipe has developed so as to reach the reservoir, its dimensions will grow in a short time into a cavity. Cave-ins and breaching will follow and, in most cases, the process will end only when all the water has been lost and, in presence of large impoundments, destruction of the dam may ensue. With due limitations, we may say that while average conditions, applicable to quite large masses, usually govern stability and deformation problems, small scale stratigraphic details and soil characteristics matter when dealing with seepage arid seepage-related problems. This is reflected in the fact that the factor of safety against piping FSP varies along each segment of a given pipe path. Because stratigraphical and textural details, as well as localized construction anomalies, acquire vital importance in relation to seepage, the level and detail of the designer's insight must be proportional to the size of the potentially unstable volumes.

Gradients as low as 4 % may be already too high, in a fine grained foundation, when combined with unfavorable local conditions and when magnified by incongruous design like the use of a heat-bonded geotextile as filter.

Discrete, modular monitoring may result totally inadequate to provide proper warning if weak spots, in relation to seepage, are not understood beforehand, duly localized and properly controlled. Numerical analyses of seepage phenomena are not straightforward. The same problem may be seen and evaluated in different ways. Yet, quantifying a safety factor against piping FSP is imperative and should be a priority item in all dams where fine grained, particulate materials are present either in the foundation or in the embankment. A similar effort should be made to define and quantify a safety factor against hydraulic fracturing FSH, in addition to the customary safety factor against shear FSS.

The potential damage of a seepage-related failure justifies using more than one approach and exploring more than one failure mechanism. Conventional flow net and/or elemental analyses, unless performed on a model closely reproducing reality, tend to provide average values and may not be appropriate to quantify the clue of the problem. Schmertmann's approach proved to be very valuable and meaningful in this particular case and should certainly receive more attention in future.

References

Krumbein WC, Sloss LL (1963) Stratigraphy and sedimentation. W.H. Freeman & Co., New York

Lambe TW, Whitman RV (1969) Soil mechanics. Wiley, New York

Schmertmann JH (1957) Effect of seepage on scour at wall face. In: Proceedings of 6th conference on coastal engineering, Gainesville, Florida

Schmertmann JH (1980) Notes and calculations for "a quantitative piping theory". Appendix B to Special Board of Consultants Report for Florida Power & Light Company

Sembenelli P, Biondani E (1984) Analisi degli Effetti prodotti dall'Invaso sulle Dighe in Terra e Roccia. Rivista Italiana di Geotecnica, Anno XVIII, N. 2, Aprile-Giugno

Sembenelli P, Sembenelli G, Ruffini A (1997) Internal erosion around relief wells in fine sand. In: Proceeding XIX international Congress on large dams, vol II, pp 715–722

Sherard Jl, Dunnigan LP, Decker RS (1976) Identification and nature of dispersive soils. ASCE J Geotech Eng Div 102(GT4):287–301

Chapter 22
Influence of Vibrations on Installation of Bored Piles

N. Santosh Rao

22.1 Introduction

This is a case study of a forensic geotechnical investigation to assess the possible cause/s which prevented the satisfactory construction of large-diameter bored piles using the bailer technique. The investigation involved first identifying the possible causes, carrying out field and laboratory tests where necessary to assess the same, analysis of the test results, and finally arriving at the conclusion.

22.2 Project Details

The project in this case study is an elevated corridor through the Kishanganj town, Bihar, India, located parallel to the existing road as also the railway line abutting the road. The elevated corridor had been proposed to be constructed with reinforced cement concrete with the deck consisting of box girders supported by two column piers resting on 1 m diameter, 16 m long-bored piles.

22.3 Problem Defined

The piling contractor had been unable to ensure satisfactory construction of the large-diameter bored piles using the bailer technique, primarily because of frequent collapse of the sides of the bores made for the purpose of construction of the bored

N.S. Rao (✉)
Nagadi Consultants Pvt. Ltd., New Delhi, India
e-mail: nsantoshrao@nagadi.co.in

V.V.S. Rao and G.L. Sivakumar Babu (eds.), *Forensic Geotechnical Engineering*,
Developments in Geotechnical Engineering, DOI 10.1007/978-81-322-2377-1_22

piles. As per the conditions of the contract, bentonite slurry was not to be used in the pile bores and use of liner/casing had also not been envisaged on account of the nature of the subsoil as indicated by the report of the geotechnical investigations carried out earlier for this project.

22.4 Possible Causes

An evaluation of the above problem indicated that one or both of the following could be the possible causes:

1. Subsoil being unsuitable for construction of the bored piles under the constraints put up by the contract.
2. Vibrations transmitted from the trains moving on the nearby railway tracks and vehicle movement on the adjacent road.

22.5 Investigation of Possible Causes

22.5.1 Subsoil Conditions

To check whether the subsoil conditions at the site of the proposed elevated corridor are similar to those considered in the design and included in the tender documents, a few boreholes had been carried out at selected locations along the length of the elevated corridor. A comparison of the typical subsoil profile indicated by these boreholes with the typical subsoil profile indicated by the geotechnical investigation report included in the tender documents is given below:

Depth (m)	Geotechnical investigation report		New boreholes	
	Soil description	Range of *N*-values	Soil description	Range of *N*-values
0–5	Silty fine sand/ sandy silt	10–15	Fine sand with silt	10–14
5–10		15–20	Medium to coarse sand	12–20
10–20		20–25	Coarse sand	25–35
20–25	–	–	Coarse sand with gravel	>50

The above comparison clearly shows that the actual subsoil conditions differ significantly from that indicated by the geotechnical investigation report. While an unsupported pile bore may remain intact in a silty sand/sandy silt soil, the same will invariably tend to collapse in a medium to coarse sandy soil. Therefore, while based

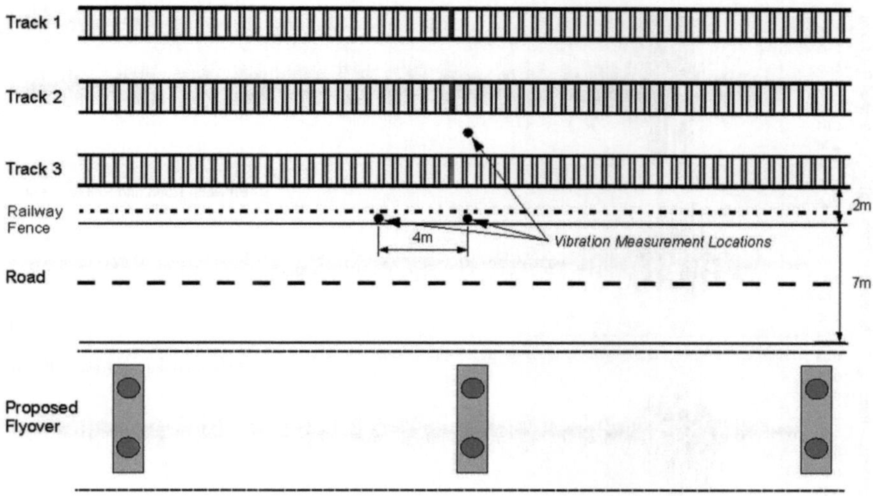

Fig. 22.1 Schematic layout plan

on the results of the geotechnical investigation report, an unsupported pile bore can be considered feasible, the same will be considered difficult to infeasible when the actual subsoil conditions as indicated by the new boreholes are taken into account.

22.5.2 Vibrations from Rail and Road Traffic

To assess the level of vibrations caused by the rail and road traffic in close proximity to the site of the proposed elevated corridor, vibration measurements have been carried out using geophones (i.e., velocity pick-ups with a sensitivity of 28.3 mv/mm/s) connected either to a storage oscilloscope or to a computer-based data acquisition system. The locations of the vibration measurement points have been indicated in the schematic layout plan of the area given in Fig. 22.1.

The recorded vibrations from different types of trains as also from trucks plying on the road have been presented in Figs. 22.2, 22.3, 22.4, 22.5, 22.6. The absorption coefficient for the subsoil at the site has been determined from the vibrations recorded during the passage of a Rajdhani express train by locating the geophones in a line perpendicular to the railway tracks on the opposite side of the road under the proposed elevated corridor.

From the recorded vibrations and the absorption coefficients determined, the values of the acceleration ratio at various distances from the tracks will be:

Distance (m)	6	10	14	18	22
Acceleration ratio	0.03	0.013	0.009	0.005	0.002

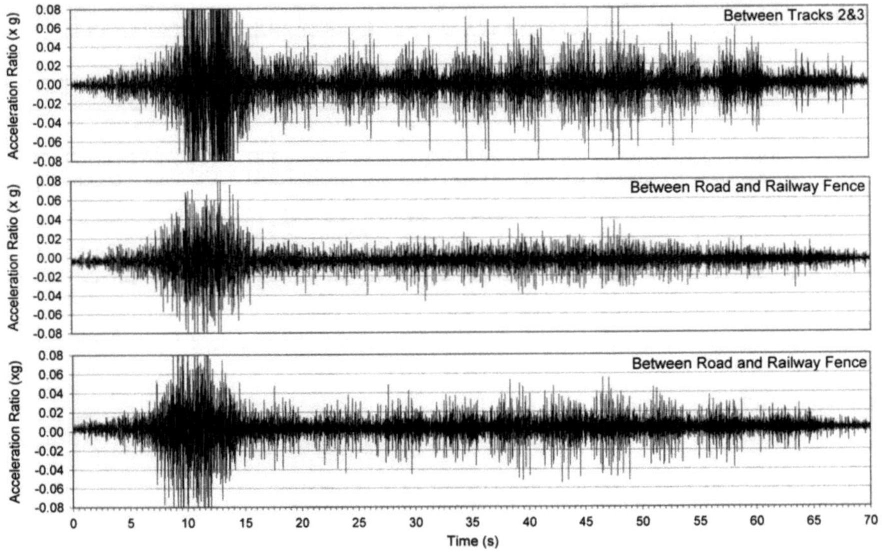

Fig. 22.2 Express train on track 3 in E-W direction

Fig. 22.3 Rajdhani express train on track 3

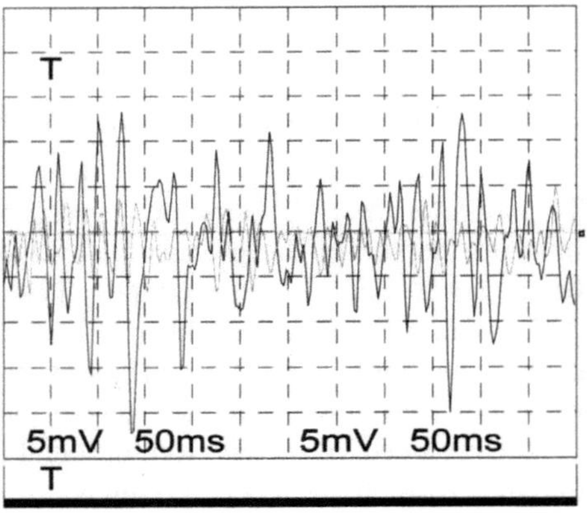

As per Tscheboterioff (1965), dynamic forces caused by pile driving operations having high accelerations in the range of 0.01–0.001 g have been shown to have caused significant soil settlements to cause distress to the nearby structures and substructures. As the levels of vibrations caused by the rail traffic are in the above range, the vibrations caused by the rail traffic can be considered as a major contributory factor in the collapse of pile bores.

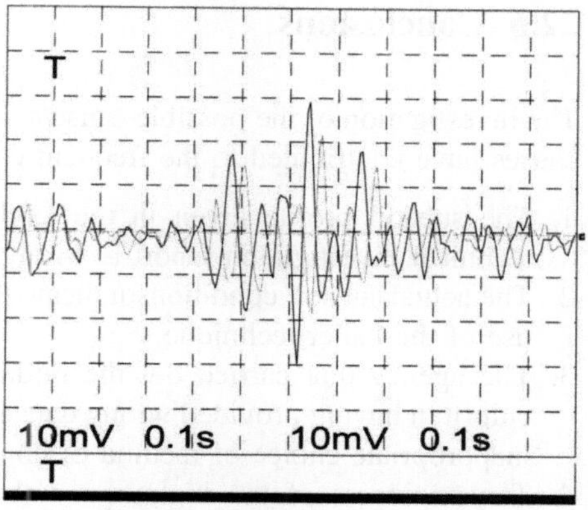

Fig. 22.4 Truck plying on the road

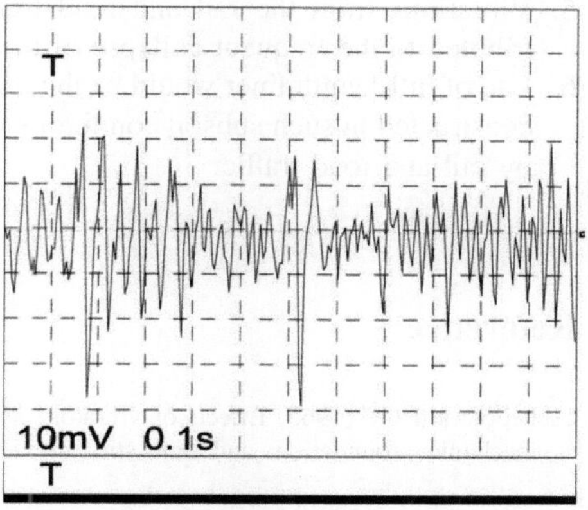

Fig. 22.5 Passenger train on track 1

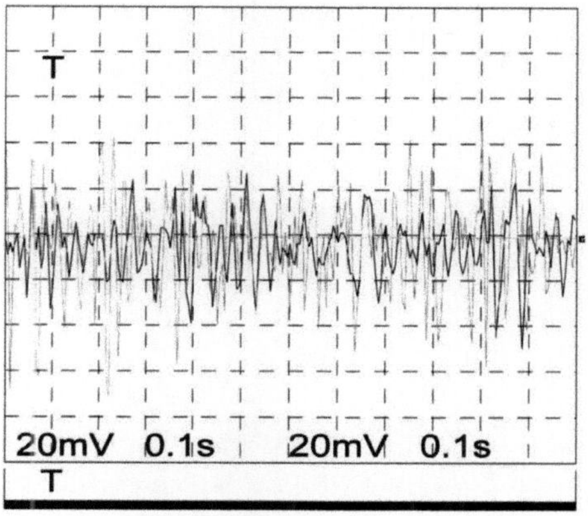

Fig. 22.6 Goods train on track 3

22.6 Conclusions

The investigation of the possible cause/s indicates that both the considered possible causes have contributed to the frequent collapse of the pile bores in that:

1. The subsoil profile given in the tender documents as indicated by the geotechnical investigation report is wrong.
2. The actual subsoil conditions indicate that the subsoil at the site is unsuitable for use of the bailer technique.
3. The agency that carried out the initial geotechnical investigation is the main culprit in having provided wrong data about the subsoil conditions leading to the inappropriate choice of method of construction of the bored piles.
4. The engineers of the highways authority should have properly verified the geotechnical investigation report before going in for tendering of the work and insisting on bored piles with restrictions on the method adopted for the same.
5. Vibrations from the rail and road traffic are significant enough to have contributed to the frequent collapse of the pile bores.
6. Use of full-length liner would be the only method by which bored piles could be constructed in such subsoil conditions where significant vibrations are generated by rail and road traffic.

Reference

Tschebotarioff GP (1965) Effects of vibratory and of slow repetitional loading of soils. In: Soil mechanics, foundations and earth structures. McGraw Hill, New York, Ch 18 pp 568–595

Chapter 23
Aspects Regarding Management of Soil Risk

Rolf Katzenbach, Steffen Leppla, Alexandra Weidle and Deepankar Choudhury

Abstract The geotechnical investigations of soil and groundwater conditions are an essential part of the design for almost every construction project. Especially at large and challenging projects, problems often occur during the construction works when the detected soil and groundwater conditions (structure, properties) differ from the result of the earlier geotechnical investigation. Time delay in the construction process, modifications of the construction, additional expenditure and in some cases even damages may be the consequences. The so-called "soil risk" occurs, when the soil and groundwater conditions are different to the results of the geotechnical investigation. In any case the soil risk is connected to the chosen construction process and type of construction ("system risk"). Most forensic geotechnical tasks deal with handling of the soil risk respectively system risk. In this paper the legal aspects of managing the soil risk respectively the system risks are explained with special focus to forensic geotechnical engineering and the qualification of publicly certified geotechnical experts. Standards are defined and the applications in engineering practice are shown through several prominent examples of retaining structures of deep excavations, foundations in organic soils and tunnel constructions in urban areas.

Keywords Observational approach · Peer review · Licence · Legal aspects

R. Katzenbach (✉) · S. Leppla · A. Weidle
Technische Universität Darmstadt, Institute and Laboratory of Geotechnics, Darmstadt, Germany
e-mail: katzenbach@geotechnik.tu-darmstadt.de

D. Choudhury
Humboldt Research Fellow at Technische Universtität Darmstadt, Institute and Laboratory of Geotechnics, Darmstadt, Germany

D. Choudhury
Department of Civil Engineering, Indian Institute of Technology Bombay, Mumbai, India

V.V.S. Rao and G.L. Sivakumar Babu (eds.), *Forensic Geotechnical Engineering*, Developments in Geotechnical Engineering, DOI 10.1007/978-81-322-2377-1_23

23.1 Introduction

According to national laws and regulations the responsibilities for the planning, construction and materials used at any kind of construction project are different around the globe. Nevertheless it is necessary to have a safety concept for risk minimisation in all planning and construction phases and for the quality of the construction material. In geotechnical engineering the soil represents the construction material. In any case the soil risk is connected to the chosen construction process and the type of construction. The soil risk and the system risk are strongly related to each other.

The so-called system risk explains that the chosen construction technique, construction material, construction process is not applicable to the soil and groundwater conditions.

Soil is used as an active or passive construction material. Soil is used as a passive construction material if it is used for the load transfer of a superstructure over a foundation system into the underground. Soil is used as an active construction material if it is used, e.g. for fillings and dams.

The difficulty of soil as a construction material is that even by an extensive soil and groundwater investigation program less than 0.1 % of the soil that is influenced by the building can be determined, as shown in Fig. 23.1.

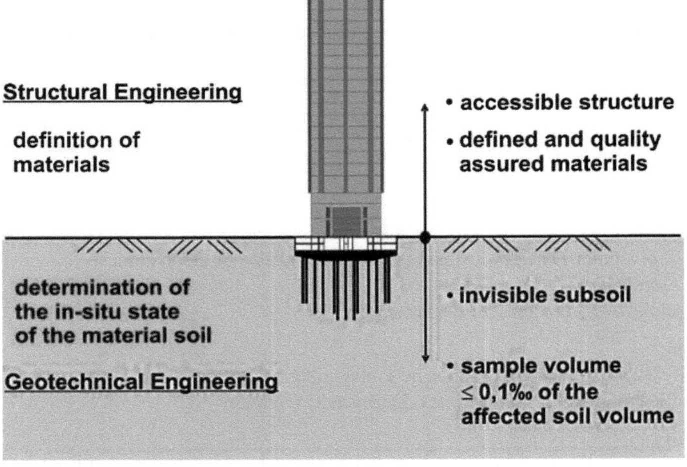

Fig. 23.1 Differences between structural and geotechnical engineering

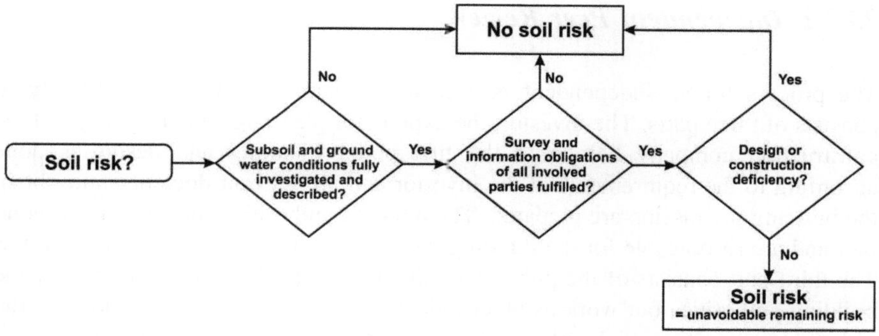

Fig. 23.2 Definition process of soil risk

Due to the random character of the soil investigation an element of risk remains. This risk means that the real soil conditions probably cannot be determined in detail. The soil risk remains as an unavoidable risk. Damages or time delays may occur, even though an extensive soil investigation was made. Figure 23.2 shows the process to the decision if soil risk is fulfilled.

It is no soil risk in the following cases:

- lack of geotechnical investigations
- wrong interpretation of the results of geotechnical investigations
- wrong investigation techniques
- mistakes in the design
- incompletion of the design
- differences between the design and the construction work, etc.

The system risk is similar. The system risk includes the construction process and the type of construction. Most forensic geotechnical tasks deal with handling of the soil risk respectively system risk (Katzenbach et al. 2012).

23.2 Concept for Safety and Quality Control

For minimisation of the soil risk a concept for safety and quality control is necessary. This concept consists mainly of:

- qualified experts for planning, design and construction
- extensive soil investigation
- independent peer review during planning, design and construction (4-Eye Principle)
- observational method

The last two aspects will be explained in detail.

23.2.1 Independent Peer Review

The process for an independent peer review is shown in Fig. 23.3. Mainly, it consists of three parts. The investor, the experts for planning and analysing and the construction company belong to the first party. Planning and design is done according to the requirements of the investor and all relevant documents to obtain the building permission are prepared. The building authorities constitute the second part and are responsible for the building permission which is given to the investor. The third part consists of the publicly certified experts. They are appointed by the building authorities but work as independent experts. They are responsible for the technical supervision of the planning, analysing process and the construction.

To achieve the licence as a publicly certified expert for geotechnical engineering by the building authorities, the following requirements should be given:

- intensive studies of geotechnical engineering in the university
- large experiences in geotechnical engineering with special knowledge about the soil structure interaction examination by a group of publicly certified experts

The independent peer review by publicly certified experts for geotechnical engineering makes sure that all information including the results of the soil investigation including the laboratory and field tests and the boundary conditions defined for the geotechnical design are complete and correct.

In the case of damage, the publicly certified expert for geotechnical engineering can be involved as an independent expert to find out the reasons for the damage and to develop a concept for the reconstruction.

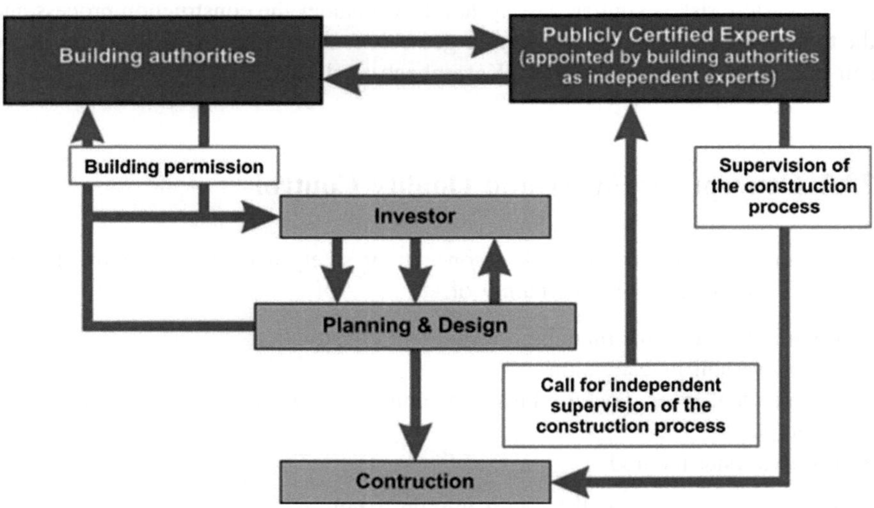

Fig. 23.3 Independent peer review process

23.2.2 Observational Method

If the identification of the mechanical behaviour of soil is difficult, it is necessary to apply the observational method to review the design during the construction time and (if necessary) during service time. For example in the European Standard Eurocode EC 7 (CEN 2008) the effect and the boundary conditions of the observational method are defined.

The application of the observational method is recommended for the following types of construction projects (Katzenbach et al. 2010, 2012):

- very complicated/complex projects,
- projects with a distinctive soil structure interaction, e.g. mixed shallow and deep foundations, retaining walls for deep excavations, Combined Pile-Raft Foundations (CPRFs),
- projects with a high and variable water pressure,
- complex interaction situations consisting of ground excavation and neighbouring buildings and structures,
- projects with pore water pressures reducing the stability,
- projects on slopes.

The observational method is always a combination of the common geotechnical investigations before and during together with the theoretical modelling and a plan of contingency actions (Fig. 23.4). Only monitoring to ensure the stability and the

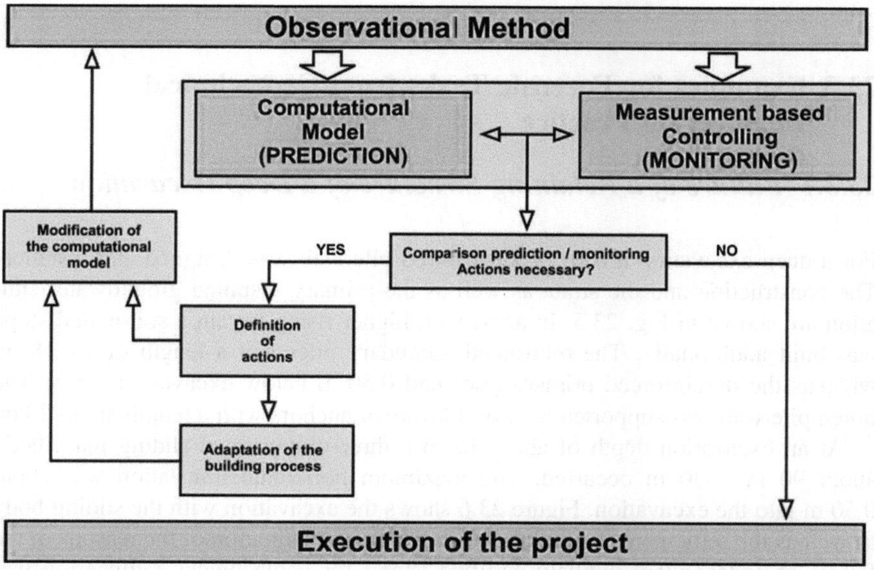

Fig. 23.4 Principle of the observational method

serviceability of the structure is not sufficient and, according to the standardisation, not permitted for this purpose.

Overall the observational method is an institutionalised controlling instrument to verify the soil and rock mechanical modelling (Katzenbach et al. 1999; Rodatz et al. 1999).

The identification of all potential failure mechanisms is essential for defining the measure concept. The concept has to be designed in that way that all these mechanisms can be observed. The measurements need to be reliable and should be done in a direct and defined way. The measurement devices need to be of an adequate accuracy to allow the identification of critical tendencies. The required accuracy as well as the boundary values need to be identified within the design phase of the observational method.

Contingency actions need to be planned in the design phase of the observational method. The construction method or design method depends on the ductility of the system. It is essential that contingency action can be applied.

The observational method must not be seen as a potential alternative for a comprehensive soil investigation campaign. A comprehensive soil investigation campaign is in any way of essential importance.

To establish the above-mentioned boundary values before the construction has started is a challenging task and requires powerful calculation tools which reply on suitable constitutive laws for the soil.

As a secondary effect the observational method is a tool of quality assurance and allows the verification of the parameters and calculations applied in the design phase. The observational method helps to achieve an economic and save construction.

23.3 Examples for Forensic Tasks from Geotechnical Engineering Practice

23.3.1 Failure of a Retaining Structure of a Deep Excavation

For a deep excavation a 0.90 m thick bored pile wall was designed and installed. The construction and the strata as well as the primary assumed groundwater situation are shown in Fig. 23.5. In areas with higher rising terrain a soil-nailed slope was built additionally. The reinforced secondary piles had a length of 14–18 m, whereas the unreinforced primary piles end 0.50 m below excavation depth. The bored pile wall was supported by 2 or 3 layers of anchors with a length about 20 m.

At an excavation depth of about 15 m a three-dimensional sliding mass body about 90 m × 30 m occurred. The maximum horizontal translation was about 0.50 m into the excavation. Figure 23.6 shows the excavation with the sliding body as well as the settlements behind the wall. For the investigation of the reasons of the failure an independent publicly certified expert for geotechnical engineering was introduced.

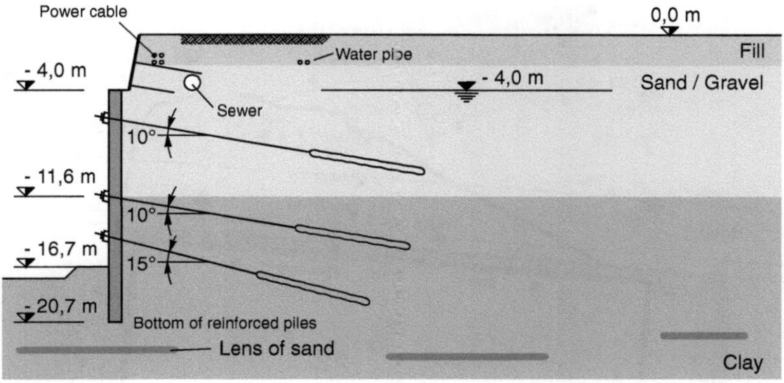

Fig. 23.5 Cross section of the retaining structure

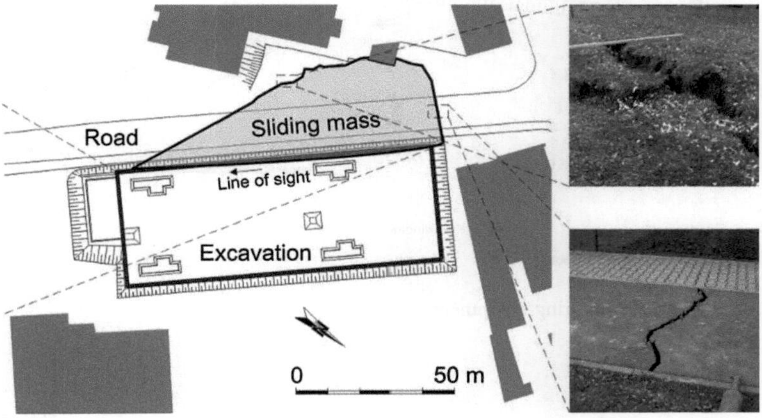

Fig. 23.6 Sliding mass body

As emergency procedure to avoid a total collapse of the retaining wall a berm and four groundwater wells were installed. The drillings for the groundwater wells showed that in the tertiary clay lenses of sand containing artesian groundwater with a water pressure level of 3–4 m above the surface exist. On the basis of new analyses taking into account the detected groundwater conditions the excavation was refilled up to a height of 7.5 m.

An extensive monitoring program was installed to observe the retaining wall and the adjacent buildings (Katzenbach et al. 2007). The monitoring equipment is shown in Fig. 23.7.

To reduce the water pressure on the retaining wall as well as on the sliding mass body and to prevent the excavation base and the refill from buoyancy a groundwater lowering was installed. To the existing anchors six anchor levels with a length of 35 m were added so the sliding mass body was stabilised and a new unaffected soil mass body was activated (Fig. 23.8).

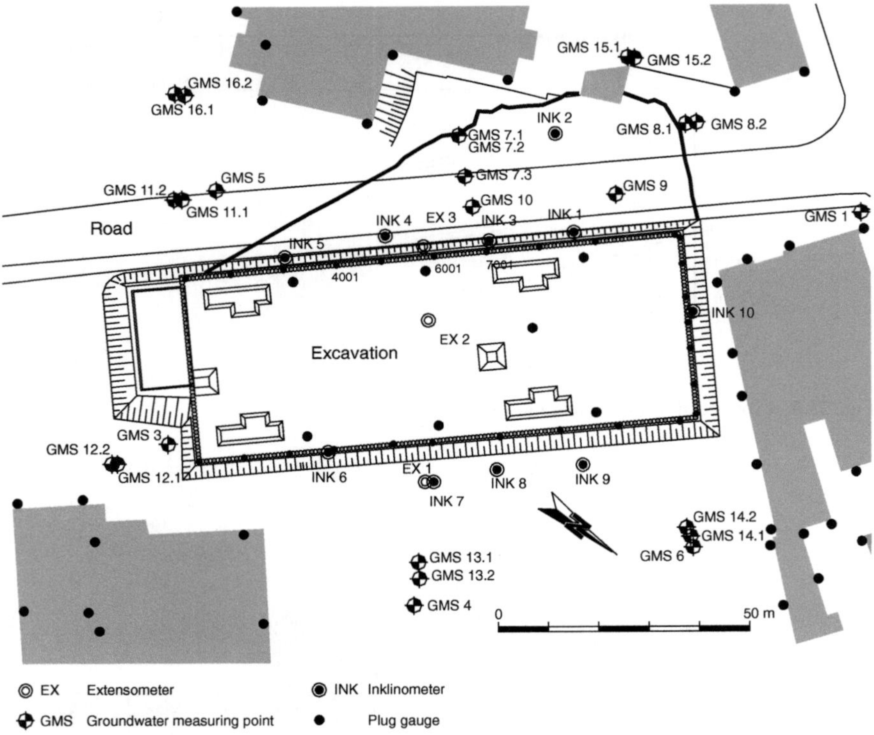

Fig. 23.7 Installed monitoring equipment

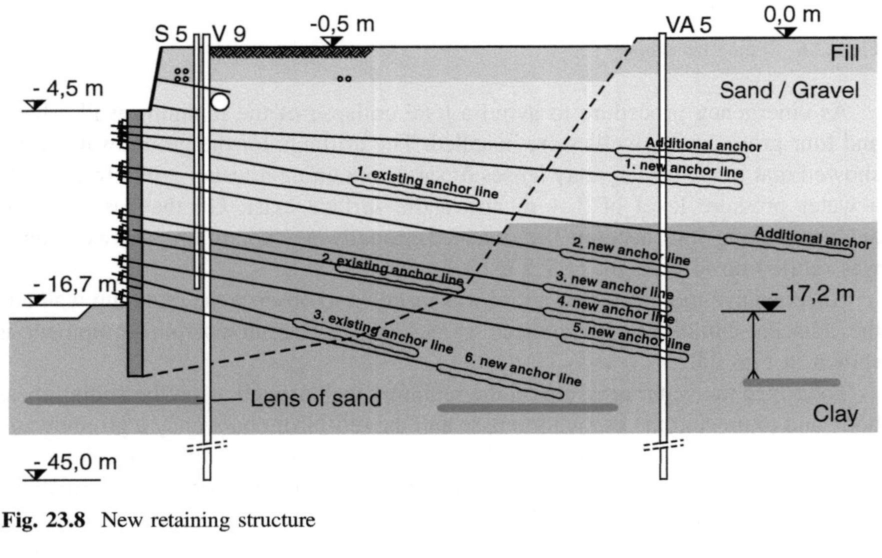

Fig. 23.8 New retaining structure

For the intact piles on the long opposite side of the sliding mass body no strengthening had to be installed. Only the two short sides of the excavation had to be stabilised by steel bracings because anchors could not be installed due to neighbourly constraints.

The fast and resolute intervention managed by a publicly certified expert for geotechnical engineering avoided a total collapse of the excavation and led back to a stable equilibrium.

23.3.2 Large Settlements of a Foundation in Organic Soil

The building shown in the ground view in Fig. 23.9 and the cross section in Fig. 23.10 is founded north of a little river at the toe of a slightly rising slope. The building has a basement, a ground floor and a top floor. The ground view is 13 m × 15 m.

The soil and groundwater conditions are shown in Fig. 23.10. The soil was investigated by four 6 m deep borings from the excavation depth. In the north up to a depth of 4 m sandy, soft ooze is detected. In the south the sandy, soft ooze reaches up to a depth of 6 m. The ooze in the south is superposed by a 1.2 m thick layer of pasty mud. The groundwater table is in the depth of the excavation. The foundation consists of a 25 cm thick reinforced concrete raft and up to 75 cm high strip foundations. The stiffness of the strip foundations including the walls is about EI = 10,000–30,000 M Nm^2 depending on the holes for doors and windows. Because of the relatively small and simple structure, no publicly certified expert for geotechnical engineering was consulted.

After one year the construction was finished. At that time differential settlements of the building from north to south of about 11 cm were recognised. One year later total settlements of 1 cm in the north and of 14 cm in the south were measured. The settlement difference increased up to 13 cm.

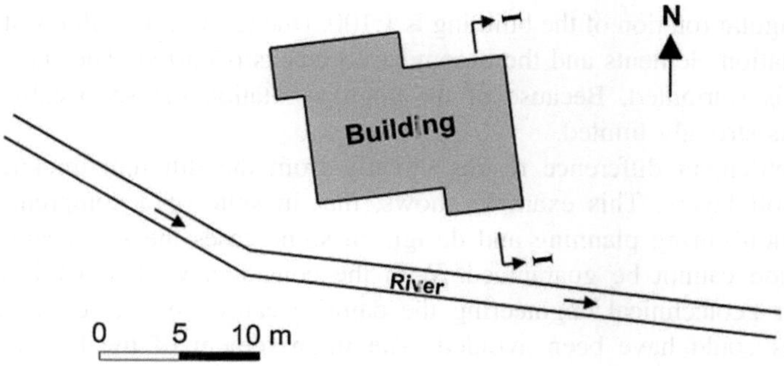

Fig. 23.9 Ground view

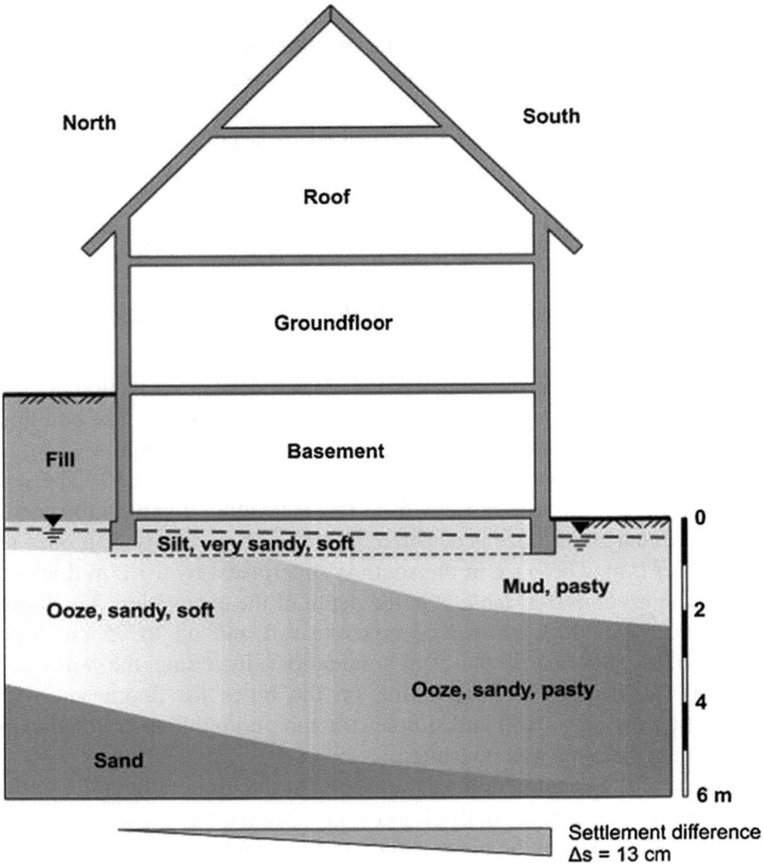

Fig. 23.10 Cross section, soil and groundwater conditions

The angular rotation of the building is 1:100. Due to the very stiff construction of the foundation elements and the basement no cracks occurred. The stability of the building is warranted. Because of the angular rotation the serviceability of the building is strongly limited.

The settlement difference results directly from the different thickness of the organic soil layers. This example shows, that in spite of a comprehensive soil investigation during planning and design, in some cases the serviceability of the construction cannot be guaranteed. With the consultancy of a publicly certified expert for geotechnical engineering the damage caused by the complicated soil conditions could have been avoided. The improvement of the building is very difficult because it will affect the boundary conditions of the construction. In any case a new deep foundation system in the organic soil layers are necessary. In this case micropiles compared with adjusting devices are a good possibility to lift up the construction. More examples of constructions on organic soils are given in Katzenbach and Rückert (1998).

23.3.3 Tunnel Constructions in Urban Areas Under World Heritage Properties

The increasing size and population density of metropolitan areas and the along going traffic demands lead to the construction of large infrastructure projects. Due to the limited surface area in urban regions in many cases underground constructions are the best solution. In many cases these infrastructure projects are close to sensitive properties.

One example is the tunnel of the new high-speed railway line (AVE) in Barcelona, connecting Madrid and the French border. The tunnel with an outer diameter of 11.55 m and a length of 5.6 km is built under the city centre next to the World Heritage Properties of the UNESCO, the famous church of Sagrada Familia and the Casa Milà. The depth of the tunnel is about 40 m below the surface in the tertiary sand. The groundwater table is about 21 m below the surface. The tunnel is built with a tunnel boring machine (TBM). Figure 23.11 shows a cross section of the tunnel construction.

To ensure the highest possible safety for the sensitive world heritage buildings the tunnelling process and the world heritage buildings itself are monitored according to the standards by the observational method. A team of publicly certified experts for geotechnical engineering and for structural engineering designed an extensive measuring program including geotechnical and geodetic measurements to

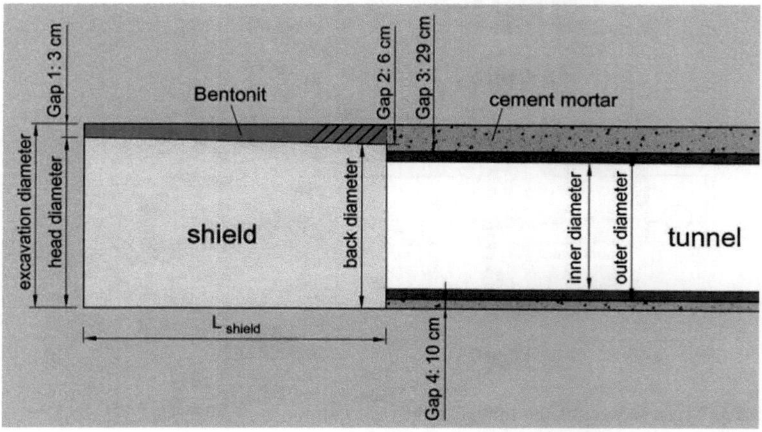

Fig. 23.11 Cross section of the tunnel construction

investigate the impact of tunnelling to the soil and the buildings. All permanent measurements were instantly analysed with a target performance comparison by the certified experts.

A cross section of the tunnel at Sagrada Familia is shown in Fig. 23.12. The TBM successfully passed Sagrada Familia in October 2010. Due to the precisely mortar injection of the gaps the measured settlements at Sagrada Familia were only about 0.1 cm and below every defined limit (Fig. 23.13). The detailed description of the project including the measured settlements at Casa Milà is printed in Katzenbach et al. (2011).

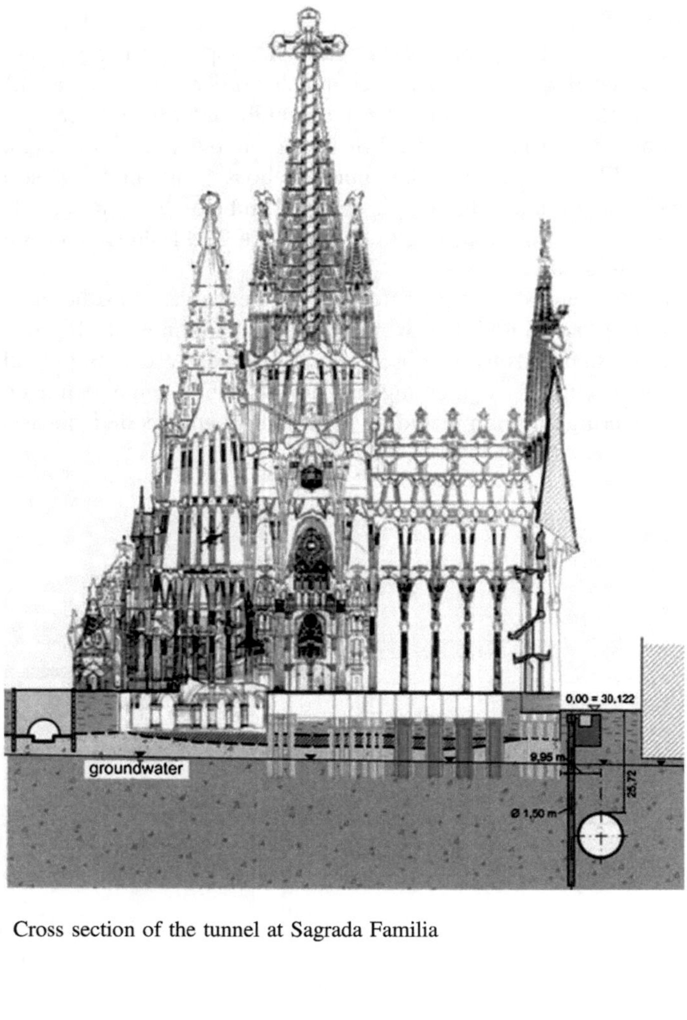

Fig. 23.12 Cross section of the tunnel at Sagrada Familia

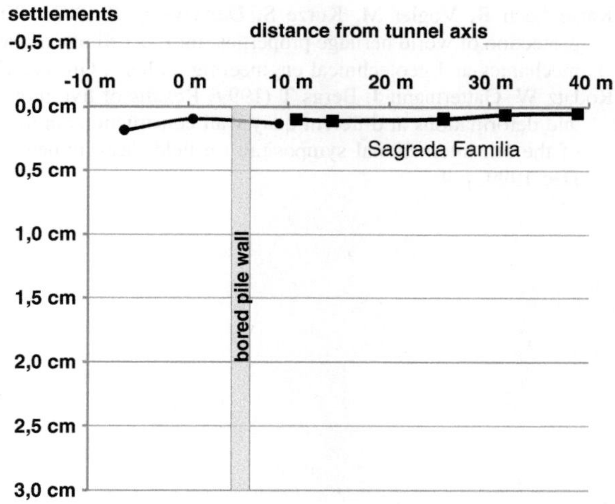

Fig. 23.13 Measured settlements in cross section at Sagrada Familia

23.4 Conclusions

How to manage the soil risk respectively system risk is explained in this paper regarding the forensic geotechnical engineering and the qualification of publicly certified expert for geotechnical engineering. However, forensic geotechnical engineering does not only deal with damages and failures but also with prevention of cases of damages and failures and the guarantee of safety and serviceability of any kind of structures.

References

CEN European Committee of Standardisation (2008) In: Eurocode 7: Geotechnical design—Part 1: General rules

Katzenbach R, Schmit A, Turek J (1999) Co-operation between the geotechnical and structural engineers—experiences from projects in Frankfurt." COST Action C7, Soil-Structure-Interaction in urban civil engineering, Thessaloniki, Greece, 1–2 Oct 1999, pp 53–65

Katzenbach R, Bachmann G, Leppla S, Ramm H (2010) Chances and limitations of the observational method in geotechnical monitoring. In: Proceedings of the 14th danube-european conference on geotechnical engineering, Bratislava, Slovakia, 2–4 June 2010 p 13

Katzenbach R, Rückert A (1998) Damages with foundations in organic soils. *Darmstadt Geotechnics, No. 3*, pp 109–120

Katzenbach R, Weidle A, Kurze S (2012) Baugrund und Grundwasser Erkungsungsproblematik, Baugrundrisiko und technische Risiken. 39. Baurechtstagung der Arge Baurecht des Deutschen Anwaltvereins, Berlin, Germany 16–17 Mar 2012, p 22

Katzenbach R, Wachter S, Vogler M, Seip M (2007) Reconstruction of a deep excavation and its numerical analysis. Darmstadt Geotechnics, No 15, pp 255–266

Katzenbach R, Vogler M, Kurze S, Dunaevskiy R (2011) Enhanced Geotechnologies for the protection of world heritage properties. In: Proceedings of the 15th european conference on soil mechanics and geotechnical engineering, Athens, Greece, 12–15 Sept 2011, pp 1859–1864

Rodatz W, Gattermann J, Bergs T (1999) Results of five monitoring networks to measure loads and deformations at different quay wall constructions in the port of Hamburg. In: Proceedings of the 5th international symposium on field measurements in geomechanics, Singapore, 1–3 Dec 1999, p 4

Chapter 24
Forensic Studies for Failure in Construction of an Underground Station of the Kaohsiung MRT System

Richard N. Hwang, Kenji Ishihara and Wei F. Lee

Abstract Several buildings collapsed as a result of leakage on diaphragm wall during the excavation for constructing O1 Station of the Kaohsiung MRT System. Resistivity image profiling was carried out to check the quality of diaphragm walls and the effectiveness of ground treatment using CCP. In addition, undisturbed soil samples were taken by using GP-75S sampler and laboratory tests were conducted for determining the characteristics of soils. This paper presents the findings of these studies.

Keywords Underground station · Building collapse · Resistivity image profiling · Defective panels

24.1 The Project

The construction for Kaohsiung MRT System (KMRTS) was commenced in October 2001. The system comprises two lines, i.e. the Red Line in the NS direction and the Orange Line in the EW direction with a total of 37 stations and three depots and is expected to be open for revenue services at the end of year 2007.

The station of interest, i.e. O1 Station (Sizihwan Station), is a terminal station at the western end of the Orange Line. It is located on the north bank of the Kaohsiung Harbour as depicted in Fig. 24.1. Figure 24.2 shows the layout of O1 Station which is 287 m in length and 16 m in width and is connected to a section of tunnel at its eastern end.

R.N. Hwang (✉)
11F, 3, Dunhwa S. Road, Sec. 1, 10557 Taipei, Taiwan
e-mail: wspc@wspc.com

K. Ishihara
Chuo University, Hachioji, Japan
e-mail: kenji-ishihara@e-mail.jp

W.F. Lee
Taiwan Construction Research Institute, Taipei, Taiwan
e-mail: weilee@mail.ntust.edu.tw

V.V.S. Rao and G.L. Sivakumar Babu (eds.), *Forensic Geotechnical Engineering*, Developments in Geotechnical Engineering, DOI 10.1007/978-81-322-2377-1_24

Fig. 24.1 Location of O1 station, Kaohsiung MRT system

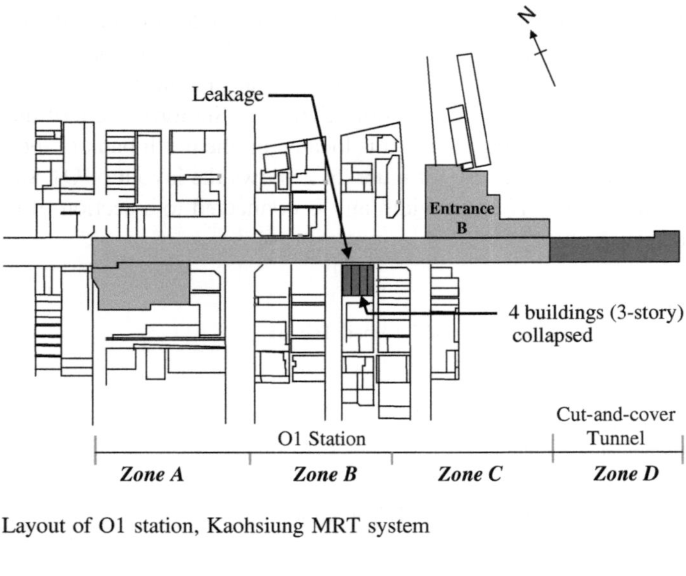

Fig. 24.2 Layout of O1 station, Kaohsiung MRT system

24.2 The Excavation

This 2-level station with side platforms is buried at a depth of 5.12 below ground surface. Excavation for the station was carried out to a depth of 20 m by using the cut-and-cover method of construction. Diaphragm walls of 800 mm in thickness were installed to a depth of 39 m in Zones A and B and walls of 1000 mm in thickness were installed to a depth of 37.5 m in Zones C and D. The pit was propped by six levels of struts as depicted in Fig. 24.3.

On 9 August 2004, a sinkhole was formed behind Panel S60M on the south side when excavation already reached a depth of 15 m and 4 buildings (3-storey) collapsed due to leakage of the diaphragm wall. Excavation was halted and measures were taken to improve the watertightness of diaphragm walls. Excavation in Zone B was resumed in July 2005 but a sinkhole was formed shortly at the same location for the second time. Additional ground improvement work was carried out and excavation resumed at the end of November 2005. The excavation was successfully completed and substructure erected in June 2006.

Fig. 24.3 Excavation scheme and retaining system

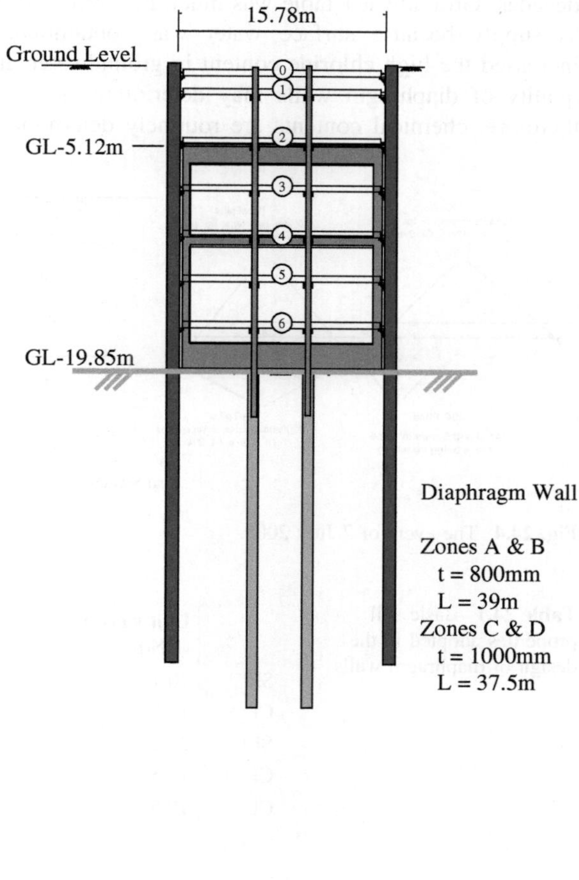

24.3 Ground Conditions

Figure 24.4 shows the soil profile obtained at Borehole WB-11. Although the thicknesses of sublayers vary from place to place, the sequence shown in the figure is quite typical. Soil properties adopted in the designs are listed in Table 24.1.

The subsoils on the west coast of Kaohsiung City are young sediments with high water contents and low plasticity. Such soils can easily be softened once disturbed, or liquefied when subjected to steep hydraulic gradients and, therefore, failures of trenches were quite common during installation of diaphragm walls. In many cases, mini-piles and/or micro-piles were installed to prevent trench from collapsing. Even so, necking still frequently occurred and reduced the sectional areas of diaphragm walls. One row of CCP piles was indeed used along the perimeter of the area to be excavated for maintaining the stability of trenches prior to the installation of the diaphragm walls.

In some cases, the soil pockets were continuous from back-to-back of the walls and became water paths. Ground treatment had to be applied behind defective diaphragm walls to stop the ingress of groundwater into pits.

The site was a salt pan a century ago. It was later used for fish farming for decades. Groundwater table was much lowered as fish farmers relied groundwater as supply because surface water was contaminated. Ingression of sea water increased the high chloride content in groundwater. It has been suspected that the quality of diaphragm walls may deteriorate as a result of chloride attack and, therefore, chemical contents are routinely determined in major projects to check

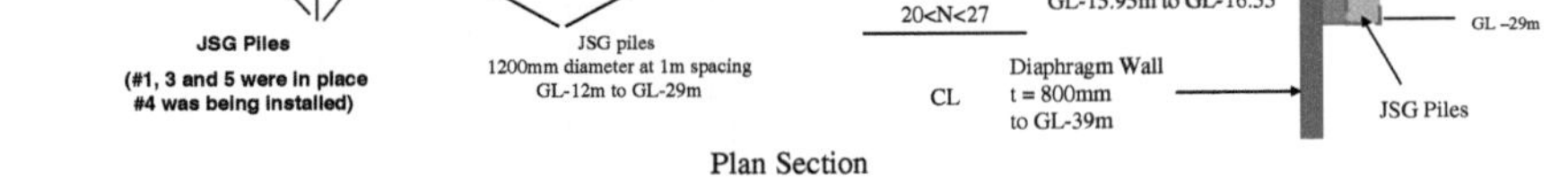

Fig. 24.4 The event of 7 July 2005

Table 24.1 Basic soil properties adopted in the design of diaphragm walls

	Unit weight (KN/m^3)	C′ (kPa)	Φ′ (°)	Su (kPa)
SF	20.1	0	28	
CL	18.9	0	27	18
SM	20.2	0	32	
CL	19.5	0	29	85
CL	19.5	0	28	113

Table 24.2 Results of chemical tests on soils

Hole	Depth (m)	pH	Organic	Cl^- (%)	SO_4^{-2} (mg/kg)	SO_3^{-2} (mg/kg)
OA-1	20	7.7	7.38	0.053	2176	5.1
OA-3	13	9.5	3.96	0.038	99.6	11.9
OA-4	15	8.3	5.29	0.11	324.7	7.5
OS-1	16	8.3	1.8	0.049	341.4	159
OS-2	16	8.3	1.8	0.049	323.3	2.6

Table 24.3 Results of chemical tests on groundwater

Hole	Depth (m)	pH	Cl^- (mg/l)	SO_4^{-2} (mg/l)	SO_3^{-2} (mg/l)	Ca^{+2} (mg/l)	Ca^{+2} (mg/l)
OA-1	15	7.9	180	112	<2.5	59.7	80.3
OA-3	15	7.2	5030	680	<2.5	113	2480
OS-1	17	6.8	15,800	2450	<2.5	984	5530
OS-2	15	7.7	17,300	2450	<2.5	1030	4780

Table 24.4 Classification of aggressiveness of groundwater

	Cl^- (mg/l)	SO_4^{-2}/SO_3^{-2} (mg/l)
Mild	0–1,000	0–150
Moderate	1,000–5,000	150–1,500
Severe	>5,000	1,500–10,000
Very severe		>10,000

whether groundwater is aggressive. Table 24.2 shows the results of chemical tests on soils and Table 24.3 shows the results of tests on groundwater in boreholes in the section of route between O1 and O2 stations. The aggressiveness of chemical attack is classified as "severe" in accordance with Table 24.4.

24.4 The Event of 9 August 2004

Water spurted from the bottom of the excavation in front of Panel S60M, refer to Fig. 24.2 for location, on the southern wall at 13:20 on 9 August 2004 as excavation reached a depth of 15 m. A sinkhole of 3 m maximum depth was formed behind the diaphragm wall and the area affected was around 500 m^2. Four 4-storey buildings collapsed in less than an hour and were demolished overnight. Several low-rise shops were severely damaged and were demolished sometime later.

One row of 11 bored piles was installed behind Panels S58–S60M. Pumping tests were performed in November 2004 to see if there were other defective panels. A total of 3,285 m^3 of water was drawn from 6 wells and water levels inside the excavation closely monitored at 60 wells. The groundwater table inside the excavation dropped;

on an average of 2.2 m as a result of pumping. The recovery of water levels was monitored for more than 10 days, however, the desired purpose was not achieved as the locations of leakages could not be identified (Ho et al. 2007).

24.5 The Event of 7 July 2005

To be on the safe side, one row of JSG piles was added along the entire perimeter of the station. Furthermore, the joints between JSG piles were treated by using CCP piles. Pumping tests were again performed subsequently to confirm the effectiveness of these measures. The results were not satisfactory as the rate of recovery of groundwater inside the excavation was only slightly smaller than that obtained previously (Ho et al. 2007). It was decided to add more JSG piles at the back of Panel S58M. 3 new piles were installed without problem. As No. 4 pile, refer to Fig. 24.9, was installed on 7 July 2005, groundwater brought a large quantity of soil into the pit. A nearby hospital was endangered and the patients in the hospital were urgently evacuated as a precautionary measure. It however survived with only minor damage. The sinkhole was about 1 m in depth and settlement spread over an area of about 1,000 m^2. Subsoils at this site, being closer to the sea, are even worse than those at O2 Station so the incident was not a surprise.

24.6 Forensic Studies

Subsequent to the event of 9 August 2004, the contractor engaged Kaohsiung Professional Civil Engineers Association, Kaohsiung Professional Geotechnical Engineers Association and Kaohsiung Architects Association on 16 August to form a joint committee to investigate the causes of failures. After the event of 7 July 2005, Taiwan Construction Research Institute (TCRI) was engaged by the contractor in December 2005 to further investigate the causes of failures.

24.6.1 Investigation by the Professional Associations

The committee studied the field records of diaphragm wall installation and checked the structural design of diaphragm walls. Upon the recommendation of the committee, additional soil exploration was carried out to determine ground conditions and tests were carried out to determine the quality of diaphragm walls. The design was found to be adequate and the failure was attributed to defects in wall panels.

Coring was carried out on Panels S59F, S60M and S61F in September 2005 to check the integrity of the walls. Panel S60M appeared to be in good quality within the depth of coring of 16 m. Segregation of concrete was detected in S59F at depths of 15.95–16.55 m. Coring was abandoned at Panel S61M as rebars were encountered at a depth of 6 m. Laboratory tests were carried out on the cores obtained from Panels S59F and S60M and an average unconfined compressive strength of 58.7 MPa (with a standard deviation of 83.8 MPa) was obtained on the 11 specimens from Panel S59F and an average strength of 713.1 MPa was obtained on the 9 specimens from Panel S60M, while the design strength of concrete is ? MPa.

Chemical tests on the cores indicate chloride contents varying from 0.016 to 0.022 kg/m^3, less than the tolerance of 0.30 specified in CNS 3090 for ready-mixed concrete.

The thicknesses of Panels S60M and S61F and the locations of rebars were confirmed to be adequate by using radar. However, Panel S59F was found to be defective.

24.6.2 *Investigation by Taiwan Construction Research Institute*

In addition to routine tests, such as CPT and SPT, TCRI carried out resistivity image profiling (RIP) of diaphragm walls and laboratory tests on soil samples taken by using GP-75S sampler which is a tool specifically developed for undisturbed sampling of difficult soils.

Table 24.5 Resistivities of different material

Material	Resistance, ρ (Ω m)
Fresh water	10,000
Sea water	1–5
Coarse sand, gravel (dry)	20,000–80,000
Coarse sand, gravel (saturated)	1,000–5,000
Sand (dry)	5,000–20,000
Sand (saturated)	200–1,000
Silt (dry)	400–2,000
Silt (saturated)	30–200
Clay (saturated)	15–30
Sandstone	100–8,000
Granite, gneiss	7,000–15,000
Concrete, gabbro	10,000–40,000
Quartzite	5,000–10,000

Cross-hole resistivity image profiling was carried out at Panel S72MF. Basically, electrical current is applied to the ground through two current electrodes and an electrical field is generated. The resulting voltage difference between two potential electrodes is measured to determine the resistance of the ground. By moving the current electrodes and the potential electrodes, ground resistances can be measured for different situations. With the modern computer technology, the data obtained

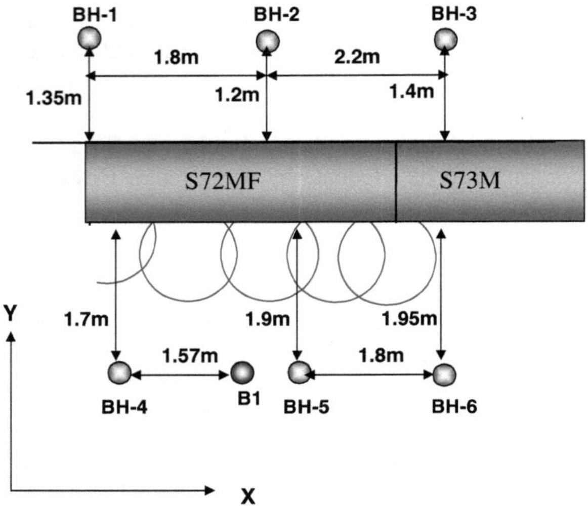

Fig. 24.5 Resistivity image profiling at panel 72

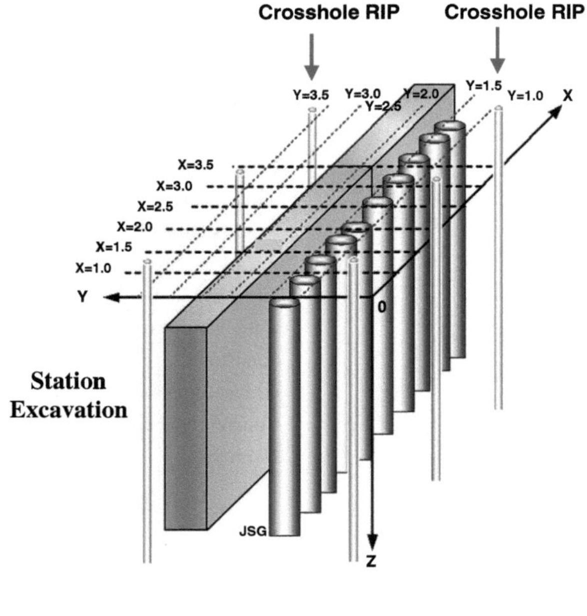

Fig. 24.6 3D presentation of resistivity image profiling

can be analysed to obtain the resistivities of underground materials at various locations and the results can be presented in a graphic form.

Table 24.5 shows the typical values of resistivities of different types of soils and rocks. As can be noted, the resistivities for concrete vary from 10,000 to 40,000 Ω m, while those for saturated sands, silts and clay vary from 15 to 1,000 Ω m.

Fig. 24.7 Side view of resistivity image profiling

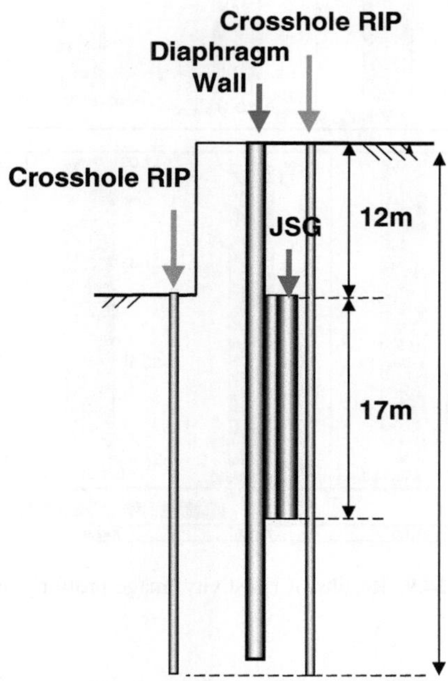

Fig. 24.8 Interpretation of results of resistivity image profiling

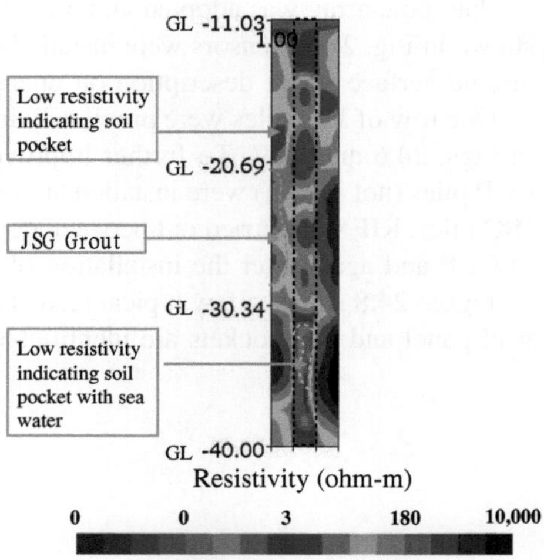

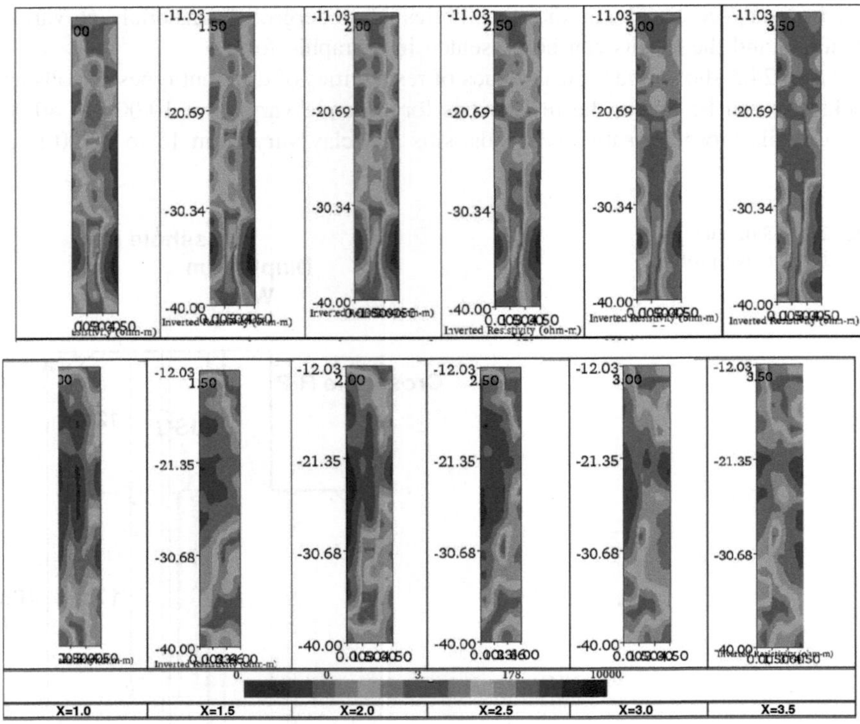

Fig. 24.9 Results of resistivity image profiling in yz sections

The resistivity of sea water is as low as 1–5 Ω m. It is therefore possible to identify defects, if any, in diaphragm walls accordingly.

Pole-pole array was adopted and 3 holes were sunk on each side of the panel as shown in Fig. 24.5. Sensors were installed at 1 m intervals to a depth of 40 m below ground surface (more description on operation).

One row of JSG piles were previously installed at the back of the panel as shown in Figs. 24.6 and 24.7. To further improve the watertightness of the wall system, CCP piles (not shown) were installed at the joints to fill up the gaps, if any, between JSG piles. RIF was carried out between depth of 11 and 40 m before the installation of CCP and again after the installation of CCP to see the differences made.

Figure 24.8 shows a few typical results obtained for a longitudinal section of the wall panel and soil pockets are identifiable (Figs. 24.9, 24.10, 24.11).

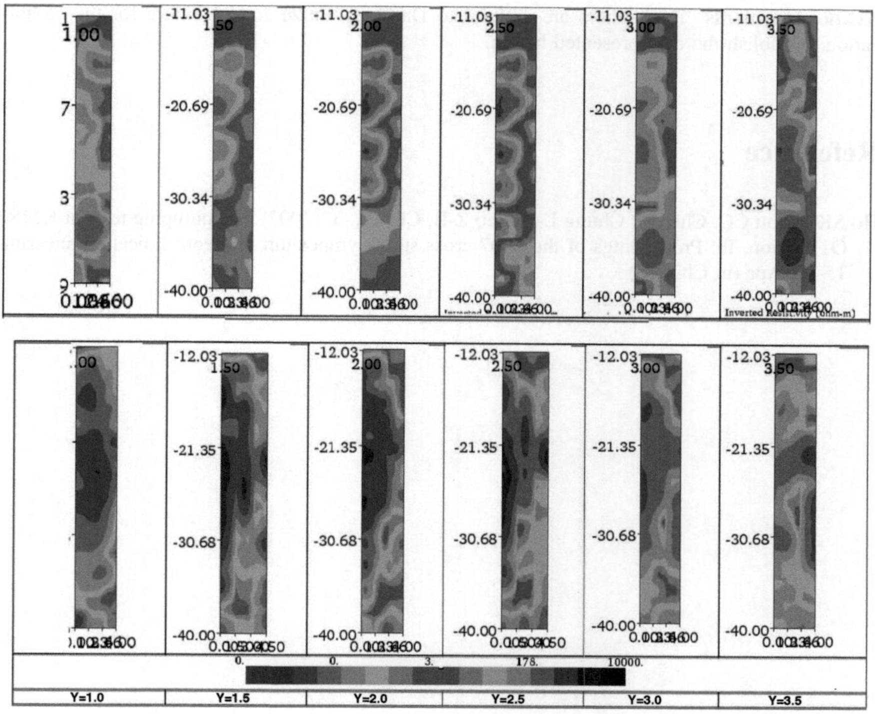

Fig. 24.10 Results of resistivity image profiling in xz sections

Fig. 24.11 3D presentation of results of resistivity image profiling

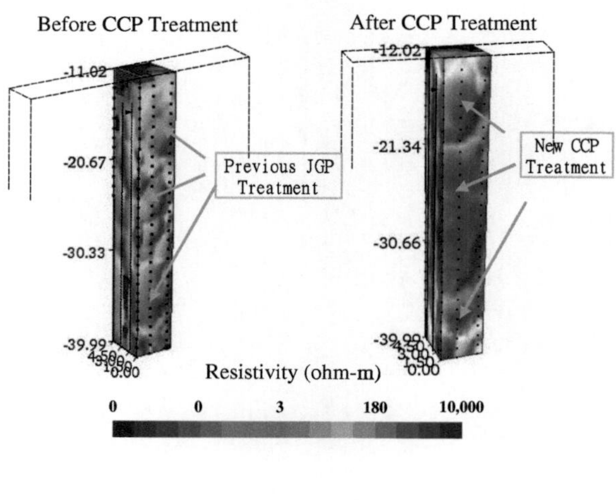

Acknowledgments The authors are grateful to Da-Cin/Shimizu Joint Venture for the authorization to publish the data presented herein.

Reference

Ho SK, Chou CC, Chen D, Chung L-J, Liao Z-B, Chen Y-Y (2007) The pumping tests at KMRT O1 Station, In: Proceedings of the 2007 cross-strait symposium on geotechnical engineering, 15–17 Apr (in Chinese)

Chapter 25
Damage to a Metro Tunnel Due to Adjacent Excavation

R.N. Hwang, B.S. Chen, T.E. Wu and S.W. Duann

Abstract Cracks were observed on the segmental linings of one of the tunnels in the Xinyi Line of the Taipei Metro, presumably, as a result of ground movements induced by a deep excavation carried out in the vicinity for constructing a high-rise building. The excavation was temporarily held and the causes of cracks were investigated. Both the contractor of the metro project and the contractor of the building project retained professionals as witnesses to provide technical supports to defend for them. The designer of the metro project also offered technical opinions to the authority in charge of the metro project. Back analyses indicated that the cracks were within tolerance and the injunction was finally lifted.

Keywords Tunnels · Segmental linings · Excavation · Instrumentation · Observations

25.1 Introduction

As shown in Fig. 25.1, an excavation was carried out in the vicinity of an existing tunnel of the Xinyi Line of the Taipei Metro for constructing the 4-level basement of a 15-storey building. It was carried out to a depth of 15.9 m in six stages as depicted in Fig. 25.2. The pit, roughly 59 m in the east-west direction and 30 m in the north-south direction, was retained by diaphragm walls of 800 mm in thickness

R.N. Hwang (✉) · T.E. Wu · S.W. Duann
Moh and Associates Inc., Oriental Technopolis Building A, 22 FL, No. 112 Xintai Wu Road, Sec. 1, Xizhi District, 22102 New Taipei City, Taiwan
e-mail: richard.hwang@maaconsultants.com

B.S. Chen
South District Project Office, Department of Rapid Transit Systems, Taipei City Government, 6F, 92, Roosevelt Road, Section 4, 10091 Taipei, Taiwan

V.V.S. Rao and G.L. Sivakumar Babu (eds.), *Forensic Geotechnical Engineering*, Developments in Geotechnical Engineering, DOI 10.1007/978-81-322-2377-1_25

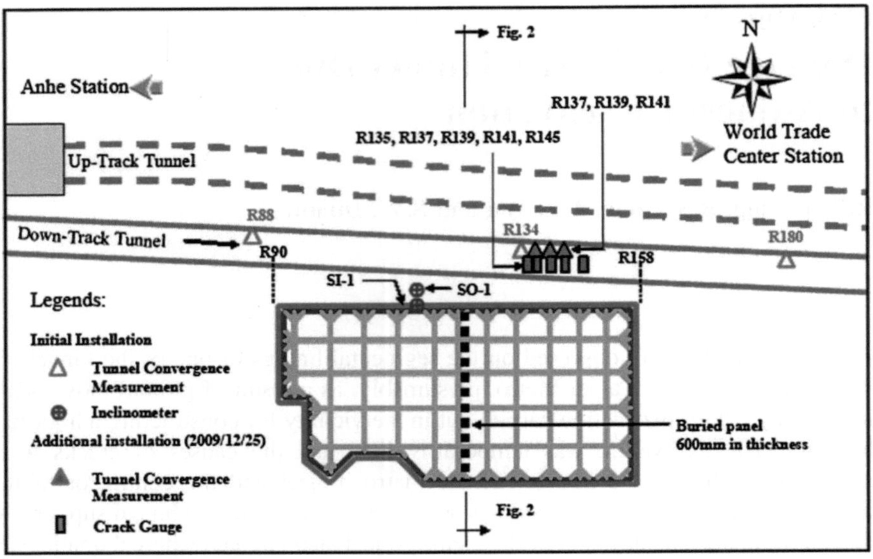

Fig. 25.1 Site plan and layout of instruments

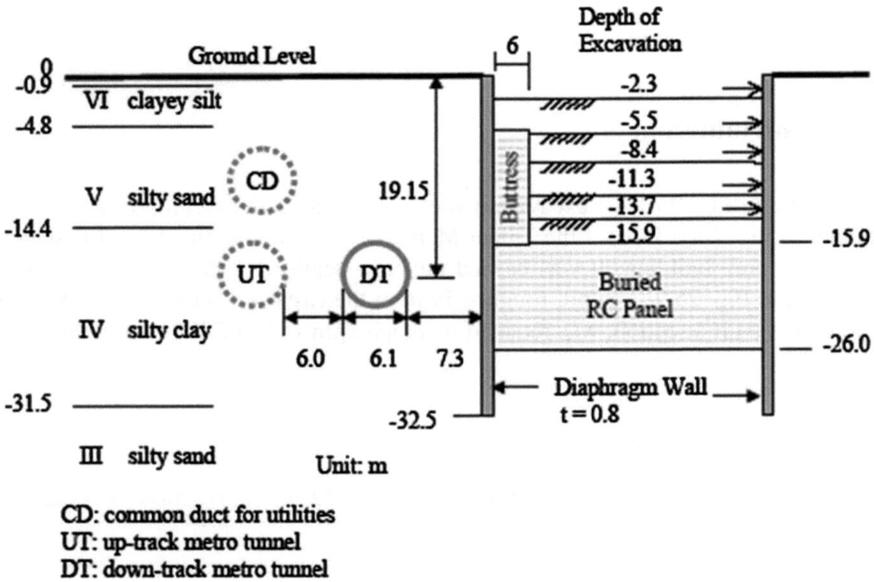

Fig. 25.2 Ground conditions and excavation scheme

and 32.5 m in length, and was braced by steel struts at five levels. To minimise wall deflections, a buried panel of 600 mm in thickness was cast near the mid-span of the site as depicted in the figures by using the diaphragm wall technique. This panel extended from the final level of excavation to a depth of 26 m below the ground surface and was structurally connected to the diaphragm walls at the two ends. At the northern end, the panel extended above the bottom level of excavation to form a buttress to further reduce wall movements.

The site is located in the TK2 Zone of the Taipei Basin (Lee 1996) and the typical six-layer sequence of soils, i.e. sublayers VI to I in a descending order, is identifiable. The bearing stratum lies at depths varying from 38 to 50 m and groundwater table was at depths of 2–3 m below ground surface. As shown in Fig. 25.2, the two tunnels are embedded in sublayer IV, which contains silty clay with SPT N-values of 2–8, at a depth of 19.15 m below ground surface.

The two tunnels are 6100 mm in outer diameter with a clearance of 6 m between and are lined with reinforced concreted segments of 250 mm in thickness. The down-track tunnel is only 7.3 m away from the pit at the most critical location. Above the up-track tunnel, a common duct is to be constructed in the future to house various types of utilities for easy maintenance. The down-track tunnel was launched from the east end of Anhe Station (R7 Station) on 27 July 2009 by using an earth pressure balance shield machine. The shield machine was driven eastward to the World Trade Center Station (R6 Station). It arrived R90 (i.e. Ring No. 90, R stands for ring number hereinafter), which corresponds to the west side of the excavation, on 6 August. Tunnelling was temporarily halted after the shield machine arrived R95 on 11 August for system transfer and was resumed on 24 August. In the meanwhile, the excavation for the basement started on 15 August. The first stage excavation had been completed and the struts at the first level were being installed when the shield machine completed the 68-m run (R90 to R158) corresponding to the frontage of the excavation on 31 August.

25.2 The Event

As depicted in Fig. 25.1, the convergences of the linings of the down-track tunnel were monitored at R88, R134 and R180. To ensure that tunnel linings would not be damaged, an alert level of 15 mm and an action level of 25 mm were imposed so mitigation measures could be taken in time to limit the deformations of the tunnel. The initial readings of tunnel convergence for R134 were established on 29 August. The ring was found to have lengthened by 13.7 mm in the horizontal direction and shortened by 19.9 mm in the vertical direction on 26 October. The squashing of the ring continued and the changes in diameters increased to 20.9 mm in the horizontal direction and 23 mm in the vertical direction on 11 November. The former already exceeded the pre-specified Alert Level. On 18 December, the lengthening of the diameter in the horizontal direction reached 27.5 mm and, however, the shortening of the diameter in the vertical direction reduced to 21.4 mm.

A joint inspection was made on 24 December with all the parties participating and cracks were observed on several rings. Since the tunnel convergence already exceeded the Action Level, an injunction was issued on 25 December by the authority incharge of the metro construction and the excavation was halted. Crack gauges were mounted on five rings next to R134 as depicted in Fig. 25.2 for long-term monitoring of the changes in widths of the cracks. In addition to R134, the convergences of R137, R139 and R141 were also monitored daily. To avoid deterioration of the situation, temporary struts were installed at the B4 level in the excavation to limit wall movements and internal bracings were installed in the tunnel to limit the deformations of R99 to R155. These measures were completed on 25 January 2010 and were found effective as the deformations of the tunnel did not increase further subsequently. The temporary internal bracings in the tunnel were kept in place until the completion of the up-track tunnel. Since the possibility for the down-track tunnel to deform could not be eliminated during the construction of the up-track tunnel and the common duct, the excavation contractor made the commitment that he would install steel linings once the changes in diameter reach 34 mm. With this commitment, the injunction was lifted and the excavation was resumed.

25.3 Instrumentation and Observations

In accordance with "the Ordinance on Constructions of Public or Private Buildings and Advertising Structures along the Routes of the Mass Rapid Transit System", the excavation scheme and instrumentation and monitoring programme were reviewed and approved by the metro authority before the excavation was allowed to commence. Therefore, sufficient instruments were available for evaluating the performance of the excavation. Wall movements were monitored by using inclinometers and ground movements in the vicinity of the excavation were monitored by using settlement markers. Besides, piezometers were available to enable ground settlements to be correlated with the changes in groundwater table. Inside the pit, strain gauges were installed on struts to measure the loads in struts. Heave markers were available for measuring the ground heaves as the excavation preceded.

The readings directly relevant to the movements of the tunnel are the movements of the diaphragm wall on the north of the site. There were two inclinometers available, i.e. SI-1 and SO-1 of which the locations are shown in Fig. 25.1. The former was installed in the diaphragm wall to a depth of 32.5 m and the latter was installed to a depth of 35 m in the ground immediately behind the wall. The readings obtained by these two inclinometers on 29 March 2010 are shown in Fig. 25.3. The maximum wall deflections recorded were 55.4 mm for SI-1 and 65.8 mm for SO-1.

Fig. 25.3 Observed and computed movements

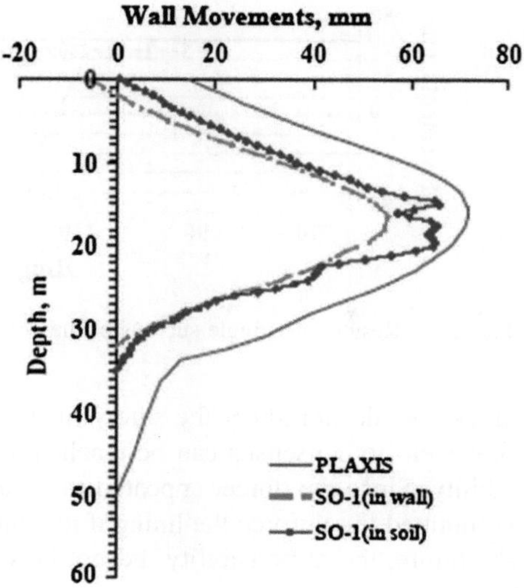

25.4 Investigations

Figure 25.4 shows the tunnel convergences measured jointly by the both contractors on 18 January and it can be noted that squashing of tunnel was significant between R55 and R150. The movements of the tunnel were monitored by wriggle survey and, as depicted in Fig. 25.5, a maximum movement of about 25 mm was observed in the transverse direction in the period between 9 September and 25 December 2009.

Both contractors engaged their own expert witnesses in mid-January to investigate the causes of tunnel deformations and evaluate the damage to the tunnel. Measurements in a joint inspection on 22 January, 2010 showed that the cracks were less than 0.2 mm in width. It was concluded by both parties that such minor

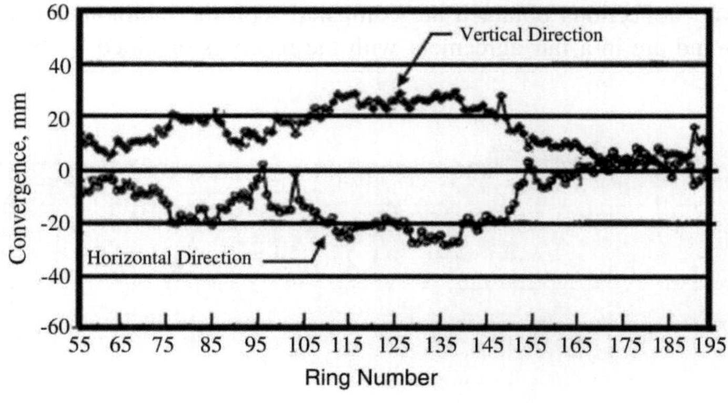

Fig. 25.4 Convergence of tunnel measured on 18 January 2010

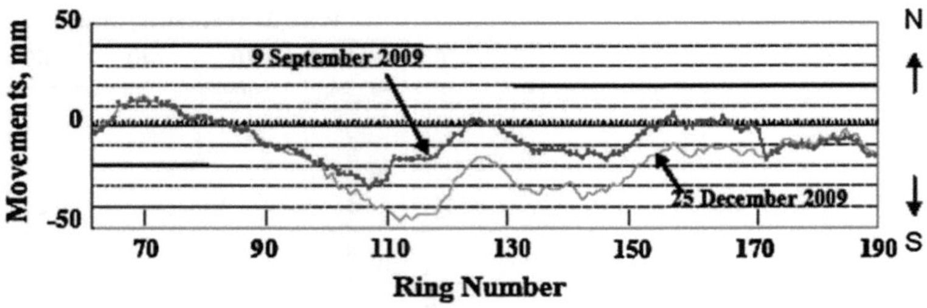

Fig. 25.5 Results of wriggle surveys of the down-track tunnel

cracks would not affect the safety of the linings. Regarding the causes of cracks, however, no consensus can be reached as both contractors disclaimed the responsibility. Since the tunnel appeared to be undamaged and the excavation contractor committed to reinforce the lining if the tunnel convergence exceeds the tolerance in the future, the responsibility did not become an issue to be argued.

25.5 Back Analyses

At the request of the authority incharge of the metro, the response of the tunnel to the excavation was analysed by the designer of the metro project. The computer software package PLAXIS was used and the finite element model adopted is shown in Fig. 25.6. The site is located in the TK2 Zone of the Taipei Basin and as can be noted from Fig. 25.2 that the subsoils comprise of an alternation of silty clay and silty sand sublayers. The Mohr-Coulomb model was adopted to simulate soil behaviour and the soil parameters adopted in the analyses are given in Table 25.1. The Young's moduli of E = 600 Su (Su = undrained shearing strength) were assumed for clayey soils and E = 2,000 N (kPa), in which N = SPT below counts, were assumed for sandy soils.

The wall deflections obtained are compared with the inclinometer readings in Fig. 25.3 and are in a fair agreement with the readings obtained by Inclinometers

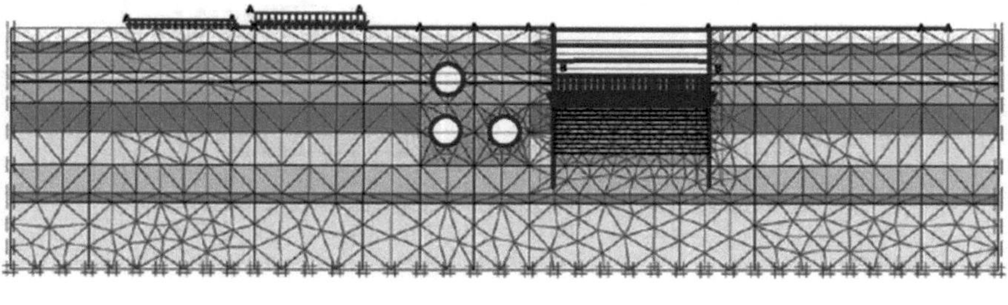

Fig. 25.6 Finite element model adopted in PLAXIS analyses

Table 25.1 Soil parameters adopted in the analyses

Stratum	Depth (m)	SPT-N	γ_t (kN/m^3)	ϕ'	s_u (kN/m^2)	ν	E (kN/m^2)
SF	3.0	3	18.9	30	0	0.30	6,000
CL1	5.0	2	18.8	31	30	0.35	18,000
SMI	9.2	6	19.5	31	0	0.30	12,000
SM2	11.0	6	21.8	33	0	0.30	12,000
CL2	21.2	4	18.5	31	60	0.35	36,000
CL3	28.2	6	18.8	30	70	0.35	42,000
CL4	34.1	17	19.8	30	92	0.35	55,200
SM3	36	25	20.8	33	0	0.30	50,000
GM	50.0	50	22.1	40	0	0.35	100,000

SI-1 and SO-1. A maximum wall deflection of 72 mm was computed in comparison with 55.4 mm obtained by Inclinometer SI-1 and 65.8 mm obtained by Inclinometer SO-1. Since inclinometer readings were calculated by assuming the toes were fixed without any movements, they have to be corrected to account for toe movements. A toe movement of 22.8 mm was obtained in the PLAXIS analyses and this amount is in an excellent agreement with the value proposed for the central areas in the Taipei Basin (Hwang et al. 2007) as depicted in Fig. 25.7. However, the chart proposed by Hwang et al. (2007) is for excavations in green field without ground improvement or auxiliary supports to walls. The RC panel and buttress (refer to Fig. 25.2) appear to be not as effective as expected in reducing wall movements for reasons which are not readily clear. In any case, the results of the PLAXIS analyses are deemed representative of the response of the northern wall to the excavation.

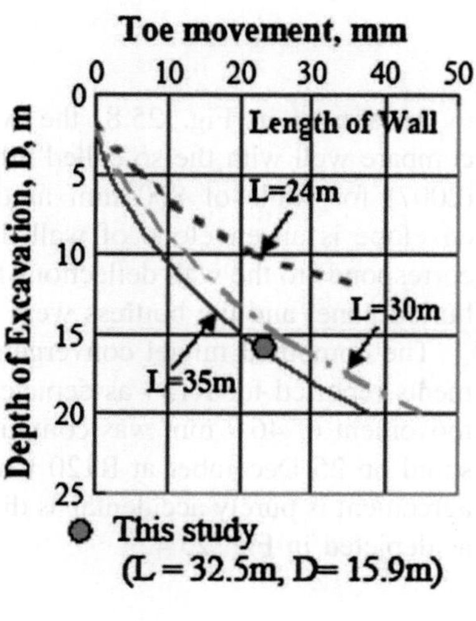

Fig. 25.7 Movements of toes of the diaphragm walls (Hwang et al. 2007)

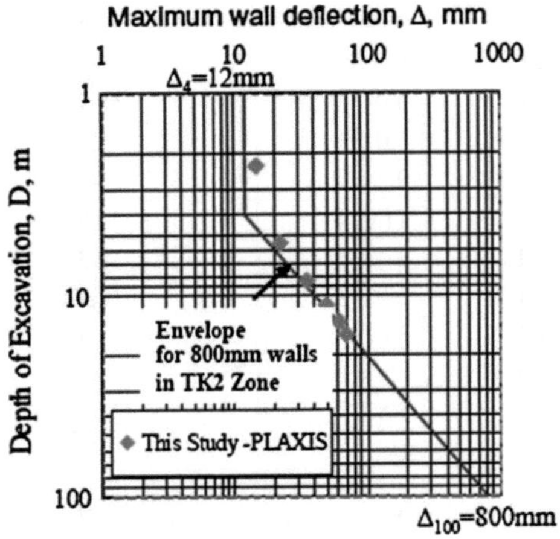

Fig. 25.8 Wall deflections obtained as compared to reference envelope

Table 25.2 The convergence of tunnel lining

Stage	Measurements for R134		PLAXIS analyses	
	Horizontal (mm)	Vertical (mm)	Horizontal (mm)	Vertical (mm)
2nd stage excavation	−0.3	−0.3	1.70	−1.55
3rd stage excavation	−0.3	−0.3	5.42	−5.45
4th stage excavation			12.37	−12.53
5th stage excavation	19.9	−13.6	20.86	−20.57
Excavation completed	20.9	−23.0	27.60	−27.03
Base slab cast	22.5	−24.6	27.59	−27.02
B4 slab cast	27.5	−21.4	27.35	−26.84

As shown in Fig. 25.8, the wall deflections obtained in PLAXIS analyses compare well with the so-called reference envelope proposed by Hwang and Moh (2007) for walls of 800 mm in thickness in the TK2 Zone. Again, since this envelope is an envelope of wall deflections in the TK2 Zone and, theoretically, corresponds to the wall deflections for excavations in green field, it appears that the buried panel and the buttress were not as effective as expected.

The computed tunnel convergences also compare very well with the measurements reported for R134 as depicted in Table 25.2. It is amazing that a horizontal movement of 46.9 mm was computed and this amount is exactly the mount measured on 25 December at R120 where the movement was the maximum. Such an agreement is purely accidental as different movements were measured at other rings as depicted in Fig. 25.4.

25.6 Conclusions

Excavations in the vicinity of tunnels are potentially dangerous and should be carried out with great care. Damage to tunnels can be avoided if early warning can be obtained and mitigation measures are taken in time.

Acknowledgments The authors are grateful to the Department of Rapid Transit Systems for the permission to publish the data presented herein. Appreciation is due to Mr. C.R. Chou for the assistance in numerical analyses.

References

Hwang RN, Moh ZC, Wang CH (2007) Toe movements of diaphragm walls and correction of inclinometer readings. J GeoEngineering 2(2):61–70

Hwang RN, Moh ZC (2007) Deflection paths and reference envelopes for diaphragm walls in the taipei basin. J GeoEng 2(1):1–12

Lee SH (1996) Engineering geological zonation for the Taipei City. Sino-Geotechnics 54:25–34

Chapter 26
A Case Study of Floor Heaving of a Condominium RC Building with Seven Stories

Yoshi Iwasaki and Junsuke Haruna

Abstract Heaves of floors inside a reinforced concrete structure with seven stories appeared one year after its construction. Apparent repair works were applied but did not improve the situation. Year by year, the heaves had increased and it was decided to bring the problem to the court. The cause was due to the use of steel slag mixed in filled soil under the structure. Free lime in the steel slag was identified as main cause of the problem by chemical analysis.

Keywords Heave · Free lime · X-ray diffraction · Numerical analysis

26.1 Introduction

A condominium building with seven stories was built and sold in April 1994 in a town west of Kobe within 30 km. After living for one year or so, residents realized deformation of the floors of the condominium. The builder explained that the deformation was caused by Kobe earthquake and repaired the distortions. After the repair was completed, however, the deformation, especially, heaves of floors, was found to continue.

Photo 26.1 shows buckling of a vertical door frame, the base of which was displaced upward by the heave of the floor of the room.

Horizontal deflection was about 80 mm with vertical heave of about 20–30 mm at the floor.

Heavy deformations were found within the structures as well as outside. The heaves of floors were found concentrated at the center of base slab between foundation piles. Uneven heaves resulted in inclination as large as 1/50. Ten years

Y. Iwasaki (✉)
Geo-Research Institute, Osaka, Japan
e-mail: yoshiiw@geor.or.jp

J. Haruna
Taisei Giken Ltd. Co., Nagoya, Japan
e-mail: junsukeharuna@yahoo.co.jp

V.V.S. Rao and G.L. Sivakumar Babu (eds.), *Forensic Geotechnical Engineering*,
Developments in Geotechnical Engineering, DOI 10.1007/978-81-322-2377-1_26

Photo 26.1 Buckling of a vertical door frame

after the construction, the resident union had decided to bring the problem to the judicial district court.

26.2 Preliminary Consideration

Heaving of the floors of the building is usually caused by expansion of ground underneath the base slab. Possible mechanisms are as follows:

1. expansive soil of clays with high IP, cotton soils, expansive tuff layers and stones,
2. release of overburden load by deep excavation,
3. soil improvement by mixing lime, and
4. steel slags mixed in filled soil that may induce ettringite or contains free lime. Near the site, steel factories have been running and providing slags as by-products.

26.2.1 Geotechnical Condition

The dimension of building is 30 × 15 m of basement supported by piles that were installed with distance of 6 m as shown in Fig. 26.1. The site locates at a small flat

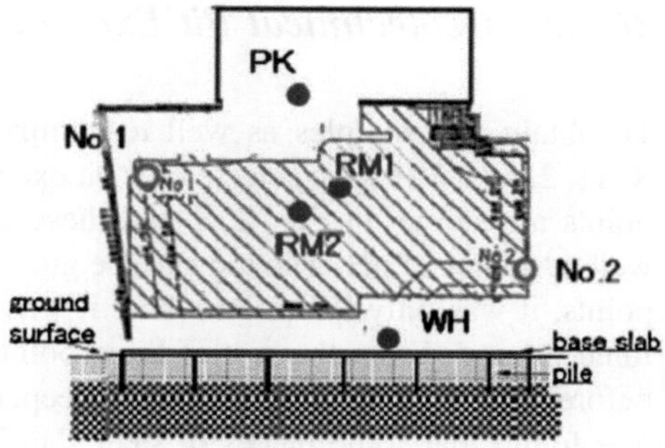

Fig. 26.1 Plan of the building section of the foundation

plain that had been cultivated for agriculture use. Before constructing the building, two geotechnical borings were performed. The both boring results were almost the same and the boring log for No.1 is shown in Fig. 26.2.

It was found that filled soil was found from the top surface to 1.6 m. The natural soil layer consists of clayey gravel and those deeper than about 5–7 m is in dense condition. The building was designed by pile foundation as the supporting bearing layer about 5–7 m below the ground surface.

Most of the natural ground consists of clayey gravel that is not likely to cause heaving. Among the above four, the latter two remained as possible causes.

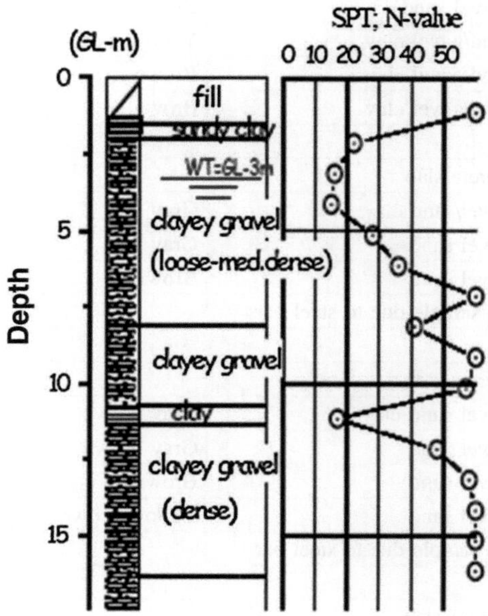

Fig. 26.2 Boring log for No.1

26.2.2 Geotechnical Pit Excavation

To obtain soil samples as well to confirm the soil condition, four points at PK, RM1, 2, and WH were selected for pit excavation of some 30 cm in diameter. These points are shown in Fig. 26.1 and these sampled soils are listed in Table 26.1 as well as values of pH. The top surface ground was filled with soil. Among these four points, it was only PK point where it was possible to excavate down to below the filled ground. From the rest of three points, some steel bars were found in the pit before reaching the natural soil. No deeper excavation in the pit was performed. It was found that some pieces of steel slag is shown as Photo 26.2. Generally, steel slag is angular shaped with small air bubbles that were produced by sudden decrease of temperature during the solidifying process. Steel slag usually shows very high alkaline potential Hydrogen. Potential Hydrogen (pH) was measured for every samples and the results are shown in Table 26.1.

Table 26.1 Soil sampled from excavated pits

PK (parking lot)				
GL (m)				pH
0.00–0.10	Gravel sand with concrete	Dark brown	Fill	11
0.10–0.30	Gavel sand	Gray		11
0.30–0.50	Gavel sand	Gray brown		11
0.50–0.65	Gravel sand	Brown		11
0.65–1.30	Sand/gavel clay	Yellow gray		9
1.30–1.70	Sand/gavel clay	Brown	Natural soil	7
1.70–	Sand/gavel clay	Brown		7
RM1				
From Bottom of concrete slab				pH
0.00–0.25 m	Gavel/sand clay	Gray	Fill	12
0.25–0.35	Gavel sand	Gray		11
0.35–0.73	Gavel sand	Brown		8
0.73–	No sample due to steel bar			
RM2				
From Bottom of concrete slab				pH
0.00–0.25 m	Gavel sand clay	Gray brown	Fill	12
0.25–0.35	Gavel sand	Gray brown		12
0.3 5–0.68	Gavel sand	Brown		8
0.68–0.77	Gavel sand	Yellow brown		8
0.77–	No sample due to steel bar			
WH ware house				
From Bottom of concrete slab				pH
0.00–0.25 m	Gavel sand	Yellow brown	Fill	11
0.25–0.45	Gavel sand	Yellow brown		11
0.45–	No sample due to steel bar			

Photo 26.2 Pieces of steel slag found in the filled soil

Most of the fills showed very high pH ≥ 11 and the natural soil showed pH = 6–7, which shows possible mixture of the steel slag with filled soil or lime.

26.2.3 X-Ray Diffraction Analysis

Since the steel slag had been identified by visual inspection of soils and the pH value, the soil fills from the surface to a depth of 0.65 m show high pH values were considered as the potential heaving layer.

Possible mechanism of heaving at the site was considered by either ettringite or by free lime. As the first stage, X-Ray diffraction analysis was performed for powder of the sample at PK site 0.10–0.30 m from the surface. The results are shown in Fig. 26.3 with characteristics refracted angles for typical chemicals of quartz, ettringite, and slacked lime.

The quartz is major mineral component of soils of gravel and sand and has a characteristic diffraction angle at $2\theta = 26.7$. Ettringite is a hexacalcium aluminate trisulfate hydrate that is major component in Portland cement for volume expansion. The characteristic diffraction angle is at $2\theta = 9.1$ for the ettringite. The obtained results do not show any response at this angle and consequently do not support the existence of the ettringite. The slacked lime is a byproduct of hydration of lime as follows, $CaO + H_2O = Ca(OH)_2$

The X-ray diffraction analysis shows the matching angle of the slacked lime at $2\theta = 18.1$, which shows positive support of the chemical reaction of the lime.

Chemical components of the sampled soils by X-Ray Fluorescence Analysis were mainly used to determine atomic components of the samples. Samples of the depths of GL-0.0–0.3, 0.5–0.65, and 1.3–1.7 at PK were selected as representatives in the site. These samples were crushed into fine grains and were applied to XFA.

For F-CaO (free lime) and T-Fe (total iron), quantitative chemical analyses were performed and listed in Table 26.2. Table 26.2 shows comparison of analyzed results of sampled soils and those of steel slag and andesite rock as natural geology. The difference between steel slag and natural rock is found to consists of the chemical components CaO, SiO_2, FeO, and P_2O_5. Generally, steel slag is rich in CaO, FeO, and P_2O_5 compared to natural rock. The analyzed results show the content of silica increases with its

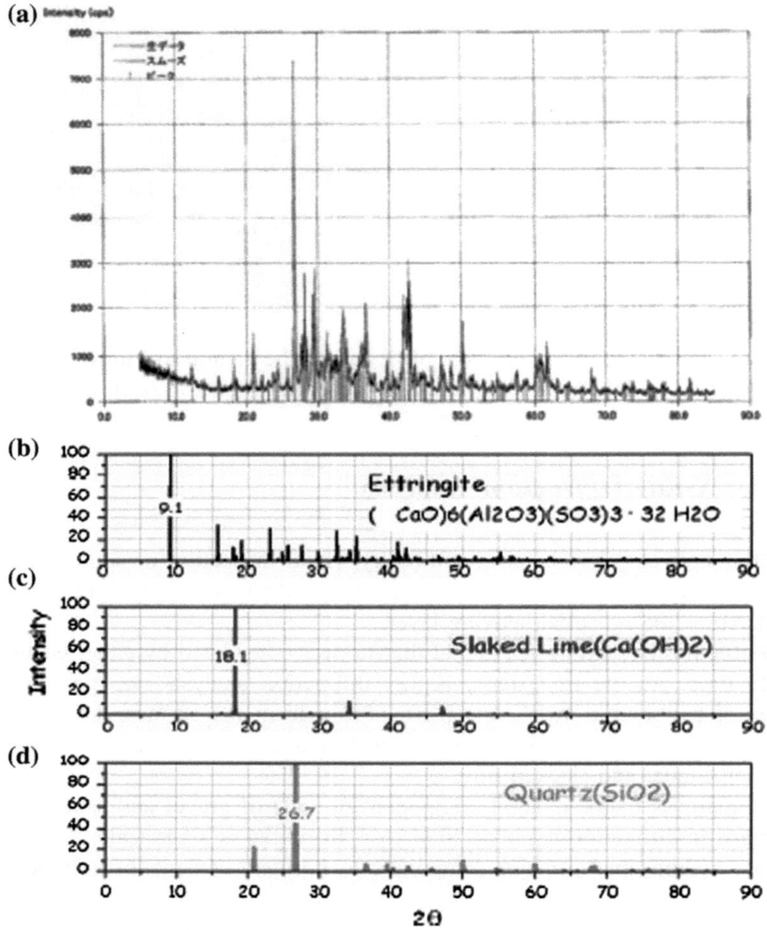

Fig. 26.3 Comparison of the Intensity by X-Ray diffraction for the sampled soil and chemical components of ettringite, $Ca(OH)_2$, and SiO_2

depth as shown in the Table 26.2. At the same time, the content of lime is higher in the upper part of the fill and becomes lower in the lower filled soil. The chemical components for the sampled soil that was considered as the mixed soil with steel slag shows a tendency of gradual change from top soil as close to steel slag to the bottom soil as close to rock. Among the limes, the existence of free lime is the positive proof of steel slag and has very active tendency of getting united with acidic atoms. When the free lime CaO meets with H_2O, it is easily produced with volume expansion.

$$CaO + H_2O \rightarrow Ca(OH)_2$$

Table 26.2 Chemical components of steel slag and soil samples

	Steels Iug	Rock (andesite)	Samples at PK depth (m)		
			0.10–0.30 (%)	0.50–0.65 (%)	1.30–1.70 (%)
CaO	45.8	5.8	28.2	14.3	2.02
F–CaO	2–5		0.52	0.23	<0.01
SiO_2	11.0	59.6	27.9	55.6	78.6
$A_2 0_3$	1.9	17.3	6.37	7.94	9.22
MgO	6.5	2.8	8.76	4.79	1.01
FeO	17.1	3.1	8.39	4.1	2.57
$P_2 0_5$	1.7	–	0.106	0.07	0.035

The contents of CaO 0.5 % are rather low compared to the 3–5 % as known common value. One reason of this low value is hydration already took place at the site. The ratio of the slag to natural soil was estimated by comparing content ratios of CaO and SiO_2 as shown in Fig. 26.4 based Table 26.2. Table 26.3 shows estimated slag contents for three layers-based assuming free lime contents of 3.5 % for steel slag.

26.2.4 Electron Probe Micro Analysis for Steel Slag

The sample of steel slag at PK was selected and the sample surface was polished and tested by electron probe microanalysis (EPMA). In EPMA, a beam of electrons is fired at a sample. The beam causes each element in the sample to emit X-rays at a

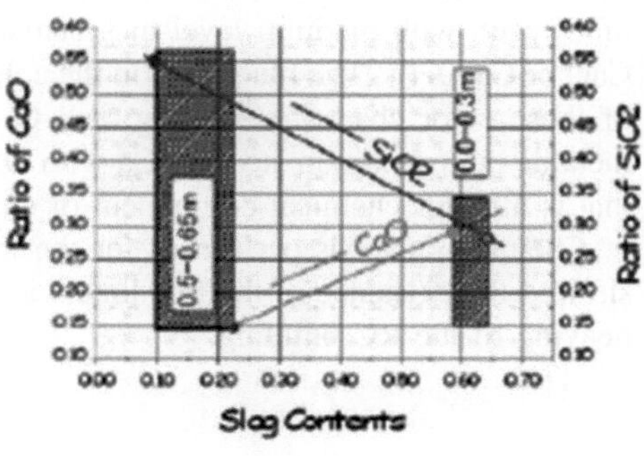

Fig. 26.4 Slag Contents versus CaO and SiO_2 (free lime for slag is assumed as 3.5 %)

Table 26.3 Slag contents and free lime of the fill below the floor

	Slag contents (%)	Free lime (%)
Depth		
0.0–0 3 m	62	2.2
0.3–0.5 m	39	1.4
0.5–0.65 m	16	0.6
(Slag)	100	3.5

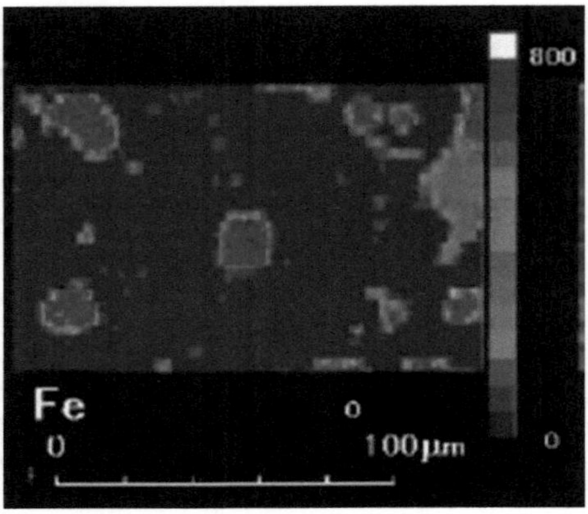

Fig. 26.5 Example of EPMA analysis color map for Fe component

characteristic frequency of the specific atom on the spot. The size of the electron beam determines accuracy of the testing size that is the trade-off between resolution and scan time is the order of 1 μm in the present study.

The electron beam was fired with scanning from edge to edge of the area. X-ray was emitted from the target element and was filtered for specific frequencies of different atoms. The number of the received X-ray with the specific frequency was counted for a unit time and the result is shown as color presentation that is called color map in Fig. 26.5. The scale of the color is shown at the right-hand side of each color maps in the Fig. 26.5. Within the frame of the scanned area, the red to orange color in Oxygen indicates high density existence of the atom at theses element area.

Figure 26.6 shows a comparison of distribution four atoms of Fe, Mg, Si, and Ca, S as well as Oxygen atom. The orange to red color pattern of Ca indicates overall existence in the frame with high density that corresponds to the same pattern of Oxygen with medium level in greenish color, which suggests the existence of CaO or $Ca(OH)_2$. The blue spots in the color chart of Ca correspond to the red spots of those color charts for Fe, Mg, and O, which indicates that Fe and Mg of the metallic atoms exist in stead of Ca with Oxygen. The very little existence of sulfur that is a major chemical component of ettringite is also confirmed by EPMA.

Chemical analysis performed for the foundation of the soils and steel slag has showed the existence of the free lime was the chemical component that has caused heaving of the RC building.

26.2.5 Heaving Characteristics of Steel Slag by Free Lime

In Japan, when the slag has become widely used for construction materials as road and highways in 1980, steel makers have noticed the swelling and expansion characteristics of steel slag. They developed and adapted a process of aging steel slag providing water has become a common practice in the middle of 1990–2000.

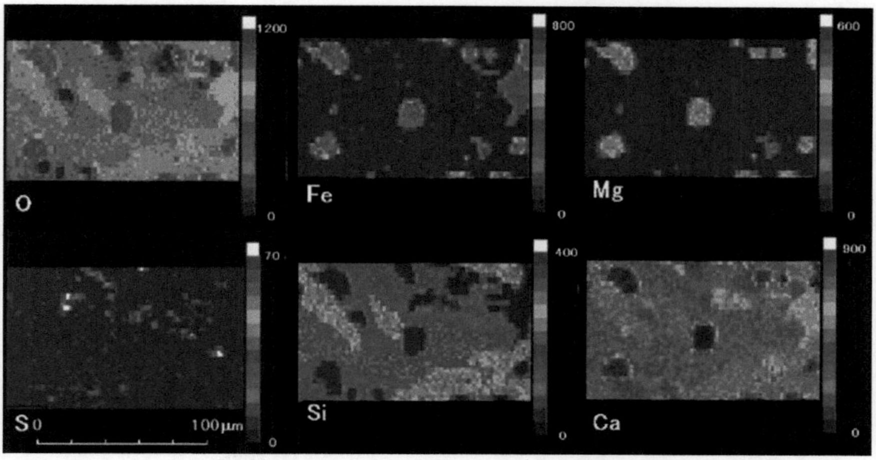

Fig. 26.6 Shows an example of the result of EPMA analysis. The size of the area of the selected surface of the sample for the analysis was 80 μm × 120 μm

Table 26.4 Steel slag arranged for expansion test

Item		S. S.3	M. S.3
Max dia	(mm)	5	5
Raw slug without aging	(%) in weight	100	80
Aged slug	(%) in weight	0	20
Optimum water contents	Wopt (%)	10.4	11.6
The max density	g_{max} (g/cm^3)	2.53	2.28
Free lime	(%) in weight	3.8	3.04

The time when the building was constructed was 1993–1994 that is transient period of adapting aging process by steel companies.

Fundamental characteristics of the heaving of steel slag were tested in a study in Japan. They have tested two kinds of slag products for road construction material with the maximum grain size of 5 mm.

One is raw steel slag without aging denoted as SS3. Another material is mixed slag of aged and without aging. The mixed ratio is 80–20 as shown in Table 26.4. Testing apparatus was specially designed as shown in Fig. 26.6 to obtain expansive pressures of the slag material under confinement with water provided. The equipment consists from two steel boxes with a hollow space of spherical shape with a radius of 50 mm. Four holes were drilled for each box to equip pipes with 5 mm in diameter for providing water into testing material under test (Fig. 26.7 and Table 26.5).

The material was compacted in the hollow space to reach 90 % of the maximum standard density of Proctor compaction method. Two hollow boxes with compacted soil were attached together as to form a spherical ball specimen and used for expansion tests. The test was performed by supplying hot water with temperature of 80 °C through pipes. The test results are shown in Fig. 26.8.

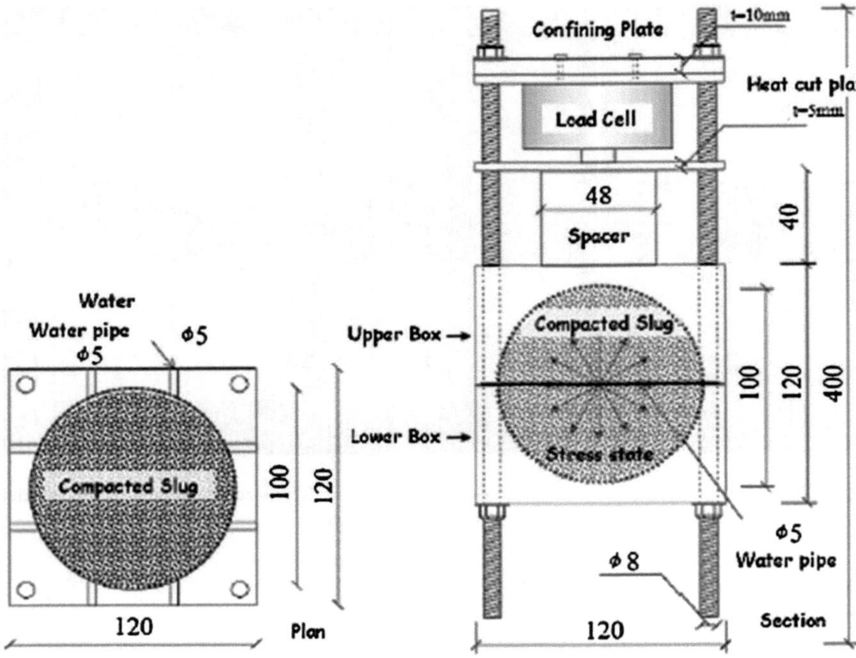

Fig. 26.7 Testing apparatus for heaving pressure for steel slag

Table 26.5 Assumed equivalent Young's moduli

Slag contents (%)	Max pressure (kPa)	Max heave (%)	Equivalent E (kPa)
100	1,150	8	14,375
60	690	4.8	14,375
40	460	3.2	14,375
20	230	1.6	14,375

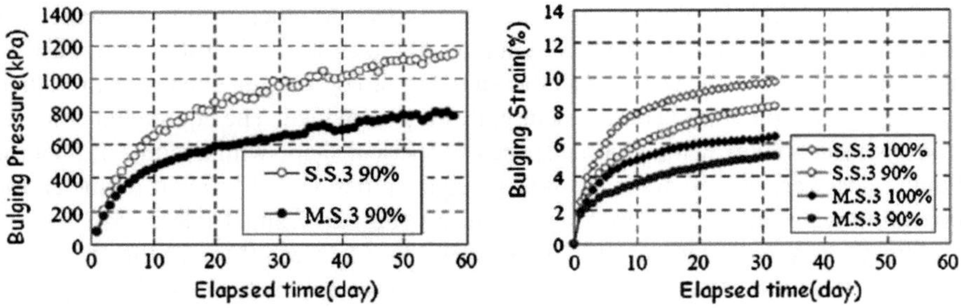

Fig. 26.8 Bulging pressure and strain versus elapsed time for steel slag

26.2.6 Simulation of Floor Heaves by Bulging Steel Slag

Assuming the bulging characteristics of the slag mixed filled layers shown in Table 26.3 based upon Fig. 26.8, expansive characteristics were assumed as adapting an equivalent Young's modulus of E = 12,000 kPa for slag mixed fill with slag content of 100 %.

The base slab of the building was supported by piles. The concrete slab was 150 mm in thickness and 6,000 mm in width between the piles as shown in Fig. 26.2. Plain strain 2D linear elastic analysis was executed introducing equivalent bulging stresses at boundaries as shown in Fig. 26.9. Upward stress equivalent to the required stress to confine material from bulging.

Figure 26.10 shows the result of deformation with two times in its scale. The maximum heave on the top surface was obtained as 44 mm that is larger than observed value of 10–30 mm at the site and might suggest the possibility of further heaving.

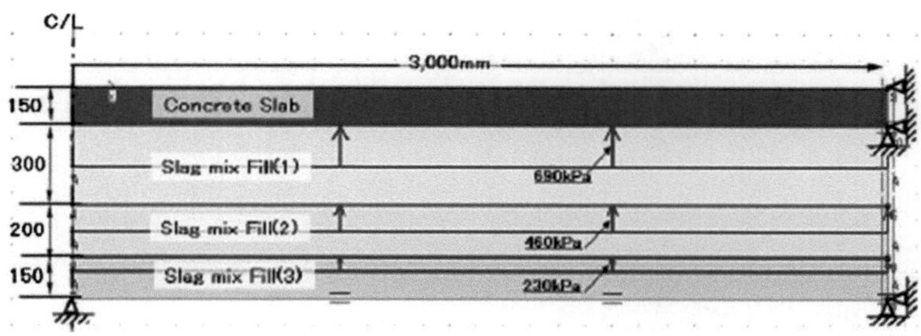

Fig. 26.9 Configuration and boundary conditions for heaving problem

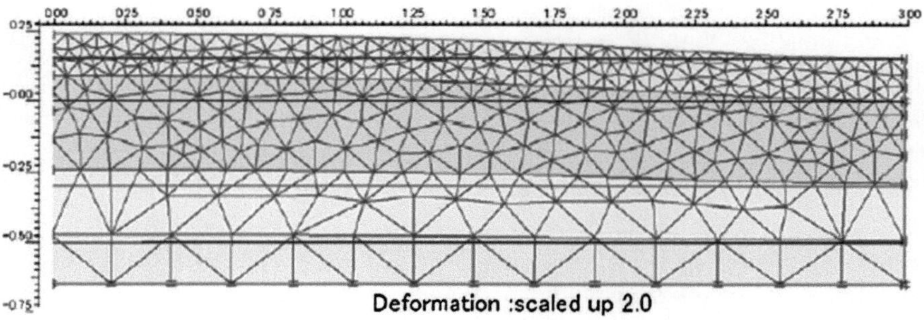

Fig. 26.10 Deformed mesh

26.3 Conclusion

A case study on floor heaves shows the cause was by the steel slag that was mixed with fill under the ground floor of RC building with seven stories. Chemical analyses including X-ray diffraction as well as EPMA were useful to identify the chemical components. Numerical simulation showed the same order of heaving of the ground floor at the site.

Chapter 27
Failure Case Study of Tieback Wall in Urban Area, Korea

S.S. Jeong, Y.H. Kim and M.M. Kim

Abstract In this study, numerical analysis was performed to reproduce the sequential behavior of an anchored retaining structure in an urban area. The numerical analysis was verified through comparisons between the prediction and a field failure case. Emphasis was placed on the wall behavior and the location of the sliding surface based on elasto-plastic method and shear strength reduction FEM method. Through the comparison study, it was found that coupled analysis using shear strength reduction method can be effectively used to perform back calculation analysis to find a critical surface in the anchored wall structures, whereas uncoupled analysis by elasto-plastic method can be applicable to the preliminary design of a retaining wall with a suitable safety factor.

Keywords Anchored retaining structure · Strength reduction method · Failure · Back analysis

27.1 Introduction

South Korea suffers from a serious lack of land space due to its high population of about 47 million people on a little less than 100,000 km^2 of land and the fact that 75 % of the land space is mountainous. Since 1980, a number of huge excavation projects such as underground space for constructing power stations, subways, high-speed railways, and many lifeline constructions have been performed in urban areas. Retaining structures such as the slurry wall, C.I.P wall, and H-pile are frequently used in these areas as temporary excavation proceeds.

S.S. Jeong · Y.H. Kim
Department of Civil Engineering, Yonsei University, Seoul, Korea

M.M. Kim (✉)
Department of Civil and Environmental Engineering, Seoul National University, Seoul, Korea

V.V.S. Rao and G.L. Sivakumar Babu (eds.), *Forensic Geotechnical Engineering*, Developments in Geotechnical Engineering, DOI 10.1007/978-81-322-2377-1_27

The analysis technique for analyzing the sequential behavior of anchored retaining wall has been developed for decades (Haliburton 1968; Clough 1971; Clough and Tsui 1974; Briaud and Kim 1998; Jeong and Seo 2004). Although these methods make slightly different assumptions, they can generally be classified into two main groups: (1) elasto-plastic method and (2) finite element method (shear strength reduction).

The objective of this study is to suggest the conceptual methodology of anchored retaining wall design by considering the coupling effect between soil and structure. Comparisons are made between the soil–wall behavior computed by elasto-plastic, and finite element analysis which considers the mechanism of shear strength reduction using the material properties of real failure case.

27.2 Analytical Methods for Anchored Retaining Wall

27.2.1 Elasto-Plastic Method

The elasto-plastic method is based on beam-column theory, which has been studied for many applications in engineering practice. The governing equation for horizontal beam modeling of anchored retaining wall is as follows:

$$EI\frac{\mathrm{d}^4y}{\mathrm{d}x^4} + \frac{A \cdot E'}{L} \cdot x = p_i - k_s \cdot x \tag{27.1}$$

where E is the elastic modulus of retaining wall; I is the inertia moments of retaining wall; A is the area of bore hole; E' is the elastic modulus of bore hole; L is the length of bore hole; p_i is the initial earth pressure; k_s is the horizontal subgrade reaction; x is the depth; y is the horizontal wall deflection. This method is developed for rigid beam with infinite length and it is assumed that the soil is rigid and perfectly plastic. Thus, this method may not represent the behavior of actual wall in the field: this method does not take into account the actual behavior of finite flexible beams, soil arching and soft soil, etc. (Hassiotis 1997; Cai et al. 2000) Schematic drawings of elasto-plastic modeling for a retaining wall is shown in Fig. 27.1.

27.2.2 Finite Element Method (Shear Strength Reduction)

The finite element method has been used to investigate the soil-structure system, which is analyzed as a continuous elastic or elasto-plastic medium using finite element formulations. This method provides coupled solutions in which the wall and slope response are considered simultaneously.

For retaining wall, the factor of safety F is traditionally defined as the ratio of the actual soil shear strength to the minimum shear strength required to prevent failure.

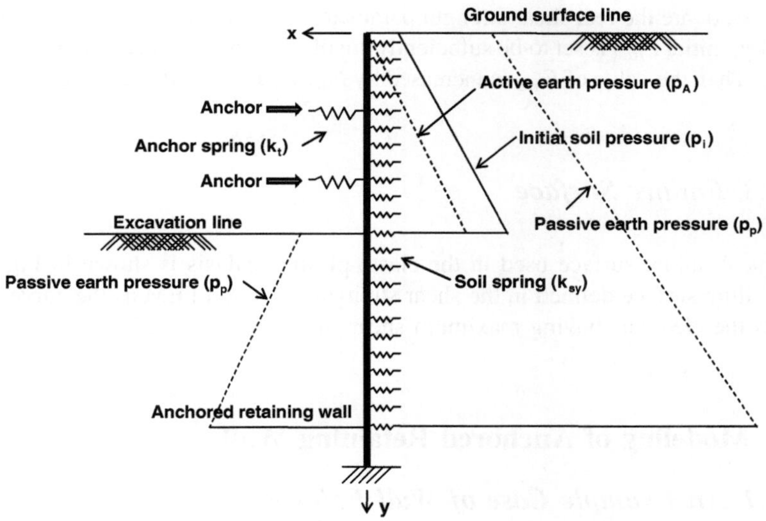

Fig. 27.1 Schematic drawing of elasto-plastic modeling

The "shear strength reduction technique" was used as early as by Zienkiewicz et al.1975, and has been applied by Naylor (1981), Donald and Giam (1988), Won (2005), etc. In the shear strength reduction FEM method, the shear strength of soil is reduced and the factor of safety is recalculated until collapse occurs.

To calculate the factor of safety of a retaining wall structure defined in the shear strength reduction technique, a series of stability analyses are performed with the reduced shear strength parameters c'_{trial} and ϕ'_{trial} (Fig. 27.2) defined as follows:

$$c'_{\text{trial}} = \frac{1}{F^{\text{trial}}} c' \tag{27.2}$$

$$\phi'_{\text{trial}} = \arctan\left(\frac{1}{F^{\text{trial}}} \tan\phi'\right) \tag{27.3}$$

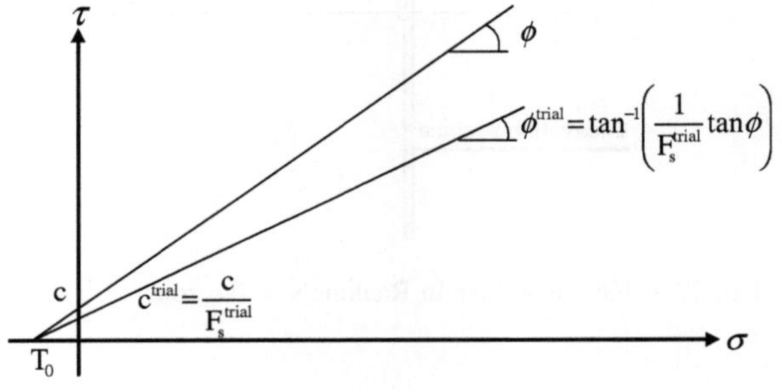

Fig. 27.2 Relationship between the actual strength and the strength reduced by a trial factor of safety

where c', ϕ' are the real shear strength parameters and F_{trial} is a trial factor of safety. Usually, initial F_{trial} is set to be sufficiently small so as to guarantee that the system is stable. Then the value of F_{trial} is increased by F_{inc} values until the retaining wall fails.

27.2.3 Failure Surface

A typical failure surface used in the elasto-plastic analysis is shown in Fig. 27.3. The failure surface defined in the shear strength reduction FEM as the curve which passes the elements having maximum shear strain ratio (Fig. 27.4).

27.3 Modeling of Anchored Retaining Wall

27.3.1 An Example Case of Wall Failure

In this study, an analysis section was adopted from a construction field in urban area where the failure accident took place. As shown in Fig. 27.5, the retaining wall consisted of 18.3 m CIP (cast-in place pile) wall plus 21.7 m H-pile, and was braced with nine earth anchors (various free length) and 13 rock bolts downwards. Based

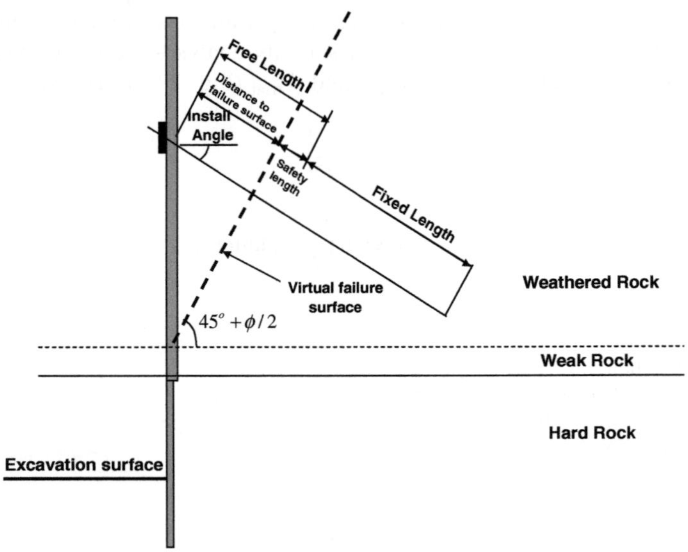

Fig. 27.3 Failure surface in Rankine's active zone

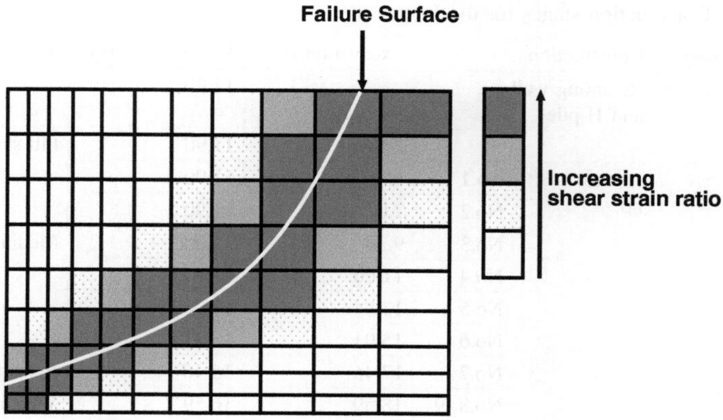

Fig. 27.4 Failure surface in finite element method

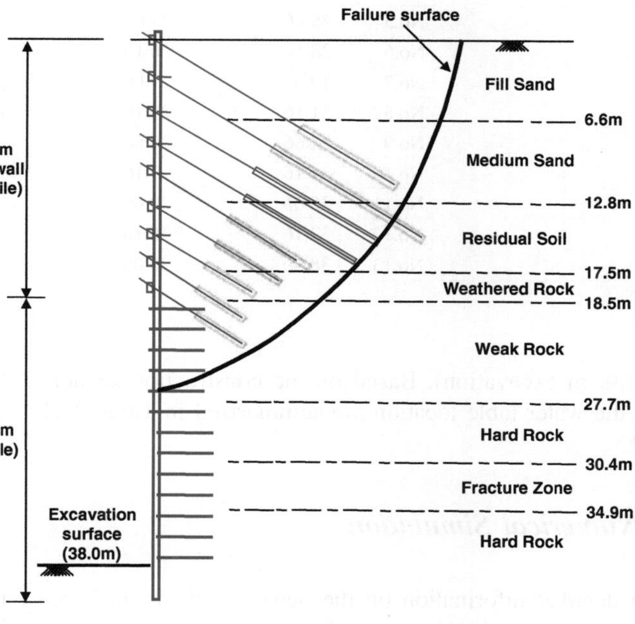

Fig. 27.5 Analysis section of anchored retaining wall

on the in situ and laboratory soil tests, soil profile was constructed as being composed of medium to dense sand in the upper 17.5 m underlain by weathered rock and weak to hard rock. Especially, the fracture zone was noted in the hard rock layer at depths below 30.4 m. Sequential excavation proceeded until the wall failure

Table 27.1 Construction stages for the failure site

Excavation stage	Construction		Excavation (m)	Water table (m)	Remark
–	Retaining wall and H-pile		–	11.90	–
1st	–		2.04	11.90	Fill sand
2nd	Anchor no.	No.1	4.58	11.90	
3rd		No.2	7.01	11.90	
4th		No.3	9.30	11.90	Medium sand
5th		No.4	11.60	11.90	
6th		No.5	13.84	11.90	
7th		No.6	15.91	13.91	
8th		No.7	17.46	15.46	Residual soil
9th		No.8, 9	18.59	16.59	Weathered rock
10th	Rock—Bolt no.	No.1	20.66	18.66	
11th		No.2	22.16	20.16	Weak rock
12th		No.3	23.66	21.66	
13th		No.4	25.16	23.16	
14th		No.5	26.66	24.66	
15th		No.6	28.16	26.16	
16th		No.7	29.66	27.66	
17th		No.8	31.16	29.16	Hard rock
18th		No.9	32.66	30.66	
19th		No.10	34.16	32.16	Fracture zone
20th		No.11	35.66	33.66	
21th		No.12	37.16	35.16	Hard rock
22th		No.13	38.00	36.00	

occurred (38 m excavation). Based on the construction sequence, the excavation stage and the water table location are summarized in Table 27.1.

27.3.2 Numerical Simulation

To obtain detailed information on the behavior of the anchored retaining wall, a series of numerical analyses were performed for the site of failure. The response of an anchored retaining wall was analyzed in parallel by the elasto-plastic method and by the 2D nonlinear finite element method.

First, the elasto-plastic analysis (SUNEX 2002) was executed which is based on an iterative and incremental procedure to take into account the changes in the earth pressures and wall displacements at each construction stages. Second, the Finite Element Method modeled the site under the plane strain condition. The finite element mesh for a typical case is shown in Fig. 27.6. The mesh consisted of 2D

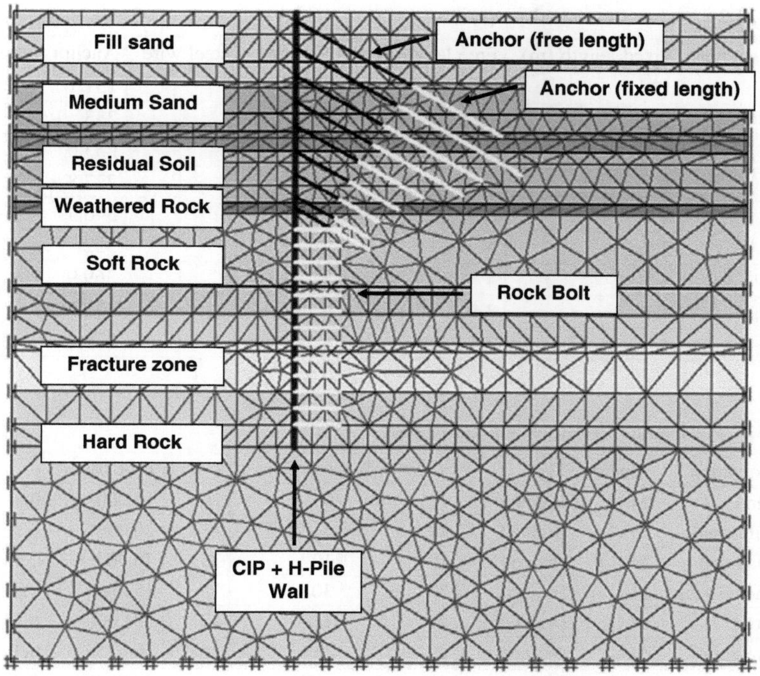

Fig. 27.6 Element mesh and boundary conditions

six-noded solid triangular elements and was assumed to be resting on a rigid layer. The wall element remained elastic at all times, while the surrounding soil was idealized as a Mohr–Coulomb elasto-plastic material. The interface element was composed of 1D quadratic three-node elements. The Coulomb's frictional criterion was used to distinguish between elastic behavior, where small displacements can occur within the interface, and plastic interface behavior (slip). The interface properties were taken from the associated soil properties surrounding the wall using the strength reduction factor (R_i) as follows:

$$c_i = R_i \times c_{\text{soil}}, \ \tan\phi_i = R_i \times \tan\phi_{\text{soil}} \tag{27.4}$$

where c_i and ϕ_i are the cohesion and friction angle of the interface, c_{soil} and ϕ_{soil} are the cohesion and friction angle of the soil mass. This model was selected in the element library of PLAXIS (2005), the commercial finite element package used for this work.

Table 27.2 shows the anchor forces and their dimensions, and Table 27.3 shows material properties used in numerical analyses.

Table 27.2 Anchor forces and properties

Earth anchor	Fixed length (m)	Free length (m)	Number of steel wire	Anchor force (kN)
No.1	12	9	3	131.1
No.2	10.5	12.5	3	233
No.3	9	10	3	255.6
No.4	8	9	3	277.8
No.5	6.5	6.5	4	346.4
No.6	5.5	5	4	352
No.7	4.5	4	4	369.9
No.8	4	3.5	4	375.2
No.9	4	3.5	4	375.2

Table 27.3 Material properties and geometries

Soil condition	Model	Type	γ_t (kN/m^3)	c (kN/m^2)	φ (°)	E (kN/m^2)	ν	R_i
Fill sand	M.C	Drained	18	0	26	10,000	0.3	–
Medium sand	M.C	Drained	18	5	31	20,000	0.3	–
Residual soil	M.C	Drained	19	10	33	40,000	0.3	–
Weathered rock	M.C	Drained	20	30	35	151,900	0.3	–
Soft rock	M.C	Drained	21	120	42	381,600	0.25	–
Facture zone	M.C	Drained	20	71	26	266,750	0.3	–
Hard rock	M.C	Drained	23	1074	63	5,000,000	0.2	–
CIP wall	Beam	Nonporous	50	–	–	8.2 × 107	0.2	0.8
H-pile	Beam	Nonporous	50	–	–	5.6 × 107	0.2	0.6

Reinforcements	Model	Horizontal spacing (m)	Size (mm)	Angle (°)	EA (kN)	ν
Earth anchor (No.1–3)	Node-to-Node Anchor and Geogrid	2.7	Φ12.7 × 3	30	19,742	0.2
Earth anchor (No.4)		1.8	Φ12.7 × 3	30	32,968	0.2
Earth anchor (No.5–9)		1.8	Φ12.7 × 4	30	236,223	0.2
Rock bolt (No.1–13)	Geogrid	1.8	Φ25	0	272,180	0.2

27.4 Comparison of Elasto-Plastic Method with Finite Element Method

27.4.1 Lateral Deflection of Retaining Wall

Figure 27.7 shows the predicted and measured lateral deflection profiles from initial to final construction stages. The measured deflection profiles were obtained from field inclinometer data.

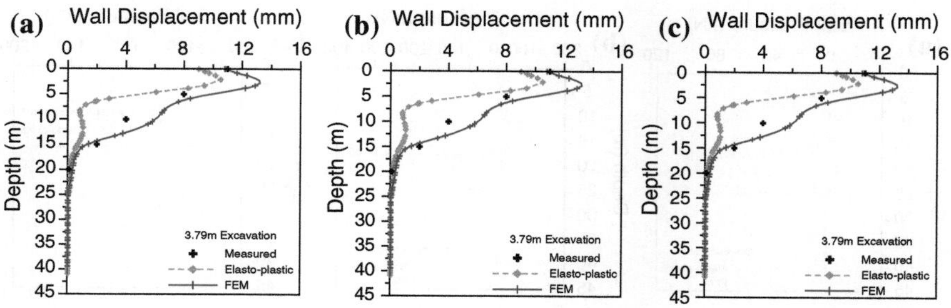

Fig. 27.7 Measured and predicted lateral deflection of anchored retaining wall. **a** 3.79 m excavation. **b** 17.41 m excavation. **c** 28.24 m excavation

It is seen in the figure that the finite element method predicts the wall deflection relatively well compared with those of the elasto-plastic method.

27.4.2 Bending Moment and Earth Pressure

Figures 27.8 and 27.9 show the predicted bending moments and earth pressures mobilized in the retaining wall. The overall distributions are similar for the finite element method and the elasto-plastic method until the middle construction stage. As the construction stage proceeds to the final stage, the distributions become dissimilar to each other, especially at the lower part of the excavation.

27.4.3 Failure Surface

Figure 27.10 compares the measured failure surface with the failure surface predicted by the FEM after the final excavation stage. From this comparison, it is

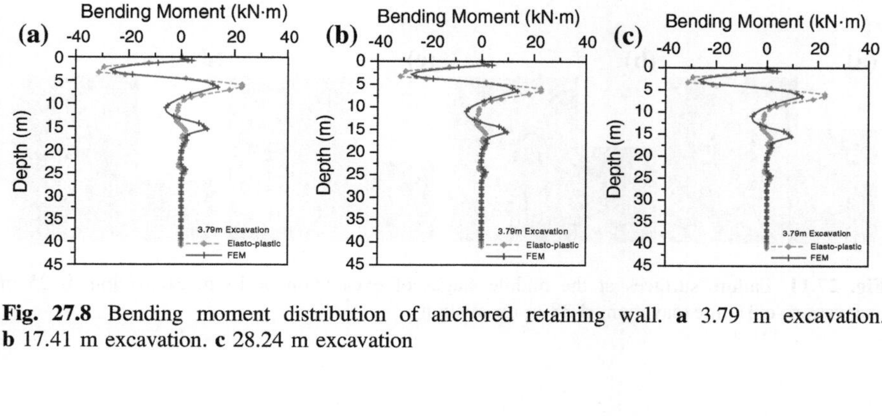

Fig. 27.8 Bending moment distribution of anchored retaining wall. **a** 3.79 m excavation. **b** 17.41 m excavation. **c** 28.24 m excavation

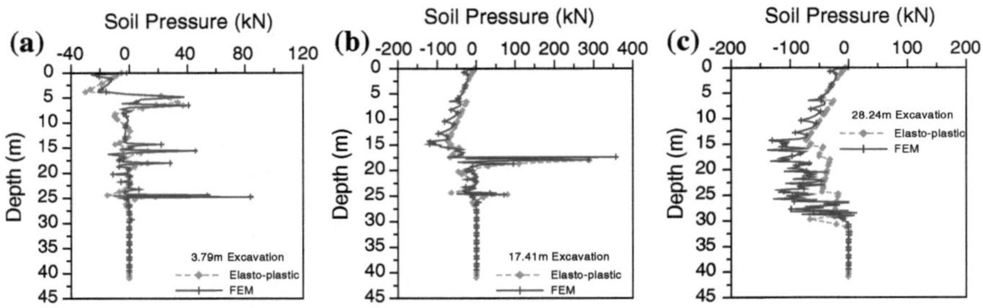

Fig. 27.9 Earth pressure distribution of anchored retaining wall. **a** 3.79 m excavation. **b** 17.41 m excavation. **c** 28.24 m excavation

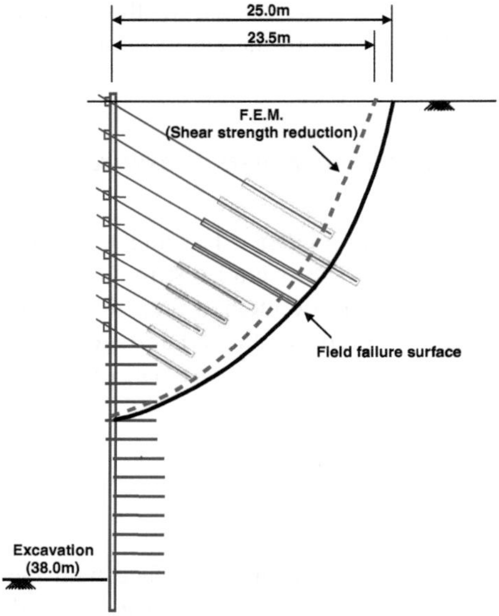

Fig. 27.10 Comparison of failure surfaces

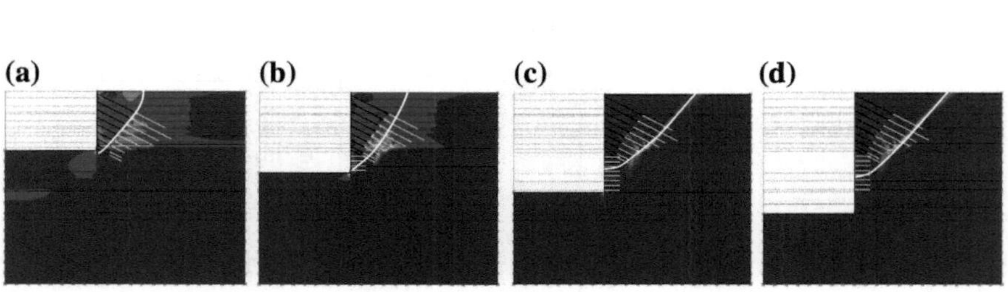

Fig. 27.11 Failure surfaces at the middle stages of excavation. **a** 18 m excavation. **b** 25 m excavation. **c** 31 m excavation. **d** 38 m excavation

confirmed that the shear strength reduction method simulates well the field failure surface. In Fig. 27.11, the possible failure surfaces are shown for the middle stages of excavation.

27.5 Conclusions

The main objective of the analysis described herein was to compare the sequential behavior of anchored retaining wall predicted by the finite element method and elasto-plastic method with the observed behavior in the field. From the comparisons, the following conclusions are drawn:

1. The finite element method utilizing the shear strength reduction technique predicted wall deflections closely compared with the measured values. However, the result of elasto-plastic method showed large discrepancy.
2. The finite element method predicted well the location of the field failure surface, which is not possible for the elasto-plastic method.
3. In overall, it was confirmed that the finite element method utilizing shear strength reduction technique can be effectively used to perform the back calculation analysis of the anchored wall structures.

References

Briaud JL, Kim NK (1998) Beam-column method for tieback walls. J Geotech Geoenviron Eng ASCE 124(1):67–79

Cai F, Ugai K (2000) Numerical analysis of the stability of a slope reinforced with piles. Soils Found Jpn Geotech Soc 40(1) 73–84

Clough GW, Duncan JM (1971) Finite element analyses of retaining wall behavior. J Geotech Eng ASCE 97(12):1657–1673

Clough GW, Tsui Y (1974) Performance of tieback walls in clay. Proc J Geotech Div ASCE 12(100):1259–1273

Donald IB, Giam SK (1988) Application of the nodal displacement method to slope stablilty analysis. In: Proceedings of the 5th Australia–New Zealand conference on geomechanics, Sydney, pp 456–460

Haliburten TA (1968) Numerical analysis of flexible retaining structure. Proc ASCE 94(6): 1233–1251

Hassiotis S, Chameau JL, Gunaratne M (1997) Design method for stabilization of slopes with piles. J Geotech Geoenviron Eng ASCE 123(4):314–323

Jeong S, Seo D (2004) Analysis of tieback walls using proposed P-y curves for coupled soil springs. Comput Geotech 31:443–456

Naylor DJ (1981) Finite Element and Slope Stability. Num Meth Geomech. In: Proceedings of the NATO Advanced Study Institute, Lisbon, Portugal, pp 229–244

PLAXIS (2005) Program manual, version 8, PLAXIS Inc

SUNEX (2002) Program manual, 8th edition for Ver w5.3, Geogroup Inc

Won J, You G, Jeong S, Kim S (2005) Coupled effects in stability analysis of pile–slope systems. Comput Geotech 32:304–315

Zienkiewicz OC, Humpheson C, Lewis RW (1975) Associated and nonassociated visco-plasticity and plasticity in soil mechanics. Geotechnique 25(4):671–689

Chapter 28
Forensics of Pile Foundations Subjected to Earthquake Induced Liquefaction and Lateral Spreading

Gopal SP. Madabhushi

Abstract Recent earthquakes in New Zealand and Haiti have shown the widespread damage suffered by pile foundations owing to soil liquefaction and lateral spreading. Several failure mechanisms of piles in liquefiable soils such as pile buckling and excessive settlement of piles in liquefied soil layers were developed at Cambridge University based on extensive centrifuge experiments. It will be shown that some of these anticipated failure mechanisms of pile foundations are confirmed by the observations in New Zealand and Haiti. The value of dynamic centrifuge modelling in deciphering the pile behaviour through observation of failure mechanisms is established. Further it is proposed that dynamic centrifuge modelling can be a valuable tool for forensic engineering in the area of geotechnical earthquake engineering.

Keywords Pile foundations · Liquefaction · Lateral spreading · Centrifuge modelling · Failure mechanisms

28.1 Introduction

Earthquakes often result in extensive damage to infrastructure. As such post-earthquake surveys act as a main tool to investigate the extent and types of damage caused by a particular earthquake. Forensic engineering is the main motivation behind such surveys and the aim is often to learn any deficiencies in our designs or identify lack of understanding either in soil behaviour or soil-structure interaction. In this paper the main focus will be on pile foundations in liquefiable soils. Two post-earthquake surveys, one in New Zealand and the other in Haiti will be considered in detail. The failure mechanisms for pile foundations observed in these

G.SP. Madabhushi (✉)
Department of Engineering, University of Cambridge, Trumpington Street, Cambridge CB2 1PZ, UK
e-mail: mspg1@cam.ac.uk

V.V.S. Rao and G.L. Sivakumar Babu (eds.), *Forensic Geotechnical Engineering*, Developments in Geotechnical Engineering, DOI 10.1007/978-81-322-2377-1_28

earthquakes will be identified. Recent research at Cambridge will be used to understand and replicate the failure mechanisms observed in these earthquakes. It will be argued that dynamic centrifuge modelling is a powerful tool that can be used in forensic engineering to replicate known failures or to anticipate future problems particularly when dealing with difficult soil conditions or unfamiliar loading scenarios.

28.2 Christchurch Earthquakes of 2010/11

Christchurch is the second major city in New Zealand. It was rocked by two powerful earthquakes in September 2010 and February 2011. Both events led to ground liquefaction, however, the second event caused much more widespread liquefaction and resulted in extensive damage to hundreds of buildings both in the city centre and the suburbs of the city. The extent of liquefaction during the February 2011 event was investigated by Cubrinovski and Taylor (2011) who produced the map of liquefaction damage based on a drive-through reconnaissance as shown in Fig. 28.1.

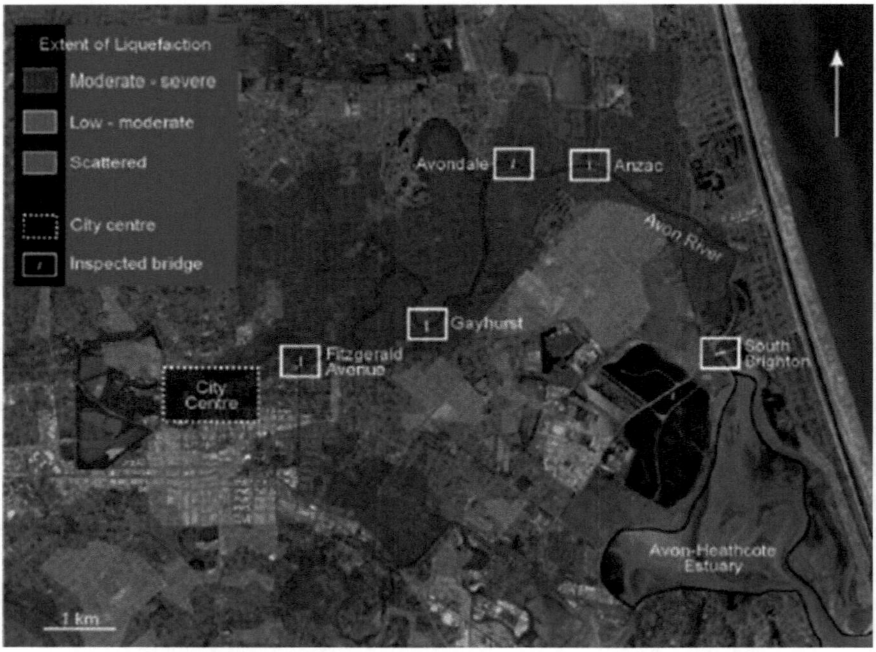

Fig. 28.1 Liquefaction extent in Christchurch, Cubrinovski and Taylor (2011)

On the ground extensive liquefaction-induced sand boils were evident in all the zones identified in Fig. 28.1. An example of a sand boil is presented in Fig. 28.2. When buildings are situated on liquefiable soils they have suffered settlement and/or rotation. An example of a single storey residential structure that suffered rotation is presented in Fig. 28.3. These structures were supported on a ring beam founded on short piles that penetrate about 1 m below the ground surface. Following liquefaction the piles were able to settle into the ground and any differential settlements between piles resulted in the rotation of the building. Such widespread liquefaction has also induced lateral spreading of even gently sloping ground. An example of lateral spreading of an embankment adjacent to the Avon River is presented in Fig. 28.4.

Fig. 28.2 Sand boils due to soil liquefaction

Fig. 28.3 Settlement/rotation of a residential building

Fig. 28.4 Lateral spreading following soil liquefaction

When infrastructure such as retaining walls or bridge abutments are present in lateral spreading ground they attract additional loading that can cause their failure. This was also observed following the Christchurch earthquake as illustrated in Fig. 28.5 that shows a failed retaining wall due to underlying lateral spreading towards the river on the right-hand side. The most significant failures in this earthquake due to the liquefaction and lateral spreading were the piles supporting bridge abutments. There were many examples of these and only a few will be considered here. In Fig. 28.6 the 'back rotation' of an abutment wall of a bridge is presented. The lateral spreading of the backfill towards the river caused the bridge abutment to suffer back rotation by a few degrees. This mechanism of failure occurred due to the high axial stiffness of the bridge deck with very little scope for axial shortening due to the lateral spreading of the river banks. The RC square piles supporting the abutment have failed in shear as seen in Fig. 28.6.

Similar observations were made in the case of a bridge abutment supported on steel piles. In Fig. 28.7 the back rotation of a bridge abutment supported on steel H piles is presented. Again lateral spreading of the backfill towards the river caused the abutment to suffer rotation inducing large shear and moment loads on the pile heads. The bridge deck has high axial stiffness and therefore did not allow any axial shortening of the bridge. Steel H piles supporting the abutment are quite strong in shear and therefore they failed by forming plastic hinges at the pile heads. However, the overall failure mechanism for this bridge abutment is similar to the previous case of abutment on RC square piles.

A final example of pile failure following lateral spreading is presented for the case of a bridge supported on battered pile groups. This was a substantial highway bridge but again the axial stiffness of the bridge deck was high due to its integral construction across spans with very little provision of axial shortening, i.e. no

Fig. 28.5 Failure of a retaining wall

expansion joints were present. This meant that the lateral spreading of backfill caused the back rotation of the bridge abutment as seen in Fig. 28.8. The abutment was supported on battered piles which failed by forming plastic hinges at the pile heads. This was evident by the spawling of concrete seen at the back of the front leading pile. Also closer inspection revealed tension cracks at the front of the leading pile.

Liquefaction-induced lateral spreading was the main reason for the back rotation of the bridge abutments. The integral bridge construction with high axial stiffness of the bridge decks and no provision for axial shortening has contributed to this failure mechanism. For all the three bridges presented here, the failure mechanism seems to be common despite the different types of pile foundations that were used in each case. The failure of the pile foundation itself was determined by the type of pile, i.e. an RC square pile failed in shear, while the steel H pile formed a plastic hinge and battered pile group failed by forming a plastic hinge at the pile heads. As mentioned before, only three bridges are considered here, although there were several other bridges with similar type of failure mechanism.

Fig. 28.6 Lateral spreading induced abutment rotation and shear failure of a square RC pile supporting the abutment

Fig. 28.7 Back rotation of a bridge abutment supported on steel H piles that formed plastic hinges at pile heads

Fig. 28.8 Back rotation of a bridge abutment supported on battered pile group

28.3 Haiti Earthquake of 2010

On 12 January 2010, a powerful earthquake with a moment magnitude Mw = 7.2 hit Haiti that has caused extensive damage to civil engineering infrastructure and led to a death toll of nearly 250,000. This earthquake led to an unprecedented effort of using satellite and pictometric imaging to evaluate both the scale of damage and in damage assessments, Saito et al. (2010). While the detailed analysis of such imagery is beyond the scope of this paper, specific use of high-resolution pictometric and pre- and post-earthquake imaging to assess the liquefaction damage at the port in Port-au-Prince will be considered in this paper. More details of the EEFIT mission to Haiti that was undertaken with support from EPSRC, UK and support from Institution of Structural Engineers, London can be found at http://www.istructe.org/knowledge/EEFIT.

Soon after the earthquake, it became clear that extensive liquefaction has occurred in the vicinity of the main port in the capital city, Port-au-Prince. Excessive settlement of container crane structure as shown in Fig. 28.9 made the use of port to receive emergency relief from rest of the world impossible. This situation highlighted the importance of designing critical facilities such as ports and harbours to withstand earthquake loading.

Comparison of the pre- and post-earthquake images reveals the extent of damage to the port facilities as shown in Figs. 28.10 and 28.11. Comparing these two satellite images it can be seen that several sections of the south wharf have collapsed and the entire section of the north pier has collapsed.

In addition to the satellite images high-resolution pictometric images were also available for the port. Pictometric images are aerial photographs taken in-flight akin to the street view images of Google Maps. These offer an oblique view of the target

Fig. 28.9 Settlement of container crane in Port-au-Prince

Fig. 28.10 Pre earthquake satellite image of the port

Fig. 28.11 Post-earthquake satellite image of the port

and give a perspective of damage suffered by the structures. An example of these pictometric images is shown in Fig. 28.12 that clearly depicts the damage to container crane seen in Fig. 28.9.

In addition to the above aerial imagery, the EEFIT field mission was also able to obtain valuable data on performance of the port structures. The island seen near the top left-hand corner in Fig. 28.12 has suffered excessive settlement following liquefaction. As a result, connecting bridge and wharf structures supported on pile foundations showed severe damage. In Fig. 28.13 the damage suffered by the bridge is shown. As the pier structure and the approach abutment on the island side settled, the bridge decks cracked to allow for these settlements. This caused the 'self-articulation' of the bridge. Similarly, a wharf structure suffered excessive damage as the pile foundations supporting the wharf settled into liquefied and laterally spreading soil. This is shown in Fig. 28.14 which depicts the formation of plastic hinges at the pile heads.

Fig. 28.12 Pictometric image showing the damaged crane structure

Fig. 28.13 Settlement of piers cause self-articulation of the bridge deck

Fig. 28.14 Hinging of piles supporting a wharf structure

Again this is similar to the formation of plastic hinges at the top of pile heads that was discussed in Sect. 28.2 during the Christchurch earthquake. This type of failure mechanism was predicted by Madabhushi et al. (2009a) as shown in Fig. 28.15, before either of these two earthquake events, based on a series of extensive centrifuge tests that were conducted to understand settlement of piles into liquefied sand overlying dense sand strata. More details of this study are discussed later in this paper.

Comparing Figs. 28.14 and 28.15 and previous cases of bridge abutments in Sect. 28.2, it can be concluded that the failure mechanism of piles in liquefiable soils derived from dynamic centrifuge tests was validated from the observations following the Christchurch and Haiti earthquakes.

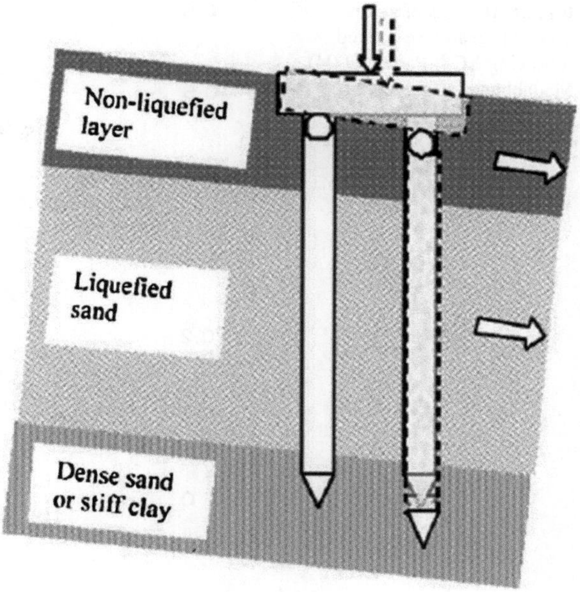

Fig. 28.15 Failure mechanism of hinging due to pile settlements, Madabhushi et al. (2009a)

28.4 Current Design Practices

In earthquake geotechnical engineering the liquefaction potential of a given site is first established. The maximum credible earthquake that is possible at the site may be obtained from the seismic zonation maps for the region. The peak ground acceleration ($a_{\max}$) that can occur may be determined for the maximum credible earthquake by using an appropriate attenuation relationship, for example Ambraseys et al. (2005). The cyclic shear stress τ_e generated by this earthquake can be calculated following Seed and Idriss (1971) as;

$$\tau_e = 0.65 \frac{a_{\max}}{g} \sigma_0 r_d \tag{28.1}$$

where σ_0 is the total stress and r_d is a stress reduction factor. Field test data from either SPT or more preferably CPT tests can be used to establish the liquefaction potential of the site. The corrected SPT values (N160) can be used, for example. In Fig. 28.16 the liquefaction lines that demarcate 'liquefaction' and 'no liquefaction' are shown for three types of sands, following Eurocode 8 Part 5 (2004) as re-plotted by Madabhushi et al. (2009b). It is preferable to obtain CPT data compared to SPT testing. If CPT data is available for the given site, then similar relationships can be found between CPT data and the liquefaction potential, for example as proposed by Robertson and Wride (1998).

Fig. 28.16 Liquefaction potential charts, re-plotted following EC8: Part 5 (2004), Madabhushi et al. (2009b)

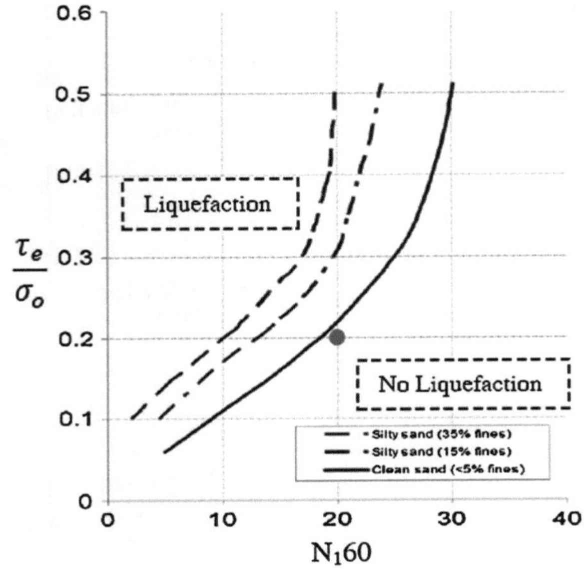

While progress has been made in the area of determining whether a given site is liquefiable there are several uncertainties associated with soil liquefaction. For example, if the cyclic shear stress value from maximum credible earthquake and the corrected SPT value for a layer give points very close to liquefaction potential lines, then it is not clear whether liquefaction should be expected or not. For example if the cyclic stress ratio (CSR) is 0.2 and the N160 is 20 for clean sand site, such a point will lie close to but below the liquefaction potential line. However, good engineering judgement requires that significant liquefaction be expected at such a site. So even when the N160 data suggests that the site will fall into 'no liquefaction' category, significant excess pore pressures can be generated during sufficiently strong earthquake events. In such events, the structures found on the site can suffer excessive settlement and/or rotations brought about by lowering of the effective stress and a consequent degradation in soil stiffness. Further details of this argument can be found elsewhere, Madabhushi and Haigh (2012).

It is both important and interesting to understand the behaviour of structures on liquefiable soils. With increased pressure on land, the geotechnical engineers have to look at designing foundations of structures on sites which may be susceptible to liquefaction. Also geotechnical engineers may have to design retrofitting measures to existing historic structures that may be located on liquefiable soils as is the case in southern European countries like Italy and Greece. A large EU funded project called NEMISREF that has concluded recently was aimed at developing novel liquefaction resistance measures for existing foundations. (For further details, please see http://www.soletanche-bachy.com/nemisref and Mitrani and Madabhushi (2008)).

In order to investigate earthquake geotechnical engineering problems, it is necessary to conduct dynamic centrifuge tests. These tests can be conducted on small-scale physical models in the enhanced gravity field of the centrifuge in which earthquake loading can be simulated. The technique is now well established and produces very valuable insights into failure mechanisms of foundations without the need to wait for large earthquake events to happen. Similarly non-linear FE analyses can be carried out to investigate these problems, but it is essential that such analyses include realistic soil models and incorporate the soil plasticity accurately.

In this paper the emphasis will be on the use of dynamic centrifuge modelling as a forensic tool to investigate failure mechanisms. After a brief introduction to the technique itself, the paper will present the case of pile behaviour in liquefiable soils and how centrifuge modelling helped to clarify the possible failure mechanisms of pile foundations.

28.5 Dynamic Centrifuge Modelling

Physical modelling in earthquake engineering with reduced scale experiments in the centrifuge is now widely considered as 'the' established experimental technique of obtaining data in controlled conditions to help engineers and researchers to understand the mechanisms involved in the response of soil-structure systems to seismic loading. This experimental approach recreates the stress state in soils which is a fundamental requirement to observe realistic soil behaviour. In Fig. 28.17 a view of the 10 m diameter Turner beam centrifuge at Cambridge is presented.

The success of earthquake geotechnical engineering at Cambridge depended to a large extent on the simple mechanical actuators that have been used for more than 30 years. First attempts of centrifuge shaking tables were mechanical 1D harmonic devices based on leaf spring device (Morris 1979), bumpy road tracks (Kutter 1982).

The current earthquake actuator at Cambridge relies on stored angular momentum (SAM) to deliver powerful earthquakes at high gravities was developed and is in operation for 14 years, Madabhushi et al. (1988). In Fig. 28.18 the front view of the SAM actuator is shown, while in Fig. 28.19 a view of the SAM actuator loaded

Fig. 28.17 The Cambridge 10 m diameter Turner beam centrifuge

Fig. 28.18 A view of the SAM earthquake actuator

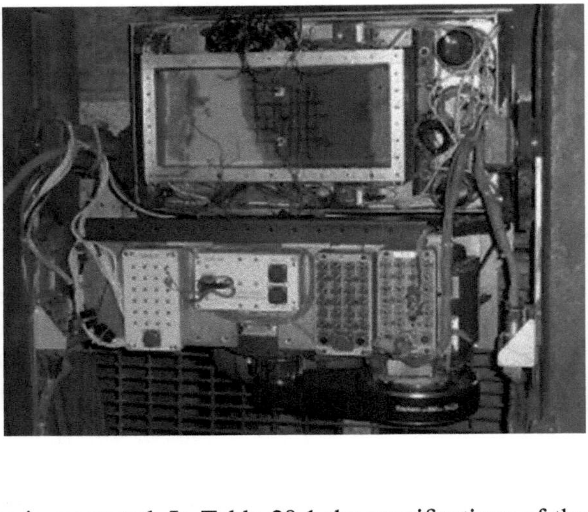

Fig. 28.19 Plan view of a centrifuge model of sloping ground with square and circular piles loaded on the centrifuge

onto the end of the centrifuge is presented. In Table 28.1 the specifications of the SAM actuator are presented. The model seen in Fig. 28.19 was from an investigation carried out by Haigh and Madabhushi (2005), on lateral spreading of liquefied ground past square and circular piles.

The SAM earthquake actuator is a mechanical device which stores the large amount of energy required for the model earthquake event in a set of flywheels. At the desired moment this energy is transferred to the soil model via a reciprocating

Table 28.1 Specifications of SAM actuator

Parameter	Value
Maximum g-level of operation	100 g
Dimension of the soil models	56 m (L) × 25 m (B) × 22 m (H)
	80 m (L) × 25 m (B) × 40 m (H)
Earthquake strength of choice	Up to 0.4 g of bed rock acceleration
Earthquake duration of choice	From 0 to 150 s
Earthquake frequency of choice	From 0.5 to 5 Hz
	Swept sine wave capability

Note All parameters above are in prototype scale

rod and a fast acting clutch. When the clutch is closed through a high pressure system to start the earthquake, the clutch grabs the reciprocating rod and shakes with an amplitude of ±2.5 mm. This is transferred to the soil model via a bell crank mechanism. The levering distance can be adjusted to vary the strength of the earthquake. The duration of the earthquake can be changed by determining the duration for which the clutch stays on. Earthquakes at different frequency tone bursts can be obtained by selecting the angular frequency of the flywheels.

With the capabilities of the SAM actuator and the geotechnical centrifuge facilities described above, it is possible to investigate a wide range of earthquake geotechnical engineering problems including soil liquefaction. In this paper, the particular example of pile behaviour in liquefiable soils will be considered. It will be demonstrated how dynamic centrifuge modelling can help understand the failure mechanisms involved. Further it will be shown how the thinking about pile behaviour in liquefiable soils has evolved as more information is deciphered from the valuable centrifuge test data, culminating in development of simplified procedures for estimating settlements of pile groups.

More recently, a new servo-hydraulic earthquake actuator was developed to enhance earthquake testing capabilities at Cambridge, Madabhushi et al. (2012). This device is able to complement the capabilities of SAM actuator with the ability to simulate desired realistic earthquake motions such as a Kobe earthquake motion or Northridge earthquake motion. This device was custom-built to adapt to the existing services and configuration of the 10 m Turner beam centrifuge shown in Fig. 28.20. Its maximum operational g-level is limited to 80 g. The specifications of this actuator are presented in Table 28.2.

An early example of the performance of the new servo-hydraulic earthquake actuator is presented in Fig. 28.21. The shaker was commanded to perform a Kobe motion and it responded satisfactorily as seen in Fig. 28.21. The higher frequency components still need further amplification as can be seen in the Fourier transform of the demand and achieved traces in Fig. 28.22.

Fig. 28.20 A view of the new servo-hydraulic earthquake actuator developed at Cambridge University

Table 28.2 Specifications of the new servo-hydraulic actuator

Parameter	Value
Maximum g-level of operation	80 g
Dimension of the soil models	56 m (L) × 25 m (B) × 22 m (H)
	80 m (L) × 25 m (B) × 40 m (H)
Earthquake strength of choice	Up to 0.6 g of bed rock acceleration
Earthquake duration of choice	From 0 to 80 s
Earthquake frequency of choice	From 0.5 to 5 Hz
	Realistic earthquake motion capability

Note All parameters above are in prototype scale

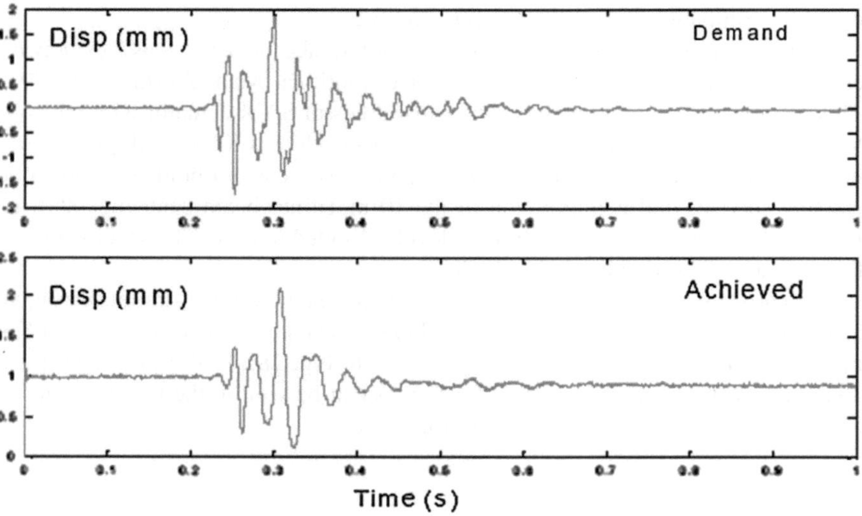

Fig. 28.21 Simulation of the Kobe motion

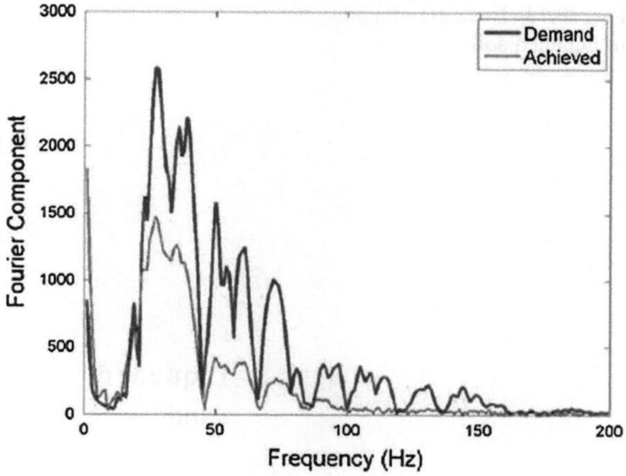

Fig. 28.22 Fourier spectra of the demand and achieved traces

28.6 Modelling of Single Piles in Liquefiable Soils

Piles are long, slender members that carry large axial loads as they transfer superstructure loads into competent soil strata. In normal soils the slenderness is not an issue, as lateral soil pressures support the pile in the radial directions. However, in soft soils such as marine clays, piles can suffer buckling failure. This problem of pile buckling in soft clays has been investigated earlier by many researchers, for example Siva Reddy and Valsangkar (1970). In sands the problem of pile buckling does not arise as large lateral pressures that offer lateral support to the pile can be easily generated. However, when the sandy soils are loose and saturated, they can suffer liquefaction. Under those circumstances piles can lose all the lateral support and can suffer buckling failure as illustrated in Fig. 28.23 by forming a plastic hinge close to the base of the pile. Madabhushi et al. (2009a) describe several possible failure mechanisms for single piles. It must also be pointed out that the piles need to be 'rock socketed' at the base so that no settlement of the pile itself is possible. This type of base condition can occur when piles are transferring the load from the ground surface onto the bed rock.

This mechanism of failure for single piles was evaluated using dynamic centrifuge modelling by Bhattacharya et al. (2004). Tubular model piles made from aluminium alloy were placed in loose, saturated sand. The pile tips were fixed to base of the model container allowing no displacement or rotation. Axial load on these single piles was modelled using brass weights. These loads were calculated using Euler's critical load formula for slender columns given by;

$$P_{\mathrm{cr}} = \pi^2 \frac{\mathrm{EI}}{(\beta L)^2} \tag{28.2}$$

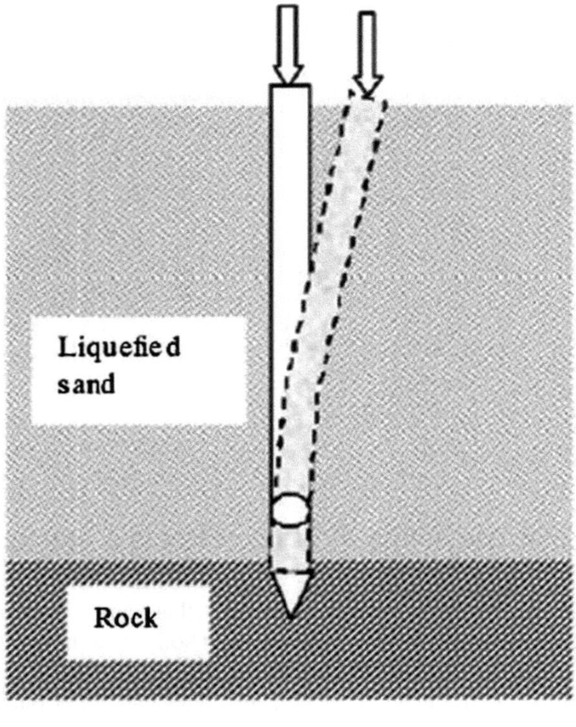

Fig. 28.23 Buckling of a pile in liquefied soil

where *EI* is the flexural rigidity of the pile, *L* is the length of the pile and β is an effective length factor that depends on the end fixity conditions. For the case of pile that is fixed at the base and free at the top, β is 2. For a pile group, where both ends of the pile may be considered as fixed, β is 0.5.

The centrifuge model was tested with 50 g and subjected to a base acceleration of 0.2 g at the bedrock level. This caused the loose sand to liquefy fully which was confirmed by excess pore pressure measurements as shown in Fig. 28.24. The piles have suffered buckling failure as seen in Fig. 28.25. After the test the model piles were extracted from the soil model to examine the location of plastic hinge formation. As seen in Fig. 28.26, the plastic hinging occurred at approximately ¼ of the length below the pile head.

Based on these series of tests, it can be concluded that pile buckling occurs in liquefied soils for the case of single piles carrying axial loads close to the Euler critical loads. Any imperfections in the pile will of course bring down the Euler critical loads significantly. However, single piles are very rarely employed in geotechnical design. One such case is the Showa Bridge foundations in which the bridge piers were extended below the water level and supported on rows of single piles. Based on the evidence of the centrifuge test data, Bhattacharya et al. (2005) argued that the failure of the Showa bridge during the 1964 Niigata earthquake could be attributed to buckling of the piles.

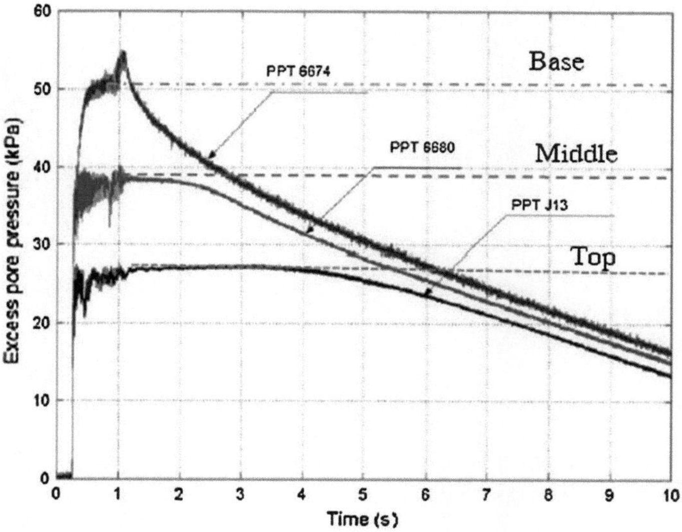

Fig. 28.24 Excess pore pressure time histories at base, middle and top of the sand layer (dashed lines indicates ru = 1)

Fig. 28.25 Centrifuge model showing buckling of three piles in liquefied soils

Interesting as the above results and their implications are, a few queries arise. Firstly, compared to Fig. 28.23 in which the plastic hinging in the pile occurs close to the base of the pile, the location of the plastic hinge in the centrifuge test occurs quite a way up as seen in Fig. 28.26. As plastic hinges form when the maximum bending moments exceed the plastic moment capacity of the pile section, clearly the liquefied soil below this level must be offering some support to the pile. This may be due to the monotonic shearing of the liquefied soil demanded by the deforming pile causes dilation and therefore a temporary reduction of the excess pore pressure.

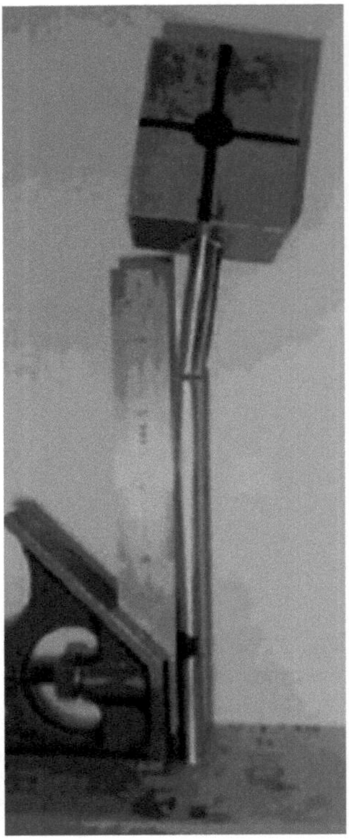

Fig. 28.26 Buckled pile

This can cause the liquefied soil to gain some strength temporarily and offer resistance to pile buckling. As a result the pile cannot buckle at the base but can only do so higher up, where the soil dilation is compromised due to inflow of water from surrounding soil.

Secondly, would pile buckling be an issue for pile groups as the effective length of a pile in a group can be significantly smaller due to the end fixity conditions (fixed at base and top and therefore $\beta = 0.5$ in 28.2). This aspect is considered next.

28.7 Pile Groups in Liquefied Soils

Piles are most often used in groups with a pile cap connecting all the pile heads. As a result there is considerable fixity of pile heads that provide resistance to rotation. In addition the pile heads are all forced to translate together. Madabhushi et al. (2009a, b) describe several possible failure mechanisms for pile groups by forming plastic hinges. Typical examples for liquefiable level ground are shown in Fig. 28.27 with four-hinge and three-hinge mechanisms.

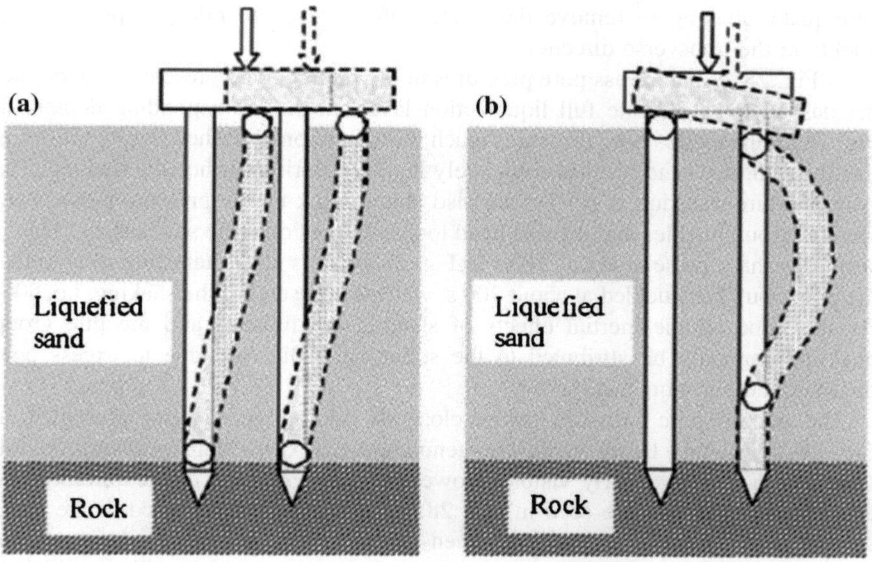

Fig. 28.27 Possible failure mechanisms for pile groups

In order to investigate the failure mechanisms for pile groups, a series of dynamic centrifuge tests were undertaken as described by Knappett and Madabhushi (2009a). As with the single piles described in Sect. 28.6, 2 × 2 model pile groups made from aluminium alloy were tested in loose saturated sands. A typical cross-section of the centrifuge models is shown in Fig. 28.28. It must be noted that the brass weights on the pile caps are restrained in the direction of

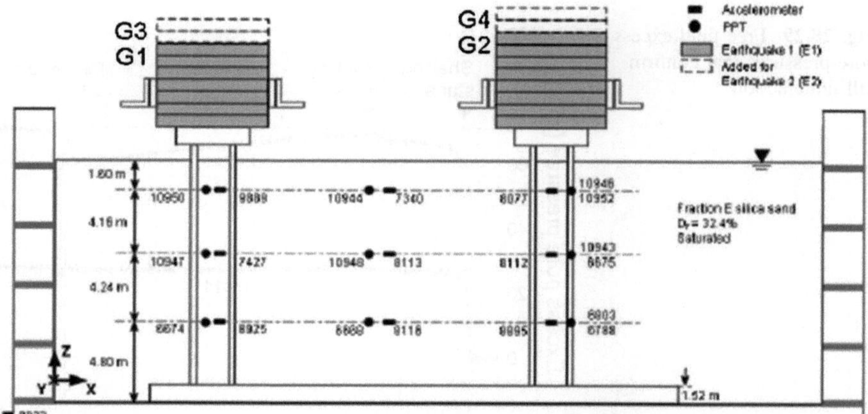

Fig. 28.28 Typical cross-section of the centrifuge model of the pile groups

earthquake shaking to remove the inertial effects, i.e. the pile groups can only buckle in the transverse direction.

In Fig. 28.29 the excess pore pressures in the free field are presented. As before, the dashed lines indicate full liquefaction levels at the corresponding depths. In Fig. 28.29 the excess pore pressures reach these horizontal dashed lines confirming that the saturated sand bed has completely liquefied during earthquake shaking. The pore pressure transducers (PPTs) are also able to pick up the pressure pulse when the pile group buckles and the pile head load collapses onto the soil surface. This is seen as a sharp pulse at about 210 s in Fig. 28.29. It is also interesting to note that the pile group has buckled at about 200 s, well after the end of the shaking, i.e. 60 s. By this time all the inertial effects of shaking are finished and the pile group buckling can only be attributed to the softening of the soil due to excess pore pressure and soil liquefaction.

The excess pore pressure traces close to pile group G4 are presented in Fig. 28.30. In this figure again the generation of excess pore pressures to full liquefaction level is clearly visible. However, during shaking cyclic variations of excess pore pressures are seen in Fig. 28.30. This may be attributed to the shear stresses induced during the shaking when the liquefying soil tries to shear past the piles. These shear stresses imposed on liquefied soil cause it to dilate, which is manifested as a drop in excess pore pressure and a temporary stiffening of the soil. PPT 10952 shows a progressive dilation on top of the cyclic variations as the pile group suffers increased deformations until the end of the earthquake at 60 s. After this the excess pore pressures recover as the pore fluid migrates from the free field into the zone next to the pile group. This softens the soil in the region next to the pile group and this softening continues until 200 s when the pile group finally buckles. Again the two PPTs are able to pick up the impulse from the pile head load impacts onto the soil surface.

Fig. 28.29 Free field excess pore pressures that confirm full liquefaction

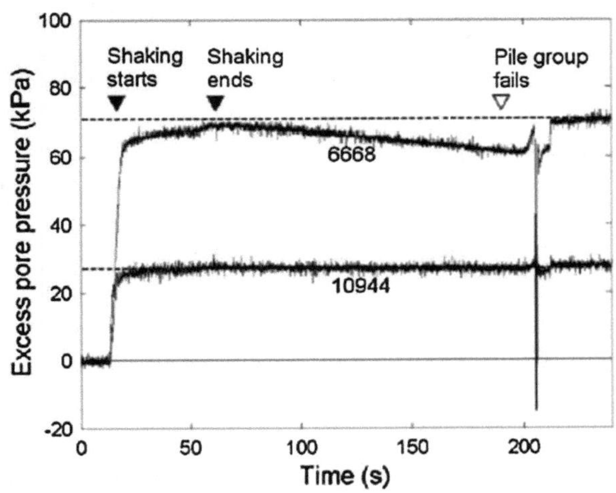

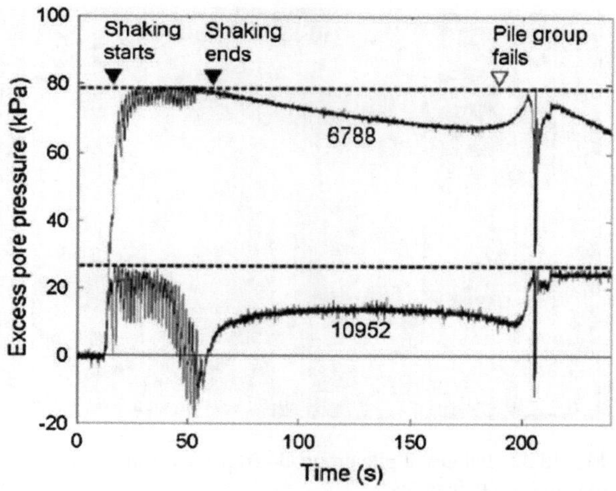

Fig. 28.30 Excess pore pressures close to the pile group G4

The lateral displacement of the pile group G4 is measured using long stroke LVDTs. The pile head displacement (in the transverse direction) with time is presented in Fig. 28.31. In this figure the evolution of lateral displacement of the pile group is clearly seen. However, the LVDT runs out of range just before the end of the earthquake shaking at about 58 s.

A view of the centrifuge model in plan is shown in Fig. 28.32. In this figure it can be seen that the pile group G4 has fully buckled and collapsed onto the side of the model container. The other pile group G1, carrying smaller axial load has also suffered significant transverse displacements. The deformed shape of the pile group G4 that has been extracted after the centrifuge test is also shown as an inset in Fig. 28.32. As in the case of the single pile, the pile group also suffered buckling failure due to liquefaction of the soil supporting the piles. However, the location of the plastic hinges is somewhat different from the anticipated positions for plastic hinges shown in Fig. 28.27a that are based on sway frame mechanisms. The

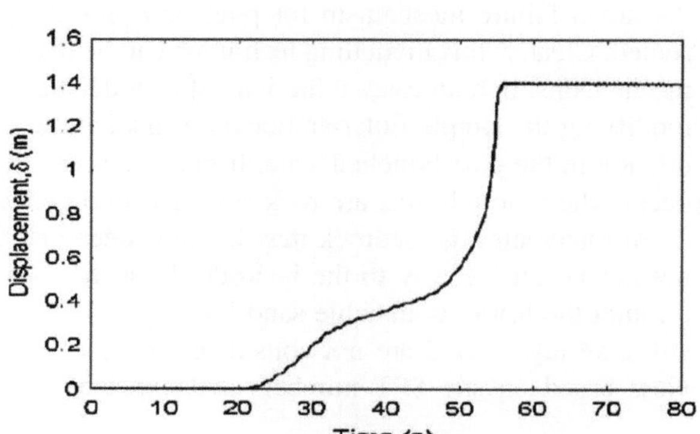

Fig. 28.31 Lateral displacement of the pile group G4

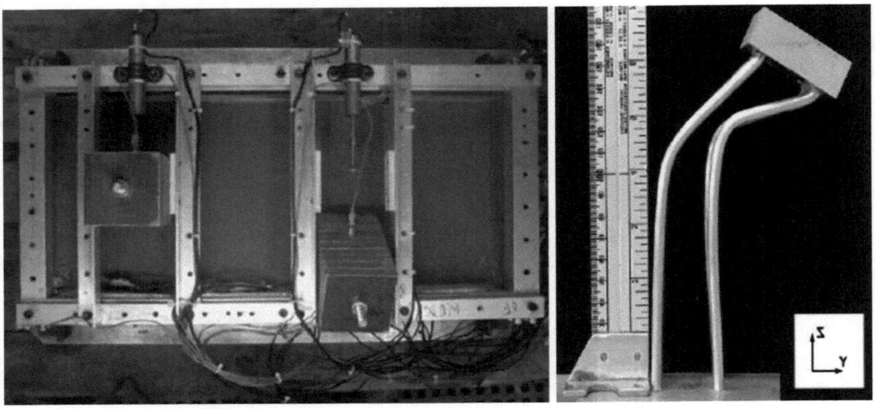

Fig. 28.32 Failure of pile group G4 (*right-hand side inset* shows the extracted pile group after the dynamic centrifuge test)

location of the plastic hinges in the centrifuge model are much higher suggesting that the liquefied soil is able to offer significant resistance to pile group buckling. Due to this resistance, the plastic hinges form much higher compared to the fully constrained pile tips. Knappett and Madabhushi (2009b) have analysed this problem using the finite element code ABAQUS that can capture large strain deformations in the post-buckled phase using Ritz's algorithm. In addition they used non-linear p-y soil springs to model the soil surrounding the pile. These FE analyses were able to capture the location of the plastic hinges in the pile group to match those seen in Fig. 28.32.

28.8 Settlement of Pile Groups

In Sects. 28.6 and 28.7, the role of dynamic centrifuge modelling in investigating the novel failure mechanism for piles and pile groups in liquefied soils was presented. Clearly, this modelling technique can be used to verify the perceived failure mechanisms. In both cases it has helped in understanding the role of liquefied soil in modifying the simple Eulerian buckling first anticipated and the importance of soil dilation in the post-liquefied state. It must be noted that this failure mechanism can occur when the pile tips are rock socketed into bedrock.

At many sites the bedrock may be quite deep and the liquefiable soil strata may not extend all the way to the bedrock. In such cases, the piles are usually driven through the loose, liquefiable sand layers into dense sand layers that lie below (or stiff clay layers) and are not considered to be at risk from a liquefaction point of view based on say SPT numbers and expected cyclic stress ratios as shown in

Fig. 28.16. This more common case of pile foundations was investigated using dynamic centrifuge modelling as described by Knappett and Madabhushi (2009c).

A series of centrifuge tests were carried out to investigate the settlement of pile groups in soil strata where a loose, saturated sand layer overlies a dense sand layer. Bedrock is considered to be below the dense sand layer.

The piles are driven through the loose sand layer into the dense sand layer. A pile length of about ten pile diameters (≈10 Do) is embedded into the dense layer to provide sufficient fixity length for the pile in dense sand. A schematic cross-section of the centrifuge model is shown in Fig. 28.33. A typical data set from this series of centrifuge tests for the numbered locations in Fig. 28.33 are presented in Fig. 28.34.

In Fig. 28.34, the input earthquake acceleration is given at the bottom as trace 7. The centrifuge test data show that significant excess pore pressures are developed at the pile tip locations as shown by trace 6 and full liquefaction levels (ru = 1) are reached quickly. It is interesting to note that such large excess pore pressures are generated even in the dense sand layer in which the pile tips rest. The horizontal response of the pile cap is given by trace 1 and this shows that after the first few cycles the pile cap oscillations are reduced and liquefaction sets in and the pile group is isolated from bedrock motion. The vertical settlement of the pile cap is given by trace 2, which shows almost continuous settlement during the whole shaking with an ultimate settlement of about 0.5 m. Traces 4 and 5 show the base and shaft capacities,

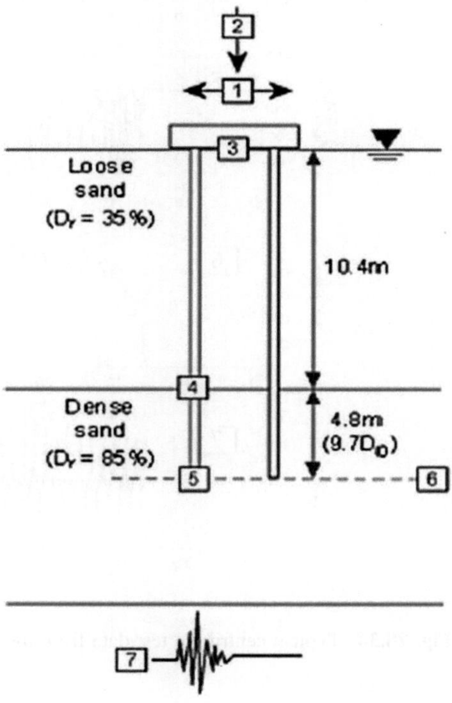

Fig. 28.33 Schematic diagram of centrifuge models with piles founded in layered soils (liquefiable sand layer overlying dense sand layer)

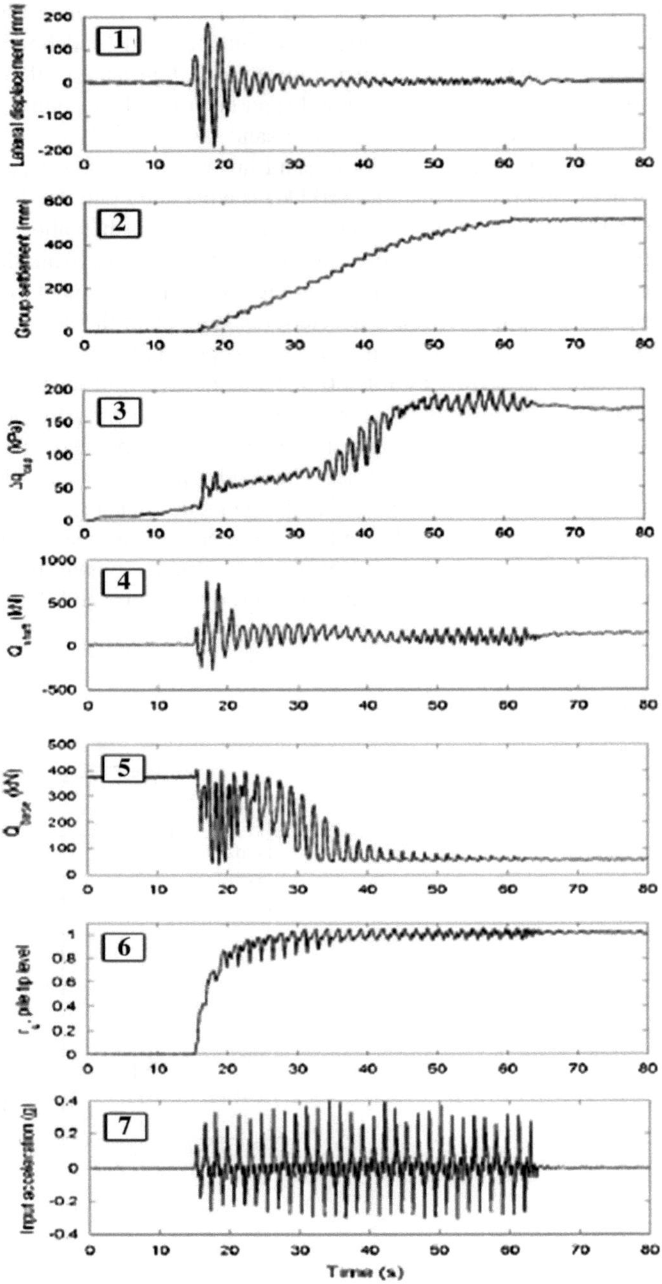

Fig. 28.34 Typical centrifuge test data from the pile group in two layered model

respectively, for one of the piles. As the earthquake loading progresses and liquefaction sets in, it can be seen that the base capacity decreases significantly. The shaft capacity of the pile also reduces. As a result the bearing pressure of the pile cap increases as shown by trace 3. This type of high-quality data and the associated insights into soil behaviour can only be obtained from dynamic centrifuge tests.

Using the centrifuge test data as outlined above, Knappett and Madabhushi (2008, 2009c) have proposed a simplified design procedure for pile groups in liquefiable soils. For example, if the designer wishes to limit the pile settlement in this type of layered soil strata to a given value, they suggest that a minimum factor of safety can be estimated as,

$$\mathrm{FOS} = 1 + A\left(r_{\mathrm{u,\,base}}\right)^{B} \tag{28.3}$$

where $r_{\mathrm{u,base}}$ is the excess pore pressure ratio at the tip of the pile, A and B are constants. For example, if the designer wishes to limit the pile settlement to 0.1 Do (10 % of pile diameter), then constants A and B are 5.5 and 3.5, respectively. The above equation is shown to provide a very satisfactory fit to the available experimental data. For example, for the case of the pile group settlement to 0.1 Do the available experimental data fits Eq. 28.3 very well as shown in Fig. 28.35, which also includes the data obtained by De Alba (1983).

More complex design cases, such as pile groups in liquefied and laterally spreading soils are considered extensively by Madabhushi et al. (2009b). This methodology for design considers the pile group settlements that a designer can allow based on the superstructure tolerance for settlement and other criterion. Of course it is necessary to be able to estimate the amount of excess pore pressure that will be generated during the design earthquake at the base of the pile. This can be done from site response analysis programmes such as CYCLIC-1D or more research based but generalised finite element codes like SWANDYNE, Chan (1988).

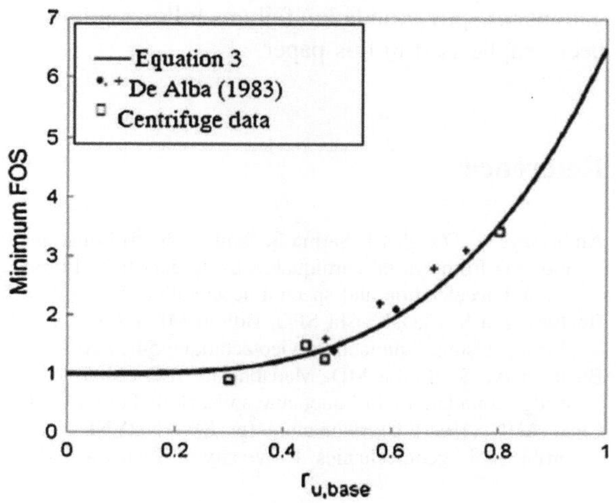

Fig. 28.35 Minimum factor of safety for a given liquefaction ratio at the base of the pile group

28.9 Conclusions

Powerful earthquakes continue to cause severe damage to pile foundations, particularly when they are located in liquefiable soil strata. The Christchurch earthquakes of 2010/11 and the Haiti earthquake of 2010 provided their share of examples of liquefaction damage. River crossings and bridge structures were found to be particularly vulnerable to liquefaction damage. Examples of damage of bridge abutments supported on three different types of pile foundations from the Christchurch earthquake were presented. Also examples of use of satellite and pictometric images were presented that can be used in rapid damage assessment immediately after the Haiti earthquake was provided. In addition ground-based investigations were very useful in recording the damage to pile foundations supporting bridge piers and wharf structures.

Dynamic centrifuge modelling offers a unique opportunity to gain insights into complex soil behaviour under earthquake loading and can be used as a powerful tool in Forensic Engineering. In this paper the problem of single piles and pile groups located in liquefiable soils is used to demonstrate the usefulness of centrifuge modelling. It was shown that using this modelling technique, failure mechanisms can be identified correctly and as result established ideas can be changed. Buckling of single piles and pile groups was investigated but the centrifuge tests revealed changes from the anticipated classical buckling. As a result, the importance of dilation of liquefied soil on shearing was recognised and incorporated into appropriate numerical models. The knowledge gained with single piles and pile groups located on bedrock was extended for the case of two-layered soil strata. The centrifuge test data again provided valuable insights into the changes in the pile base capacity and pile shaft capacity as the surrounding soil suffered liquefaction. An example of simplified design procedure was presented. This procedure considers the requirement of limiting settlements to satisfy performance-based criteria. The important role of dynamic centrifuge modelling to carry out forensic investigations into pile foundation failures following liquefaction and lateral spreading has been emphasised in this paper.

References

Ambraseys N, Douglas J, Sarma S, Smit P (2005) Equations for the estimation of strong ground motions from crustal earthquakes using data from Europe and the middle east: vertical peak ground acceleration and spectral acceleration. Bull Europ Earthq Assoc 3:55–73

Bhattacharya S, Madabhushi SPG, Bolton MD (2004) An alternative mechanism of pile failure during seismic liquefaction. Geotechnique 54(3):203–213

Bhattacharya S, Bolton MD, Madabhushi SPG (2005) A reconsideration of the safety of piled bridge foundations in liquefiable soils. Soils Found 45(4):13–26

Chan AHC (1988) Users manual for SWANDYNE—A finite element code for generalised problems in geomechanics. University of Swansea, UK

Cubrinovski M, Taylor M (2011) Liquefaction map of Christchurch based on drive-through reconnaissance after the 22 February 2011 earthquake, University of Canterbury

De Alba PA (1983) Pile settlement in liquefying sand deposit. J Geotech Engng 109(9):1165–1179

EN 1998-5 (2004) Eurocode 8: design of structures for earthquake resistance. Part 5: foundations, retaining structures and geotechnical aspects. CEN European Committee for Standardisation

Haigh SK, Madabhushi SPG (2005) The effects of pile flexibility on pile loading in laterally spreading slopes, invited paper, ASCE geotechnical special publication No. 145, ISBN 0-7844-0822-X, pp 24-37

Knappett JA, Madabhushi SPG (2008) Liquefaction induced settlement of pile groups in liquefiable and laterally spreading soils. ASCE J Geotech Geo-Environ Eng 134(11):1609–1618

Knappett JA, Madabhushi SPG (2009a) Influence of axial load on lateral pile response in liquefiable soils. Part I: Phys Model Geotech 59(7):571–581

Knappett JA, Madabhushi SPG (2009b) Influence of axial load on lateral pile response in liquefiable soils. Part II: Numer Model Geotech 59(7):583–592

Knappett JA, Madabhushi SPG (2009c) Seismic bearing capacity of piles in liquefiable soils. Soils & Found 49(4):525–536

Kutter BL (1982) Centrifugal modelling of the response of clay embankments to earthquakes, PhD thesis, Cambridge University, UK

Madabhushi SPG, Schofield AN, Lesley S (1988) A new stored angular momentum (SAM) based earthquake actuator. In Kimura et al (eds), Centrifuge 98, Tokyo, Japan, Balkema, Rotterdam, pp 111–116

Madabhushi SPG, Haigh SK (2012) How well do we understand earthquake induced liquefaction? Indian Geotech J 42(3):150–160. doi:10.1007/s40098-012-0018-2

Madabhushi SPG, Haigh SK, Houghton NE, Gould E (2012) Development of a servo-hydraulic earthquake actuator for the cambridge turner beam centrifuge. Int J Phys Model Geotech 12 (2):77–88

Madabhushi SPG, Knappett JA, Haigh SK (2009a) Design of pile foundations in liquefiable soils. Imperial College Press, ISBN 1848163622

Madabhushi SPG, Thusyanthan NI, Lubkowski Z, Pecker A (2009b) Design of shallow foundations. In: Elghazouli AY (ed) Seismic design to Eurocode 8. Spon—Taylor and Francis. ISBN 9780415447621

Mitrani H, Madabhushi SPG (2008) Centrifuge modelling of inclined micro-piles for liquefaction remediation of existing buildings. Geomech Geoengineering: an Int J 3(4):245–256

Morris DV (1979) The centrifuge modelling of dynamic behaviour, PhD Thesis, Cambridge University, UK

Robertson PK, Wride CE (1998) Evaluating cyclic liquefaction potential using the cone penetration test. Can Geotech J 35(3):442–459

Saito K, Spence R, Booth E, Madabhushi SPG, Eguchi R, Gill S (2010) Damage assessment of Port-au-Prince using pictometry. In: Proceedings of the 8th international conference on remote sensing for disaster response. Tokyo Institute of Technology, Tokyo, Japan

Seed HB, Idriss IM (1971) Simplified procedure for evaluating soil liquefaction potential. J Soil Mech Found Div ASCE 97(SM9):1249–1273

Siva Reddy A, Valsangkar AJ (1970) Buckling of fully and partially embedded piles. J Soil Mech Found Div ASCE 96(SM6):1951–1965

Chapter 29
Piping Failure of a Metro Tunnel Construction

Wei F. Lee and Kenji Ishihara

Abstract In this paper, the forensic investigation of a subway tunnel construction failure occurred in Kaohsiung, Taiwan is presented. The studied construction failure occurred during a cross-pass tunnel excavation of a shield tunnel construction work of the Kaohsiung Mass Rapid Transit system, and had resulted in severed tunnel collapse and extensive ground failure that even reached to ground surface 30 m above the tunnel depth. In order to investigate such a serious construction failure, particular soil testing programs and geophysical testing methods were carried out to access the possible failure scenario of the investigated case. Information presented in this paper is hoped to be helpful to improve engineers' knowledge for preventing similar construction risks.

Keywords Tunnel construction failure · Resistivity image profiling · Internal erosion · Critical hydraulic gradient

29.1 Outline of the Incidence

Subway construction had been underway in east–west direction in the central part of Kaohsiung. Upon finishing the tunnel construction by the method of earth-balanced shield, the corridor connecting the two tunnels (up-line and down-line) was constructed by means of what is called the NATM method involving open excavation with the help of steel-framed support and injection. Then, a vertical shaft 3.3 m in diameter was excavated to provide a sump for water collection in the middle of the corridor in open dry conditions with the support of the H-shaped circular beams. When the shaft excavation reached a level 4.95 m from the floor of the corridor, a chunk of wet soil tumbled out from the southern wall of the shaft at the bottom.

W.F. Lee (✉)
National Taiwan University of Science and Technology, Taipei, Taiwan
e-mail: weilee@mail.ntust.edu.tw

K. Ishihara
Chuo University, Hachioji, Japan

V.V.S. Rao and G.L. Sivakumar Babu (eds.), *Forensic Geotechnical Engineering*,
Developments in Geotechnical Engineering, DOI 10.1007/978-81-322-2377-1_29

The small collapse was followed by steadily increased outflow of mud water. The amount of water increased minute after minute. This breakage led to undermining the already built-up tunnels, accompanied by an inflow of soil and water from the rips opened at the junction on the ceiling between the corridor and tunnel. This breakage culminated eventually in a large-scale collapse of the tunnel structure involving formation of two cave-ins on the ground surface.

29.2 Description of the Site Conditions

The location of the collapse is shown in Fig. 29.1 and the feature of the cave-in on the ground surface is displayed in Fig. 29.2. A bird's-eye view over the cave-in is displayed in Fig. 29.3. The plain and side views of the tunnel are shown in Fig. 29.4. It is to be noted that there was an underground roadway called Chung Cheng underpath just above the subway tunnel. The dual tunnels were constructed by the tunnel boring machine (TBM) which can advance by rotating a large steel disk equipped with cutting blade, while the cutting face is balanced by the mud pressure.

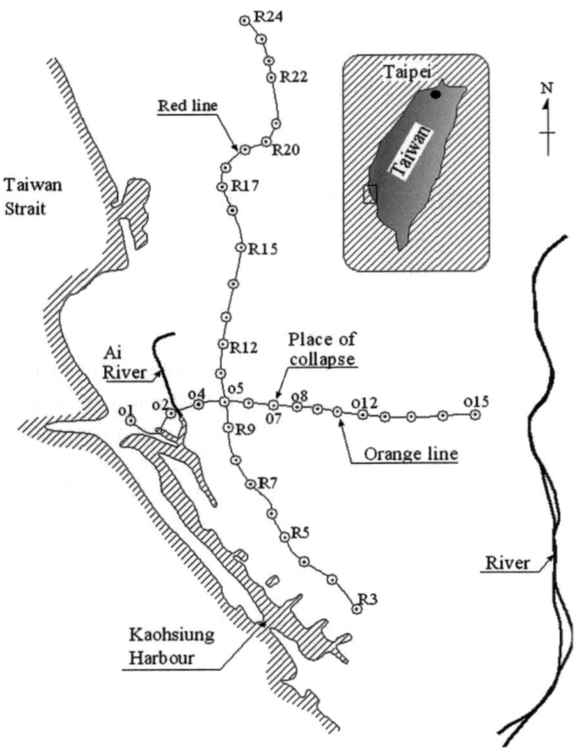

Fig. 29.1 The location of the collapse

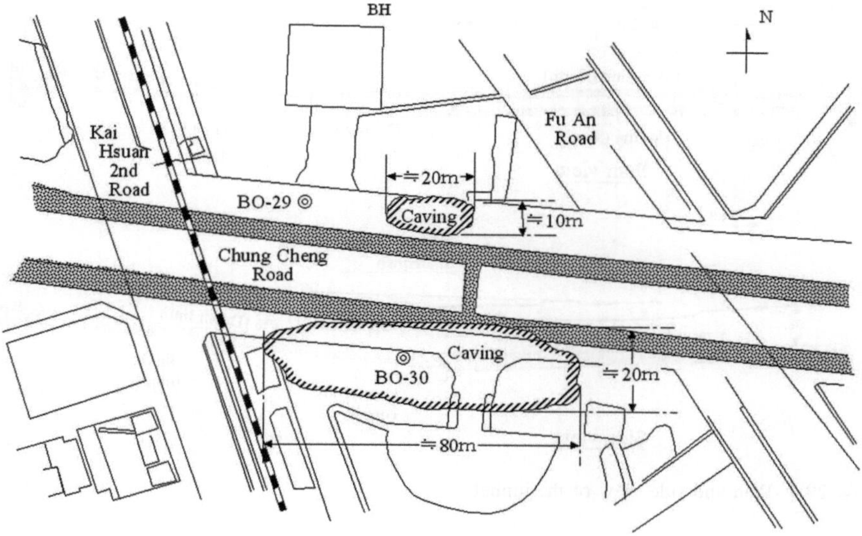

Fig. 29.2 Features of the cave-in in plan view

Fig. 29.3 A bird's-eye views over the cave-in in Kaohsiung

The soil profiles at the locations BO29 and BO30 shown in Fig. 29.2 are indicated in Fig. 29.5, where it can be seen that the deposits are comprised predominantly of silty sand with occasional layers of low-plasticity clay (CL) to a

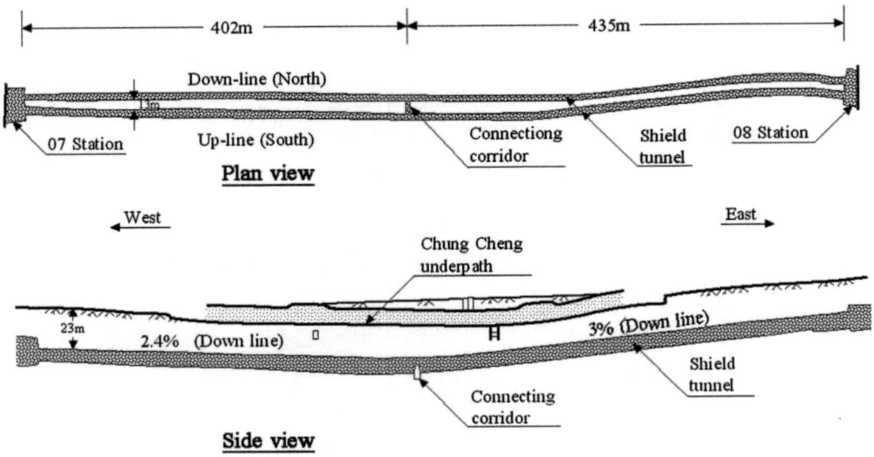

Fig. 29.4 Plan and side view of the tunnel

Fig. 29.5 Soil profile at the site of collapse

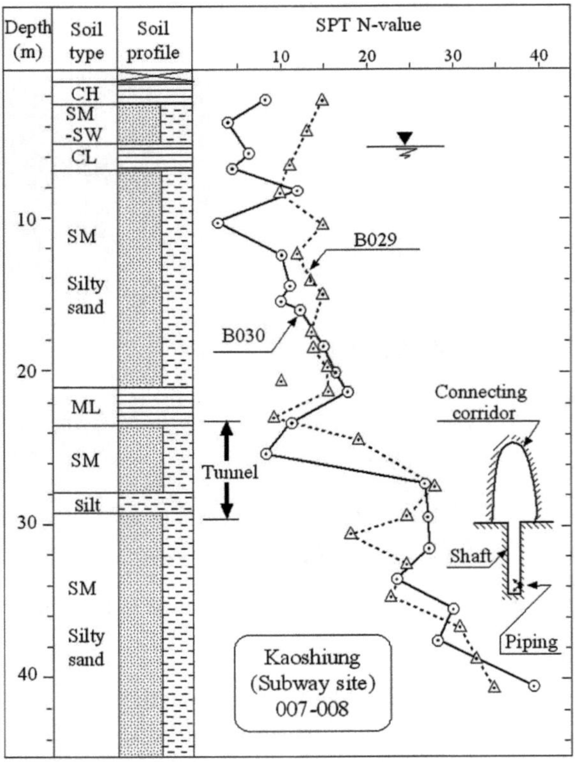

depth of 40 m. The blow count, N-value, in the standard penetration test is shown to increase with depth and to have a value of 20–30 at the depth where the sump for water collection was excavated.

29.3 Immediate Retrofit Work

29.3.1 Ground Stabilization

The first stage of immediate retrofit work is mainly focused on backfilling the sinkholes. Total volume of 9000 m^3 sand, concrete, and aggregates was dumped from ground surface to fill the caves developed from the deep-seated corridor. In addition to backfill, remediation grouting and curtain grouting were also applied from both north and south sides of the car pass to try to stabilize the underground structures and further confine the damaged ground. Figure 29.6 shows the layout and the total injection volume of grout for the immediate work. In the meantime, double-layered concrete walls were also constructed inside the damaged tunnels to

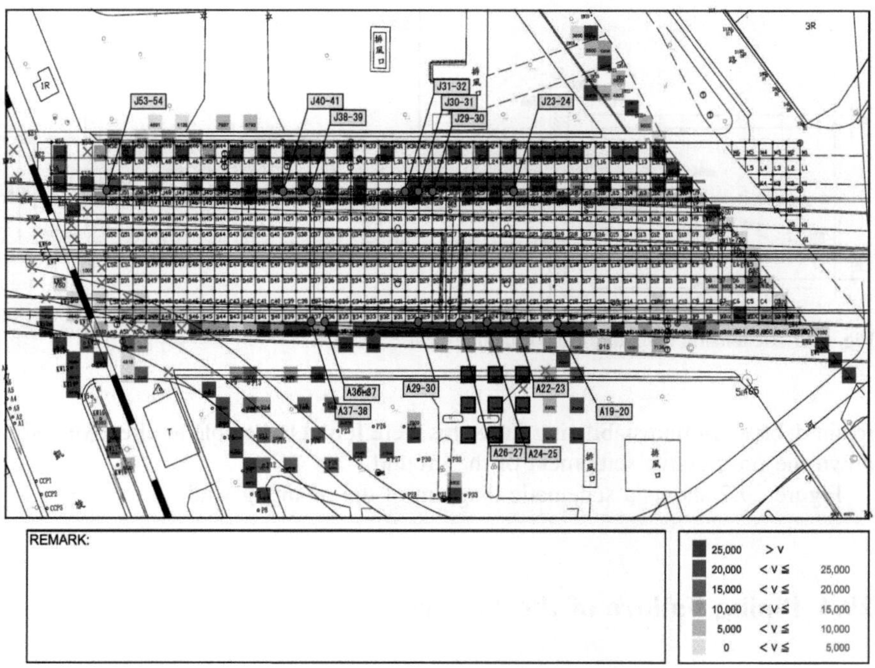

Fig. 29.6 Layout and total volume of grout applied (TCRI 2006)

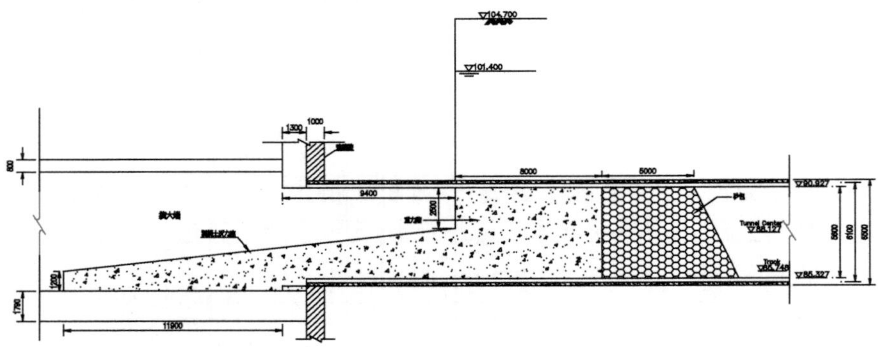

Fig. 29.7 Schematic diagram of the retaining wall

retain debris. Ground stabilizing measures were found to be able to effectively slow down the progressive settlement of the ground.

Figure 29.7 shows a schematic diagram of the retaining wall.

29.4 Piping Failure of the Tunnel

On December 4, 2005, the piping occurred at 3:30 p.m. at the very last stage of excavation of the sump. A block of silty sand was detached from the south wall at the bottom of the excavation which is located at a depth of 35 m from the ground

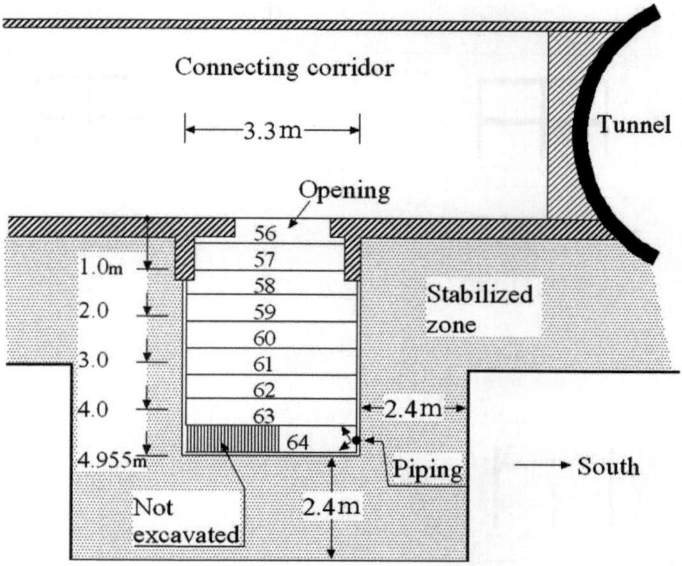

Fig. 29.8 Details of the excavation for the sump and initiation of piping

surface. The circumstance at this time at the bottom of the sump excavation is illustrated in details in Fig. 29.8.

A sequence of the events after triggering the piping which are likely to have occurred is illustrated in the cross sections in Fig. 29.9. The blackish mud water continued to come out in the sump with increasing volume. Two men at work at the bottom strove in vein to clog the hole by dumping sand bags from the corridor. About 1 h later, the sand bags were seen moving around in the sump as illustrated in Fig. 29.9b. At this stage, the workmen closed the entrance to the sump by placing a steel lid and several bars for its support as illustrated in Fig. 29.9a.

It is believed that the mud water flowed into the sump from the bottom through a single hole probably with a diameter of about 30 cm. This assumption would hold true in the light of the fact that the zone stabilized by the jet grouting could not so easily broken down in the horizontal direction, and a weak zone might have existed in the form of vertical pipe.

About 2 h later, engineers at spot heard squeaking sounds of breaks at segment joints of the tunnel on the south, accompanied by mud water cascading from the ripped joints of segments at the ceiling. It appears likely that because of the underscoring of the bottom portion, the stabilized zone subsided together with the tunnel body, resulting in the stepwise settlement in the longitudinal direction. The probable feature of the breakage at this stage is illustrated in Fig. 29.9a.

Around 10:20 p.m., the falloff of the soil above the ceiling had spread upwards and, assisted by the slipping along the vertical wall of the motorway structure, soil mass fell down into the tunnel. This resulted in creation of a large cave-in on the

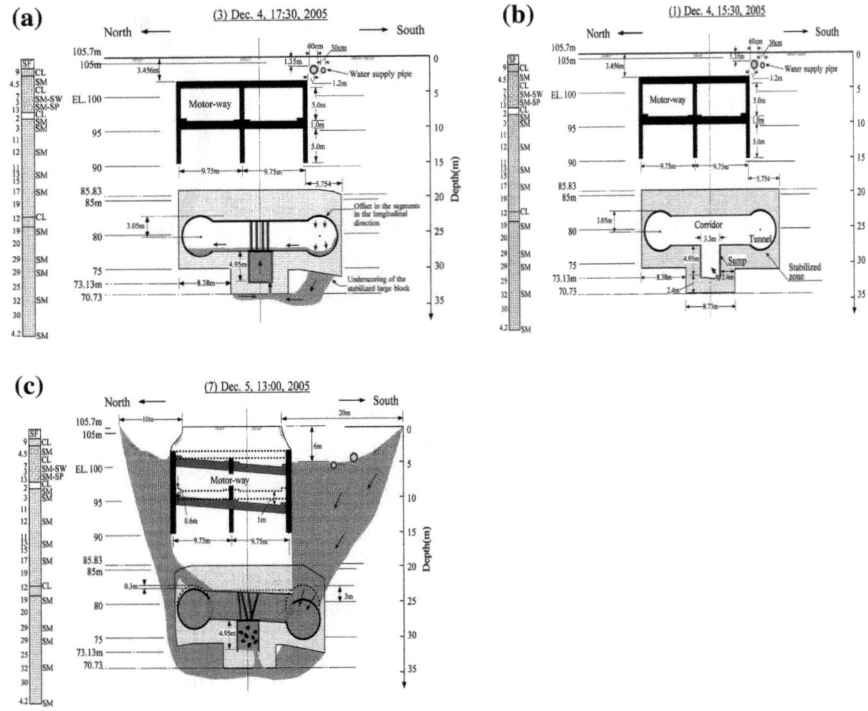

Fig. 29.9 A sequence of likely scenario for progressing the piping failure

ground surface on the south side. At this time, two mains 60 and 30 m in diameter for water supply were broken releasing a large amount of water for a prolonged time. This breakage of buried pipes appears to have transferred the collapsed soil deposits into more flowable debris.

The feature of the collapse propagation as above would be understood more vividly by visualizing the sequence of events that might have occurred in the longitudinal direction. These features are described in Fig. 29.10a–c. Figure 29.10a, b, shows the progress of the collapse from 3:30 to 5:30 p.m. on December 4, 2005. Figure 29.10b indicates the breakage of the tunnel, segment by segment, resulting in the falloff of the debris from the offsets of the segments in the ceiling on the east side. On the north side undermining of the tunnel body did occur as well as shown in Fig. 29.9c, resulting in the sinking of the tunnel together with the large block of the stabilized zone. This led to the breakage of the segment joints on the north side and allows the mud water cascading mainly through the opening at the junction between the portal and the adjacent segment. The length of the tunnel in which the debris packed in the entire cross section was as long as 130 m on the west side and

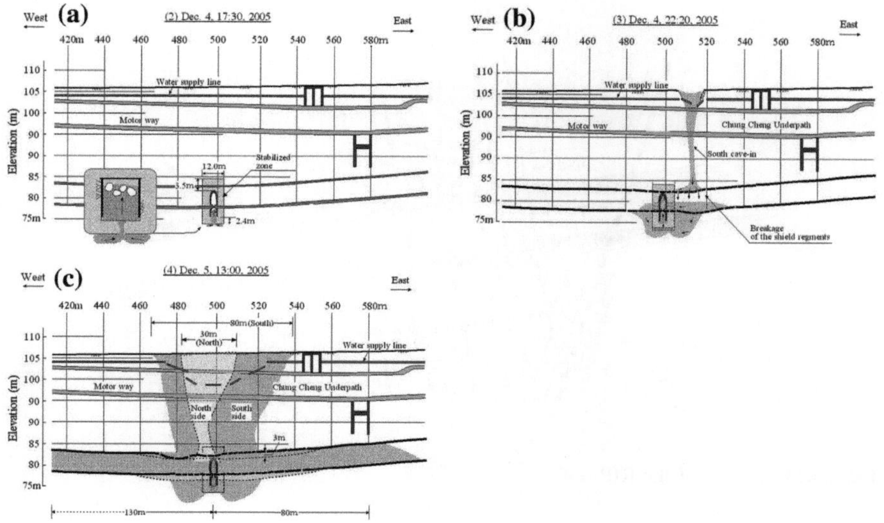

Fig. 29.10 A sequence of likely scenario in the longitudinal direction

80 m on the east side in the up-line on the south, as shown in Fig. 29.13. Thus, another cave-in occurred on the north side, as seen in Fig. 29.2. The amount of soils and concrete dumped to fill the caves was 12,000 m^3 and the volume of water inflow from the 60 cm diameter water main lasting about 18 h until 11:50 p.m. next day was approximately 2,000 m^3. Thus, the total amount of debris was as much as 14,000 m^3.

29.5 Resistivity Profile Imagine (RIP) Testing

29.5.1 RIP Method

In this study, surface 2D RIP testing was conducted to probe ground condition, and to check the effectiveness of remediation grouting right after the accident. Figure 29.11 shows the layout of the 2D RIP testing. As depicted in the figure, six survey lines in north–south direction and one in east–west direction were installed in an attempt to cover the damaged area; and hopefully to identify different soil conditions and underground structures. As described in previous paragraphs, the corridor was surrounded by grout material which was installed to stabilize the ground prior to excavation. Large numbers of sand bags and backfill material were

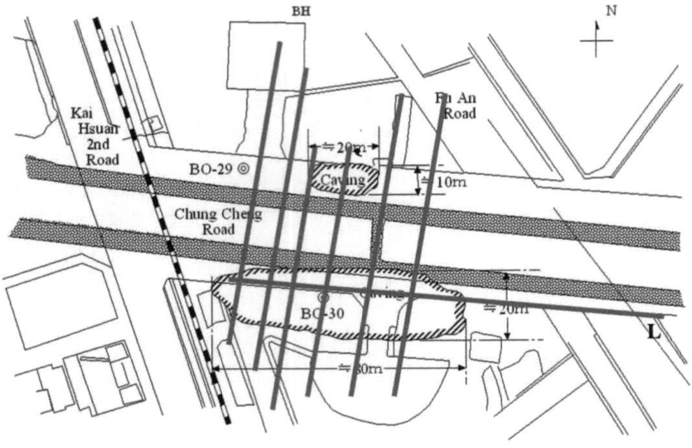

Fig. 29.11 Layout of the RIP test

dumped into the surface sinkholes to stabilize the disturbed ground after the event. Ground conditions were fairly complicated when the 2D RIP survey was conducted.

29.5.2 Test Results

Figure 29.12 shows the results of 2D RIP in cross-sectional profiles. In the figures, cold colors, from green to deep blue, represent low resistivity materials as the disturbed ground with larger volume of groundwater content. On the contrary, warm colors, from yellow to red and brown, represent high-resistivity material such as grout materials or cemented backfill. Underground structures shown in imagines of 2D RIP tend to be distorted from rectangular shapes into upward shell shapes. As depicted in Fig. 29.12, despite the imagine distortion, area in between the underground car pass and the subway corridor were identified as in seriously disturbed condition. At locations of Line L2 and Line E, which are the east and west boundaries of the soil improvement block of the damaged corridor, soil beneath the subway tunnels was also found to be severely agitated by the piping failures. This observation implied that the subway tunnels outside the grouted area might subside and resulted in dislocations of tunnel ring segments. Imagines shown in Fig. 29.12 also indicate that immediate remediation grout treatments have effectively reached both sides of the corridor as displayed in warm color. By interpreting the 2D RIP results, failure schemes shown in Fig. 29.9 were possibly pictured to pursuit causes of the failure. Borehole probing was also conducted at the same time when 2D RIP survey was performed in an effort to validate the scale of the damaged tunnel portion.

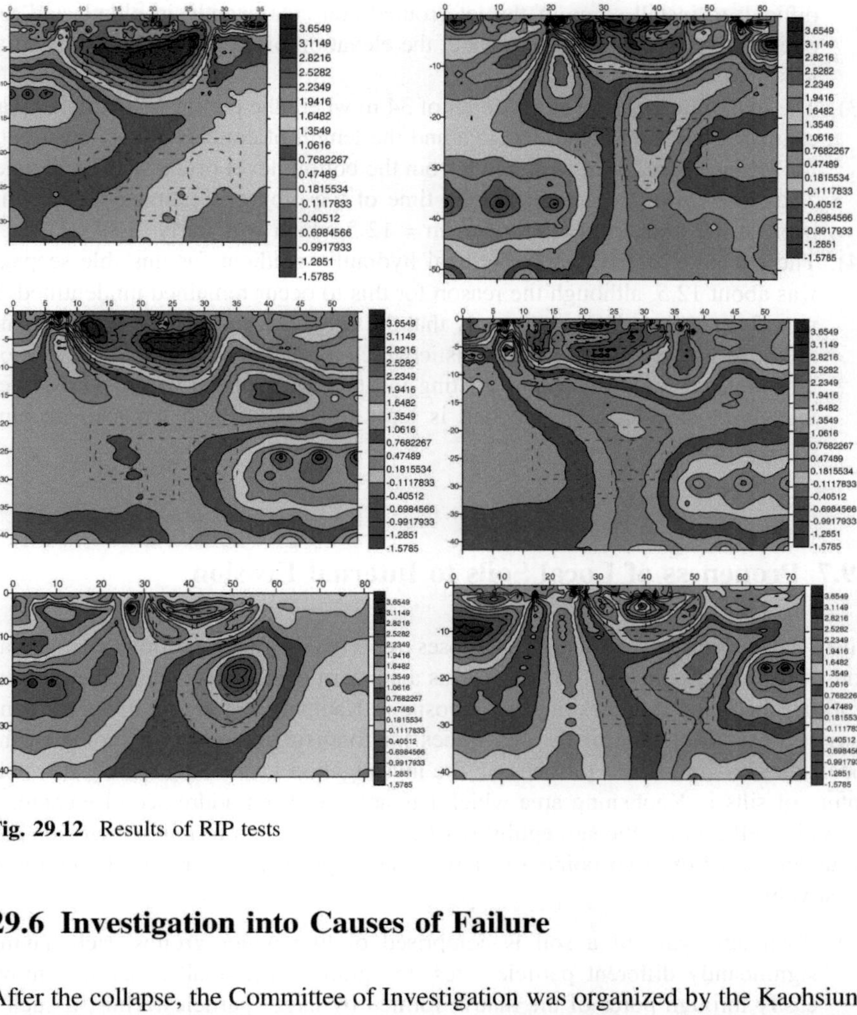

Fig. 29.12 Results of RIP tests

29.6 Investigation into Causes of Failure

After the collapse, the Committee of Investigation was organized by the Kaohsiung Mass Transit Authority and its work commissioned to National Taiwan Construction Institute in Taipei. Even after comprehensive studies and discussions, it appeared difficult to single out definitive causes with proven evidences. However, there were several points agreed upon by the committee which could be summed up as follows:

(1) It was obvious from the testimony of the two men at work that the phenomenon of piping was responsible for triggering the collapse.
(2) The reasons why the piping was initiated were conceived variously. One of them was existence of seepage-prone weak zones resulting from imperfect overlapping in the arrangement of the soil-cement columns created by the jet-grouting method. In fact, because of the obstructions in the ground, it was

difficult to install some of the jet-grouted columns exactly in plumb and the overlapping of adjacent columns at the elevation of the piping is estimated at less than 60 cm.

(3) The hydraulic pressure at the depth of 34 m where the piping was initiated was estimated as being about 300 kPa and the length of conceivable shortest path for seepage was 2.4 m as counted from the bottom level of the stabilized zone. Thus, the hydraulic gradient at the time of the piping is estimated as having been of the order of i = 30 m/2.4 m = 12.5 which was fairly high.

(4) Thus, it is expected that the critical hydraulic gradient for unstable seepage was about 12.5, although the reason for this to occur remained unidentified. It was also suspected rather strongly that the silty sand in the area of Kaohsiung might be possessed of a characteristic which is vastly prone to internal erosion as compared to soil deposits existing in other parts of the world. Thus, once the piping develops, the deposit is least self-healing and tends to become easily unstable.

29.7 Proneness of Local Soils to Internal Erosion

There have been several small collapses reported in other sections of subway construction in Kaohsiung. In fact, it was a concern among geotechnical engineers working there that the fines in sand deposits in Kaohsiung is highly nonplastic and easy to flow. Thus, one of the key issues for forensic diagnosis for piping of the local soil was considered to focus on the investigation into the extremely flowable nature of silts in Kaohsiung area which has not ever been addressed elsewhere.

As is well known, the susceptibility of a given silty sand to internal erosion can be understood from two points of views, that is, generic reason and durability or resistively.

(1) If an aggregate of a soil is comprised of two major groups each having significantly different particle sizes, the grains with smaller size can move easily through pores of the matrix formed by larger particles. Thus, if such a soil with a gap grading is subjected to seepage flow, the smaller particles can be easily detached and washed away. This is the generic concept underlying the criterion for the design of filters in rock-fill dams. However, it is normally applied to the type of soils in a range of particle sizes which are larger than those of silt sands encountered in Kaohsiung area.

(2) The condition as to whether or not a given soil actually suffers the piping collapse is expressed in terms of the critical hydraulic gradient. In fact, the experiments by Skempton and Brogan (1994) showed that the critical hydraulic gradient, ic, for the occurrence of piping in gap-graded sand–gravel mixture could be as low as 0.2–0.3 as against i = 0.9–1.0 for clean sands. This would hold true for soils with the range of particle size coarser than the silty sand in Kaohsiung. To the author's knowledge, there appears to be no study

ever performed to clarify the vulnerability to piping or internal erosion for silty sands or sandy silts subjected to the seepage flow. Thus, this issue was taken up as a new problem area deserving further scrutiny to identify causes of the collapse in deep-seated excavation.

29.8 Laboratory Tests on Internal Erosion

With respect to the piping or erosion, there would be three types of problems to be distinguished depending upon the range of grain size of soil materials in question. These are shown in a graphical form in Fig. 29.13. The segregation piping has been the target of extensive studies in the past in association with the filter criteria for the design of rock-fill dams. There are many studies reported in this context such as those by Terzaghi (1939), Kenny and Lau (1985), Skempton and Brogan (1994) and Sherard et al. (1984a, b).

Erosion in clay–silt materials has been investigated by Sherard et al. (1984a, b) in response to the problem emerging from occasional failure of low-height earth dams which occurred upon water filling in Australia and U.S. The grain sizes associated with erosion or gully formation has been known to lie in the range indicated in Fig. 29.13. It is known that dispersive nature of clayey soils is a dominant factor for inducing gulley or erosion tunnels in the clay fills. Tests for chemical analysis and pinhole tests were suggested as means to identify the dispersive nature of the fine-grained soils.

The outcome of the grain size analyses for the silty sand at Kaohsiung has shown that the soil there belongs neither to the broadly graded gravelly sands for the segregation piping nor to the dispersive clay related to the erosion or gulley formation. In fact, the grain size in Kaohsiung soil lies midway in the range of silt and sand, as shown in Fig. 29.13, which is coarser than the dispersive clay and finer than the soils related with segregation piping.

Thus, the collapse of the silty sand in the subway construction in Kaohsiung was considered to have posed a new challenge and addressed a novel problem area. The solution for this is not yet settled, but performing some tests was regarded as a useful attempt to shed some light on this problem. In this context, a set of pinhole tests and measurements of specific surface test were performed as described below.

(a) *Pinhole Tests*

In order to examine the vulnerability of the silty sands at Kaohsiung to the internal erosion or piping, what might be called "Pinhole test" was carried out. The layout of the test system is displayed in Fig. 29.14. The sample 4 cm thick and 5.17 cm in diameter was sandwiched by highly pervious gravel layers at the top and bottom with filter meshes placed between the sample and gravel. A vertical hole 3 mm in diameter was drilled as shown in Fig. 29.14. Water was then circulated through the sample. In this type of test, water is supposed to flow mainly through the

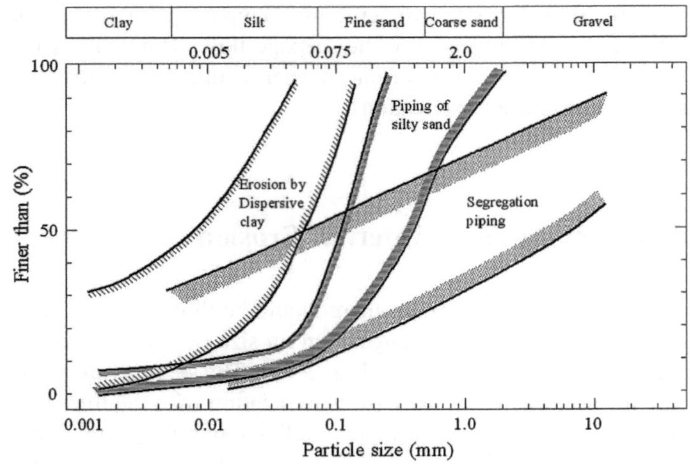

Fig. 29.13 Three groups of soils in terms of grain size range associated with erosion and seepage instability

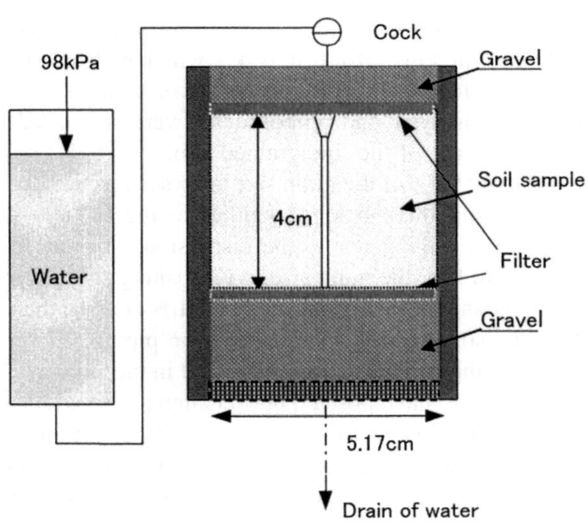

Fig. 29.14 Layout of the pinhole test

pinhole. If the silty portion is erodible, the water coming out is expected to be muddy, but otherwise the water transparent.

The silty sand from Kaohsiung tested had a grain size distribution as shown in Fig. 29.15. For comparison sake, another material from a site in Chiba, Japan was secured and used for the same testing. The grading of this Japanese soil was almost the same as indicated in Fig. 29.15. It is to be noticed that both silty sands had an average grain size of about $D_{50} = 0.075$ mm and fines content was about 50 %. The

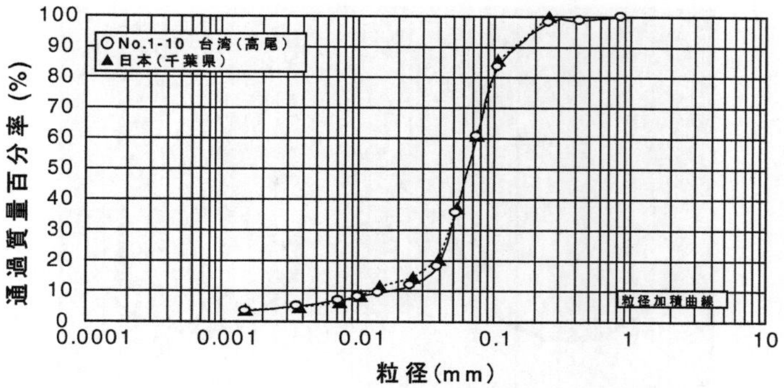

Fig. 29.15 Grain size distribution curves of silty sands used for the pinhole tests

Table 29.1 Grading of two materials used for the pinhole tests

Material	Kaohsiung sand	Japanese sand from Chiba
Dry density ρ_d (g/cm^3)	1.902	1.902
D_{60} (mm)	0.0651	0.0654
D_{50} (mm)	0.0746	0.0743
D_{10} (mm)	0.0168	0.0124
Uc	4.443	2.51
Specific gravity Gs	2.733	2.708

physical characteristics of these two materials are shown in Table 29.1. The two samples were compacted so as to have the same wet density of 1.902 g/cm^3. A sample from Kaohsiung as prepared for the pinhole test is displayed in the middle of Fig. 29.16 and gravel layers to be placed on the top and at the bottom are shown on both sides of Fig. 29.16. The procedures for testing were as follows:

1. Water was circulated first slowly with a low pressure through the sample to ensure saturation.
2. Water was then circulated under a pressure of 98 kPa and color of water coming out of the sample was observed. Pictures were taken every 1 min. Water percolation was continued for 5 min.

In the first series, the tests were conducted on disturbed samples, one from Kaohsiung and another from Japan. The result of the pinhole tests on a disturbed sample is presented in Fig. 29.17. For the disturbed silt containing sand sample from Kaohsiung, the pinhole is seen enlarging to a diameter of 5 mm from 3 mm as displayed in Fig. 29.17a. Correspondingly, the drained water from the bottom was muddy for the first one-minute period of water percolation as seen in the water color in the left-side cup of Fig. 29.17b. The result of the pinhole test on the Japanese silty sand is shown in Fig. 29.18. In contrast to the Kaohsiung soil, the circulation of water through the Japanese soil did not exhibit any appreciable change in color in

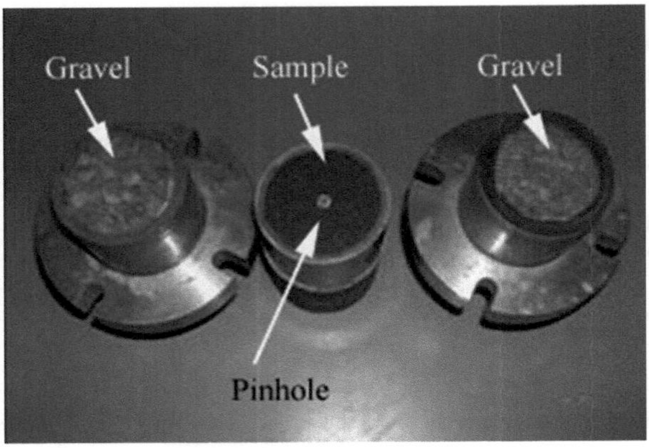

Fig. 29.16 A samples from Kaohsiung for the pinhole test (*the middle*)

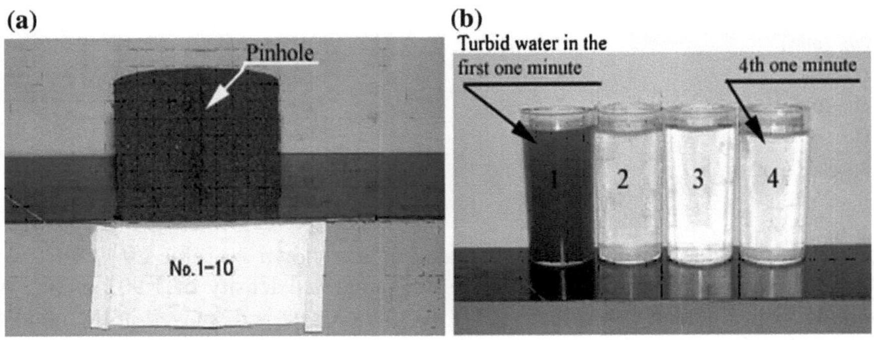

Fig. 29.17 Silt-contained sample from Kaohsiung. **a** Pinhole enlarged from 3 to 5 mm, visible erosion. **b** In the first one minute percolation, water was mud as seen in the *left* cup

Fig. 29.18 The results of the pinhole test on the Japanese silty sand. **a** Diameter of the pinhole remained unchanged except local collapse at the *top*. **b** In the first one minute percolation, water was seen almost transparent as observed in the *left* cup

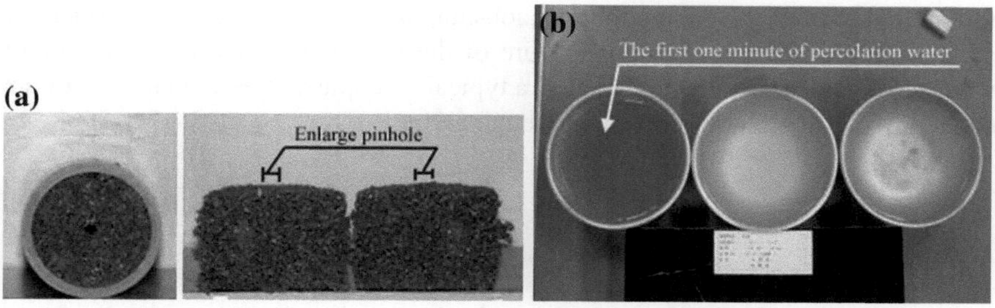

Fig. 29.19 Pinhole tests on undisturbed samples from Kaohsiung. **a** Diameter of the pinhole enlarged from 3 to 4 mm. **b** In the first one minute percolation, water was seen becoming muddy in the *left* bowl

the first 1 min of percolation as seen in Fig. 29.18b. There was no change in the pinhole diameter for the Japanese soil before and after water permeation except for a local collapse at the top. Thus, it may be conclusively mentioned that the silty sand from Kaohsiung is potentially susceptible to internal erosion as compared to the soil from Chiba, Japan, although they had almost the same grading. This is believed to stem from the highly nonplastic nature of the Kaohsiung silt contained in the sand.

In the subsequent series, similar pinhole tests were conducted on undisturbed samples. The undisturbed sample from the depth of 14 m in Kaohsiung also demonstrated dirty blown color in the first 1 min of percolation accompanied by the enlargement of the pinhole from 3 to 4 mm, as displayed in Fig. 29.19.

Two fine sand samples from a deposit at depth 22 m in Kaohsiung were tested likewise with the result that it is also vulnerable to internal erosion.

From the series of the pinhole tests as described above, it may be concluded that, no matter whether the soil is disturbed or undisturbed, the silt ingredient contained in the Kaohsiung sand is not only nonplastic but also highly erodible being susceptible to piping as compared to other silts with the same gradation such as those in Japan.

29.9 Lessons Learned from the Incident in Kaohsiung

In the practice of subway construction in Kaohsiung, it has been known that the silty sand in that area is of peculiar nature being highly susceptible to erosion due to seepage. Although the silt portion is known to be nonplastic, there has been no way further on to scrutinize the nature of the silt. In an effort to grope for some gauge, what is called pinhole test and specific surface test were conducted for the Kaohsiung silt and also for silt from Japan. The results of these tests have shown that the silt from Kaohsiung is more erodible and has a smaller value of specific surface area than the silt in Japan. This fact is indicative of the tendency of the Kaohsiung silt to be more liable to erosion as compared to other silts. The piping failure in the subway construction in Kaohsiung as described above is considered as a consequence of such a peculiar nature of the local soil which had not explored well. This incident is to be regarded as a typical example addressing an issue of new challenge in the area of geotechnical engineering.

References

Kenny TC, Lau D (1985) Internal stability of granular filters. Can Geotech J 22:215–225

Sherard JL, Dunnigan LP, Talbot JR (1984a) Basic properties of sand and gravel filters. J Geotech Eng, ASCE 110:684–700

Sherard JL, Dunnigan LP, Talbot JR (1984b) Filters for silts and clays. J Geotech Eng, ASCE 110 (6):701–718

Skempton AW, Brogan JM (1994) Experiments on piping in sandy gravels. Geotechnique 44:449–460

Terzaghi K (1939) Soil mechanics—a new chapter in engineering science. J Inst Civ Eng (UK) 12(7):106–141

Chapter 30
Forensic Analysis of Failure of Retaining Wall

G.L. Sivakumar Babu, J. Raja, B. Munwar Basha and Amit Srivastava

Abstract In this paper result of forensic analysis of a cantilever retaining wall for road approach embankments which showed distress in the form of translation, vertical settlements and rotation is presented. Extensive geotechnical soil investigation and field measurements of distress are collected. The paper presents prominent causes of failure of cantilever retaining wall using forensic geotechnical investigation. Back analysis of the cantilever retaining wall is performed using classical conventional methods and finite elements analysis. From the results of the conventional analysis and finite element methods, it is observed that the retaining wall designs based on prescriptive guidelines may not lead to satisfactory designs and considerations of deformations are important. Back analysis of failure showed that the mechanism of failure is a combination of sliding and overturning and the deformations are in conformity with the predictions obtained from the numerical analysis.

Keywords Forensic analysis · Retaining wall · Distress patterns

---

G.L. Sivakumar Babu (✉) · J. Raja
IISc, Bangalore, India
e-mail: glsivakumar@gmail.com; gls@civil.iisc.ernet.in

J. Raja
e-mail: jraja@civil.iisc.ernet.in

B. Munwar Basha
Department of Civil Engineering, IIT, Delhi, India
e-mail: basha@civil.iitd.ac.in

A. Srivastava
Department of Civil Engineering, JP University of Engineering and Technology, Guna, MP, India

V.V.S. Rao and G.L. Sivakumar Babu (eds.), *Forensic Geotechnical Engineering*, Developments in Geotechnical Engineering, DOI 10.1007/978-81-322-2377-1_30

30.1 Introduction

Forensic geotechnical engineering is the investigation of the possible causes and processes of distress in a structure that is not able to perform the intended use. Such a critical analysis provides answers to understand the causes of failure. It also gives strong inputs to rehabilitation of structure. The normally adopted standard procedures of testing, design and construction are not adequate for the forensic analysis in majority of cases. The forensic geotechnical engineering differs in many ways from design geotechnical engineering. Generally, design geotechnical engineering involves preliminary investigation, standard field and laboratory investigation, design, construction and evaluation of performance. The forensic investigations involve fresh field and laboratory tests apart from collection of all available data as well as distress measurements. The test parameters and design assumptions in the forensic analysis will have to be representative of the actual conditions encountered at site. The design procedures are mostly stress based, based on limit equilibrium of failure modes such as shear failure, sliding overturning, etc. On the other hand, the forensic analysis has to be deformation based to understand the distress pattern. It often needs to include: collection of data, characterization of distress, development of failure hypothesis, diagnostic tests and back analysis. In the present case forensic analysis of cantilever retaining wall for ROB approach embankment is analysed to determine the possible cause of distress in the structure.

There are many types of rigid retaining walls like gravity retaining wall, semi-gravity retaining wall, cantilever retaining wall and counter fort/buttressed retaining wall. However, the use of cantilever retaining wall in road approaches is popular with some road construction authorities and agencies due to its simplicity in design and execution. In classical conventional methods, the stability of retaining wall is checked for four prominent failure modes like overturning, sliding, basal heave and drainage requirements behind the wall.

30.2 Literature Review

Previous studies done by researchers (Clough and Duncan 1971; Goh 1993) on retaining wall analysis using finite element analysis indicated that horizontal earth pressures acting on the wall are in excess of the active earth pressures and also highlighted the importance of taking into account the wall movement in designing the retaining wall. In the literature the failure of many retaining walls are reported and are attributed due to improper selection of backfill in the retaining wall, temporary rise of pore water pressures due excess rainfall in the active zone increasing loading on the wall and causing the wall to translation or rotation, etc. Some of the common causes of failure are long-term increase in pressure of back fill, cyclic freeze-thaw pressures, water pressure behind retaining wall, compaction pressure.

Most of the failures of retaining wall occurred when the back fill is clay. Clean, free draining, granular sand or granular gravel is recommended as a back fill material. In clayey backfill, swelling pressures, high pore water pressures may substantially increase the thrust on the wall and tends to the failure of the structure. Marsh and Walsh (1996) reported that many cases of distress in retaining wall occurred because the walls were constructed outside the design criteria. He also reported that the retaining wall ceases to perform the intended use as design soil parameters obtained during preliminary soil investigation could be different from the values at the time of construction. Day (1997) presented two case studies of retaining wall failures. For the first case, the retaining wall failed as a result of a reduction in the lateral support. In the second case, the wall was damaged as a result of the increase in the lateral pressure from the earthquake. The above suggest the importance of considering the deformation behaviour of retaining wall using finite element analysis while designing a structure to perform its function over life of the structure using the parameters that are representative of the field conditions. Many researchers studied the failure mechanisms of mechanically stabilised walls (Wu and Chou 2012; Hossain et al. 2011; Scarborough 2005). However, there are limited studies on the forensic geotechnical investigation of cantilever retaining wall and this paper highlights the forensic geotechnical investigation of cantilever retaining wall to determine the possible causes for the distress in the structure. It also presents the importance of using finite element analysis with consideration of deformation behaviour while designing cantilever retaining wall. Forensic geotechnical investigation of retaining walls that were constructed for a bypass road project in India is presented. Extensive geotechnical soil investigation and field measurements of distress are collected. Geotechnical soil investigations are performed on eight bore holes along the centre line of the approach road. The paper presents prominent causes of failure of cantilever retaining wall using forensic geotechnical investigation. Back analysis of the cantilever retaining wall is performed using classical conventional methods and finite elements analysis.

30.3 Distress Patterns

The distresses observed in the field are in the form of lateral movements, vertical settlements and rotations. However, the cantilever retaining wall made of concrete showed no sign of distress or cracks and acted like a rigid structure. The magnitudes of distress patterns of tilt and settlements along the approach road are shown in Figs. 30.1 and 30.2. From Figs. 30.1 and 30.2, it is observed that the lateral displacement of the wall increases as the height of the wall increases. A maximum lateral displacement of 630 mm is observed on the right of east side at 1+208.7 chainage as shown in Fig. 30.3. Figure 30.4 shows visual cracks (more than 60 mm wide) on the flexible pavement due to the excessive lateral displacement and vertical

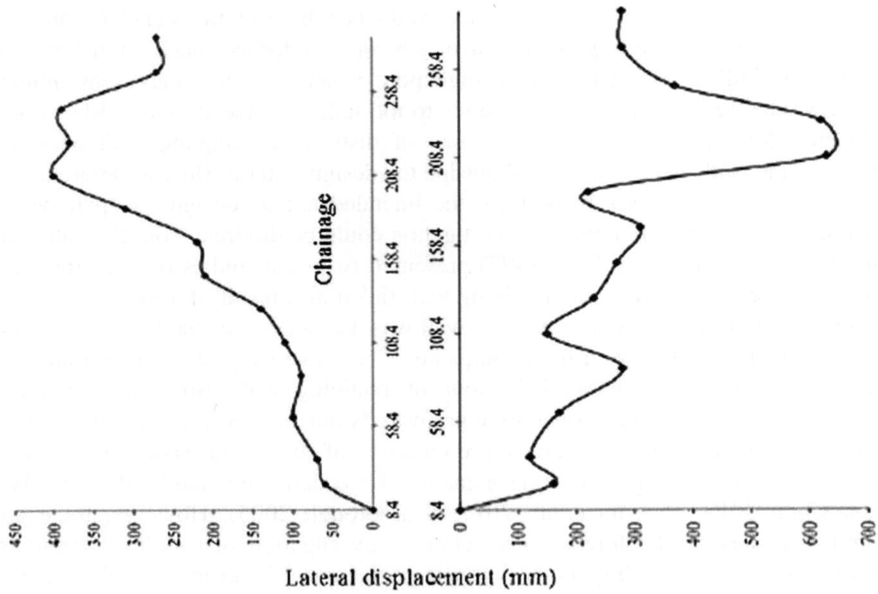

Fig. 30.1 Distress patterns of retaining wall on the *left* and *right* side of east end

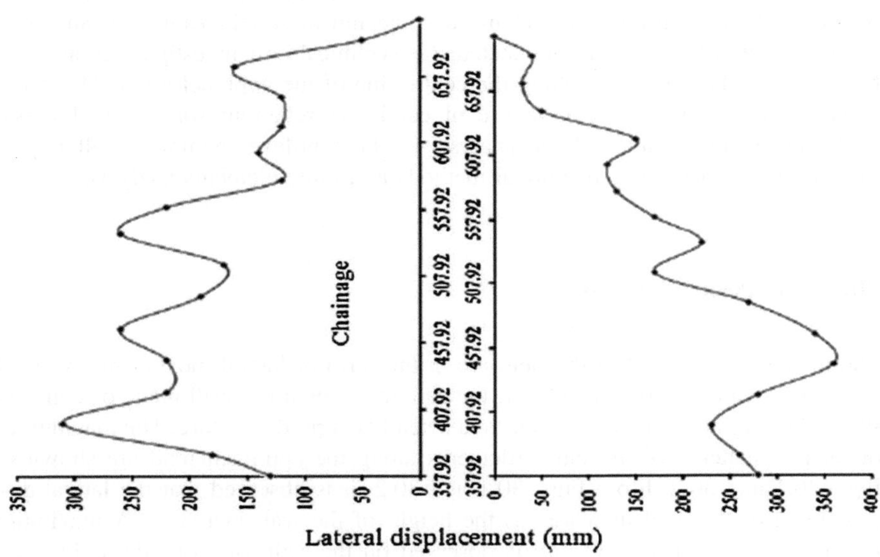

Fig. 30.2 Distress patterns of retaining wall on the *left* and *right* side of west end

Fig. 30.3 Maximum lateral displacement of wall

Fig. 30.4 Distress on the flexible pavement surface

settlement of the embankment fill beneath the flexible pavement. Figures 30.5 and 30.6 show the vertical settlement of the pavement measured with respect to final road level (FRL) on east and west side, respectively. From field measurements of vertical displacement, it is observed that a maximum of 508 mm is observed at chainage 1 +397.9 on the west side. The structure showed clear signs of distress immediately after completion and hence could not be opened for traffic.

30.4 Collection and Interpretation of Test Data

This section describes the initial stage of forensic geotechnical investigation. This stage allows identifying the causes of failure or distress in a structure. In order to carry out back analysis of distressed retaining wall, information is required on the design parameters as well as distress. For this essential data is required for the

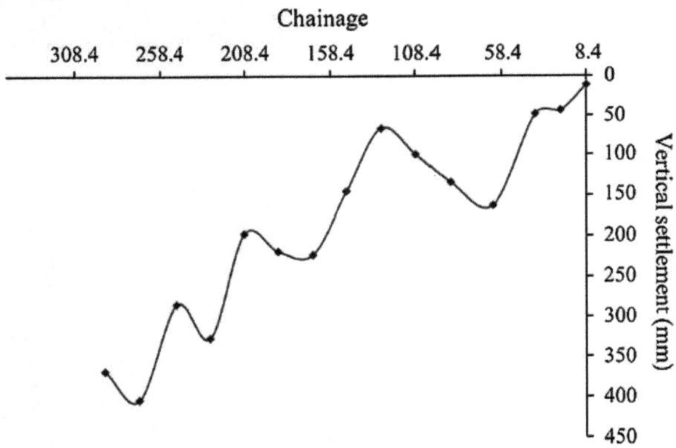

Fig. 30.5 Vertical settlement on east side wrt to FRL

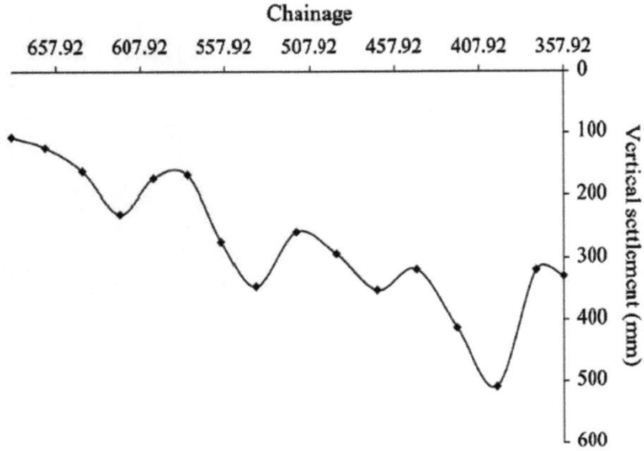

Fig. 30.6 Vertical settlement on west side wrt to FRL

analysis which depends on the amount of distress observed from the field observations. From the amount of field observations of lateral displacements and settlement of the cantilever retaining wall, it is noted that the maximum observed values of distress are located near both sides of the bypass bridge as shown in Figs. 30.1 and 30.2. So, the spacing between the boreholes varied from a minimum of 54 m near bypass bridge to a maximum of 140 m at either ends. Geotechnical properties are determined from eight boreholes along the centre line of the approach road starting from chainage 1+8.4 at east end to chainage 1+688.4 at west end. Extensive geotechnical soil investigation and field measurements of distress are collected. Standard penetration tests, classification tests, grain size distribution tests were performed on the samples collected from eight bore holes. All the boreholes were bored using conventional rotary drilling rigs. Soil samples were collected at every 1.5 m depth of boring. Both disturbed and undisturbed samples were collected from each bore hole. The SPT-samples collected were used as disturbed soil samples. These samples were used for visual and physical identification and for conducting laboratory classification tests. Representative undisturbed samples were collected at depth intervals indicated in Table 30.1. Values of SPT N value (see Table 30.2) for all bore holes greater than 100 are reported as 100. These samples were used to determine bulk densities, cohesion and angle of friction using direct shear test. Summary of soil data of cohesion, angle of internal friction, bulk density and SPT N60 value of the backfill and foundation soil from eight boreholes with depth from FRL is tabulated in Tables 30.1 and 30.2. From the grain size distribution the embankment fill have the properties of silty sand in majority of the portion of the back fill and foundation soil. Tables 30.3 and 30.4 show the limits of the gradation of the backfill material according to FHWA and TxDOT guidelines. Natural moisture content is determined from the samples collected from eight bore holes. From the observation of the variation of natural moisture content in the eight bore holes, it is observed that natural moisture content varies from a minimum of 10 % to a maximum of 18 %.

30.5 Back Analysis

Back analysis provides a means to analyse the failures that frequently occur in various technical fields including geotechnical engineering. It is of prime importance that back analysis must represent the in situ conditions to the extent possible. Choice of back analysis methodology must be based on the technical inputs and data available from the failure site and detailed laboratory investigations. Back analysis was conducted on distressed retaining wall using conventional methods and by performing rigorous finite element analysis using a computational code (Plaxis V8). Various section of retaining wall with different heights (11.18 and 12.4 m) at chaniages 1+228.7 and 1+397.47 were analysed independently. Section 30.1

Table 30.1 Borehole log data

S. no.	Barehnle no.	Depth of sample (m)	Cohesion (kPa)	Angel of internal friction (°)	Bulk density (kN/m^3)
1.	BH-01	1.5	–	–	21.7
		3.0	–	–	23.0
		4.5	–	–	25.0
		6.0	–	–	22.2
		9.0	–	–	22.4
2.	BH-02	1.5	14.0	20.2	19.4
		3.0	–	–	20.7
		4.5	–	–	21.4
		6.0	–	–	20.8
		9.0	–	–	22.3
		10.5	–	–	21.7
		12.0	–	–	22.2
		13.5	–	–	22.5
3.	BH-03	1.5	5.0	23.0	18.8
		3.0	10.0	30.2	19.6
		4.5	10.0	32.0	19.6
		6.0	–	–	19.6
		7.5	–	–	20.4
		12.0	–	–	22.2
		13.5	–	–	20.5
		15.0	–	–	21.9
		16.5	–	–	19.6
		18.0	–	–	22.2
4.	BH-04	1.5	6.0	27.0	18.4
		3.0	12.0	28.90	18.9
		4.5	12.0	30.30	18.9
		6.0	15.0	29.00	18.9
		7.5	–	32.20	19.5
		9.0	16.0	31.00	19.5
		12.0	15.0	22.00	19.9
		15.0	–	–	20.9
		16.5	–	–	21.40
		18.0	–	–	21.4
5.	BH-05	1.5	4.0	23.4	18.7
		3.0	7.0	30.3	19.4
		4.5	12.0	31.00	20.7
		6.0	–	–	21.2
		7.5	15.0	32.00	19.8
		9.0	14.0	30.30	19.9
		10.5	22.0	30.30	29.9
		12.0	20.0	31.00	18.9
		13.5	–	–	21.6
		15.0	–	–	20.9
		18.0	–	–	21.9
		19.5	–	–	20.9

(continued)

Table 30.1 (continued)

S. no.	Barehnle no.	Depth of sample (m)	Cohesion (kPa)	Angel of internal friction (°)	Bulk density (kN/m^3)
6.	BH-06	1.5	4.0	22.30	18.9
		3.0	12.0	27.30	19.6
		4.5	–	–	21.2
		7.5	–	30.30	20.20
		9.0	–	–	21.9
		10.5	–	–	19.6
		13.5	–	–	22.2
		15.0	–	–	22.2
		16.5	–	–	22.4
7.	BH-07	1.5	5.0	31.3	19.4
		3.0	–	–	20.4
		6.0	–	–	20.9
		9.0	–	–	22.2
		10.5	–	–	22.5
		16.0	–	–	21.4
8.	BH-08	1.5	18.0	29.3	19.9
		3.0	–	–	20.2
		4.5	–	–	20.4
		10.5	–	–	22.2

(11.0 m height of wall) has lateral displacement of 380 mm and vertical settlement of 327 mm as shown in Figs. 30.1 and 30.5. Section 30.2 (12.4 m height of wall) has horizontal tilt of 280 mm and maximum vertical tilt of 508 mm at the centre as shown in Figs. 30.2 and 30.6. The following section presents the result of the analysis from conventional analysis and numerical analysis.

30.6 Conventional Analysis

The stability checks of cantilever retaining wall include overturning about the toe, condition of no tension at the base, sliding stability, maximum pressure on the foundation soil and other checks. For the present case two sections (Sects. 30.1 and 30.2) stability of the retaining wall is determined with respect to overturning, sliding and basal heave. The results of the stability checks for Sects. 30.1 and 30.2 are presented in this section. For Sect. 30.1, the height of the wall is 11.18 m with base width of 8 m and a thickness of 0.3 m of stem as the top width. For Sect. 30.2 the height of the wall is 12.4 m with base width of 8.5 m and a thickness of 0.3 m of stem as the top width. The geometry of the Sects. 30.1 and 30.2 are at chainages 1 +228.7 and 1+397.47 on the east end and west end, respectively, are shown in Figs. 30.7 and 30.8. The cantilever retaining wall is constructed with concrete of unit weight 25 kN/m^3. Figure 30.9 shows the prescriptive dimensions of the

Table 30.2 SPT N60 values

Depth of sample below FRL (m)	BH-01	BH-02	BH-03	BH-04	BH-05	BH-06	BH-07	BH-08
2	68	24	9	26	8	11	59	37
3.5	100	76	16	13	18	55	32	45
5	30	100	38	14	36	53	51	34
6.5	100	20	65	26	38	15	42	88
8	100	100	41	29	16	100	29	100
9.5	100	44	32	17	16	26	100	100
11	100	60	100	19	11	69	69	–
12.5	–	100	61	54	100	100	100	–
14.0	–	100	48	61	48	81	100	–
15.5	–	100	100	63	81	100	100	–
17.0	–	–	77	100	100	100	–	–
18.5	–	–	64	44	53	100	–	–
20.0	–	–	100	61	100	100	–	–
21.5	–	–	100	100	100	100	–	–
23.0	–	–	–	100	100	100	–	–

Table 30.3 Gradation limit of backfill (FHWA)

U. S sieve size	Percent passing
102 mm (4 in.)	100
0.425 mm (no. 40)	0–60
0.075 mm (no. 200)	0–15
Plasticity index should not exceed 6 %	

cantilever retaining wall given in Chen (1997) and Das (2002) followed in general in many countries. From the observations of the dimensions, it is noted that Sects. 30.1 and 30.2 follow the prescriptive guidelines for the retaining wall designs.

The soil properties of BH-03 which is close to the chainge 1+228.7 for Sect. 30.1 and BH-05 for Sect. 30.2 are considered for the analysis. The average soil properties considered for the analysis for backfill and foundation soil are presented in Tables 30.7 and 30.8. From the grain size analysis the embankment fill is classified as silty sand. Silty sands will have higher Poisson's ratio (0.3–0.35) when compared to sands and gravel. The low modulus of elasticity and high Poisson's ratio allows the backfill to deform higher than the required to develop active pressures compared to clean sands and gravels. From the results of the conventional methods presented in Table 30.5, it is observed that the safety factors with respect to overturning and basal heave are higher than the critical values specified for Sects. 30.1 and 30.2. However, the safety factor with respect to sliding stability is less than the critical factor of safety 1.5. From the above, it is evident that the retaining wall will suffer excessive lateral displacement due to failure with

Table 30.4 Gradation limit of backfill (TxDOT)

Type A		Type B	
Sieve size	Percent passing	Sieve size	Percent passing
76 mm (3 in.)	100	152 mm (6 in.)	100
0.425 mm (no. 40)	0–60	76 mm (3 in.)	75–100
0.075 mm (no 200)	0–15	0.075 mm (no. 200)	0–15
Plasticity index should not exceed 6 %			

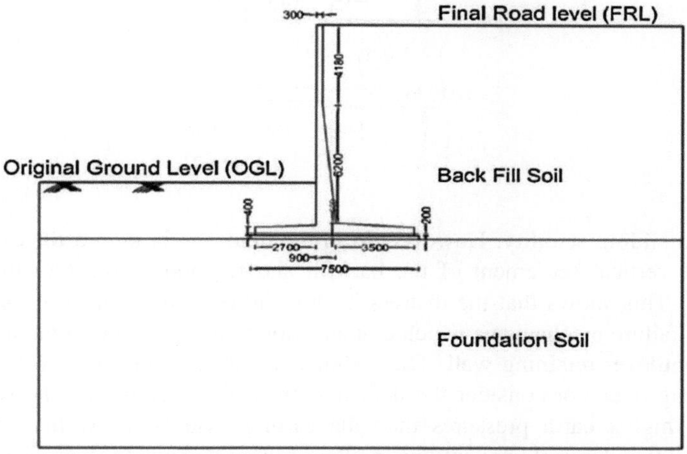

Fig. 30.7 Geometry of Sect. 30.1 at chainage 1+228.7

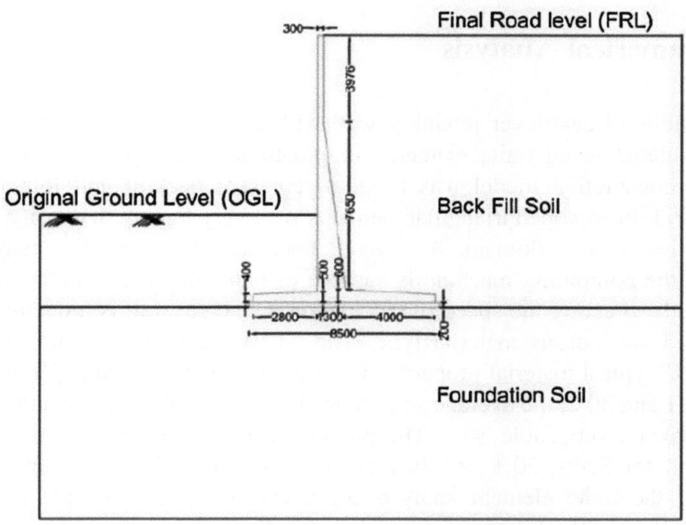

Fig. 30.8 Geometry of Sect. 30.2 at 1+397.47

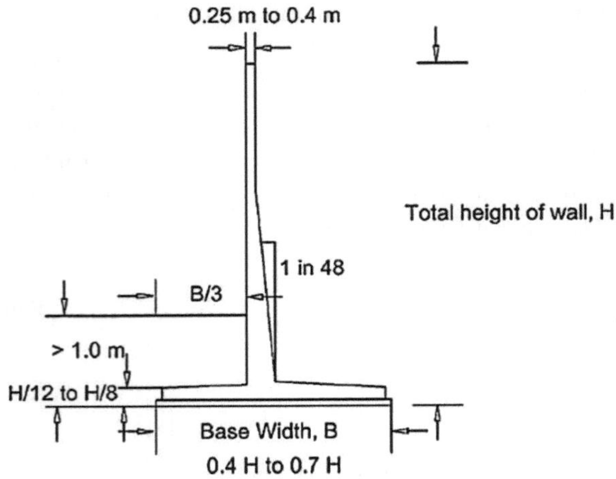

Fig. 30.9 Prescriptive cantilever wall dimensions

respect to sliding stability. However, from the field observation a tilt with combination of vertical settlement of the backfill and rotation of the retaining wall is observed. This shows that the distress in the cantilever retaining wall is a combination of failure mechanisms which contains horizontal tilt, translation and rotation of the cantilever retaining wall. The stability of the retaining wall with respect to overturning does not consider the deformation of the backfill and the wall which develops higher earth pressures than the earth pressures for vertical wall. This shows that the classical conventional method cannot be applied to predict the actual failure modes.

30.7 Numerical Analysis

Two sections of cantilever retaining wall with design height (11.18 and 12.4 m) were simulated using finite element computational code (plaxis V8). The mohr coulomb constitutive model was used to represent backfill and foundation soil behaviour. Fifteen noded triangular elements with very fine mesh density were used to discretise the soil domain. An overall fineness set to very fine may be used provided the computing machine is capable of handling such simulations without significantly affecting the speed of calculation. Concrete wall is modelled as linear elastic and non-porous material type. The unit weight of the concrete is set as 25 kN/m^3. Typical material properties for concrete are presented in Table 30.6. For Sects. 30.1 and 30.2, the average soil properties for backfill material and foundation soil are taken from Table 30.1. The properties of the backfill material and foundation soil for Sects. 30.1 and 30.2 are presented in Tables 30.7 and 30.8. The results of the finite element analysis for Sects. 30.1 and 30.2 are presented in Tables 30.9 and 30.10. Form Tables 30.9 and 30.10, it is observed that the global

Table 30.5 Results from conventional analyses

Failure mode	Section 30.1	Section 30.2	Required minimum safety factor
Overturning about toe	3.6	3.95	2.0
Sliding stability	1.0	1.05	1.5
Basal heave failure	6.4	5.36	3.0

Table 30.6 Concrete properties

Parameter	Value
Material model	Linear elastic
Unit weight, γ_c	25.0 kN/m^3
Elasticity modulus, E_c	40 MPa
Poisson's ratio, ν	0.15

Table 30.7 Properties of backfill soil

Parameter	Name	Section 30.1	Section 30.2
Material model	Model	Mohr-Coulomb	Mohr-Coulomb
Type of material	Type	Drained	Drained
Behaviour unit weight	γ_{unsat}	20.03 kN/m^3	21.1 kN/m^3
Saturated unit weight	γ_{sat}	21 kN/m^3	22 kN/m^3
Horizontal permeability	k_x	1.0 m/day	1.0 m/day
Vertical permeability	k_y	1.0 m/day	1.0 m/day
Young's modulus	E_{ref}	15 MPa	11 MPa
Poisson's ratio	ν	0.3	0.3
Cohesion	c_{ref}	2.0 kPa	2.0 kPa
Friction angle	Φ	28.4	29.75

Table 30.8 Properties of foundation soil

Parameter	Name	Section 30.1	Section 30.2
Material model	Model	Mohr-Coulomb	Mohr-Coulomb
Type of material	Type	Drained	Drained
Behaviour unit weight	γ_{unsat}	21.05 kN/m^3	21.3 kN/m^3
Saturated unit weight	γ_{sat}	21 kN/m^3	22 kN/m^3
Horizontal permeability	k_x	1.0 m/day	1.0 m/day
Vertical permeability	k_y	1.0 m/day	1.0 m/day
Young's modulus	E_{ref}	42 MPa	27 MPa
Poisson's ratio	ν	0.3	0.3
Cohesion	c_{ref}	2.0 kPa	2.0 kPa
Friction angle	Φ	28.4	29.75

Table 30.9 Results of numerical modelling for Sect. 30.1

Parameter	Section 30.1	Field measurement
Global factor of safety	0.4	–
Lateral displacement (mm)	342	380
Vertical settlement (mm)	342	327

Table 30.10 Results of numerical modelling for Sect. 30.2

Parameter	Section 30.2	Field measurement
Global factor of safety	0.34	–
Lateral displacement (mm)	300	280
Vertical settlement (mm)	600	508

factor of safety using phi-c reduction technique is less than 1. The deformations measured in the field are in good agreement with finite element results. Although conventional analysis for Sects. 30.1 and 30.2 shows that the safety factor for over turning and basal heave are greater than the critical safety factor, failure occurred as a combination of overturning and sliding. Back analysis of failure showed that the mechanism of failure is a combination of sliding and overturning and the deformations are in conformity with the predictions obtained from the numerical analysis. This can be better represented by the global safety factor which considers the deformation behaviour of the soil to calculate the global safety factor.

30.8 Conclusions

This paper presents the results of forensic geotechnical investigation of a distressed cantilever retaining wall for ROB approach embankment. From the results of the conventional analysis and finite element analysis, it is observed that the retaining wall designs based on prescriptive guidelines may not lead to satisfactory designs and considerations of deformations are important. Back analysis of failure showed that the mechanism of failure is a combination of sliding and overturning and the deformations are in conformity with the predictions obtained from the numerical analysis.

References

Chen WF (1995) The civil engineering handbook. CRC Press, Inc.

Clough GW, Duncan JM (1971) Finite element analysis of retaining wall behaviour. J Soil Mech Found ASCE 97(12):1657–1673

Das BM (2002) Principles of geotechnical engineering, 7th edn. Cengage Learning

Day RW (1997) Design and construction of cantilevered retaining walls. Pract Periodical Struct Des Constr 2(1):16–21

Goh ATC (1993) Behaviour of cantilever retaining walls. J Geotech Eng 19(11):1751–1770

Hossain MS, Kibria G, Khan MS, Hossain J, Taufiq T (2011) Effect of backfill soil on excessive movement of MSE wall. J Perform Constr Facil (in press)
Marsh T, Walsh RK (1996) Common causes of retaining-wall distress: case study. J Perform Constr Facil 10(1):35–38
Scarborough JA (2005) A tale of two walls: case histories of failed MSE walls. In: Proceedings of slopes and retaining structures under seismic and static conditions, GSP 140, Geo-Frontiers 2005, ASCE, Reston, VA
Wu JY, Chou NN (2012) Forensic studies of geosynthetic reinforced structure failures. J Perform Constr Facil (in press)

Chapter 31
Role of Reliability in Forensic Geotechnical Engineering

K.K. Phoon, G.L. Sivakumar Babu and M. Uzielli

Abstract This chapter discusses some possible roles for reliability and risk in forensic geotechnical engineering. A preliminary statistical framework is presented to quantify the difference between expected and observed performance in the presence of unavoidable and potentially significant geotechnical variabilities. The probable ranges of geotechnical variabilities are reviewed. Other potentially useful results in the recent reliability and risk literature are highlighted. The intention of this chapter is to stimulate further discussions and research in this important but somewhat overlooked area.

Keywords Reliability · Variability · Factor of safety · Risk-based criteria

31.1 Introduction

Forensic geotechnical engineering (FGE) is a relatively new field. ISSMGE TC40 was formed only in 2005 under Dr. V.V.S. Rao as chair. Subcommittee 6 was convened under Dr. K.K. Phoon to explore the potential role of reliability in FGE.

K.K. Phoon (✉)
Department of Civil Engineering, National University of Singapore,
Singapore 117576, Singapore
e-mail: cvepkk@nus.edu.sg

G.L. Sivakumar Babu
Department of Civil Engineering Indian Institute of Science,
Bangalore 560012, India
e-mail: gls@civil.iisc.ernet.in

M. Uzielli
Georisk Engineering S.r.l, Florence, Italy
e-mail: muz@georisk.eu

V.V.S. Rao and G.L. Sivakumar Babu (eds.), *Forensic Geotechnical Engineering*,
Developments in Geotechnical Engineering, DOI 10.1007/978-81-322-2377-1_31

A cursory Google search using keywords "reliability, forensic, and geotechnical" shows that this concept is entirely new. This chapter seeks to explore if reliability/risk concepts are potentially useful to forensic geotechnical engineering. This study is very preliminary given the dearth of previous literature. Brown (2006) noted that: "The concept of reliability as 'the likelihood that a system will perform in an acceptable manner' (Bea 2006) is important in forensic geotechnical engineering". No details were discussed.

Engineering with natural materials such as soils and rocks is difficult and challenging owing to the variability and the uncertainties associated with engineering decisions. In fact, one key aspect that distinguishes geotechnical engineering from structural engineering is the large variability (natural/intrinsic variability, testing errors, and transformation uncertainties introduced when measured parameters are converted to engineering parameters) related to naturally occurring geomaterials. In the case of quality of concrete, variability (in terms of coefficient of variation or COV) can be classified as follows (Réthàti 1988 citing 1965 specification of the American Concrete Institute):

COV < 10 %	Excellent
COV = 10–15 %	Good
COV = 15–20 %	Satisfactory
COV > 20 %	Bad

The COVs related to the natural variability of geomaterials as well as those pertaining to measurements errors and transformation uncertainties can be much larger and do not fall within a narrow range. Phoon and Kulhawy (1999a, b) compiled an extensive database of soil statistics for calibration of simplified reliability-based foundation design equations. Details are given in Sect. 31.3. One key observation that emerged from their calibration studies is that geotechnical variability of common design parameters can be broadly grouped using the following qualitative categories:

Geotechnical parameter	Property variability	COV (%)
Undrained shear strength	Low	10–30
	Medium	30–50
	High	50–70
Effective stress friction angle	Low	5–10
	Medium	10–15
	High	15–20
Horizontal stress coefficient	Low	30–50
	Medium	50–70
	High	70–90

These observations are not entirely surprising given that the volume of geomaterials investigated by direct or indirect means is extremely small in comparison to the volume of interest. Chilès and Delfiner (1999) cited volume fractions investigated at Brent field, North sea, to be 10^{-9} for cores and cuttings and 10^{-6} for logging. The immediate impact of these observations is that Leonards (1982) definition of "failure"—*unacceptable difference between expected and observed performance*—cannot be evaluated in a meaningful way using deterministic methods. To elaborate, "expected performance" must vary given the backdrop of potentially significant geotechnical variability. We can argue that this variability is "inherent", in the sense of being irreducible, in geotechnical engineering based on the prevailing standards of best practice. Some reduction in some of the components of variability is always possible by carrying out more tests, but it is generally not realistic to expect orders of magnitudes reduction. In view of this widely acknowledged and widely established reality in geotechnical engineering, it is more realistic and perhaps more credible to quantify "unacceptable difference" in a statistical sense. In broad terms, forensic engineering is related to the investigation of failures with the view of rendering an opinion regarding *responsibility*. Hence, an objective statistical measure of "unacceptable difference" (specifically, a difference not explainable by underlying variability) should provide useful additional information in the formulation of such an opinion. It should be emphasized, however, that the statistical criteria tentatively proposed in the following sections should never replace—nor downplay the importance of—preliminary qualitative assessments of the quality of design based on geotechnical knowledge and experience. Rather, they should serve as a complementary tool that could improve the degree of objectivity in forensic geotechnical assessments.

31.2 Statistical Framework

31.2.1 Rejection Criteria for Factor of Safety

A preliminary statistical framework can be constructed using the widely used global factor of safety (FS) as an indicator of performance. It is plausible to imagine FS as a lognormal random variable with mean = μ, standard deviation = σ and coefficient of variation = σ/μ = θ. Silva et al. (2008) provided some empirical evidence to support this lognormal assumption for slope problems. The lognormal distribution satisfies the most basic constraint that most load and strength parameters are non-negative. It is implicit in this model that one speaks of the "likelihood" of failure, rather than failure in our usual absolute deterministic sense. Figure 31.1 illustrates a lognormal distribution for $\mu = 3$ and $\theta = 0.3$. It can be shown that ln(FS) is normally distributed with mean (λ) and standard deviation (ξ) given by:

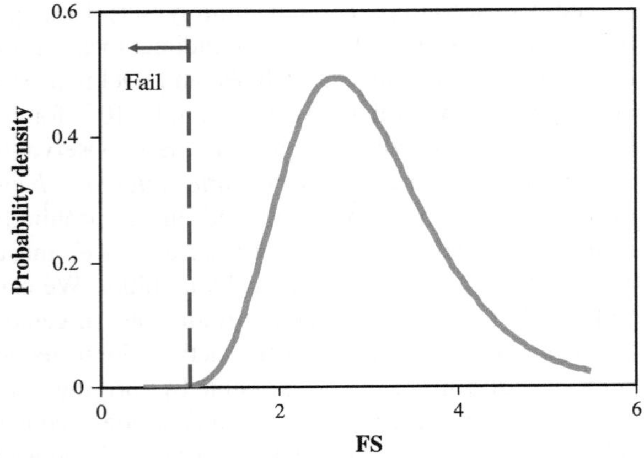

Fig. 31.1 Lognormal distribution of FS

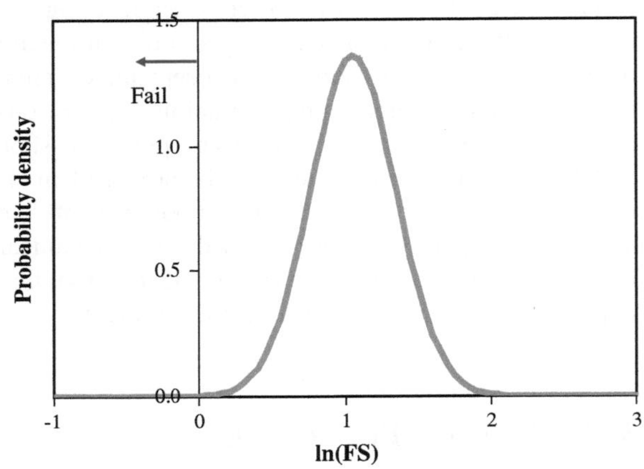

Fig. 31.2 Equivalent normal distribution

$$\lambda = \ln(\mu) - 0.5\xi^2 \text{ and } \xi^2 = \ln(1 + \theta^2). \tag{31.1}$$

The equivalent normal distribution for Fig. 31.1 is shown in Fig. 31.2 with $\lambda = 1.056$ and $\xi = 0.29$. Note that the standard deviation of ln(FS) (ξ) is approximately equal to the COV of FS (θ) up to θ of about 0.5.

The observed performance, $\overline{\text{FS}}$, can be viewed as a sample mean (of sample size 1) from a *population* of factors of safety such as that shown in Fig. 31.1. It is obvious that a low value of $\overline{\text{FS}}$ near to 1 is not impossible, although it is "unlikely" if the population follows a lognormal distribution with $\mu = 3$ and $\theta = 0.3$. A naïve interpretation is that "failure" (as in $\overline{\text{FS}} < 1$) is always possible and geotechnical variability is responsible, rather than human errors. Sowers (1993) noted that the majority of foundation failures were due to human shortcomings. This naïve interpretation essentially misses the key principle of hypothesis testing described below.

Our proposed approach is to assume that: (a) "expected performance" can be described by the reliability index and (b) a conventional hypothesis test can be performed to ascertain if the target reliability index had been achieved in the original design based on the observed sample mean, $\overline{\text{FS}}$. To do this, it is necessary to define "expected performance". If FS follows a lognormal population, the reliability index is given by:

$$\beta = \frac{E[\ln(\text{FS})]}{\sigma[\ln(\text{FS})]} = \frac{\lambda}{\xi} \tag{31.2}$$

Numerous reliability calibration studies (e.g. Phoon et al. 1995) have shown that existing foundations are typically designed to achieve a target reliability index (β) of about 3 (corresponding to a probability of failure of about one in a thousand). Target reliability indices for other types of geotechnical design are not as well established presently.

In Fig. 31.2, the reliability index is β = 3.64 or probability of failure = $\Phi^{-1}(-\beta)$ = 0.000136, where Φ is the standard normal cumulative distribution function. In EXCEL, Φ(.) and Φ^{-1}(.) are computed using the NORMSDIST (.) and NORMSINV(.) functions, respectively. Based on the definition of the reliability index and β = 3, the following null and alternate hypotheses on the population mean [actually, the mean of ln(FS)] can be formulated as:

$$H_0{:}\lambda = 3\xi$$

$$H_1{:}\lambda < 3\xi$$

Assuming that ξ is known and a sample size of 1, the null hypothesis is rejected at the customary 5 % level of significance if:

$$\frac{\ln\left(\overline{\text{FS}}\right) - \lambda}{\xi} = \frac{\ln\left(\overline{\text{FS}}\right) - 3\xi}{\xi} < \Phi^{-1}(0.05) = -1.645 \tag{31.3}$$

$$\overline{\text{FS}} < \exp(1.355\xi) = \exp\left[1.355\sqrt{\ln\left(1+\theta^2\right)}\right] \tag{31.4}$$

The above rejection criterion provides an example of a simple numerical yardstick to evaluate "unacceptable difference between expected and observed performance" in the presence of potentially significant geotechnical variability. A rejection means that the observed factor of safety does not support the claim that β = 3 at the 5 % level of significance. Different values of the threshold factor of safety can be computed for different significance levels. It is also possible for a rejection to arise because the underlying "true" geotechnical variability was grossly

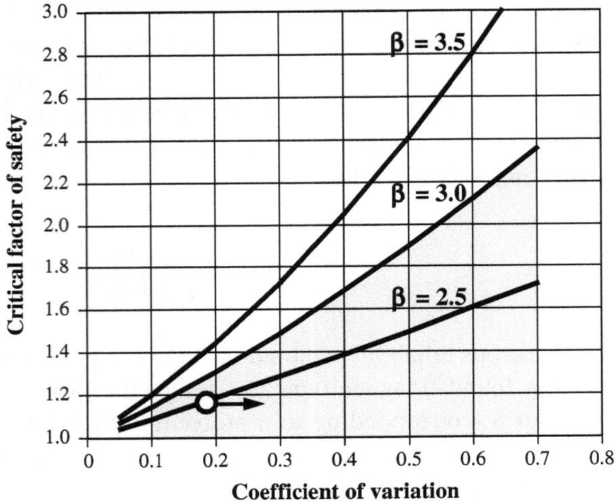

Fig. 31.3 Rejection criteria for different expected performance levels—a possible role for risk in FGE based on one observed factor of safety

overestimated. A "do not reject" scenario means that the observed factor of safety is not unreasonably "low" and failure may be caused by geologic "surprises", limitations in the existing factor of safety, critical failure mechanism not identified, etc. Again, it is possible for "do not reject" to arise because we have grossly underestimated the underlying geotechnical variability.

The above framework is by no means perfect and comprehensive, given the diversity and complexities of actual failures. However, it does provide an example of an objective framework to perform an initial evaluation of the observed factor of safety, particularly to eliminate the more obvious claim that it is an "unfortunate" realization caused by geotechnical variability.

Figure 31.3 presents the critical values of factor of safety below which the difference between expected and observed performance is "unacceptable". For example, if the expected performance corresponds to $\beta = 3$ and the underlying variability corresponds to $\theta = 0.3$, an observed factor of safety of 1.4 is too "low" and not explainable as a random outcome from a population of FS with $\beta = 3.0$ and $\theta = 0.3$. It will be useful to interpret actual case studies using this chart to evaluate its usefulness in forensic geotechnical engineering and to fine-tune it if necessary. In the general case where we observe n factors of safety, the rejection criterion for β will become:

$$\begin{aligned}\overline{\mathrm{FS}} &< \exp\left(\beta\xi - 1.645\xi/\sqrt{n}\right)\\ &= \exp\left[\sqrt{\ln(1+\theta^2)}\,\left(\beta - 1.645/\sqrt{n}\right)\right]\end{aligned} \tag{31.5}$$

The rejection curves for $\beta = 3.0$ corresponding to various sample sizes (n) are shown in Fig. 31.4. If this statistical framework proves to be useful, more

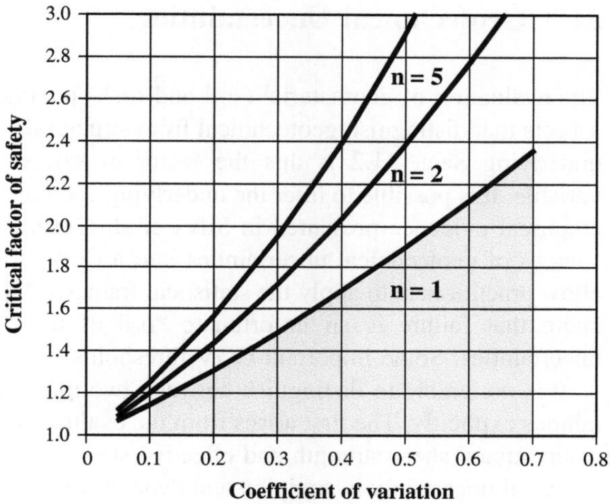

Fig. 31.4 Rejection criteria for different sample sizes of observed factors of safety based on expected performance of $\beta = 3$

sophisticated rejection criteria can be developed based on a sample estimate of ξ (rather than the population version used in the above equations). It suffices to note here that they do not follow the t-distribution in standard statistical texts.

31.2.2 Example

An example of the use of Fig. 31.3 can be illustrated with reference to the results presented in Duncan (2000) wherein a case study of underwater slope failure was reported. The failure took place entirely within San Francisco Bay mud, a normally consolidated, slightly organic clayey silt or silty clay of marine origin. Previous experience in the area indicated that 1(H):1(V) slopes with a factor of safety of 1.25 were satisfactory. To reduce the quantity of excavation, slopes of 0.875(H):1(V) having a factor of safety of 1.17 were excavated which failed subsequently. A risk-based back analysis indicated a probability of failure of 18 % which is unacceptable though the factor of safety is 1.17. Duncan (2000) mentioned that the coefficient of variation of shear strength parameters was high and hence the probability of failure was high. This case study can be plotted in Fig. 31.3 as an open circle. It is clear that an observed $\overline{\text{FS}} = 1.17$ cannot support the claim that $\beta = 2.5$ for any COV larger than about 0.2. Following the same argument, we can conclude that $\beta = 3$ is not supported for any COV larger than about 0.1. Because COV of 0.1 is the lower bound for most geotechnical problems, it is reasonable to say that $\beta = 3$ was not achieved in the original design with a fair degree of confidence.

31.3 Geotechnical Uncertainties

The evaluation of geomaterial (soil and rock) parameters is one of the key design aspects that distinguish geotechnical from structural engineering. The basic premise underlying Sect. 31.2 is that the factor of safety (FS) is a lognormal random variable. It is possible to infer the underlying coefficients of variation of FS from the empirical evidence presented in Silva et al. (2008). This approach of *lumping* all sources of geotechnical uncertainties into a COV of FS is quite crude, but it does allow practitioners to apply the statistical framework in Sect. 31.2 rapidly to test the claim that failure is an unfortunate "roll of the dice" caused by geotechnical uncertainties. Some important COV thresholds for FS are given in Sect. 31.3.1.

It is preferable to distinguish between two main sources of geotechnical uncertainties explicitly. The first arises from the evaluation of design soil properties, such as undrained shear strength and effective stress friction angle. This source of geotechnical uncertainty is complex and depends on inherent soil variability, degree of equipment and procedural control maintained during site investigation, and precision of the correlation model used to relate field measurement with design soil property. A total variability analysis that lumps all these components produces only site-specific statistics which should not be generalized to other sites. Realistic statistical estimates of the variability of design soil properties have been established by Phoon and Kulhawy (1999a, b) and presented in Sect. 31.3.2. The second source arises from geotechnical calculation models. Although many geotechnical calculation models are "simple", reasonable predictions of fairly complex soil structure interaction behaviour still can be achieved through empirical calibrations. Model factors, defined as the ratio of the measured response to the calculated response, usually are used to correct for simplifications in the calculation models. Recent literature includes estimation of model statistics for the calibration of deep foundation resistance factors for AASHTO (American Association of State Highway and Transportation Officials) (Paikowsky 2002). None of these studies addresses the applicability of model statistics beyond the conditions implied in the database. This question mirrors the same concern expressed previously on the possible site-specific nature of soil variabilities. More rigorous model statistics for both ultimate and serviceability limit states are presented in Sect. 31.3.3.

It is crucial to highlight that the COV of FS is indeed exceedingly crude, because it lumps parametric and model uncertainties into a single number.

31.3.1 Coefficient of Variation of Factor of Safety

Silva et al. (2008) proposed several relationships between the annual probability of failure and the factor of safety based on 75 projects (zoned and homogeneous earth dams, tailings dams, natural and cut slopes, and some earth retaining structures) and

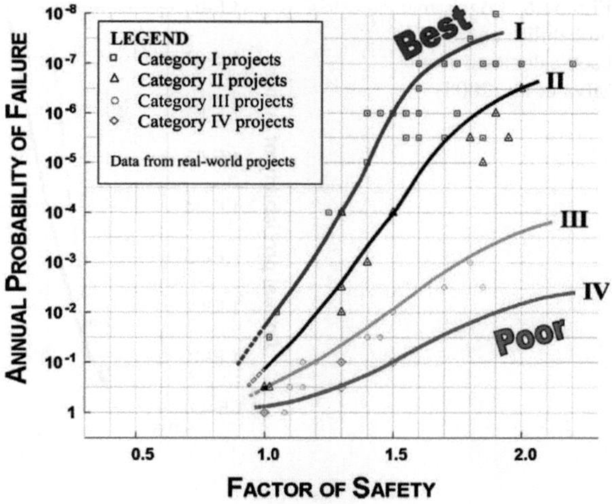

Fig. 31.5 Annual probability of failure versus factor of safety for earth structures (Silva et al. 2008)

expert judgment. As shown in Fig. 31.5, there are four categories of earth structures. They are defined by Silva et al. (2008) as:

- Category I—facilities designed, built and operated with state-of-the-practice engineering. Generally these facilities have high failure consequences;
- Category II—facilities designed, built and operated using standard engineering practice. Many ordinary facilities fall into this category;
- Category III—facilities without site-specific design and substandard construction or operation. Temporary facilities and those with low failure consequences often fall into this category;
- Category IV—facilities with little or no engineering.

Silva et al. (2008) compared the empirical data in Fig. 31.5 with some theoretical curves, but did not provide any mathematical details. Their theoretical curves can be easily reproduced using the following procedure:

1. Assume that FS is lognormally distributed with parameters λ and ξ.
2. The horizontal axis of Fig. 31.5 is the *mean* factor of safety, μ.
3. If the COV of FS (θ) is sufficiently small, $\lambda \approx \ln(\mu)$ and $\xi \approx \theta$.
4. The vertical axis of Fig. 31.5 is the probability of failure, given by:

$$p_f = \text{Prob}(\text{FS} < 1) = \text{Prob}[\ln(\text{FS}) < 0] = \Phi\left(\frac{0-\lambda}{\xi}\right) = \Phi\left[\frac{-\ln(\text{FS})}{\theta}\right]$$

The theoretical lognormal curves from Silva et al. (2008) are reproduced in Fig. 31.6 using the above procedure with $\theta = 0.072$ (Category I), 0.109 (Category II), 0.174 (Category III), 0.316 (Category IV). Based on Silva et al. (2008), it would

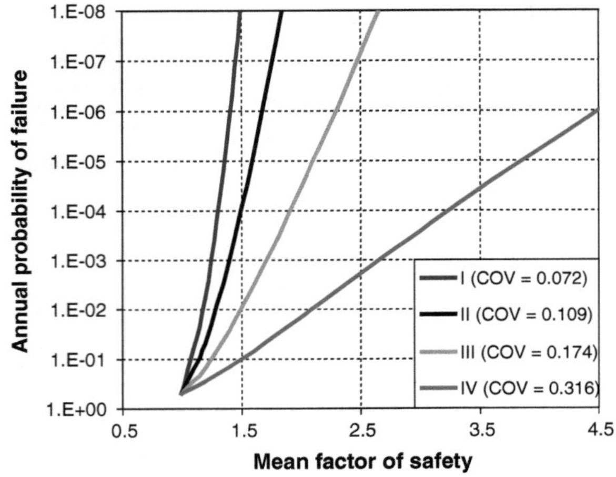

Fig. 31.6 Lognormal probability curves back-calculated from Silva et al. (2008)

Table 31.1 COV thresholds for factors of safety corresponding to different categories of earth structures (back-calculated from Silva et al. 2008)

Categories of earth structures	COV (FS)
I	0.072
II	0.109
III	0.174
IV	0.316

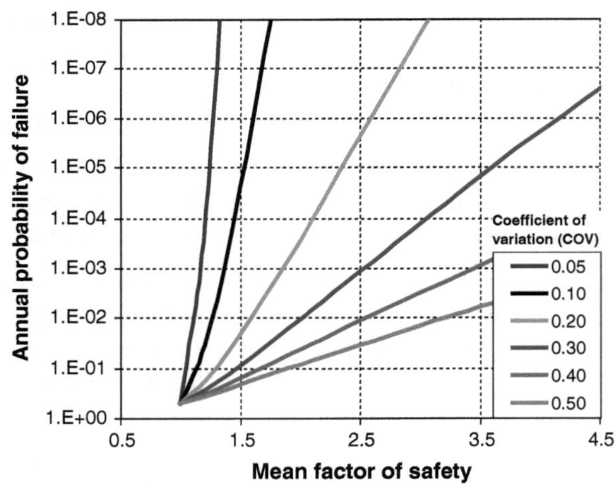

Fig. 31.7 Lognormal probability curves for higher COVs of FS

appear that the COV of FS for standard earth structures is between 0.1 and 0.2. These COV thresholds are important for FGE and they are summarized in Table 31.1. Lognormal probability curves for a more complete and more systematic range of COVs are shown in Fig. 31.7.

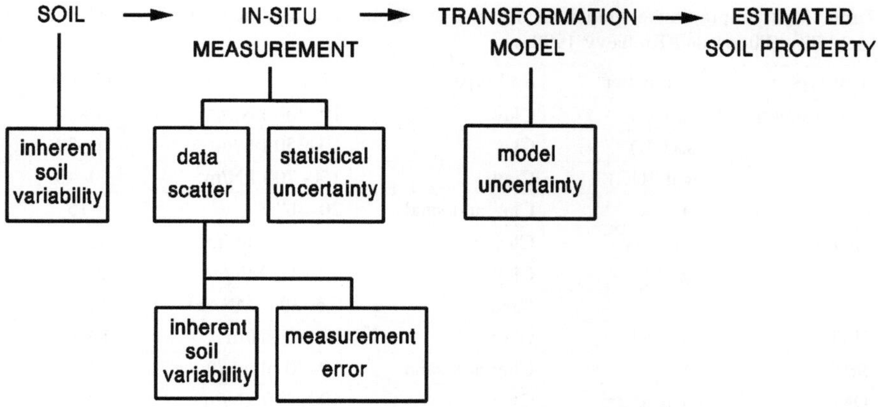

Fig. 31.8 Sources of uncertainties contributing to overall uncertainty in design soil parameter

31.3.2 Parametric Uncertainties

There are three primary sources of geotechnical uncertainties as illustrated in Fig. 31.8: (a) inherent variability, (b) measurement error and (c) transformation uncertainty (Phoon and Kulhawy 1999a). A comprehensive effort was undertaken to provide realistic geomaterial statistics of sufficient generality to underpin current and future developments of practical reliability-based design codes (Phoon and Kulhawy 1999a, b; Kulhawy et al. 2000). The results of this compilation effort are also useful for FGE and are summarized below. Others are reported by Jones et al. (2002) and Uzielli et al. (2006).

31.3.2.1 Inherent or Natural Variability

Inherent variability results primarily from the "real" heterogeneity which is inherent to geomaterials. Such heterogeneity stems from natural geologic and physical/chemical/biological processes that produced and continually modified the soil/rock mass in situ. Uzielli et al. (2006) provided a state-of-the-art review of methods for quantifying inherent variability in geotechnical engineering analyses and design. Approximate guidelines for inherent soil variability are given in Table 31.2. A more detailed presentation containing the mean, standard deviation (S.D.) and range of the COV are shown in Table 31.3 with the total number of data groups per test (m). The statistics for rock are given for comparison.

Table 31.2 Approximate guidelines for ranges of mean values and COVs of inherent soil variability (Phoon and Kulhawy 1999a)

Test type	Parameter[a]	Soil type	Mean	COV(%)
Lab strength	s_u(UC)	Clay	10–400 kN/m^2	20–55
	s_u(UU)	Clay	10–350 kN/m^2	10–30
	s_u(CIUC)	Clay	150–700 kN/m^2	20–40
	ϕ	Clay and sand	20–40°	5–15
CPT	q_T	Clay	0.5–2.5 MN/m^2	<20
	q_c	Clay	0.5–2.0 MN/m^2	20–40
		Sand	0.5–30.0 MN/m^2	20–60
VST	s_u(VST)	Clay	5–400 kN/m^2	10–40
SPT	N	Clay and sand	10–70 blows/ft	25–50
DMT	A reading	Clay	100–450 kN/m^2	10–35
		Sand	60–1300 kN/m^2	20–50
	B reading	Clay	500–880 kN/m^2	10–35
		Sand	350–2400 kN/m^2	20–50
	I_D	Sand	1–8	20–60
	K_D	Sand	2–30	20–60
	E_D	Sand	10–50 MN/m^2	15–65
PMT	p_L	Clay	400–2800 kN/m^2	10–35
		Sand	1600–3500 kN/m^2	20–50
	E_{PMT}	Sand	5–15 MN/m^2	15–65
Lab index	w_n	Clay and silt	13–100 %	8–30
	w_L	Clay and silt	30–90 %	6–30
	w_P	Clay and silt	15–25 %	6–30
	PI	Clay and silt	10–40 %	b
	LI	Clay and silt	10 %	b
	γ, γ_d	Clay and silt	13–20 kN/m^3	<10
	D_r	Sand	30–70 %	10–40[c]
				50–70[d]

[a] s_u undrained shear strength; *UC* unconfined compression test; *UU* unconsolidated-undrained triaxial compression test; *CIUC* consolidated isotropic undrained triaxial compression test; ϕ effective stress friction angle; q_T corrected cone tip resistance; q_c cone tip resistance; *VST* vane shear test; *N* standard penetration test blow count; *A* and *B* readings, I_D, K_D and E_D = dilatometer *A* and *B* readings, material index, horizontal stress index, and modulus; p_L and E_{PMT} pressure meter limit stress and modulus; w_n natural water content; w_L liquid limit; w_P plastic limit; *PI* plasticity index; *LI* liquidity index; γ and γ_d total and dry unit weights; D_r relative density

[b] COV = (3–12 %)/mean

[c] total variability for direct method of determination

[d] total variability for indirect determination using SPT values

Table 31.3 Inherent soil variability for soil and rock (Kulhawy et al. 2000)

Test type	Parameter	Material type	Coefficient of variation (%)			
			m	Mean	S.D.	Range
Index	γ, γ_d	Fine-grained	14	7.8	5.8	2–20
	w_n	Fine-grained	40	18.1	7.9	7–46
	w_P	Fine-grained	23	15.7	6.0	6–34
	w_L	Fine-grained	38	18.1	7.1	7–39
	PI—all data	Fine-grained	33	29.5	10.8	9–57
	– ≤20 %	Fine-grained	13	35.0	11.4	16–57
	– >20 %	Fine-grained	20	26.0	9.0	9–40
	γ, γ_d	Rock	42	0.9	0.7	0.1–3
	n	Rock	25	25.9	19.4	3–71
Strength	ϕ, tan ϕ	Sand, clay	48	13.9	10.4	4–50
		Sand	32	9.0	3.0	4–15
		Clay	16	23.5	13.0	10–50
	s_u	Clay	100	31.5	14.2	6–80
	q_u	Rock	184	14.2	11.7	0.3–61
	$q_{t\text{-brazilian}}$	Rock	74	16.6	10.4	2–58
Stiffness	E_{t-50}	Rock	32	30.7	15.0	7–63
CPT	q_c	Sand, clay	65	36.6	15.5	10–81
		Sand	54	38.2	16.3	10–81
		Clay	11	28.4	6.8	16–40
	q_T	Clay	9	7.9	4.9	2–17
VST	s_u(VST)	Clay	26	25.3	6.5	13–36
SPT	N	Sand, clay	23	38.0	10.8	19–62
DMT	A, B	Sand, clay	56	27.9	11.9	12–59
		Sand	30	34.8	11.3	13–59
		Clay	26	19.9	6.2	12–38
	I_D—all data	Sand	30	40.7	21.6	8–130
	– w/o outliers		29	37.7	14.2	8–66
	K_D—all data	Sand	31	41.2	19.2	15–99
	– w/o outliers		29	37.6	13.5	15–67
	E_D—all data	Sand	31	42.7	19.6	7–92
	– w/o outliers		30	41.1	17.6	7–69

31.3.2.2 Measurement Error

Inherent variability is caused primarily by the natural geologic processes that are involved in soil formation. Measurement error, on the other hand, arises from equipment, procedural/operator, and random testing effects. A summary of total measurement error is given in Table 31.4 for laboratory tests and in Table 31.5 for field tests.

Table 31.4 Summary of total measurement error of some laboratory tests (Phoon and Kulhawy 1999a)

Parameter[a]	Soil type	No. of data group	No. of tests/group		Property value (units[b])		Property COV (%)	
			Mean	Range	Mean	Range	Mean	Range
s_u(TC)	Clay, silt	11	–	13	7–407	125	8–38	19
s_u(DS)	Clay, silt	2	13–17	15	108–130	119	19–20	20
s_u(LV)	Clay	15	–	–	4–123	29	5–37	13
ϕ(TC)	Clay, silt	4	9–13	10	2–27°	19.1°	7–56	24
ϕ(DS)	Clay, silt	5	9–13	11	24–40°	33.3°	3–29	13
	Sand	2	26	26	30–35°	32.7°	13–14	14
tan ϕ(TC)	Sand, silt	6	–	–	–	–	2–22	8
tan ϕ(DS)	Clay	2	–	–	–	–	6–22	14
w_n	Fine-grained	3	82–88	85	16–21	18	6–12	8
w_L	Fine-grained	26	41–89	64	17–113	36	3–11	7
w_P	Fine-grained	26	41–89	62	12–35	21	7–18	10
PI	Fine-grained	10	41–89	61	4–44	23	5–51	24
γ	Fine-grained	3	82–88	85	16–17	17.0	1–2	1

[a]s_u undrained shear strength; ϕ effective stress friction angle; *TC* triaxial compression test; *UC* unconfined compression test; *DS* direct shear test; *LV* laboratory vane shear test; w_n natural water content; w_L liquid limit; w_P plastic limit; *PI* plasticity index; γ total unit weight

[b]units of s_u kN/m^2; units of w_n, w_L, w_P, and *PI* %; units of γ kN/m^3

Table 31.5 Summary of measurement error of common in situ tests (Kulhawy and Trautmann 1996)

Test	Coefficient of variation, COV (%)				
	Equipment	Procedure	Random	Total[a]	Range[b]
Standard penetration test (SPT)	5[c]–75[d]	5[c]–75[d]	12–15	14[c]–100[d]	15–45
Mechanical cone penetration test (MCPT)	5	10[e]–15[f]	10[e]–15[f]	15[e]–22[f]	15–25
Electric cone penetration test (ECPT)	3	5	5[e]–10[f]	7[e]–12[f]	5–15
Vane shear test (VST)	5	8	10	14	10–20
Dilatometer test (DMT)	5	5	8	11	5–15
Pressuremeter test, pre-bored (PMT)	5	12	10	16	10–20[g]
Self-boring pressuremeter test (SBPMT)	8	15	8	19	15–25[g]

[a]$\text{COV(Total)} = [\text{COV(Equipment)}^2 + \text{COV(Procedure)}^2 + \text{COV(Random)}^2]^{0.5}$

[b]Because of limited data and judgment involved in estimating COVs, ranges represent probable magnitudes of field test measurement error

[c,d]Best to worst case scenarios, respectively, for SPT

[e,f]Tip and side resistances, respectively, for CPT

[g]It is likely that results may differ for p_o, p_f and p_L, but the data are insufficient to clarify this issue

31.3.2.3 Transformation Uncertainty

The third component of uncertainty is introduced when field or laboratory measurements are transformed into design parameters using empirical or other correlation models (e.g. correlating the standard penetration test N value with the undrained shear strength) as shown in Fig. 31.9. Obviously, the relative contribution of these components to the overall uncertainty in the design parameter depends on the site conditions, degree of equipment and procedural control, and quality of the correlation model. Therefore, geomaterial statistics that are determined from total variability analyses only can be applied to the specific set of circumstances (site conditions, measurement techniques, correlation models) for which the design parameters were derived. In other words, the COV of geomaterials cannot be viewed as an intrinsic statistical property.

For each combination of soil type, measurement technique and correlation model, the uncertainty in the design soil property is evaluated systematically by combining the appropriate component uncertainties using a simple second-moment probabilistic approach:

$$s_{\xi d}^2 \approx \left(\frac{\partial T}{\partial w}\right)^2 s_w^2 + \left(\frac{\partial T}{\partial e}\right)^2 s_e^2 + \left(\frac{\partial T}{\partial \varepsilon}\right)^2 s_\varepsilon^2 \quad (31.6)$$

in which $\xi_d = T(\xi_m, \varepsilon)$, $T(.)$ = correlation function between test measurement (ξ_m) and design parameter (ξ_d), ε = transformation uncertainty, w = inherent variability, e = measurement error and s^2 = variance.

The above equation refers to the variance of the design parameter at a point in the soil mass. For foundation design, it is not uncommon to evaluate the spatial average of the design parameter over some depth interval, rather than using the value of the design parameter at a point. The spatial average of ξ_d is defined as:

$$\xi_a = \frac{1}{L_a}\int_{z_t}^{z_b} \xi_d(z)\,dz \quad (31.7)$$

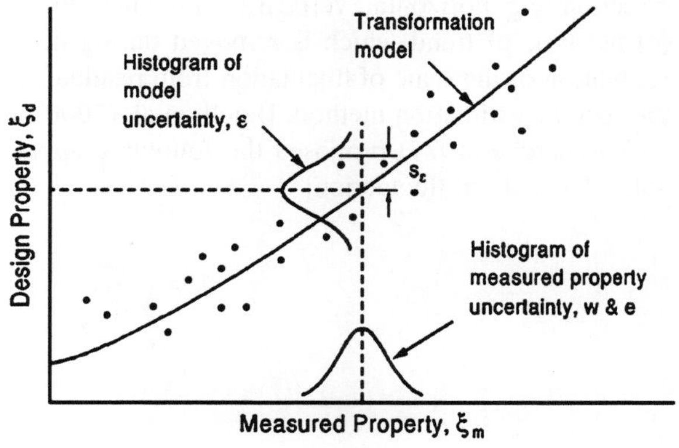

Fig. 31.9 Transformation uncertainty (Phoon and Kulhawy 1999b)

in which ξ_a = spatial average, z_t and z_b = top and bottom coordinates of a depth interval, respectively, and $L_a = z_b - z_t$ = averaging length.

If t and $\partial T/\partial w$ are constants, it can be shown that the variance of the spatial average ($s^2_{\xi a}$) is given by (Phoon and Kulhawy 1999b):

$$s^2_{\xi a} \approx \left(\frac{\partial T}{\partial w}\right)^2 \Gamma^2(L_a) s^2_w + \left(\frac{\partial T}{\partial e}\right)^2 s^2_e + \left(\frac{\partial T}{\partial \varepsilon}\right)^2 s^2_\varepsilon \qquad (31.8)$$

in which $\Gamma^2(.)$ = variance reduction function, which depends on the length of the averaging interval, L_a. The following approximate variance reduction function is proposed for practical application (Vanmarcke 1983):

$$\Gamma_2(L_a) = 1 \qquad \text{for } L_a \le \delta_v$$
$$\Gamma_2(L_a) = \delta_v / L_a \quad \text{for } L_a > \delta_v$$

in which δ_v = vertical scale of fluctuation. It can be seen that the variance reduction function decreases as the length of the averaging interval increases. Therefore, the effect of averaging is to reduce the uncertainty associated with inherent variability (s^2_w).

The scale of fluctuation quantifies the spatial extension in which a property of interest can be considered significantly autocorrelated. Within separation distances smaller than the scale of fluctuation, the deviations from the trend function are expected to show relatively strong correlation. When the separation distance between two sample points exceeds the scale of fluctuation, it can be assumed that little correlation exists between the fluctuations in the measurements. From a geological and geomorphological perspective, it is intuitive that soil formation and modification processes, as well as factors contributing to the definition of the in situ state (e.g. stress) would result in a greater heterogeneity of soil properties in the vertical direction and, hence, in a weaker spatial correlation. Hence, the scale of fluctuation of a given soil property in the vertical direction is generally much smaller than in the horizontal direction.

The scale of fluctuation is not an inherent property of a soil parameter. Numerical values of the scale of fluctuation depend at least on: (a) the spatial direction [e.g. horizontal, vertical]; (b) the measurement interval in the source data; (c) the type of trend which is removed during decomposition; (d) the method of estimation of the scale of fluctuation from residuals and (e) modelling options from the specific estimation method. Uzielli et al. (2006) discussed these issues in detail.

Vanmarcke (1977) proposed the following approximate relationship for evaluating the scale of fluctuation:

$$\delta \approx \sqrt{\frac{2}{\pi}}\bar{\Delta} \qquad (31.9)$$

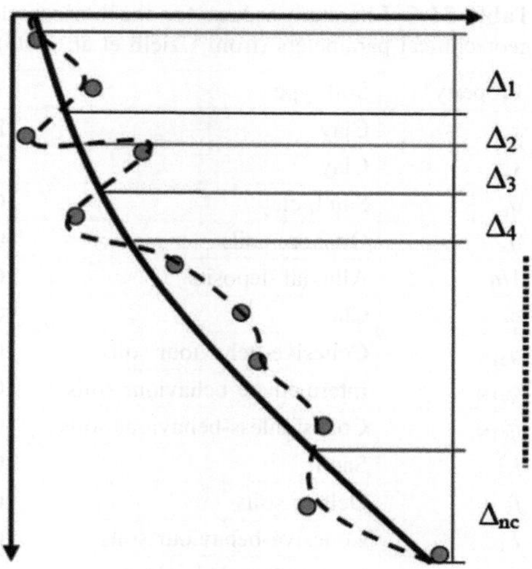

Fig. 31.10 Simplified estimation of the scale of fluctuation as proposed by Vanmarcke (1983)

where

$$\overline{\Delta} = \frac{1}{n_c} \sum_{i=1}^{n_c} \Delta_i \qquad (31.10)$$

is the average distance between the intersections of the fluctuating component and the trend of a given profile (see Fig. 31.10).

Other methods to estimate the scale of fluctuation have been implemented in the geotechnical literature. Uzielli et al. (2006) provided an overview of such methods. Uzielli (2004) compared the estimates of the scale of fluctuation of profiles of normalized cone tip resistance using different estimation methods. None of the models were shown to rank consistently above or below the others, though however, in some cases the scatter among estimates from different methods is significant. Ranges of δ_v for a number of geotechnical parameters are given in Table 31.6.

Useful guidelines on typical “total” coefficients of variation of many common design soil strength properties have been summarized by Phoon and Kulhawy (1999b) and are given in Table 31.7 for reference. Once again, tabulated values should not be applied uncritically, i.e. if it has not been assessed that the variability related to site conditions, testing methods and transformation models are significantly similar to those used to obtain specific literature values.

Table 31.6 Literature values for the horizontal (δ_η) and vertical (δ_v) scale of fluctuation of geotechnical parameters (from Uzielli et al. 2006)

Property[a]	Soil type	Testing method[b]	δ_h (m)	δ_v (m)
s_u	Clay	Lab. testing	–	0.8–8.6
s_u	Clay	VST	46.0–60.0	2.0–6.2
q_c	Sand, clay	CPT	3.0–80.0	0.1–3.0
q_c	Offshore soils	CPT	14–38	0.3–0.4
$1/q_c$	Alluvial deposits	CPT	–	0.1–2.6
q_t	Clay	CPTU	23.0–66.0	0.2–0.5
q_{c1N}	Cohesive-behaviour soils	CPT	–	0.1–0.6
q_{c1N}	Intermediate-behaviour soils	CPT	–	0.3–1.0
q_{c1N}	Cohesionless-behaviour soils	CPT	–	0.4–1.1
f_s	Sand	CPT	–	1.3
f_s	Deltaic soils	CPT	–	0.3–0.4
F_R	Cohesive-behaviour soils	CPT	–	0.1–0.5
F_R	Intermediate-behaviour soils	CPT	–	0.1–0.6
F_R	Cohesionless-behaviour soils	CPT	–	0.2–0.6
I_c	Cohesive-behaviour soils	CPT	–	0.2–0.5
I_c	Intermediate-behaviour soils	CPT	–	0.6
I_c	Cohesionless-behaviour soils	CPT	–	0.3–1.2
N	Sand	SPT	–	2.4
w	Clay, loam	Lab. testing	170.0	1.6–12.7
w_L	Clay, loam	Lab. testing	–	1.6–8.7
γ'	Clay	Lab. testing	–	1.6
γ	Clay, loam	Lab. testing	–	2.4–7.9
e	Organic silty clay	Lab. testing	–	3.0
σ'_p	Organic silty clay	Lab. testing	180.0	0.6
K_S	Dry sand fill	PLT	0.3	–
$\ln(D_R)$	Sand	SPT	67.0	3.7
n	Sand	–	3.3	6.5

[a]s_u undrained shear strength; q_c cone tip resistance; q_t corrected cone tip resistance; q_{c1N} dimensionless, stress-normalized cone tip resistance; f_s sleeve friction; F_R stress-normalized friction ratio; I_c CPT soil behaviour classification index; N SPT blow count; w water content; w_L liquid limit; γ' submerged unit weight; γ unit weight; e void ratio; σ'_p preconsolidation pressure; K_S subgrade modulus; D_R relative density; n porosity

[b]*VST* vane shear testing; *CPT* cone penetration testing; *CPTU* piezocone testing; *SPT* standard penetration testing; *PLT* plate load testing

31.3.3 Model Uncertainties

31.3.3.1 Ultimate Limit State

Robust model statistics can only be evaluated using: (1) realistically large scale prototype tests, (2) a sufficiently large and representative database, and (3)

Table 31.7 Approximate guidelines for design soil parameter variability (Phoon and Kulhawy 1999b)

Design parameter[a]	Test[b]	Soil type	Point COV (%)	Spatial avg. COV[c] (%)	Correlation equation[f]
s_u(UC)	Direct (lab)	Clay	20–55	10–40	–
s_u(UU)	Direct (lab)	Clay	10–35	7–25	–
s_u(CIUC)	Direct (lab)	Clay	20–45	10–30	–
s_u(field)	VST	Clay	15–50	15–50	14
s_u(UU)	q_T	Clay	30–40[e]	30–35[e]	18
s_u(CIUC)	q_T	Clay	35–50[e]	35–40[e]	18
s_u(UU)	N	Clay	40–60	40–55	23
s_u^d	K_D	Clay	30–55	30–55	29
s_u(field)	PI	Clay	30–55[e]	–	32
ϕ	Direct (lab)	Clay, sand	7–20	6–20	–
ϕ (TC)	q_T	Sand	10–15[e]	10[e]	38
ϕ_{cv}	PI	Clay	15–20[e]	15–20[e]	43
K_o	Direct (SBPMT)	Clay	20–45	15–45	–
K_o	Direct (SBPMT)	Sand	25–55	20–55	–
K_o	K_D	Clay	35–50[e]	35–50[e]	49
K_o	N	Clay	40–75[e]	–	54
E_{PMT}	Direct (PMT)	Sand	20–70	15–70	–
E_D	Direct (DMT)	Sand	15–70	10–70	–
E_{PMT}	N	Clay	85–95	85–95	61
E_D	N	Silt	40–60	35–55	64

[a]s_u undrained shear strength; *UU* unconsolidated-undrained triaxial compression test; *UC* unconfined compression test; *CIUC* consolidated isotropic undrained triaxial compression test; s_u(*field*) corrected s_u from vane shear test; ϕ effective stress friction angle; *TC* triaxial compression; ϕ_{cv} constant volume ϕ; K_o in situ horizontal stress coefficient; E_{PMT} pressure meter modulus; E_D dilatometer modulus
[b]*VST* vane shear test; q_T corrected cone tip resistance; *N* standard penetration test blow count; K_D dilatometer horizontal stress index; *PI* plasticity index
[c]averaging over 5 m
[d]mixture of s_u from UU, UC and VST
[e]COV is a function of the mean; refer to COV equations in Phoon and Kulhawy (1999b) for details
[f]Quality of correlation affects the COV of design parameters. The correlation models used are referenced by the equation numbers in Phoon and Kulhawy (1999b)

reasonably high quality testing where extraneous uncertainties are well controlled. It is common to correct for simplifications in the calculation model using the following multiplicative form (as exemplified by laterally loaded drilled shafts):

$$H_m = M\,H_u \tag{31.11}$$

in which H_m = "measured" lateral capacity (more precisely, capacity interpreted from load test), H_u = ultimate lateral capacity computed using limit equilibrium analysis, and M = model factor, typically assumed to be an independent lognormal random variable. It is well known that many different models exist for the computation of H_u.

The distributions of the model factors for H_m determined using two different criteria (H_L or H_h) and H_u computed from four different lateral soil stress models are shown in Fig. 31.11. Note that $M < 1$ implies that the calculated capacity is larger than the measured capacity, which is unconservative. If H_m is defined as the hyperbolic capacity (H_h), $M < 1$ is most likely unsafe as well since there is no reserved capacity beyond H_h and it is mobilized at very large displacements.

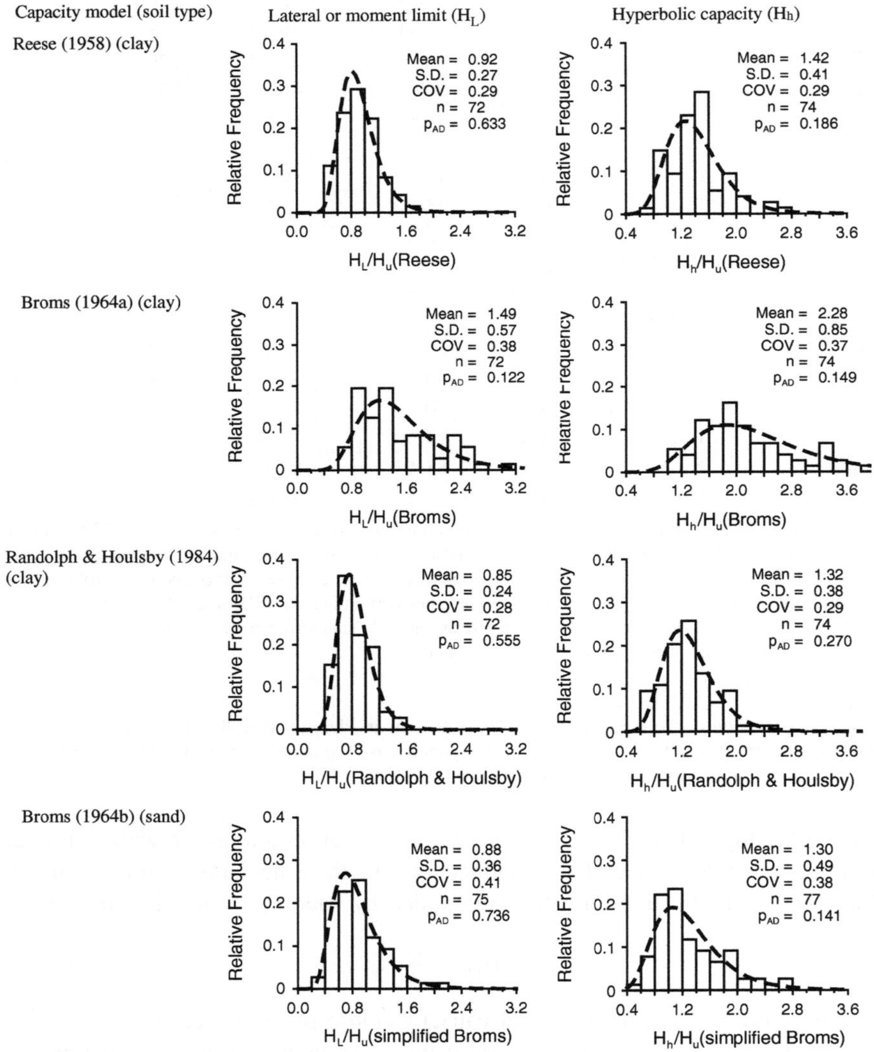

Fig. 31.11 Distribution of model factors (Phoon and Kulhawy 2005)

31.3.3.2 Serviceability Limit State

Phoon et al. (2006) reported a probabilistic characterization of load-displacement curves using an augered cast-in-place (ACIP) pile load test database. The normalized hyperbolic curve considered in their study is expressed as:

$$\frac{Q}{Q_{\mathrm{STC}}} = \frac{y}{a + by} \tag{31.12}$$

in which Q = applied load, Q_{STC} = failure load interpreted using the slope tangent method, a and b = curve-fitting parameters and y = pile butt displacement. Note that the curve-fitting parameters are physically meaningful—the reciprocals of a and b are equal to the initial slope and asymptotic value, respectively.

Each continuous load-displacement curve can be reduced to two curve-fitting parameters (plotted as a single point in Fig. 31.12). The scatter in the load-displacement curves is captured by the scatter between a and b. If the values of a are plotted as a histogram, a non-uniform distribution will be obtained (Fig. 31.13). For example, values close to 5 mm are more frequently encountered. The standard approach is to fit one of the many classical distributions (e.g. lognormal distribution) to the histogram. The crucial point to be emphasized here is that such an approach implicitly assumes that a and b are statistically independent random variables. However, Fig. 31.12 clearly shows that the variation of a (variation along y-axis) and the variation of b (variation along x-axis) are coupled. In other words, it is incorrect to assume that a can vary independently of b. The correct probabilistic model in this case is not independent random variables but a bivariate random vector. Computational details on the construction of this bivariate random vector are given elsewhere (Phoon and Kulhawy 2008).

31.4 Other Potential Roles of Risk-Based Analysis

One of the important requirements in forensic geotechnical engineering is the identification of failure mechanisms. A possible role for risk-based analysis can be discerned from the simulation studies detailed below.

Popescu et al. (1997) studied the effects of spatial variability of soil properties on soil liquefaction for a saturated soil deposit subjected to seismic excitation. They compared standard results derived from deterministic inputs and probabilistic results derived from stochastic finite element analyses. They concluded that both the pattern and the amount of dynamically induced pore water pressure buildup are strongly influenced by the spatial variability of soil parameters. For the same average values of soil parameters, more pore water pressure build-up is predicted by the stochastic model than by the deterministic model, which is attributed to a water injection phenomenon triggered by the presence of loose pockets in the spatially variable soil deposit.

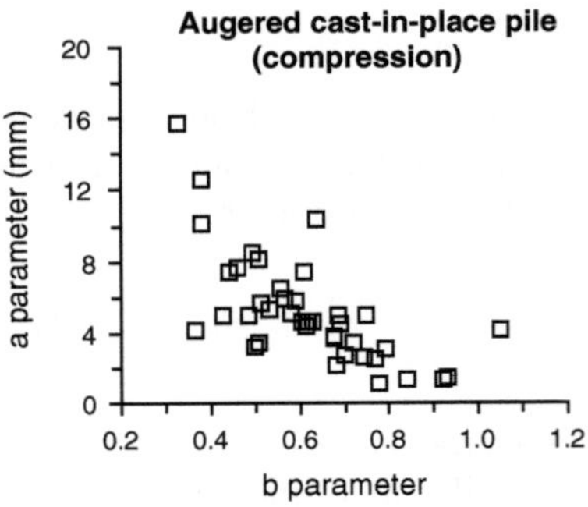

	No. of tests =	40
a	Mean =	5.15 mm
	Standard deviation =	3.07 mm
	Coefficient of variation =	0.60
b	Mean =	0.62
	Standard deviation =	0.16
	Coefficient of variation =	0.26
	Correlation =	-0.67

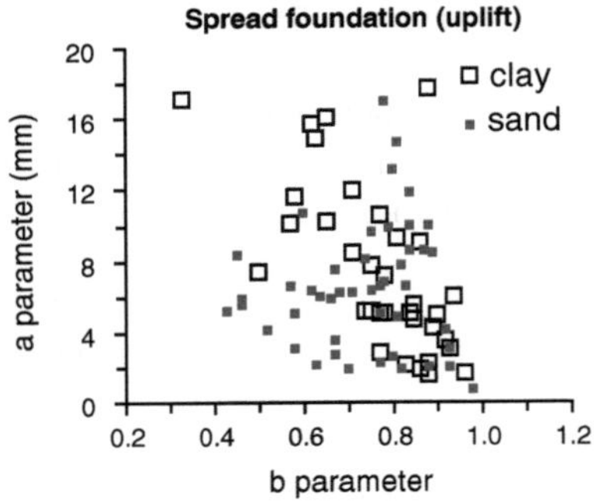

	No. of tests =	85
a	Mean =	7.13 mm
	Standard deviation =	4.66 mm
	Coefficient of variation =	0.65
b	Mean =	0.75
	Standard deviation =	0.14
	Coefficient of variation =	0.18
	Correlation =	-0.24

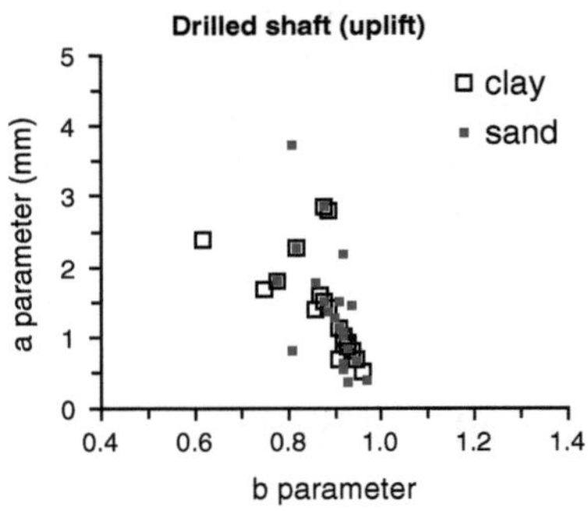

	No. of tests =	48
a	Mean =	1.34 mm
	Standard deviation =	0.73 mm
	Coefficient of variation =	0.54
b	Mean =	0.89
	Standard deviation =	0.063
	Coefficient of variation =	0.07
	Correlation =	-0.59

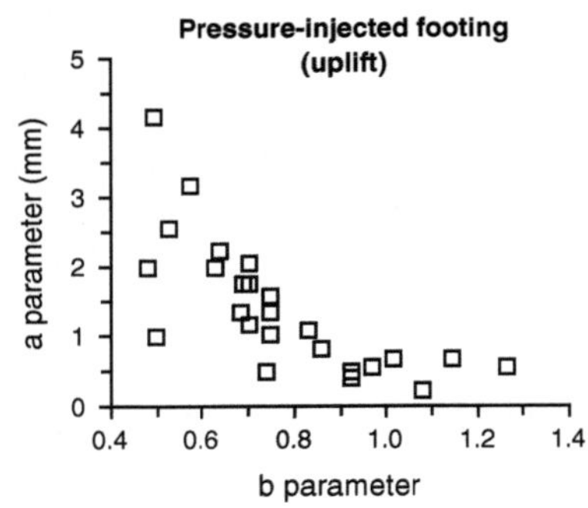

	No. of tests =	25
a	Mean =	1.38 mm
	Standard deviation =	0.95 mm
	Coefficient of variation =	0.68
b	Mean =	0.77
	Standard deviation =	0.21
	Coefficient of variation =	0.27
	Correlation =	-0.73

Fig. 31.12 Correlation between hyperbolic parameters (Phoon et al. 2006, 2007)

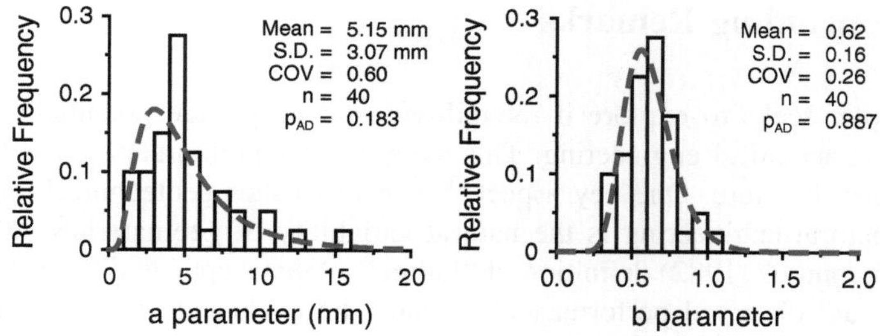

Fig. 31.13 Marginal distributions of hyperbolic parameters (Phoon et al. 2006)

Griffiths et al. (2002) demonstrated that failure mechanisms of a strip footing are considerably different in the case of uniform soils and spatially variable soils. The spatial correlation structure of the foundation soil in terms of autocorrelation structure has an important effect on the failure mechanism. More extensive studies are presented by Kim (2005).

Goldsworthy (2007) presented a fairly comprehensive risk-based study of the effect of site investigations on foundation failures. One conclusion is that an optimum site investigation minimizing the risk of failure can be planned based on the variability and autocorrelation structure of the foundation soil.

The above studies and other similar efforts demonstrate that failures can be studied in a more realistic way using spatially variable soils, rather than the traditional uniform or layered soil profiles. Notwithstanding this, a significant number of failures does arise from human failures and a lack of proper quality assurance and control plans. Bea (2006) suggested that the failure development process can be categorized into three phases viz, initiating, contributing and propagating causes and a proper risk management strategy is necessary using observational methods.

A risk-based approach to forensic geotechnical engineering may include elements such as:

i. Detailed soil investigations in the area in the form of vertical and horizontal soil profiles to formulate plausible hypotheses concerning prevalent/relevant failure mechanisms. The number and spatial location of boreholes and spacing should be adequate to provide a proper estimate of the mean, variance and autocorrelation properties of in situ soils.
ii. Analysis of the actual failure mechanism in relation to results from probabilistic studies.
iii. Reanalysis of different loading and the associated variability.
iv. Re-examination of quality assurance plans which include specifications on quality control of materials and construction, construction sequences proposed and adopted, observations of performance depending on the nature of project such as deformations, pore pressures, etc. which are expected in important projects and
v. Reliability-based computational back analysis.

31.5 Concluding Remarks

This chapter seeks to explore if reliability/risk concepts are potentially useful to forensic geotechnical engineering. This study is very preliminary given the dearth of previous literature. One key aspect that distinguishes geotechnical engineering from structural engineering is the natural variability of geomaterials. Within this context, Leonards (1982) definition of "failure"—"unacceptable difference between expected and observed performance"—cannot be evaluated in a meaningful way using deterministic methods. In broad terms, forensic engineering is related to the investigation of failures with the view of rendering an opinion regarding responsibility. Hence, a statistical measure of "unacceptable difference" (specifically, a difference not explainable by underlying variability) should provide useful additional information in the formulation of such an opinion. A preliminary statistical framework is presented to quantify the difference between expected and observed performance in the presence of unavoidable and potentially significant geotechnical variabilities. It should be emphasized that the quality of the statistical analysis can only be as precise, accurate and meaningful as the engineer's characterization of uncertainties. Other potentially useful results in the recent reliability and risk literature are highlighted. The intention of this chapter is to stimulate further discussions and research in this important but somewhat overlooked area.

References

Bea R (2006) Reliability and human factors in geotechnical engineering. J Geotech Geoenviron Eng ASCE 132(5):631–643

Brown ET (2006) Forensic engineering for underground construction. In: Proceedings of the ISRM international symposium 2006 and the 4th Asian rock mechanics symposium, Singapore, 8–10 November 2006, Chap. 1, pp 1–16

Chilès J-P, Delfiner P (1999) Geostatistics—modeling spatial uncertainty. Wiley, New York

Duncan JM (2000) Factors of safety and reliability in geotechnical engineering. J Geotech Geoenviron Eng ASCE 126(4):307–316

Goldsworthy JS (2007) Quantifying the risk of geotechnical site investigations. PhD Thesis, University of Adelaide

Griffiths DV, Fenton GA, Manoharan N (2002) Bearing capacity of rough rigid strip footing on cohesive soil: probabilistic study. J Geotech Geoenviron Eng 128(9):73–755

Jones AL, Kramer SL, Arduino P (2002) Estimation of uncertainty in geotechnical properties for performance-based earthquake engineering. PEER report 2002/16, Pacific Earthquake Engineering Research Center, University of California, Berkeley

Kim H (2005) Spatial variability in soils: stiffness and strength. PhD Thesis, Georgia Institute of Technology

Kulhawy FH, Trautmann CH (1996) Estimation of in-situ test uncertainty. Uncertainty in the geologic environment—from theory to practice (GSP 58). ASCE, New York, pp 269–286

Kulhawy FH, Phoon KK, Prakoso WA (2000) Uncertainty in the basic properties of natural geomaterials. In: Proceedings of 1st international conference on geotechnical engineering education and training, Sinaia, Romania, pp 297–302

Leonards GA (1982) Investigation of failures. J Geotech Eng Div ASCE 108(GT2):187–246

Paikowsky SG (2002) Load and resistance factor design (LRFD) for deep foundations. In: Proceedings of international workshop on foundation design codes and soil investigation in view of international harmonization and performance based design, Tokyo, Japan, pp 59–94. Balkema, The Netherlands

Phoon KK, Kulhawy FH, Grigoriu MD (1995) RBD of foundations for transmission line structures. Report TR-105000, electric power research institute (EPRI), Palo Alto

Phoon KK, Kulhawy FH (1999a) Characterization of geotechnical variability. Can Geotech J 36 (4):612–624

Phoon KK, Kulhawy FH (1999b) Evaluation of geotechnical property variability. Can Geotech J 36(4):625–639

Phoon KK, Kulhawy FH (2005) Characterization of model uncertainties for laterally loaded rigid drilled shafts. Geotechnique 55(1):45–54

Phoon KK, Kulhawy FH (2008) Serviceability limit state reliability-based design, Chap. 9. Reliability-based design in geotechnical engineering: computations and applications, Taylor & Francis, April 2008, pp 344–383

Phoon KK, Chen J-R, Kulhawy FH (2006) Characterization of model uncertainties for augered cast-in-place (ACIP) piles under axial compression. Foundation analysis & design: innovative methods (GSP 153), ASCE, Reston, pp 82–89

Phoon KK, Chen JR, Kulhawy FH (2007) Probabilistic hyperbolic models for foundation uplift movement, probabilistic applications in geotechnical engineering (GSP 170). ASCE, Reston, CDROM

Popescu R, Prevost JH, Deodatis G (1997) Effects of spatial variability on soil liquefaction: some design recommendations. Geotechnique 47(5):1019–1036

Réthâti L (1988) Probabilistic solutions in geotechnics. Elsevier, New York

Silva F, Lambe TW, Marr WA (2008) Probability and risk of slope failure. J Geotech Geoenviron Eng ASCE 134(12):1691–1699

Sowers GF (1993) Human factors in civil and geotechnical engineering failures. J Geotech Eng ASCE 119(2):238–256

Uzielli M (2004) Variability of stress-normalized CPT measurements and application to seismic liquefaction initiation assessment, University of Florence (Italy). Thesis downloadable at: http://www.georisk.eu

Uzielli M, Lacasse S, Nadim F, Phoon KK (2006) Soil variability analysis for geotechnical practice. In: Proceedings of second international workshop on characterisation and engineering properties of natural soils, Singapore. Balkema, The Netherlands, pp 1653–1752

Vanmarcke EH, (1977) Probabilistic modeling of soil profiles. J Geotech Eng ASCE 103, 1227–1246

Vanmarcke EH (1983) Random field: analysis and synthesis. MIT Press, Cambridge

Chapter 32
Important Role of Uncertainty in Forensic Geotechnical Engineering

R.B. Gilbert

Abstract Uncertainty plays a significant role in forensic geotechnical engineering. There is generally uncertainty in the available evidence and there is subsequently uncertainty in any conclusions drawn from it. Probability theory highlights the following principles when drawing conclusions from evidence:

1. If evidence can be explained by multiple hypotheses, then it is impossible to draw conclusions with certainty;
2. Additional evidence does not necessarily reduce and can actually increase uncertainty in the conclusions drawn from it; and
3. Neglecting hypotheses from the set of all possibilities can lead to erroneous conclusions.

Three case histories are presented to demonstrate these principles in practice. These case histories underscore the importance of considering uncertainty carefully and in communicating uncertainty clearly in conducting forensic analyses.

Keywords Bayes theorem · Evidence · Hypotheses · Case histories

32.1 Introduction

Uncertainty plays a significant role in forensic geotechnical engineering. There is generally uncertainty in the available evidence and there is subsequently uncertainty in any conclusions drawn from it. It is important to consider uncertainty carefully in conducting a forensic analysis and it is equally important to communicate uncertainty clearly while presenting conclusions.

R.B. Gilbert (✉)
Department of Civil, Architectural and Environmental Engineering,
The University of Texas at Austin, Austin, TX 78712, USA
e-mail: bob_gilbert@mail.utexas.edu

V.V.S. Rao and G.L. Sivakumar Babu (eds.), *Forensic Geotechnical Engineering*,
Developments in Geotechnical Engineering, DOI 10.1007/978-81-322-2377-1_32

Probability theory provides valuable insight into the role of uncertainty in drawing conclusions. Bayes' theorem is a mathematical representation of learning from evidence:

$$P(\text{Hypothesis is True}|\text{Evidence}) = P(\text{Evidence}|\text{Hypothesis is True})$$
$$P(\text{Hypothesis is True without Evidence}) \Big/ \sum\nolimits_{\downarrow}(\text{all Hypothesis}) \equiv P(\text{Evidence}|\text{Hypothesis})$$

where P(Hypothesis is True|Evidence) is the probability of interest, the probability that a conclusion is true given the available evidence; P(Evidence| Hypothesis is True) is the probability for the evidence if the hypothesis is true; and the summation in the denominator includes all possible hypotheses.

Bayes' theorem highlights three useful principles:

1. If evidence can be explained by multiple hypothesis, then it is impossible to draw conclusions with certainty;
2. Additional evidence does not necessarily reduce and can actually increase uncertainty in the conclusions drawn from it; and
3. Neglecting hypotheses from the set of all possibilities can lead to erroneous conclusions.

Consider simple examples to illustrate these principles. Figure 32.1 shows a case with two possible hypotheses, A and B. The probabilities for these hypotheses without evidence are weighted heavily toward Hypothesis B (Fig. 32.1a). However, the probability of obtaining the evidence is three times greater for Hypothesis A versus Hypothesis B (Fig. 32.1b). Subsequently, the updated probabilities for the hypotheses from applying Bayes' theorem reflect uncertainty (Fig. 32.1a); both hypotheses are still possible and a conclusion cannot be drawn with certainty given the evidence. Furthermore, since the evidence contradicts the prior expectation that Hypothesis B is more probable, the evidence actually increases the uncertainty in the conclusion; instead of the updated probabilities moving closer to certainty [P (Hypothesis A) = 0 and P(Hypothesis B) = 1 or P(Hypothesis A) = 1 and P (Hypothesis B) = 0], the updated probabilities move closer to maximum uncertainty [P(Hypothesis A) = 1/2 and P(Hypothesis B) = 1/2] when the evidence is considered (Fig. 32.1a). Figure 32.2 shows a case with three possible hypotheses, A, B, and C. If the probabilities for these hypotheses without evidence neglect the presence of Hypothesis C (Fig. 32.2a), even though the probability of obtaining the evidence is greatest with this hypothesis (Fig. 32.2b), then the updated probabilities are weighted heavily toward Hypothesis A being correct. However, if the probabilities for these hypotheses without evidence include the presence of Hypothesis C (Fig. 32.2c), then the updated probabilities indicate that Hypothesis C is more likely to be correct than either A or B (Fig. 32.2c).

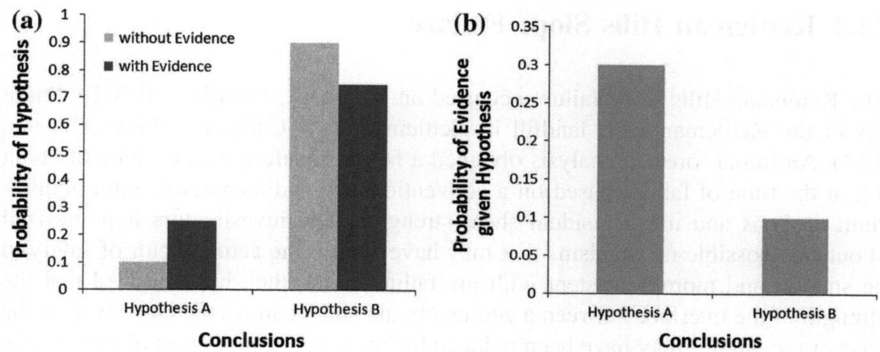

Fig. 32.1 Example illustrating that evidence does not necessarily reduce and can increase uncertainty

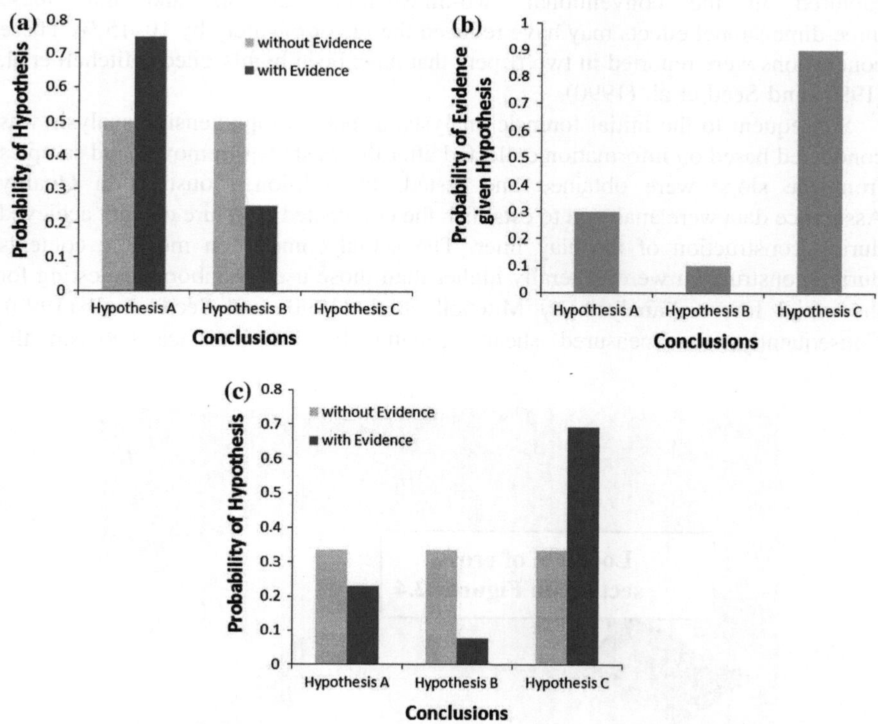

Fig. 32.2 Example illustrating that neglecting a possible hypothesis can lead to an erroneous conclusion. Three case histories are presented in the following sections to demonstrate these theoretical principles in practice.

32.2 Kettleman Hills Slope Failure

The Kettleman Hills slope failure occurred on March 18, 1988 in Cell B-19, Phase 1A of the Kettleman Hills landfill in Kettleman City, California (Figs. 32.3 and 32.4). An initial forensic analysis obtained a factor of safety for the slope of about 1.2 at the time of failure based on a conventional, two-dimensional, limit equilibrium analysis and using residual shear strengths. The investigators hypothesized about two possible mechanisms that may have caused the actual factor of safety to be smaller and more consistent with the failure. First, they hypothesized that the strength of the interface between a geomembrane and a compacted clay layer on the base of the landfill may have been reduced by "wetting," a migration of moisture in the clay to near the interface after it was constructed. They concluded that this "wetting" may have reduced the factor of safety by about 10 %. They also hypothesized that the three-dimensional geometry of the slip surface was not captured in the conventional two-dimensional analysis and that these three-dimensional effects may have reduced the factor of safety by 10–15 %. These conclusions were reported in two papers that have been highly cited, Mitchell et al. (1990) and Seed et al. (1990).

Subsequent to the initial forensic analysis, a more comprehensive analysis was conducted based on information collected after the waste was removed and samples from the slope were obtained and tested. In addition, Construction Quality Assurance data were analyzed to establish the compacted moisture content achieved during construction of the clay liner. The actual compaction moisture contents during construction were generally higher than those used in laboratory testing for the initial forensic analysis by Mitchell et al. (1990) and Seed et al. (1990). Consequently, the measured shear strengths for the interface between the

Fig. 32.3 Aerial photograph of Kettleman Hills Landfill before slope failure (*photo* provided by Chemical Waste Management, Inc.)

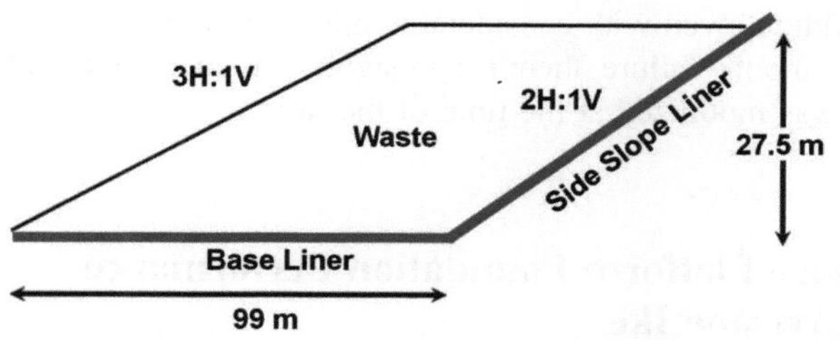

Fig. 32.4 Cross-section through Kettleman Hills Landfill before slope failure

geomembrane and compacted clay liner on the base of the landfill were less than those obtained in the initial analysis. The primary conclusion from the subsequent analysis was that the calculated factor of safety from a conventional, two-dimensional, limit equilibrium analysis was about 1.0 and consistent with the failure (Byrne et al. 1992). A second conclusion from this analysis was that three-dimensional effects did not reduce the factor of safety and, if anything, probably increased the factor of safety. A third conclusion from this analysis was that the shear strengths mobilized at the time of failure were probably close to the peak shear strength on the side slope and the residual shear strength on the base. This conclusion was further reinforced by Stark et al. (2001).

This slope failure was analyzed further to better understand the relationship between the mobilized shear strengths at failure and the peak and residual shear strengths. Numerical analyses were conducted to maintain compatibility between stresses, strains, and displacements. An analysis was then performed with all of the evidence using a formal implementation of Bayes' theorem to account for uncertainty (Gilbert et al. 1998). The conclusion from this analysis was that the average strength mobilized along the base of the landfill was probably between the peak and residual strength (Fig. 32.5).

In summary, initial forensic analyses that did not fully capture the uncertainty in the evidence led to unsubstantiated conclusions about the Kettleman Hills Landfill

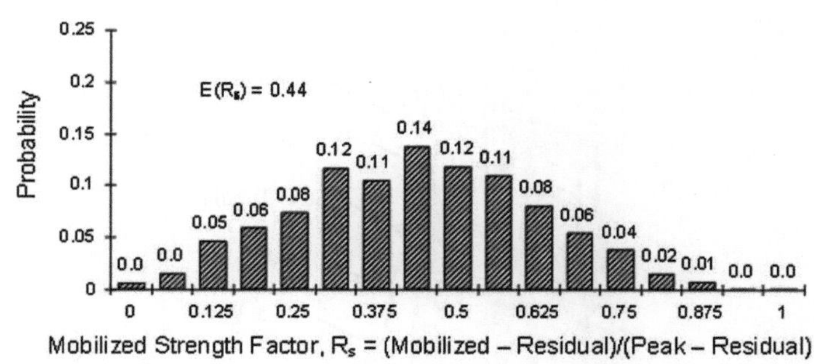

Fig. 32.5 Probability distribution for mobilized shear strength on base of Kettleman Hills Landfill based on Bayesian analysis of evidence (from Gilbert et al. 1998)

failure. In addition, even with considerable information obtained from a variety of investigations into the failure, there is still significant uncertainty in the actual shear strength that was mobilized at the time of the failure.

32.3 Offshore Platform Foundation Performance in Hurricane Ike

In 2008, Hurricane Ike loaded many offshore platforms in the Gulf of Mexico to beyond their design capacity and some platforms even to their ultimate capacity (Energo 2010). One platform loaded to its ultimate capacity was an eight-leg structure located in about 30 m of water (Fig. 32.6). The foundation system for this structure is eight steel pipe piles that are driven on a batter 52 m into the seafloor. The stratigraphy over the length of the piles consists of 50 m of soft to stiff clay with a 10-m thick layer of medium to very dense sand near the pile tips.

For purposes of continuing to use this structure after the hurricane, a forensic analysis was performed of its performance during Ike. First, the hindcast oceanic and atmospheric data from the hurricane were analyzed to develop the maximum sea state conditions at the location of the platform. Then, the entire structural system including the steel pipe piles was analyzed using a three-dimensional, nonlinear Finite Element Method model.

The results of the analysis are shown in Fig. 32.7. This figure shows an interaction curve for combinations of base shear and overturning moment on the foundation system that lead to collapse; the flatter left-hand side of this curve corresponds to a shear-dominated failure where the piles are failing laterally, while the steeper right-hand side of this curve corresponds to an overturning-dominated failure where the piles are failing axially. Figure 32.7 also shows the maximum load in Hurricane Ike estimated from the hindcast analysis. This forensic analysis indicates that the foundation system was expected to fail during the hurricane (Fig. 32.7). However, the structure was intact with only minor structural damage above the mudline when it was inspected after the hurricane. Therefore, further

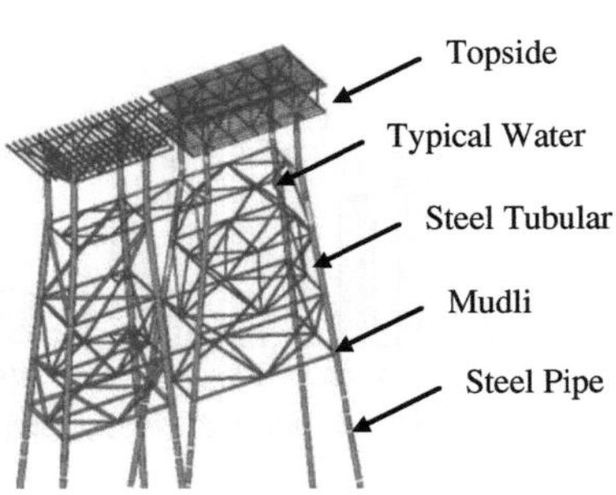

Fig. 32.6 Schematic rendering of 8-leg jacket loaded to ultimate capacity in Hurricane Ike

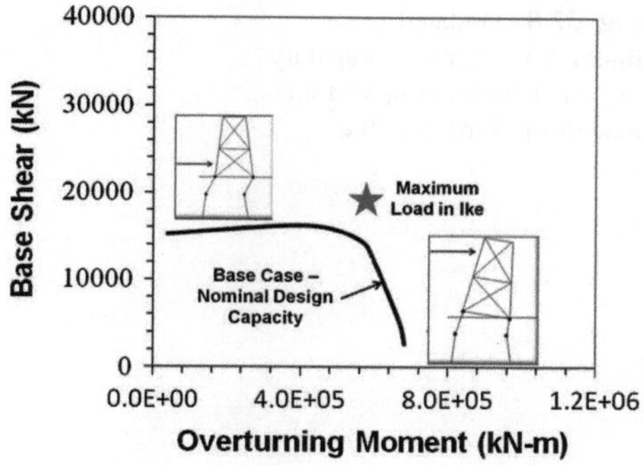

Fig. 32.7 Interaction curve for capacity of jacket foundation system loaded in Hurricane Ike

forensic analyses were required in order to explain its survival and update its capacity before the platform could be used again.

Since the "failure" of the structure in the structural analysis was taking place in the foundation system, the structural engineers analyzing the platform focused on the shear strength of the soil to explain the survival. The structural engineers incrementally increased the undrained shear strength of the soft to stiff clay until the capacity of the foundation system exceeded the load in Hurricane Ike. An increase of 300 % to the profile of undrained shear strength versus depth was required to explain the survival. While a 300 % increase in the undrained shear strength explains the survival of this platform, it is not considered to be plausible or realistic (particularly to a geotechnical engineer).

The potential failure mechanism for the foundation system in the Hurricane Ike loading is dominated by shear versus overturning (Fig. 32.7). The lateral capacity of the pile system is relatively insensitive to the shear strength of the soil (e.g., Gilbert et al. 2010), which is why a 300 % increase in the shear strength of the soil is required to produce a relatively modest increase in the capacity of the foundation system. However, the lateral capacity is sensitive to the structural capacity of the piles (i.e., the yield strength of the steel) and to the structural details at the pile heads near the mudline. In addition, when the ultimate capacity of the foundation system is mobilized, the piles will push laterally into soil that has not been degraded by previous cyclic loading during the storm. Therefore, an ultimate lateral resistance corresponding to static loading is more reasonable than one that reflects cyclic degradation. Figure 32.8 shows the capacity of the foundation system using a mean versus nominal yield strength for the steel piles, accounting for jacket leg stubs that extend several meters below the mudline and using static rather than cyclic *p*–*y* curves. With these small and plausible adjustments to the structural analysis, the survival of this foundation system in Hurricane Ike can be readily explained (Fig. 32.8).

In sum, by not considering plausible hypotheses to explain the survival of this offshore platform loaded to its capacity in Hurricane Ike, unrealistic conclusions were drawn from the evidence.

Fig. 32.8 Updated interaction curve for capacity of jacket foundation system loaded in Hurricane Ike

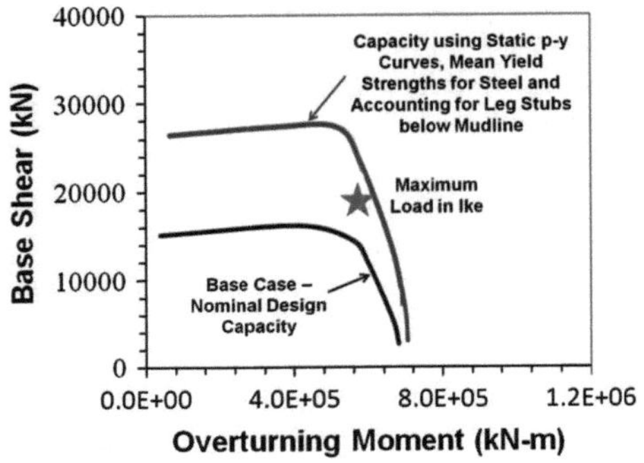

32.4 Landfill Cover Failure

A failure of a landfill cover slope occurred during construction (Fig. 32.9). A 0.6-m thick soil cover layer slid about 7 m down a 3H:1 V slope along the interface between a geocomposite drainage layer and a textured geomembrane. The geocomposite drainage layer ruptured in tension at the head scarp, and the cover soil layer bulged at the toe of the slide.

The initial forensic analysis focused on the shear strength of the failure plane. First, data from the Construction Quality Assurance testing were compiled and additional direct shear tests were conducted by the same laboratory (Laboratory A). Next, the contractor hired Laboratory B to conduct additional tests on samples from the site. These test data are summarized in Fig. 32.10. The results from Laboratories A and B were very similar and, while there is variability in individual tests results due apparently to spatial variability in the texturing of the geomembrane; there was a relatively high degree of confidence in the average shear strength for this interface. Subsequently, the design engineer hired Laboratories C and D to conduct additional shear tests (Fig. 32.10). These new data were statistically similar to one

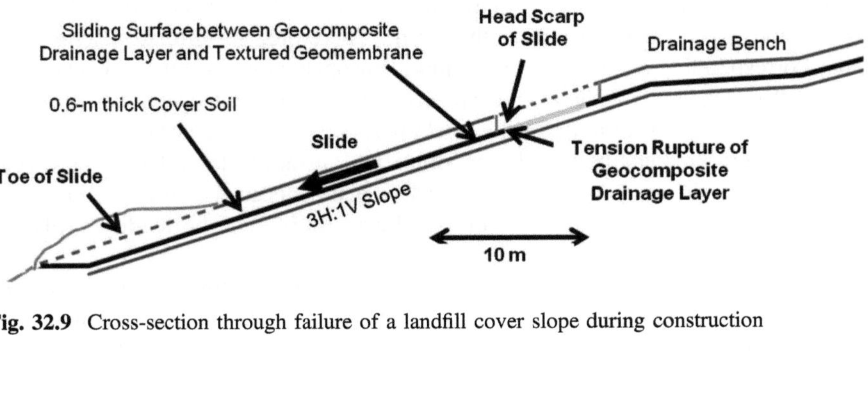

Fig. 32.9 Cross-section through failure of a landfill cover slope during construction

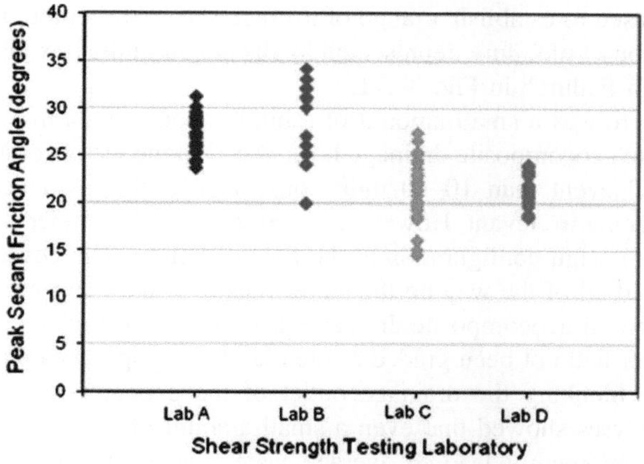

Fig. 32.10 Test data for shear strength of interface between geocomposite drainage layer and textured geomembrane from landfill cover slope failure

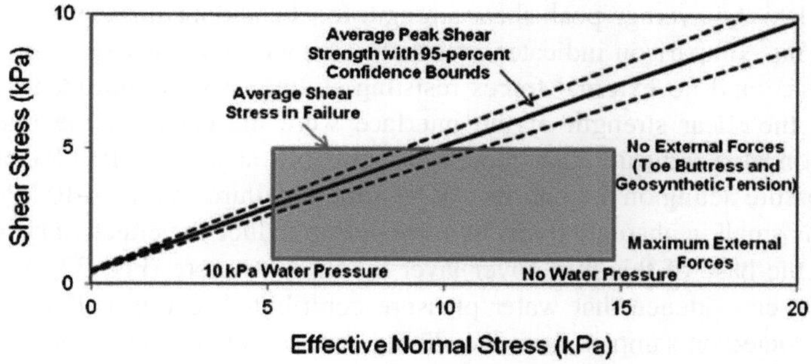

Fig. 32.11 Strength envelope for sliding surface from landfill cover slope failure

another but statistically different from those obtained from Laboratories A and B. Therefore, the uncertainty increased in the average shear strength for this interface as additional information was obtained. Furthermore, a hypothesis was formulated that spatial variability in the shear strength of the interface led to the failure. An assessment of the average shear strength envelope together with 95 % confidence bounds is shown in Fig. 32.11.

There was additional evidence beyond test data for the shear strength of the sliding surface. First, a limit equilibrium stability analysis was conducted, accounting for a range of possible contributions to sliding resistance from the tensile capacity of the geocomposite drainage layer at the head of the slide and the compressive capacity of the cover soil layer at the toe of the slide (Fig. 32.9). This analysis was used to establish a range of average shear stress mobilized on the slip surface at the onset of failure, represented by the height of the box labeled "Average Shear Stress in Failure" in Fig. 32.11.

Second, there was a small amount of rainfall, about 10–20 mm, shortly before the failure. The geocomposite drainage layer was designed to accommodate a much greater rainfall event than 10–20 mm; consequently, this evidence was initially dismissed as being irrelevant. However, the cover slope was under construction and not in its final design configuration at the time of failure. The cover soil layer had not been placed all of the way up the slope, meaning that there was a 3–5 m wide section of exposed geocomposite drainage layer at the crest of the slope. Also, the cover soil layer had not been graded at the toe of the slope, meaning that it was at least partially blocking the drainage outlet of the geocomposite drainage layer. Hydraulic analyses showed that even a small amount of precipitation under these conditions could readily lead to average water heads on the slide surface that exceeded the thickness of the cover soil. The range of possible water pressures is represented by the width of the box labeled "Average Shear Stress in Failure" in Fig. 32.11.

The mobilized average shear stress on the sliding surface is compared in Fig. 32.11 to the average peak shear strength for the sliding surface based on shear testing. This comparison indicates first that, even with no water pressure acting on the interface and no external forces resisting sliding, it was improbable that variations in the shear strength of the interface were the cause of the failure. This comparison also indicates that there was probably at least 5–10 kPa of average water pressure acting on the interface at the time of failure. While 5–10 kPa of water pressure is small in absolute terms, it is enough to reduce the effective normal stress acting at the base of this thin cover layer by 50 % or more (Fig. 32.11).

As further evidence that water pressure contributed to this failure, shear tests were conducted on samples from the sliding surface obtained after the failure. If an effective normal stress of 15 kPa had been acting on this interface in the field (i.e., no water pressure), then a residual shear strength for this effective normal stress would have developed after about 7 m of shear displacement in the field. Tests on fresh samples of the interface at this effective normal stress produced residual shear strengths that were about 60 % of the peak shear strengths. However, shear tests on the interface tested at an effective normal stress of 15 kPa after the failure showed no evidence of post-peak strength loss, consistent with shearing in the failure having occurred at a much smaller effective normal stress.

In sum, this landfill cover slope failure illustrates that uncertainty does not necessarily decrease with additional evidence. It also illustrates that it is important to consider all of the evidence in formulating and evaluating hypotheses as to the cause of a failure.

32.5 Conclusion

Probability theory highlights the following principles when drawing conclusions from evidence:

1. If evidence can be explained by multiple hypotheses, then it is impossible to draw conclusions with certainty;
2. Additional evidence does not necessarily reduce and can actually increase uncertainty in the conclusions drawn from it; and
3. Neglecting hypotheses from the set of all possibilities can lead to erroneous conclusions.

Three case histories were presented to demonstrate these principles in practice. The Kettleman Hills Landfill slope failure showed that there was a significant uncertainty in the shear strength mobilized in the failure even after numerous forensic analyses. The offshore platform survival in Hurricane Ike showed that unrealistic conclusions were drawn when plausible hypotheses were neglected. The landfill cover slope failure showed that uncertainty increased with additional evidence and that all evidence needed to be considered holistically in drawing conclusions.

In conclusion, it is important to consider uncertainty carefully in conducting a forensic analysis and it is equally important to communicate uncertainty clearly in presenting conclusions from that analysis.

References

Byrne RJ, Kendall J, Brown S (1992) Cause and mechanism of failure Kettleman Hills Landfill B-19, Phase IA. In: Proceedings, ASCE specialty conference on stability and perform. of slopes and embankments II, Berkeley, pp 1152–1187

Energo Engineering (2010) Assessment of damage and failure mechanisms for offshore structures and pipelines in Hurricanes Gustav and Ike, final report prepared for minerals management services, MMS TAR no. 642

Gilbert RB, Wright SG, Liedtke E (1998) Uncertainty in back analysis of slopes. J Geotech Eng ASCE 124(12):1167–1176

Gilbert RB, Chen JY, Materek B, Puskar F, Verret S, Carpenter J, Young A, Murff JD (2010) Comparison of observed and predicted performance for jacket pile foundations in hurricanes. In: Proceedings of offshore technology conference, paper no. OTC 20861, Houston, Texas

Mitchell JK, Seed RB, Seed HB (1990) Kettleman Hills waste landfill slope failure. I: liner system properties. J Geotech Eng ASCE 116(4):647–668

Seed RB, Mitchell JK, Seed HB (1990) Kettleman Hills waste landfill slope failure. II: stability analyses. J Geotech Eng ASCE 116(4):669–690

Stark TD, Eid HT, Evans WD, Sherry PE (2001) Municipal solid waste slope failure. II: stability analyses. J Geotech Eng ASCE 126(5):408–419

Chapter 33
Technical, Ethical, and Legal Issues with Forensic Geotechnical Engineering—A Case History

D.S. Saxena

Abstract For geotechnical engineers it is only a myth to believe that practicing perfect engineering, or conforming to normal standards of care, will provide immunity from civil liability. Unfortunately, when problems or failures occur, all parties including engineers get named in the lawsuit regardless of their innocence. In USA, attorneys sue everyone involved with a damaged project to secure compensation or claim to their liability insurance limit. A practicing geotechnical engineer cannot provide services without the fear of a lawsuit. Strategies for limiting liability range from assessing risk to securing professional liability insurance and including limitation of liability classes in consulting contracts between engineer and the client. Although experts retained by opposing parties generally disagree on issues resulting from differences in professional judgment, they are invaluable to the jurisprudence system in America. This paper presents some legal issues involving the jurisprudence system and introduces a case history that includes all elements of forensic geotechnical engineering.

Keywords Liability insurance · Jurisprudence · Lawsuit · Expert witness

33.1 Introduction

When something breaks, falls down, or otherwise fails to perform as it was intended, our society demands to know why. This inevitably becomes the job of the forensic engineer, who must be equally comfortable in the field, the office, and the courtroom while providing an objective analysis of the failure, whether the facts are helpful or harmful to the client's interests.

D.S. Saxena (✉)
Director of Forensic/Geo-Forensic Engineering, Biller Reinhart Engineering Group, 102 South Florida Avenue, Suite 102, Lakeland, FL 33803, USA
e-mail: E-dsaxena@billerreinhart.com

V.V.S. Rao and G.L. Sivakumar Babu (eds.), *Forensic Geotechnical Engineering*, Developments in Geotechnical Engineering, DOI 10.1007/978-81-322-2377-1_33

A new discipline known as forensic geotechnical engineering (FGE) has been created to deal with investigation of soil-interaction-related failures of engineered facilities or structures. Services of forensic geotechnical engineers are generally commissioned to investigate such failures. They also prepare reports, and sometimes provide expert witness testimony, in an attempt to assist their client in understanding the cause and liability for the failure. Because of the adversarial nature of most failure investigations many forensic engineers are generally under pressure to give an opinion that benefits their client's position. Accordingly, the technical facts are not presented; yielding to this kind of pressure sometimes leads to inappropriate occurrences, practicing outside one's competence, manipulation of facts, crafted testimony, inadequate or defective investigations, misrepresentation of standards of practice, or other ethical violations.

A forensic geotechnical engineer looks at a pile of rubble as a problem that needs to be solved. Each broken piece in the pile tells a story about the forces that acted upon it and somewhere in the twisted mass of intertwined building components is the first facture surface that resulted in the progressive collapse of the structure. At the end of the failure analysis process, which may include laboratory testing of critical items and hours of analysis in the office, a clear picture of what happened begins to develop, and the forensic engineer must then be able to explain the mode of failure and the forces, action, or process that caused the failure. The explanation may be a simple letter, a detailed report, or expert testimony in the courtroom (Bell and John 2007).

FGE prepares civil engineers to read, think, speak, and analyze like a lawyer. In addition, it familiarizes him with the jurisprudence system so that he is much able to understand and deal with legal issues since he has to work closely with statutes and regulations, may become involved in litigation, or may serve as an expert witness.

33.1.1 Who Wants to Know What

A forensic geotechnical engineer needs to answer questions such as:

- The building owner who wants to know why the building fell down and who is going to pay for it.
- The insurance adjuster who needs to determine if the cause of the collapse is covered under the building owner's insurance policy, and if it was the result of a defect that someone else might be liable for.
- The contractor and subcontractors, who built the building and their liability carriers want to know if the defect is construction-related and if so, which trade or trades are responsible.
- The design professionals and their liability carriers want to know if the defect is design-related.

Each of these entities may retain their own forensic engineers to assist them with understanding how they may or may not be responsible.

There will be attorneys for each entity who will evaluate and argue how the law should be applied to the insurance policies and contracts. The forensic engineer's opinion of what happened may be explored during a deposition in which the attorneys for all sides have the opportunity to probe the engineer as to the process used and the basis for the opinion. If the parties cannot reach a settlement, the forensic engineer may be called to provide expert testimony in the courtroom, with the goal of helping the judge and/or jury understand the facts and how they relate to the particular incident.

Opposing attorneys normally attempt to discredit the engineer through cross-examination with the hopes of revealing some flaw in the forensic engineer's thought process.

The forensic engineer must be able to apply the art and science of engineering during the entire process, from the extreme conditions encountered in the field, through the intense pressure experienced in the courtroom, and never lose sight that seeking the truth is all that matters. By understanding what went wrong, the forensic engineer is able to add to the body of knowledge that hopefully will prevent a similar occurrence from happening again.

33.2 Ethical Considerations

Ethics is defined as a discipline dealing with good and bad, right and wrong, moral duty, and obligation.

Why does a group need a code of ethics? One does not have to look any further than the daily newspaper to find that unethical behavior exists in the highest levels of government, in the clergy, and in virtually every segment of society. Social scientists have studied and found that in any group a certain percentage will deviate from the norm. Therefore, there is the need to codify and regulate or influence the behavior of various professions and groups. But, when we speak of a code of ethics applicable to engineers do we mean merely the business practice with respect to another professional or a client, or are there higher goals and standards to which a professional engineer should aspire?

The NSPE code of ethics states that in the fulfillment of their professional duties engineers shall:

"1. Hold paramount the safety, health and welfare of the public in the performance of their professional duties; 2. Perform services only in the areas of their competence; 3. Issue public statements only in an objective and truthful manner; 4. Act in professional matters for each employer or client as faithful agents or trustees; 5. Avoid deceptive acts in the solicitation of professional employment."

Upon close examination of the canons, it is seen that none are in conflict with the practice of forensic engineering. The forensic engineer should adhere to the NSPE Cannons not only as a member of NSPE but also as an engineer (Dixon 1992).

Engineers are recognized as the most ethical professionals above doctors, accountants, and lawyers in a survey of 200 CEOs of national companies.

While ethical core values for engineers are generally identified as integrity, honesty, faithfulness, charity, responsibility, and self-discipline the classical seven deadly sins are described as pride, greed, anger, envy, gluttony, lechery, and sloth.

It is clear that there are problems in ensuring the integrity of expert testimony. There are no minimum qualifications for becoming an expert witness, nor any means of regulating them. Code of ethics are voluntary and not often enforced (Grover et al. 2003).

In sum, the middle of the road rule for an ethical engineer is to be honest, tell the truth no matter how bad it may seem, and let your conscience be your guide (Baker and John 1997).

33.3 Legal Issues

The legal considerations of these forensic geotechnical engineering services illustrate the reality that the engineering investigation of a failure incident is a fact-finding mission that results in uncovering the probable causes of that failure. It concentrates on the identification of hidden clues. The procedures adopted for the analysis, testing, opinions, and written reports should be able to satisfy even legal scrutiny of their validity.

33.3.1 Role of Expert Witness

Experts are vitally important to contemporary American jurisprudence. Expert witness can assist retained counsel in fact gathering, request for documents, and evaluation of evidence; fashioning precise questions for opposing experts and through candid discussions of strengths and weaknesses of the case; retained counsel to properly qualify and question an expert witness to maximize his impact on the jury; and expansion of the traditional testifying roles in the litigation process.

An expert in forensic engineering is not automatically dubbed an "expert witness." An individual is not granted the privileges of an expert witness until a legal forum confers that recognition upon a properly qualified professional.

Experts must possess specialized knowledge of science, profession, business, or occupation that is beyond the experience of the lay juror and can assist the fact finder (Grover et al. 2003). Additionally, expert witnesses in the field of engineering are governed by their own code of ethics.

An experienced expert prepares a well-written, persuasive expert based on reliable accepted methodology. An expert witness can express her opinion in an unambiguous and legally sufficient manner. Expert witness can dramatically improve their performance by being good teachers, both to the lawyers who hire them, and to jurors and other fact finders. They must keep it simple with illustrations, explanation, and nontechnical language.

The four key elements of a successful expert witness assignment are (SEAK 2006).

33.3.1.1 Part I: Prevention

The most valuable experts isolate themselves from attacks and deny opposing counsel ammunition to attack their credentials and credibility. It helps to identify a detailed checklist of potential areas of attack that experts may be subject to regarding their credentials and credibility including: every word on their CVs, past testimony, their image, controversial or political associations, missing credentials, fee schedules, fee agreements, marketing materials, web page, speeches, work on past case, apparent and actual conflicts of interest, nonrelated litigation, hobbies, professional complaints or discipline, presentations, and writings.

The best experts form and express airtight opinions and opinions that hold up under the most rigorous scrutiny and cross-examination. There are many ways in which opposing counsel is able to poke holes in an expert's opinion. The expert must provide specific actions to bullet-proof his opinion including, proper case and client selection, avoiding time crunches, using careful and confident language, not overstating or understand facts or opinions, consistency, dealing with the opinions of other experts, knowing exactly what needs to be proved, testing alternative theories, taking careful and precise measurements, being well-trained and well-versed in any computer program used, verifying computer results, leaving no stone unturned, taking photographs, verifying factual assumption, gaining as much firsthand knowledge as possible, thoroughly researching the issues at hand, obtaining and carefully reviewing all relevant documents, not sharing draft reports with counsel, and avoiding "junk science."

33.3.1.2 Part II: Preparation

Peak performance requires proper disciplined preparation done correctly prior to deposition and trial. Well-prepared experts are able to deliver confident testimony, deal with cross-examination far more effectively, and are in a better position to articulate and defend their opinions. It should include advanced techniques that can and should be used to prepare for depositions, direct examination, and cross-examination.

33.3.1.3 Part III: Performance

The best experts recognize that most cases are won and lost in the discovery phase and that the expert's deposition is a crucial, often outcome determinative, component of the case. In order to excel at the highest level during a deposition, experts need to be able to recognize and defeat opposing counsel's deposition tactics and recognize how these tactics differ from those used during trial.

The best experts deliver powerful and understandable direct testimony. They also learn to explain and demonstrate numerous advanced techniques for delivering captivating, memorable, and persuasive direct expert testimony. These advance techniques include, showing—not telling, getting to the point up front and explaining later, being well-prepared and well-organized, making the complex simple, entertaining, being likeable, highlighting your most relevant qualifications, working on a smooth flow and style, getting out of the jury box early and often, using visual aids that work, aggressively self-editing, employing powerful, memorable analogies, showing your human side and bonding with the jury, using precise language, using confident language, employing short preview and review summaries, using numbered lists, citing references, speaking conversationally, conforming your testimony to the theme of the case, and reading and reacting to the jury.

33.3.1.4 Part IV: Practice

The best experts skillfully answer cross-examination questions and explain how the question could have been avoided, how they could have and should have prepared to answer the question, identifying the tactic that counsel is using and delivering a response that defeats the tactic and/or allows the expert to go on the offensive.

33.3.2 Legal Considerations

In order to fully evaluate the forensic expert's role in the legal process, the expert should understand the definitions and concepts of a dispute resolution.

The professional services provided in the investigative process and the duties expected of the expert witness are generally the same no matter what method of formal dispute resolution is selected by the client. Following is an overview of the various methods of dispute resolution that may be encountered (Task Committee on Guidelines for Failure Investigation 1989).

33.3.2.1 Civil Litigation

This process occurs in a court of justice having jurisdiction in the dispute. The court is presided over by a judge who decides on the applications of the law and procedure. These decisions usually are determined by a jury of laypersons, or by the judge in the absence of a jury. The decision of this court and jury may be subjected to appeals and retrials.

33.3.2.2 Arbitration

Instead of a judge and/or jury, an arbitrator or a panel of arbitrators are used in a private proceeding. Unlike litigation, arbitration eliminates or significantly relaxes the discovery process and is usually limited to the production of documents to an extent determined by the arbitrators.

33.3.2.3 Administrative Hearings or Special Investigations

Hearings are held by governmental authorities to resolve disputes or to gather information about public opinion prior to making an administrative decision, or for inquiries into the integrity of the structure or the cause of a failure of an engineering facility.

33.3.2.4 Private Litigation

This form of dispute resolution requires both parties to agree to resolve their differences in a private court that is created, and paid for, by both parties. A mutually respected individual, such as a retired jurist, is chosen as a judge, and civil litigation rules are followed.

33.3.2.5 Mediation

This method of "non binding" dispute resolution is used to resolve certain construction-oriented disputes. A mediator acceptable to both sides is selected and probes the disputants to determine their position relative to "minimum" settlement. He or she then strives to attain a middle ground that both can accept comfortably.

33.4 A Case History

A case history where forensic geotechnical engineering was effectively utilized to identify, investigate, and remediate the problem is presented. It assisted in litigation and resolution of legal issues.

Project specifics and forensic facts summary;

- The backyard slope of a one-story dwelling subsided abruptly, failed, and slid into the lake along with the rip-rap from lake edge. It also extended into the neighboring property on the south side.

- The lakefront backyard of a residence was damaged and stability of the structure was threatened, and removal/replacement of the completed structure was considered a viable yet costly option.
- The owner filed a claim against the developer/builder for negligence and for not informing them of a potentially unstable preexisting condition.
- The owner secured the services of a Forensic Geotechnical Engineer (FGE) to investigate, remediate, and assist in litigation as well as serve as an expert witness during the resolution of dispute and legal issues.
- The developer/builder offered a band-aid solution that the owner then rejected upon advice of this expert.

Figures 33.1, 33.2, 33.3, 33.4, 33.5, 33.6 and 33.7 illustrate the salient features of this case history.

Fig. 33.1 Close-up view of the subsided backyard

Fig. 33.2 View of subsided slope of adjoining house

Fig. 33.3 View of subsided slope looking east

Fig. 33.4 Close-up of the underdrain installation

Fig. 33.5 View of sodded and finished backyard

Fig. 33.6 Recurrence of sheared slope

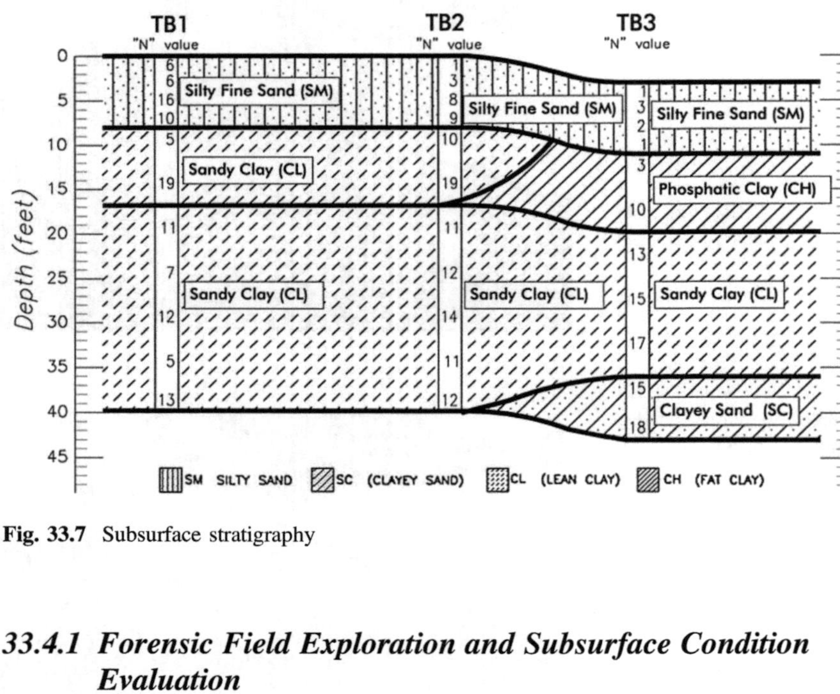

Fig. 33.7 Subsurface stratigraphy

33.4.1 Forensic Field Exploration and Subsurface Condition Evaluation

Subsurface conditions beneath the site were evaluated by advancing several test borings. Typical subsurface conditions, as illustrated in Fig. 33.7, consisted of very soft, loose low strength subsoils to 15.0 ft depth. These low strength subsoils consisting of highly plastic clays (mc 113 %, LL-107, PI-89) were unstable under the weight of the additional fill material.

Slope stability analyses were performed for the original subsurface profile as well as the contractor/developer proposed restructured slope that yielded Factors of Safety (F.S.) of 0.87 and 0.93, respectively, for the two conditions.

33.4.2 Observations and Findings

The expert's investigation at the project development site revealed that;

- Slope failure was triggered by the presence of unconsolidated sediment layers under the site fill.
- Unstable sediments would continue to consolidate resulting in continued movement of the back of the house and the backyard.
- A retaining wall should have been constructed prior to the house being built to contain and stabilize the soils beneath the house foundation.

Upon the advice of their legal counsel the owner put the developer on notice, and elected to proceed with the proper fix pending resolution of litigation and the claim.

33.4.3 Remediation

Based on findings and observations from the FGE investigation, a two-part remediation program was recommended and undertaken:

1. Installation of helical anchor piers under the exterior wall footing, along the back of the house, for the underpinning stabilization.
2. Installation of a helical anchor bulkhead to stabilize the backyard along the lake.

33.4.3.1 Part 1-Underpinning Stabilization Program

Based on the conclusions of the investigation, this part of the remediation program consisted of:

- Stabilization layout of nine (9) helical underpinning piers encompassed multiple helices.
- These piers were installed at 5.5 ft centers with a terminal depth of 15.0 ft and loaded to a capacity to completely stabilize the foundation without lift.

- They were torqued until a resistance criteria of 2500 psi for an approximate ultimate compression load of 37.5 kip was reached and monitored by a forensic engineer representative.

33.4.3.2 Part 2-Anchored Bulkhead Stabilization Program

- This part of the remediation program consisted of a total of six 10-in. (butt) diameter, and 14.0 ft long timber piles on 8.0 ft centers piles that were advanced by jetting and the embedded piles were then secured by quick setting grout.
- A total of six helical anchors with multiple helices that were installed in lengths ranging from 44.0 to 21.0 ft. since the shear plane traversed away from the east side wall at an angle as it moved from north to south.
- Anchors were installed to an approximate termination criteria and length so as to provide satisfactory embedment/resistance and capacity.
- The slope was stabilized with properly anchored geotextile that was secured and wrapped along the vertical face of the bulkhead wall.
- Fill material was placed in lifts, graded, and compacted by the construction.
- Equipment. Subsequent to final grading, the fabric was allowed to undergo tensioning for about a week under the fill material.
- Based on field monitoring of pier installation operations, anchored timber bulkhead, post stabilization walk through, and review/evaluation of the field stabilization data it was concluded that the methods and procedures utilized and installed by the specialty contractor were effective, satisfactory, and acceptable. Slope stability analyses of this remediated structure indicated a Factor of Safety (F.S.) of 1.57.

Figures 33.8, 33.9, 33.10, 33.11, 33.12, 33.13, 33.14 and 33.15 illustrate various sequences of the underpinning and bulkhead stabilization program.

Fig. 33.8 View of helical support anchor in place

Fig. 33.9 View of support cap and bracket assembly under the footing

Fig. 33.10 Setting up of the tie back helical anchor

Fig. 33.11 Hydraulic setup for achieving torque/ resistance capacity

Fig. 33.12 Close-up the helical anchor tie back with timber bulkhead

Fig. 33.13 Close-up of geotextile lay down and wrap up on timber bulkhead

Fig. 33.14 View of completed and restored backyard

Fig. 33.15 Additional fill being dozed into the anchored geotextile

33.5 Concluding Remarks

1. Following the close coordination between the forensic geotechnical engineer, the owner, and the remediation contractor and based on results of remediation monitoring it was determined that the repair of the residence structure had been achieved satisfactorily.
2. The final repair resulted in restoration of the residential structure and backyard to its originally planned and constructed stage.
3. As a result of the monitored and satisfactory remediation program, the consultant recommended acceptance of the restored structure.
4. The property owner retained an attorney who sent a notice of claim and remediation cost summary to the developer/builder. It is the author's understanding that the property owner was able to recover these remediation costs along with attorney fees during the mediation process.

33.6 Overview

A well-organized, methodical approach to failure investigation is needed because of the complexities of modern engineered facilities and the vast variety of possible failure causes. The investigative process should not employ a rigid "cookbook" approach, but there are certain steps that are common to an effective failure investigation.

An emphasis on an open-minded acceptance of all pertinent data, while recognizing the relevant information, is the fundamental basis of a failure investigation. As various hypotheses develop and change, continuous review of previously rejected data is required.

There is seldom a single cause of a failure but rather a complex interaction of components and forces. As a result, the outcome of a failure investigation seldom leads to absolutely irrefutable results but rather to a most probable cause of failure. While the collection of data and facts surrounding the failure is specific, expert opinions arising from these data may differ widely.

The most accepted failure investigation findings will be the one employing a qualified investigation team that presents the most plausible failure scenario, based on well-documented support data (Task Committee on Guidelines for Failure Investigation 1989).

References

Baker PE, John A (1997) Professional engineering ethics versus expert witness requirements, Los Angeles Section ASCE, Forensic Engineering Technical Group

Bell PE, John T (2007) What is forensic engineering? Florida Eng Soc J 8–9

Dixon EJ (1992) The NSPE code of ethics applied to forensic engineering. J Nat Acad Forensic Eng IX(1)

Grover JD, PE, JL (2003) Ethical considerations for expert witnesses in forensic engineering. Ethical Dilemmas Tech Forensic Pract 441–52

SEAK inc. (2006) Advanced testifying skills for experts—the master's program, pp 4–11

Task Committee on Guidelines for Failure Investigation (1989) Guidelines for failure investigation, American Society of Civil Engineer (ASCE)

Chapter 34
Forensic Engineering, Legal Considerations, and Property Damage Assessment from Construction Vibrations

D.S. Saxena, Prashanth Vaddu and Anu Saxena

Abstract Forensic geotechnical engineering is generally identified as a science concerned with relations between engineering and law. It does not generally embrace traditional disciplines. Other subdisciplines including geological, applied civil, hydrological, environmental, and geophysical must be considered prior to conducting an effective investigation and/or resolution of a failure. It is, therefore, obvious that the determination of the cause(s) of failure requires a thorough familiarity with an array of related disciplines and an ability to pursue different lines of investigation simultaneously. Specialized case history is presented where damage to sensitive structures from a variety of vibratory construction methods ranging from pile driving, highway construction, rock excavation and/or blasting, was considered. Twenty homeowners from a residential subdivision in the Florida Panhandle, USA, claimed damage including cracking to their residences during construction of various portions of city's storm water treatment/enhancement project. It was also alleged that during excavation and construction of drainage structures, especially four subsurface treatment vaults that required installation and extraction of sheet piles using vibratory hammer, stability of the nearby residential structures was endangered and resulted in movement/cracking. Forensic engineering analysis was effectively utilized to identify, investigate, and remediate the concerns as well as assist in litigation and in some cases avoidance. This technical

D.S. Saxena (✉)
Director of Forensic/Geo-Forensic Engineering, Biller Reinhart Engineering Group,
102 South Florida Avenue, Suite 102, Lakeland, FL 33803, USA
e-mail: E-dsaxena@billerreinhart.com

P. Vaddu · A. Saxena
Technologies, Inc, Lakeland, FL 33815, USA
e-mail: prashanth.vaddu@kci.com

A. Saxena
e-mail: anu.saxena@kci.com

V.V.S. Rao and G.L. Sivakumar Babu (eds.), *Forensic Geotechnical Engineering*,
Developments in Geotechnical Engineering, DOI 10.1007/978-81-322-2377-1_34

paper presents some legal issues related to identification, investigation, mediation, and resolution in conjunction with a case history that includes many elements of forensic engineering.

Keywords Liability insurance · Jurisprudence · Lawsuit · Expert witness

34.1 Introduction

Forensic geotechnical engineering is a relatively new discipline. By its very nature, forensic or forensic geotechnical engineering is not in itself a stand-alone traditional discipline. The analysis of geotechnical or applied civil engineering failure often involves application of an array of other disciplines. For example, the investigation of a slope and/or settlement failure would involve the application of statics, dynamics, structural configuration, geology, hydraulics, soil mechanics, and other proactive elements including historical maintenance. A failure is rarely the result of a single cause. It is, mostly a combination of a number of causes acting in concert or in a sequence. It is, therefore, obvious that the determination of the cause(s) of failures requires familiarity with a number of related disciplines and an ability to pursue different lines of investigation simultaneously.

34.2 Failures

Failure is generally defined as the inability of a component, structure, or facility to perform its intended function.

Additionally, failure does not necessarily involve a collapse or rupture and they are generally of three types.

- Safety Failure—involves death or injury or placing people at risk, examples include formwork installation during concrete placement, trench collapse, or slip and fall on wet floor.
- Functional Failure—involves compromise of intended usage of structure or facility, examples are excessive vibration of floor, roof leaks, poor acoustics, etc.
- Ancillary Failure—includes factors that perversely affect schedules, cost, or intended use. Examples include delayed construction or unexpected foundation problems.

It must also be pointed out that general causes of failure range from design deficiencies to construction errors to material defects and maintenance deficiencies.

Forensic analysis of geotechnical failure(s) must generally include a number of steps including:

- Characterization of distress/failure
- Instrumentation for monitoring the behavior of distress
- Data assimilation
- Performance of diagnostic tests
- Back analysis
- Formulating a failure hypothesis and verification of its validity
- Conclusions

An application of this process, encompassing most or all of these steps constitutes a complete forensic analysis (Rao 2009).

34.3 Expert Witness

An engineer can become an expert witness if he can objectively explain technical issues to judge, jury, arbitrator and render opinions and conclusions as to the importance of these issues as they relate to the details of dispute.

A lawyer tries to represent his/her client to the best of his ability while an engineer tries to analyze the complete issues and render opinions based upon facts.

An engineering expert must

- Avoid conflict of interest
- Take assignments he can explain
- Consult and obtain other opinions before rendering his own
- Get facts and make no assumptions
- Must establish standards of care
- Terminate the assignment if one cannot conduct the full engineering
- Terminate the assignment if the fee is being used to bias an opinion

While litigation is lawyer's game played with lawyer's ball on lawyer's field with lawyer's rules yet he does not have all the ammunition. It is an engineer's or expert witness job to make sure the attorney has the right ammunition at the right time to argue and litigate the case.

34.4 Construction Vibrations—Damage Assessment

Construction activities can affect structures and their occupants in different ways. Most types of construction activities, such as soil and rock excavation, driving of piles and sheet piles, quarry and construction blasting (often carried out in densely populated areas and/or close to sensitive structures), and installations generate noise and vibrations that may result in annoyance and be the suspected cause of cracking or other damages. Construction work as illustrated in Fig. 34.1 can often take place in densely populated areas and close to residential structures.

Fig. 34.1 Sources of construction vibrations. **a** Rock blasing at construction site. **b** Pile installation at hospital. **c** Vibratory compactor near surgical center

This aspect is enhanced by the increasing awareness of the public concern, issues of environmental issues, vibration-sensitive electronic equipment and machinery (Massarsch 2000). Accordingly, many agencies frequently require that a forensic risk analysis be performed prior to issuance of construction permit. A risk analysis requires due perusal of many factors and consequences on humans, buildings, and installations as illustrated in Fig. 34.2.

34.5 Introduction

Construction activities cause different types of damage to structures and installations which can either be related to, or attributed to, ground vibrations and transmission through soils, rock, or other types of materials. They can either be soil movements, ground distortion, cyclic effects, or dynamic effects as illustrated in Fig. 34.3.

Site characterization and establishing site-specific threshold levels from test section, or preinstallation survey, are necessary tools for implementing an effective vibration and noise monitoring program. This will assist in reducing and possibly

Fig. 34.2 Factors included in a Risk Analysis (Holmberg et al. 1984)

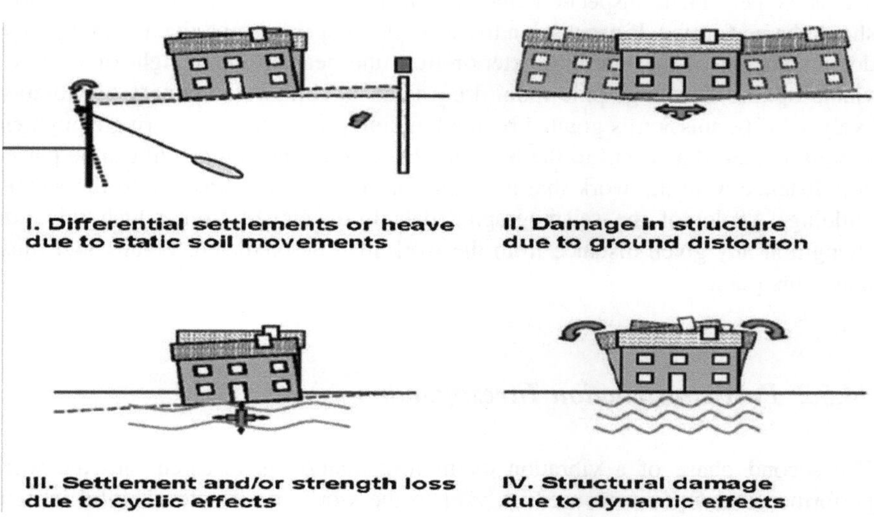

Fig. 34.3 Different types of building damage observed in connection with construction activities

eliminating claims against the contractor, consulting engineer, owners, developers, or others. Possibilities of installing vibration and/or noise attenuation measures like wave barriers (e.g., trenches, broken rock zones, sheet-pile walls, etc.) should be explored fully; likewise, initiative to switch to an alternate driving approach (e.g., change rated energy of hammer, change from driving to vibration installation, change from compressed air to diesel hammer, etc.) may be suggested to contractor by the owner. Construction activities, such as blasting, driving of piles and sheet piles, dynamic, soil compaction, or excavation generate noise and vibrations. In populated areas, these can have a negative impact on the environment, disturb inhabitants and, under unfavorable conditions, cause damage to buildings and installations. This condition is aggravated by the growing public awareness of environmental issue and the increasing use of vibration-sensitive electronic equipment and machinery in industry. In many countries, environmental regulations are enforced more stringently and limit or even prohibit the use of impact or vibratory hammers. The development has restricted the use of cost-effective foundation methods, such as driven and vibrated piles and sheet piles.

34.5.1 Preconstruction Surveys

Preconstruction surveys can play a significant role in vibration damage claims and litigation. A preconstruction survey should include the documentation of cracks and other damages in and around buildings near the intended work. When building access is permitted, inspection and documentation of the interior and exterior should be performed. If permission to access the property is not given, photographic documentation of the building exterior from the nearest public right of way is a viable option. Seismographic monitoring is an important component of vibration analysis. If permission is granted by the building owner, the monitoring equipment should be placed adjacent to the building. Otherwise, the equipment can be placed at a distance from the work that is similar in distance and direction to the nearest building. A plot of the seismographic data is developed from which vibration strength at any given distance from the work may be predicted. Figure 34.4 illustrates this point.

34.5.2 Post-construction Investigations

The second phase of a vibration claim investigation is research into the work performed and equipment used. A visit to the work site is often helpful in conjunction with the property damage inspection. The goal is to obtain information regarding dates and nature of work, construction equipment used, preconstruction surveys if available, seismographic monitoring reports, and geotechnical reports. A written request for information is typically made to the parties responsible for the construction. Finally, after the construction data has been obtained, engineering analysis is performed to estimate the vibrations and determine the potential for damage due to direct vibrations and/or settlement. The determination as to whether

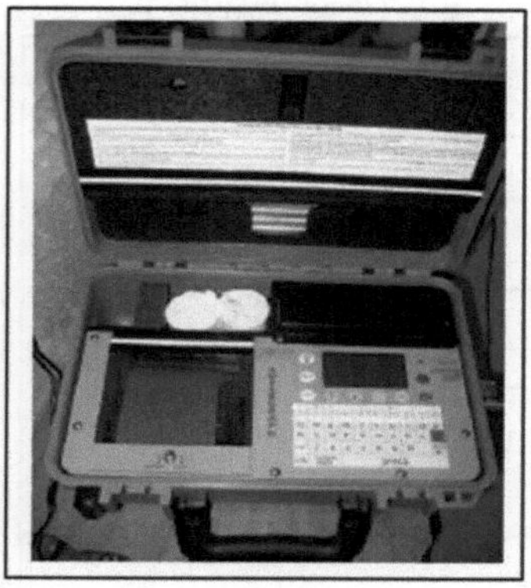

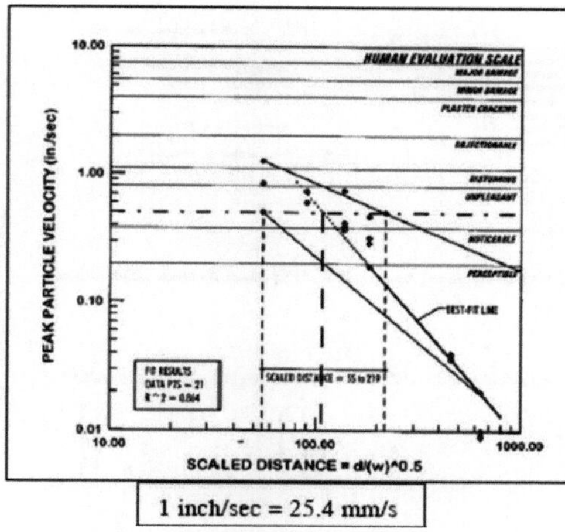

Fig. 34.4 Seismographic equipment and sample vibration attenuation band and best data fit curve

or not vibrations from construction activities were responsible for cracking or other damage requires the study of two mechanisms by which damage may occur. First, the direct vibrations or shaking of the structure as a result of the construction activities must be determined and compared to known thresholds for damage. Second, in the case where the structure is founded on a cohesionless soil such as sand, the potential for settlement of the structure as a result of vibrations can be approximated by rational analysis.

Vibration damage claims can be reduced by preconstruction surveys to document existing cracks and other conditions. Seismographic monitoring provides actual ground vibration experienced by nearby structures. In the absence of seismographic data, engineering analysis can be used to estimate the maximum vibrations at a given structure. The vibrations data is used to determine whether the vibrations could have caused the damage.

34.5.3 How Do We Measure Vibration?

The measurement of vibration is undertaken using special monitors, such instrumentation, termed a seismograph, which is capable of recording both ground and airborne vibration. Ground vibration is recorded in terms of peak particle velocity in millimeters per second in three mutually perpendicular directions (T, V & L). Airborne velocity is measured in terms of decibels (dB). Seismographs can be left unattended and set to trigger when an emission level exceeds a predetermined set level. Figure 34.5 represents a typical vibration event record produced using monitoring blast vibration and air overpressure during construction of a major project.

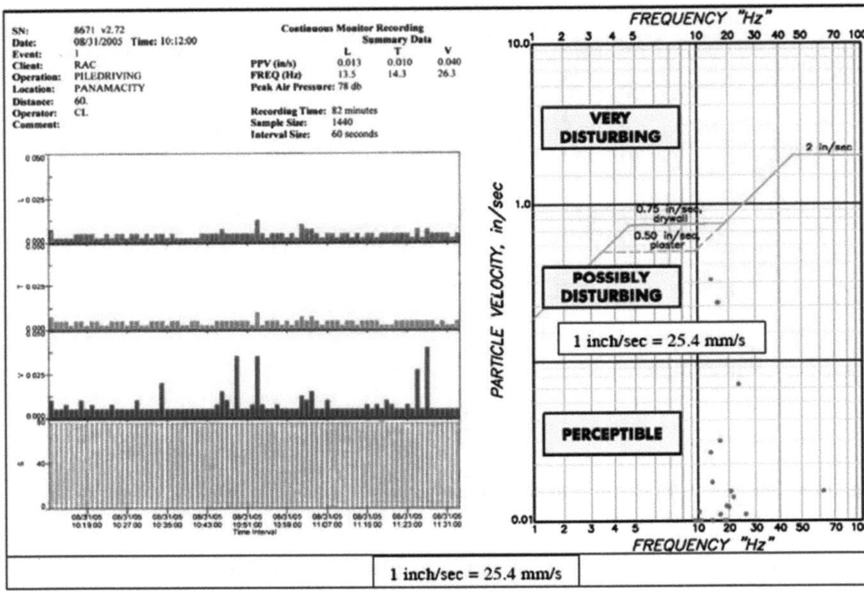

Fig. 34.5 Typical vibration event record

The ground vibration records show the variation of velocity with time. Each trace has a point where the velocity is a maximum (+ve or −ve) and this is known as the peak particle velocity (or PPV) which has units of mm/s or in/s. Monitoring usually takes place at the closest vibration-sensitive building to current operations. Where blasting/vibration generating construction takes place in more than one area within a site then more than one monitoring location may be necessary. In some situations, access to a vibration-sensitive building may not be practicable. In this case, consideration can be given to the selection of a location away from the building in a general line with the area where the activity is occurring and at which monitoring could be regularly undertaken. Such locations may be at or just within the site boundary.

When vibration limits are set, they usually refer to recordings taken at ground level at specified properties. The reason for this is that structural damage criteria from around the world specify vibration limits that apply at foundation level. Indeed, the vast majority of damage noted in studies does not occur on upper floors in structures but mainly on lower floors where the strains are greatest due to the confinement of the foundation.

Vibration standards come in two varieties: (i) those dealing with human comfort and (ii) those dealing with cosmetic or structural damage to buildings. In both instances, the magnitude of vibration is expressed in terms of peak particle velocity (PPV).

In practice, construction vibration limits may be found in planning permission conditions for large projects which are likely to generate potentially significant vibration, limits are usually set which are below the thresholds of structural and cosmetic damage in order to limit nuisance.

34.5.4 Human Response to Vibration

Human response to blast-induced vibration is a relatively complex phenomenon and is dependent upon a range of factors of which the actual vibration magnitude is only one and not necessarily the most important. It is well recognized that the human body is very sensitive to the onset of vibration albeit very poor at distinguishing relative magnitudes. Although sensitivity to vibration varies significantly between individuals, a person will generally become aware of blast induce vibration at levels of around 0.5 mm/s peak particle velocity; however, individuals are very poor at distinguishing between vibrations of differing magnitudes.

Once a received vibration is greater than an individual's perception threshold then it is possible for concern to be expressed about the blasting or construction activity involved. Such concern normally relates to the vibration's potential for causing damage to the complainant's property. Concern may be expressed that damage has already occurred due to the recent discovery of cracking that may have been present for some time or have been caused by natural processes. More often, however, concerns are based on the fear that damage will be caused at some time in the future as a result of repeated vibration. It is usually the case that adverse comments are less likely once a neighbor has become accustomed to the perceived effects of blasting.

Good communication is one of the best ways to help minimize vibration complaints around a construction site. Keeping neighbors informed of the nature of the work and progress is a great way to help alleviate unnecessary concern.

34.6 Case Study

34.6.1 Salient Features

Twenty homeowners from a residential subdivision in the Florida Panhandle (refer to Fig. 34.6), claimed damage including cracking to their properties during construction of various portions of city's storm water treatment/enhancement project. Legal filing stated that "while site work and construction was being performed and proceeding at the project the tenants and/or owners of the various residences experienced severe vibrations, shaking and rattling of the windows and other fixtures and personal property in their respective homes." They argued that their houses have been damaged as a result of the construction activities related to vault construction. These construction activities included excavation dewatering, sheet

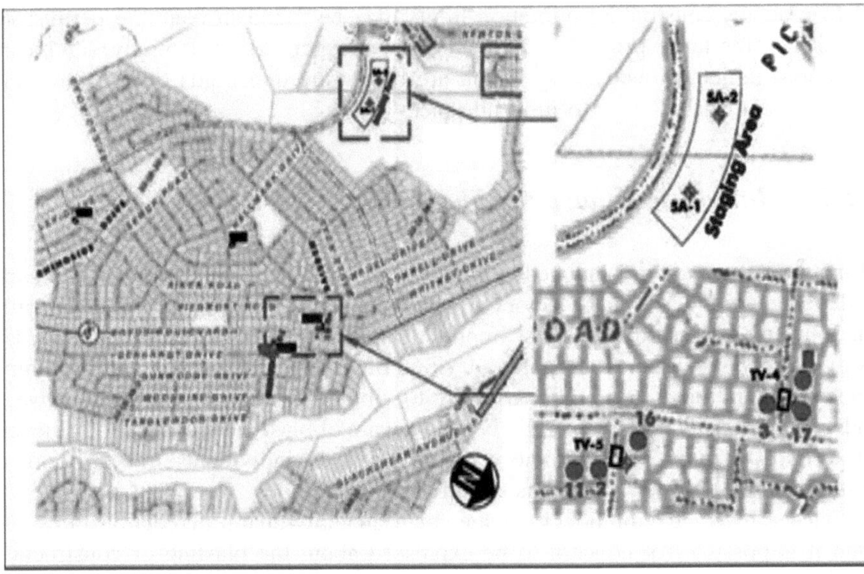

Fig. 34.6 Structure layout, with treatments vaults (TV), staging area test section (SA-1 and SA-2)

pile installation, and construction/compaction activities using vibratory equipment including vibratory hammer and compactor.

It was also alleged that during excavation and construction of drainage structures, especially four subsurface treatment vaults that required installation and extraction of sheet piles using vibratory hammer, stability of the nearby residential structures was endangered and resulted in movement/cracking.

Forensic field investigation performed for the defendant contractor included a two-stage program that consisted of (1) a detailed field exploration, data assimilation, stability analyses of the alleged sites and (2) measurement of ground vibrations on simulated tests performed on a comparable site designated as the staging area. All records of ground vibrations obtained by the contractor during construction of treatment vaults TV-1 and TV-4 were analyzed.

34.6.1.1 Stage 1 Program

It consisted of performing a forensic investigation at each of the 20 claim locations that included:

- A geotechnical exploration
- A visual condition survey
- An interview with the property owner
- A field probing and coring program
- Test pits and footing inspection

Appropriate samples were collected and stability analyses were performed for each of the 20 residential structures.

34.6.1.2 Stage 2 Program

Stage 2 program involved performing special vibration tests at a comparable site designated as the "staging area." This was done to determine effects from various construction sources on ground vibrations. Simulated tests consisted of driving two sheet piles in parallel and perpendicular mode to a line of seismograph settings, dropping a heavy collector section from a height of 4–5 ft, and movement of heavy construction equipment along seismograph setting. A total of four calibrated seismographs were placed in radial direction at distances ranging from 25 to 200 ft from the sheet pile driving location. Ground vibrations (PPV and frequency) were measured in longitudinal and transverse directions for a total of five vibration modes. Two typical test modes, installation/extraction of sheet piles in a parallel mode, and vibrations from moving of vibratory construction equipment are illustrated in Figs. 34.7 and 34.8.

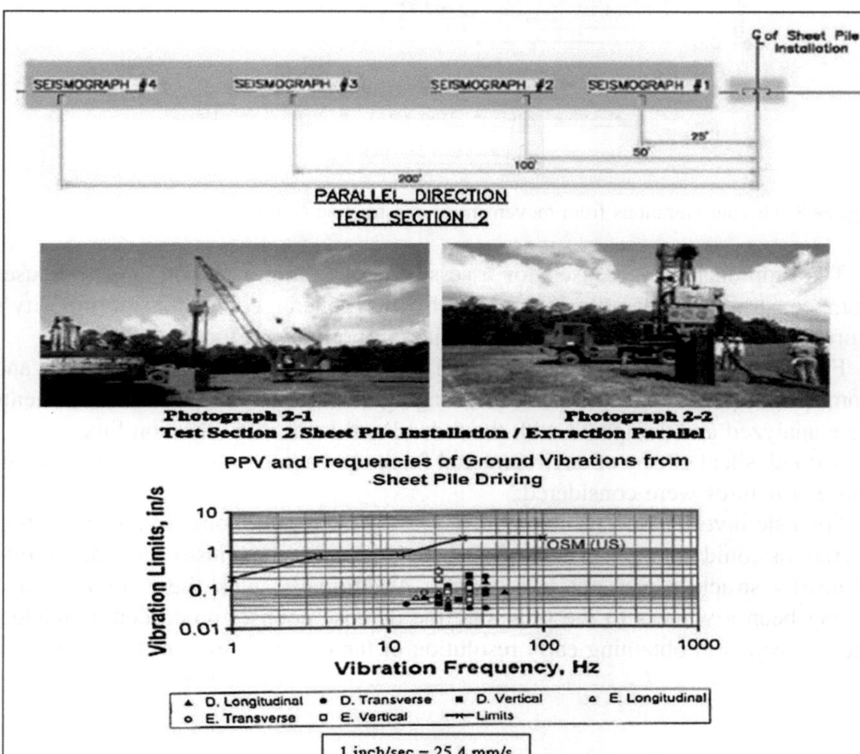

Fig. 34.7 Ground vibrations along parallel direction from pile driving

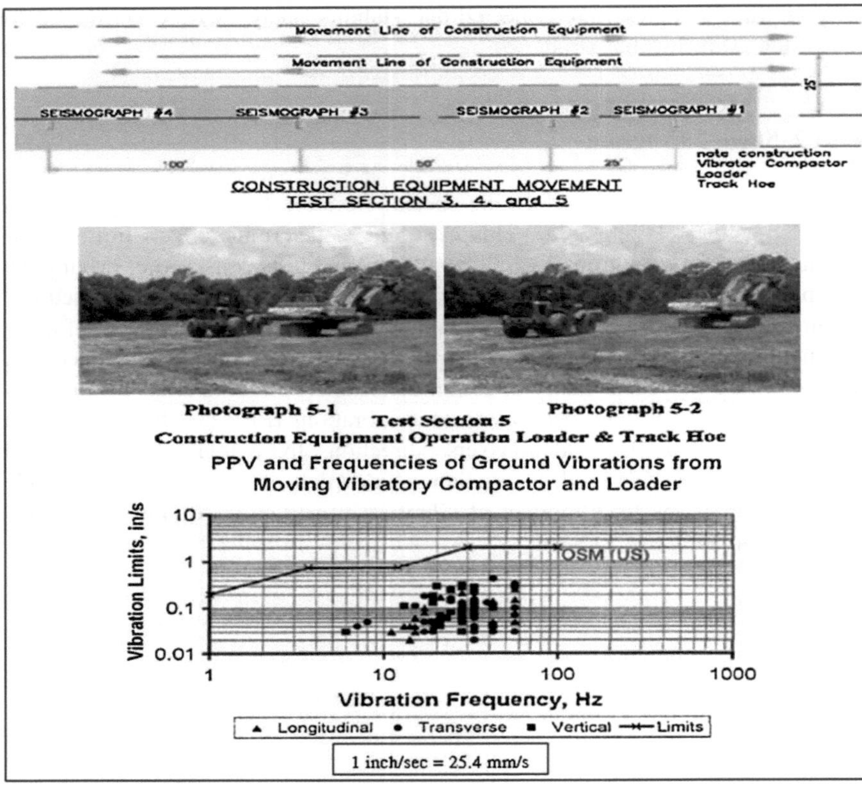

Fig. 34.8 Ground vibrations from movement of construction equipment

Two approaches were used for assessment of vibration effects on the houses from construction operations provided by the general contractor for the city's Stormwater Treatment Enhancement Project in northwest Florida, USA.

First, ground vibrations were measured during construction of TV1 and TV4, and from special vibration tests made at the staging area. Results of these measurements were analyzed and compared with the federal and local safe vibration limits.

Second, silent effects of environmental forces and aging processes in materials of house structures were considered.

Forensic investigation results revealed that ground vibrations from construction operations could not be the cause of alleged damages to the houses and that stability of nearby structure was not in jeopardy. Additionally, with these findings, and having been a witness to the tests site, the defense counsel proceeded, defended, and succeeded in obtaining early resolution of the claim in favor of the defendant.

34.6.2 Concluding Remarks

Based upon our review and analysis of all available and developed data, it was concluded with a reasonable degree of scientific certainty that:

1. Ground vibrations from construction operations at the project site and during installation/extraction of sheet piles using vibratory hammer as well as the use of vibratory construction equipment, with high degree of probability, cannot be the cause of the alleged damages to the houses.
2. The analysis presented in this report is of sufficient scope to eliminate vibration activity resulting in dynamic soil settlement as the cause of alleged damage within a reasonable degree of engineering probability.
3. Environmental forces and aging processes in house materials appear to the actual causes of the damages to the houses.
4. It was concluded that the quality of construction of the homes could play a factor in the litigation. It was highly unlikely that the damage may have been caused by the construction. The vibrations from the construction were within acceptable standards.

Forensic investigation results revealed that ground vibrations from construction operations could not be the cause of alleged damages to the houses and that stability of nearby structure was not in jeopardy. Additionally, with these findings, and having been a witness to the tests site, the defense counsel proceeded, defended, and succeeded in obtaining early resolution of the claim in favor of the defendant.

Acknowledgments The opportunity to perform the services described herein provided an interesting exercise in the planning and execution of this unique project. The information herein is from projects where the author and their firm were involved as the project's forensic engineering consultant. The author sincerely takes this opportunity to thank the other case study project team members: subconsultant, M.R. Svinkin, P.E., Vibroconsult, Inc. and David Jester, Esquire, Senior counsel with the law firm of GJTB & S, PLC.

References

Holmberg R et al (1984) Vibrations generated by traffic and building construction activities

Massarsch KR (2000) Settlements and damage caused by construction-induced vibrations. In: Proceedings, international workshop wave 2000, Bochum, Germany 13–15 Dec 2000, pp 229–315

Rao VVS (2009) Introduction to forensic geotechnical engineering Chap. I, ISSMGE TC40: forensic geotechnical engineering, Oct 2009

Chapter 35
A Case Study of Observational Method for a Failed Geotechnical Excavation in Singapore

Y. Iwasaki

Abstract Observational method (OM) was studied based on a case history of failed geotechnical excavation in Singapore. Several items including inclinometer provide vital information about the safety of geotechnical construction sites. An engineer who is specialized in OM, with adequate knowledge of geotechnical engineering as well as instrumentation, should be engaged during the design and construction phase. An ISO standardization is necessary to provide a basic process to implement OM.

Keywords Observation method · Piezometer · Inclinometer · Strain gauge · Settlement

35.1 Introduction

Observational method (OM) was proposed by Terzaghi and Peck (1967) for several reasons, including to confirm the assumed geotechnical condition during design stage as well as to avoid critical situations in geotechnical construction. One of the failed cases in Singapore that was described by Prof. Ishihara during the TC302 Osaka Symposium is discussed in terms of OM.

After the collapse of Nicoll Highway on April 20, 2004, a Committee of Inquiry (COI) was established to study the process of failure.

The COI reported their conclusion on May 13, 2005 to the Minister of Manpower, Government of Singapore, which also released the report to the public (COI report 2005).

Y. Iwasaki (✉)
Geo-Research Institute, Osaka, Japan
e-mail: yoshi-iw@geor.or.jp

V.V.S. Rao and G.L. Sivakumar Babu (eds.), *Forensic Geotechnical Engineering*, Developments in Geotechnical Engineering, DOI 10.1007/978-81-322-2377-1_35

The COI report concludes the accident was associated with two inadequate designs of "effective method of analysis for cut and cover excavation" and "wailer system."

This paper will show the results of monitoring and how the process to the failure could be traced.

35.2 Cut and Cover Section of M3 Zone

As shown in Fig. 35.1, the section is parallel to the existing Nicoll Highway on the ground surface and consists of different types of retaining walls, including M2 and M3 that failed as shown in Fig. 35.1.

As shown in Fig. 35.3, the special design was two sets of grout slabs that had been installed before the excavation started. The depth of the excavation was planned to GL-34.5 m above the lower grout slab (Fig. 35.2).

The cone penetration resistance shown in Fig. 35.3 shows very soft soil down to 35 m from the surface where stiff clayey soil continues.

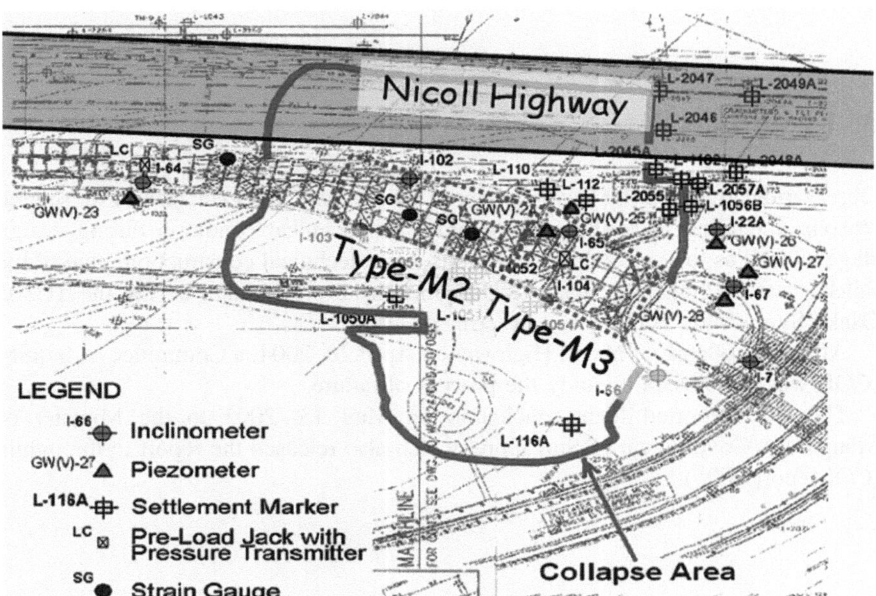

Fig. 35.1 General plan view

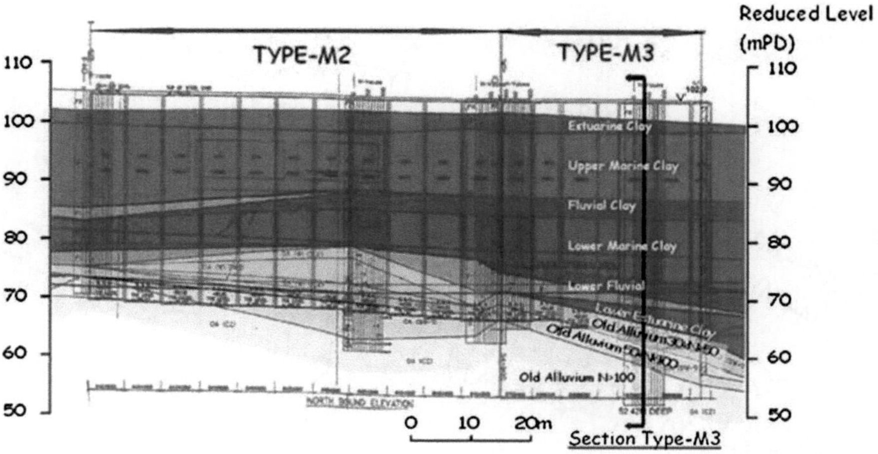

Fig. 35.2 Geotechnical section used for design phase

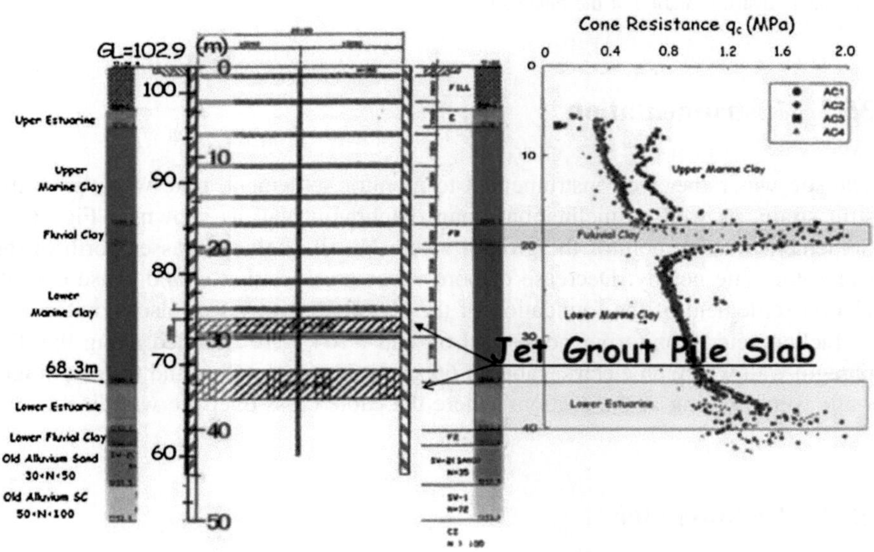

Fig. 35.3 Design of cut and cover of Type M3

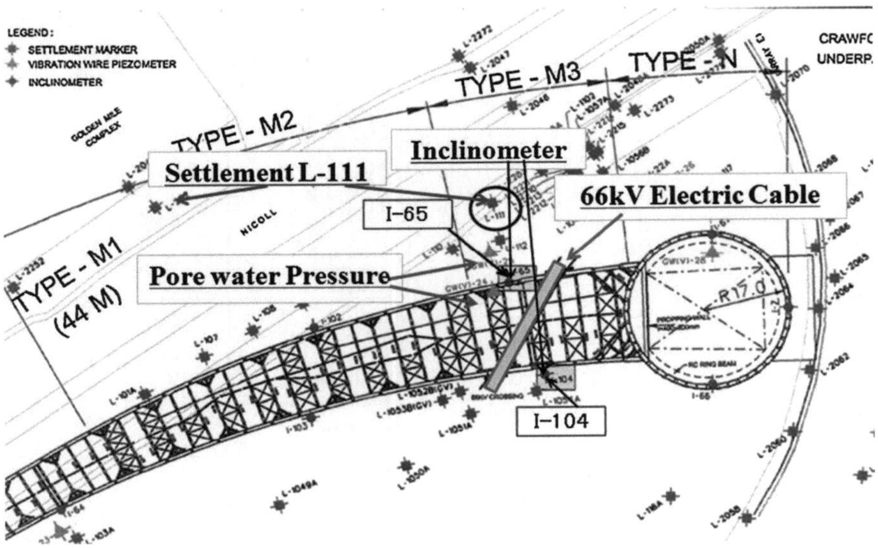

Fig. 35.4 Instrumentation at the failed site

35.3 Instrumentation

The site was rather well instrumented to monitor settlement, pore water pressure, strut strain, as well as inclination along diaphragm wall as shown in Fig. 35.4. Settlement was to monitor the ground where Nicoll Highway passes north of the excavation site nearby. Decrease of pore water pressure that would cause consolidation settlement of the foundation of the Nicoll Highway was also monitored.

Inclinometer monitoring points at I-65 and I-104 were installed along the diaphragm wall where an electric cable of 66 kV had been installed and the diaphragm walls were missing at the sections where the cable cross over the wall lines.

35.4 Inclinometer

The inclinometer guide pipes were installed at I-65 and I-104 at the failed section.

Excavation started on September 26, 2003, at the M3 Section. The monitored wall deflections from the six strut installation until before the failure took place are shown in Fig. 35.5 as well as chronological time change of excavation with strut and the maximum deflection for I-65 and I-104 in Fig. 35.6.

The allowable deflection could be estimated by mechanical properties of such related materials as diaphragm wall, soils to be excavated, and jet grout pile.

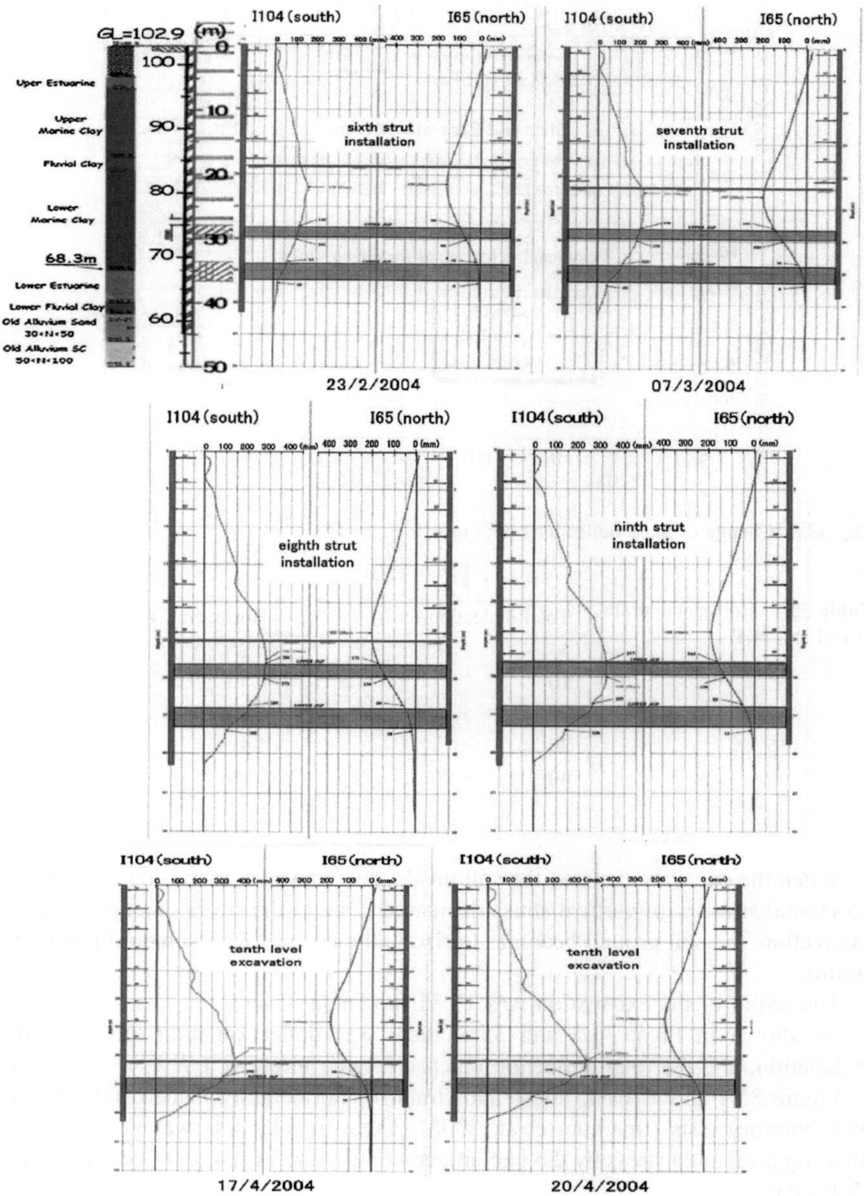

Fig. 35.5 Deflection of diaphragm wall

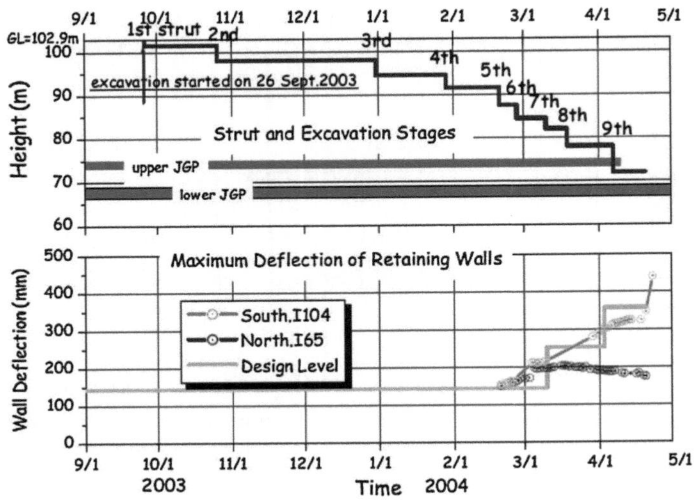

Fig. 35.6 Change of max. deflections with time

Table 35.1 Horizontal strain in soil and JGP

Total deflection (cm)	Horizontal strain (%)
5	0.25
10	0.50
20	1.00
30	1.49
40	1.99

When the excavation starts, the soil inside the excavated zone is under increased horizontal stress with vertical stress decreased. Due to the stress change caused by excavation, the soil ground becomes horizontally stressed with some critical failure strains.

The width of the excavation was 19.85 m (Table 35.1).

As shown in Figs. 35.5 and 35.6, there was deflection after the sixth strut installation, and the total deflection reached 35 cm, at about 1.7 %.

Figure 35.7 shows basic stress and strain for horizontally loaded JGP obtained by laboratory tests (Imamura et al. 2008). The vertical stress was normalized by dividing horizontal stress by the unconfined compression strength of column of JGP (500 kPa).

Figure 35.7 shows the maximum allowable horizontal strain estimated at 2 %.

If safety margin is given as FS = 2, the allowable deflection at the level of JGP is 1 %, which is 20 cm in total, or, 10 cm for each side.

As shown in Fig. 35.8, the deflection at the level of the upper JGP slab had reached 12 cm for southern and 8 cm for the northern side wall.

Fig. 35.7 Stress-strain for JGP (Imamura et al. 2008)

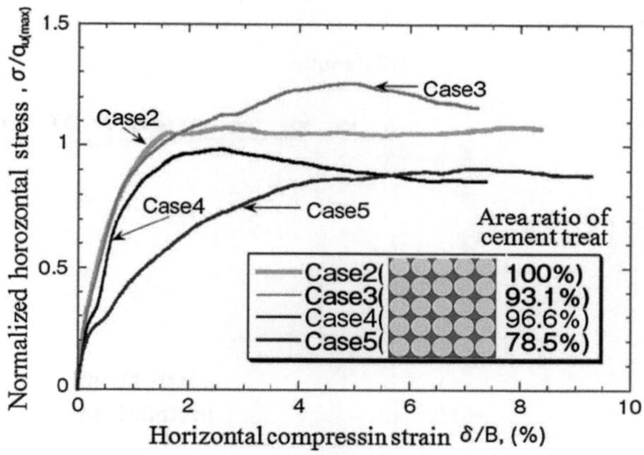

If OM was applied, this could be the one of the critical phases when some countermeasures should be considered. The possible countermeasures include additional soil improvement in the zone above the upper JGP slab and/or additional strut installation to strengthen the temporary retaining structure.

At this stage, different responses of the wall deflections for north and south sides should have been noticed. The lower JGP was observed to have moved to the center with about 5 cm for southern side but no displacement for northern side.

When the strut 8 was completed, the displacement for southern wall had increased to 35 cm (25 cm for southern and 10 cm for the northern wall) at the upper JGP slab, which was almost 1.8 % in failure.

The asymmetry of the wall deflection for north and south sides was increased with the excavation steps.

When the excavation reached the upper JGP, the asymmetry became clear and the reason that caused the phenomena might well be studied. The reason could be any difference in water pressure or earth pressure against for the north and south walls.

Many settlement points were installed along the retaining wall adjacent to Nicoll Highway.

35.5 Settlement and Pore Pressure

The Nicoll Highway is the nearby structure that should not be affected. Monitoring had been installed for evaluation. Pore water measurements were also installed near the retaining wall as shown in Fig. 35.9. One of the typical settlements installed at L111 is shown in Fig. 35.10. The level of the ground surface was GL = +102.9 m RL.

Figure 35.11 shows a comparison between the max deflection at I-65 and the settlement at L-111 with the same timescale. The pore water pressure was

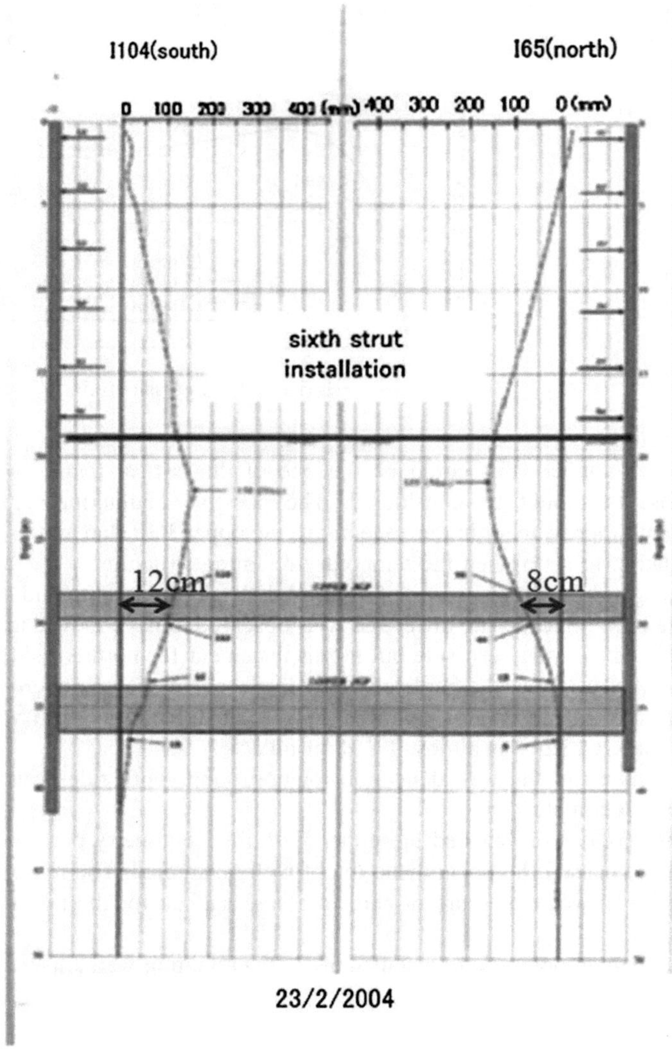

Fig. 35.8 Wall deflection monitored after the sixth strut

monitored at two depths at GL-17.5 m and −30.5 m of behind the diaphragm wall. The ground water was observed at GL-2 m in general in the area.

The ground settlement at L111 had gradually taken place and reached about 10 cm in January 2004. After this point, the settlement increased rather rapidly with time.

The pore pressure at GL-17.5 m had gradually decreased with excavation stage.

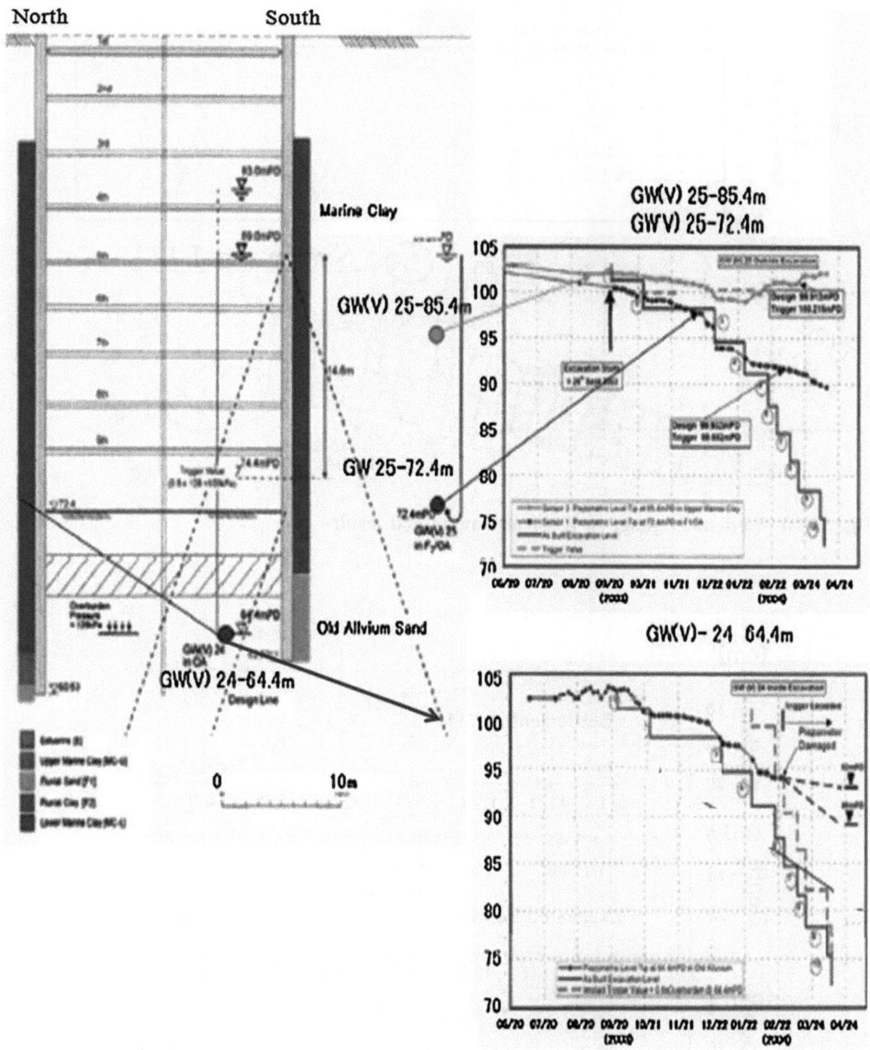

Fig. 35.9 Pore water monitoring

The decrease in the pore water pressure might have been caused by inward movement of the diaphragm wall, which resulted in volume expansion of the soil behind the wall.

The settlement at L111 was caused either by consolidation or displacement of the diaphragm wall.

Since the settlement had increased so fast, it was very likely considered by the movement of the diaphragm.

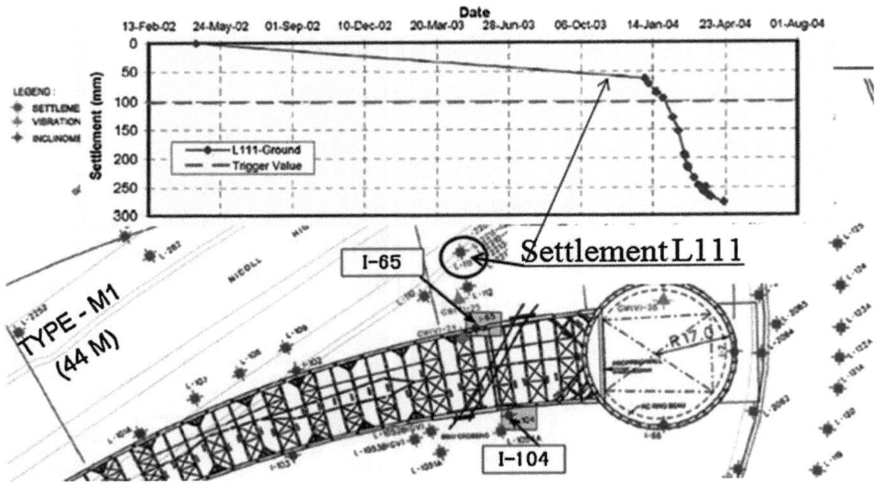

Fig. 35.10 Settlement gauge at L111 and monitored result

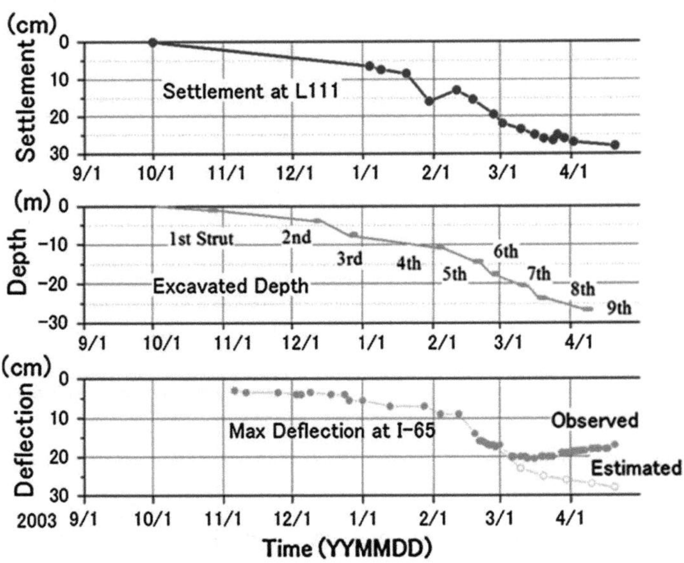

Fig. 35.11 Settlement versus deflection

35.6 Settlement and Max Deflection

The max deflection at I-65, shown in Fig. 35.6, shows its max deflection at the excavation stage after the sixth strut installation. After this point, the maximum deflection had decreased, which implies the diagram wall of the north side was moving backwards.

Figure 35.12 shows the relationship between the settlement and the maximum deflection. Until March 1, 2004, the settlement at L-111 was almost the same as the max. deflection at I-65. The direct relation of the settlement of the surface ground with the max. deflection means that the settlement is not caused by consolidation but compensation of the soil volume caused by the retaining wall. After March 1, the settlement continued to increase. On the other hand, the deflection had stopped and had even gone backwards.

The backwards movement of the diaphragm wall is not easy to understand. This might be caused by the displacement assumption for the inclination calculation. It is common practice to assume the bottom point of the inclinometer guide pipe, which is embedded within rather dense sand or hard clayey layer, to be a fixed point. It is rather realistic to consider the bottom of the inclination guide pipe had moved. Figure 35.11 shows the estimated increase in deflection assuming the linear relationship between the settlements at L-111 and the max deflection at I-65 as in Fig. 35.12, which resulted to about 28 cm.

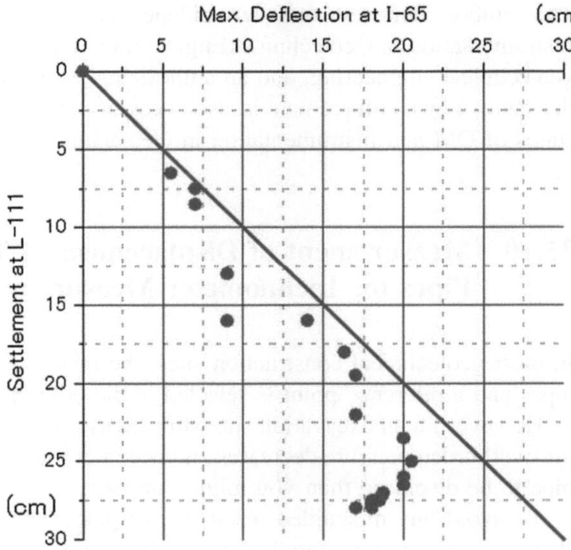

Fig. 35.12 Max. deflection at I-65 versus settlement at L-111

35.7 Lessons Learned

The author had selected several topics from the failure of excavation in Singapore that happened on April 20, 2004 from the view points of OM as well as instrumentation, which was not referred in the report of Committee of Investigation.

35.8 General Review of Applicability of Observational Method in the Failure

Among many monitoring items, displacement is much clear than stress or force. If the failure strains are known before the construction starts, displacement criteria to avoid failure of different materials might well be established. In the case under discussion, allowable horizontal displacement for jet grout pile slab was discussed.

35.9 Necessity of Specialized OM Engineer for Geoconstruction

If OM should be applied to any geotechnical construction, it is not easy to implement the process without a good quality geo-engineer, who is basically a geotechnical engineer but one who has a profound knowledge of instrumentation. The author believes that an independent professional field of "OM and Instrumentation in Geotechnical Engineering" must be separated from the common geotechnical engineering, and an authorizing system established to select and give the status. Legal enforcement is also necessary to include an engineer who holds the status of OM and Instrumentation in Geotechnical Engineering.

35.10 Measurement of Displacement of Top of Guide Pipes for Inclinometer Measurement

In most geotechnical construction sites, the inclinations are measured using guide pipes and a reference point is selected at the bottom of the pipe.

However, in an excavation site, with excavation depth approached to final level, the overburden pressure decreases and becomes much easier for bottom of the guide pipe to be displaced than you might consider.

To avoid any misguided result, the displacement of the top of the guide pipe should always be monitored when the inclinometers are measured.

GPS is a good system for automatic monitoring instead of manual survey when the OM should be applied for rather a long period.

35.11 Interpretation of Geotechnical Behavior Based on Related Monitored Data

It is important to discuss settlement data with deflection of the nearby retaining wall. The relationship among various monitored items provides important and vital information to understand the site conditions.

35.12 Effective or/and Total Stress?

Traditional soil mechanics assumes and discusses failure in terms of stress. Since the characteristics of soil changes with various factors and conditions, some simplification was necessary to obtain a conclusion.

One of the unique conditions to be considered in soil mechanics is "drained" or "undrained."

When stress path that is expected for construction process is known, the soil specimen may be tested along the stress path to obtain strength parameters for "Total Stress" approach.

When pore pressure is monitored during the test, "Effective Stress" approach may also be available. Additional information about the pore pressure that is generated during the shearing process due to dilatancy is vital for the Effective Stress approach.

Discussion of total or effective stress is for only design stage. In the design stage, various conditions are assumed as safer side. When the construction site allows "undrained condition," the total stress approach may still be applied during the construction to discuss safety in terms of stress level.

Since the deformation parameters of soil generally depend on effective confining pressure, stress path, and other several factors, effective stress approach will make it easier to deal with deformation analysis than total stress approach.

35.13 ISO Standardization for OM in Geotechnical Engineering

In the early stage of soil mechanics, big projects were not many in number and such giants like Terzaghi and Peck could work by themselves. At present, it has become rather common to implement huge geotechnical constructions.

Most failed cases are due to lack of monitoring instrumentation or lack of knowledge of how to interpret the monitored data, even though instrumented.

We need to create the standard basic principles of instrumentation and how to implement OM in geotechnical construction as one of ISO in geotechnical engineering.

It is necessary to have five member countries of ISO to make such a new item.

35.14 Conclusion

The failed excavation work in Singapore on April 20, 2004 is reviewed in terms of instrumentation and OM.

Conclusions are obtained as follows:

Observational method provides the basic approach to complete geotechnical construction with safety.

It is necessary to provide a special geotechnical engineer who understands not only geotechnical engineering but also instrumentation and OM.

It is also pointed out that key information about displacement of top of guide pipe for inclinometer should be prepared.

ISO standardization for OM is highly recommended.

Effective and total stress approach should be discussed for not only obtaining safety of the project but also for deformation.

References

Imamura S et al (2008) Lateral deformation characteristics of footing beam improved. Technical report, Nishimatsu Corp., vol 31, pp 7–12 by DMM in Earth-Retaning Support

Terzaghi K, Peck RB (1967) Soil mechanics in engineering practice. Wiley, New York

The Committee of Inquiry (2005) Report of committee of inquiry into the incident at the MRT circle line worksite that led to the collapse of Nicoll highway on 20 April, 2004, Ministry of Manpower, Government of Singapore

Appendix

TC302 (2005–2013)—Forensic Geotechnical Engineering

Aim: To prepare a guidance manual for Forensic Investigations of Geotechnical Failures

Composition:

Chairman: Dr. V.V.S. Rao (India)
Vice-chairman: Prof. M.R. Madhav (India)
Secretary: Prof. G.L. Sivakumar Babu (India)

A. Core members:

Mr. P.W. Day (S.Afrika)
Dr. R. Hwang (Taiwan)
Mr. D. Hight (UK)
Prof. M. Popescu (USA)
Dr. Yoshi Iwasaki (Japan)
Prof. J. Mecsi (Hungary)
Prof. W.F. Lee (Taiwan)
Prof. K.K. Phoon (Singapore)
Mr. David Starr (Australia)
Prof. Luiz Guilherme de Mello (Brazil)
Mr. D.S. Saxena (USA)

B. Members:

Dr. Axel Ng (Hongkong)
Dr. Suraj de Silva (Hongkong)
Mr. Robert, J. (France)
Mr. Alain Pecker (France)
Mr. Itoh, K (Japan)
Mr. Barry, T (UK)
Mr. Filatov, A.V. (Kazaksthan)
Mr. Dirk Lugar (Netherlands)

V.V.S. Rao and G.L. Sivakumar Babu (eds.), *Forensic Geotechnical Engineering*,
Developments in Geotechnical Engineering, DOI 10.1007/978-81-322-2377-1

Mr. Cichy, W. (Poland)
Prof. Malek Bouazza (Australia)
Mr. Rob Jessep (UK)
Dr. Anand J Puppala (USA)
Dr. Jan Hellings (UK)
Mr. Enrico Conte (Italy)
Mr. Vincenzo Caputa (Italy)
Mr. Murray, G (New Zealand)
Mr. Edmund Medley (USA)
Mr. Enrique Dapena (Spain)

Author Index

V.V.S. Rao and G.L. Sivakumar Babu (eds.), *Forensic Geotechnical Engineering*,
Developments in Geotechnical Engineering, DOI 10.1007/978-81-322-2377-1

MIX
Papier aus verantwortungsvollen Quellen
Paper from responsible sources
FSC® C105338

If you have any concerns about our products,
you can contact us on
ProductSafety@springernature.com

In case Publisher is established outside the EU,
the EU authorized representative is:
Springer Nature Customer Service Center GmbH
Europaplatz 3, 69115 Heidelberg, Germany

Printed by Libri Plureos GmbH
in Hamburg, Germany